SIXTH EDITION
PHYSICAL SCIENCE

bju press
Greenville, South Carolina

NOTE: The fact that materials produced by other publishers may be referred to in this volume does not constitute an endorsement of the content or theological position of materials produced by such publishers. Any references and ancillary materials are listed as an aid to the student or the teacher and in an attempt to maintain the accepted academic standards of the publishing industry.

PHYSICAL SCIENCE
Sixth Edition

Coordinating Writer
David M. Quigley, MEd

Writer
Christopher D. Coyle

Biblical Worldview
Tyler Trometer, MDiv

Academic Oversight
Jeff Heath, EdD
Rachel Santopietro, MEd

Project Editor
Rick Vasso, MDiv

Cover, Design, and Interior Concept Design
Sarah Lompe

Page Layout
Carrie Walker

Illustration
Jonathan Andrews
Sarah Lompe
Craig Oesterling
Garison Plourde

Permissions
Tatiana Bento
Lily Kielmeyer
Hannah Labadorf
Rita Mitchell
Ashleigh Schieber
Elizabeth Walker

Project Coordinators
Chris Daniels
Tony Every

Photo credits appear on pages 563–67.

All trademarks are the registered and unregistered marks of their respective owners. BJU Press is in no way affiliated with these companies. No rights are granted by BJU Press to use such marks, whether by implication, estoppel, or otherwise.

The cover photo shows a close-up of paint in water.

© 2020 BJU Press
Greenville, South Carolina 29609
Fifth Edition © 2014 BJU Press
First Edition © 1974 BJU Press

Printed in the United States of America
All rights reserved

ISBN 978-1-62856-505-8

15 14 13 12 11 10 9 8 7 6

CONTENTS

Taking Off with Physical Science!   vii

1 THE STRUCTURE OF MATTER   xii

CHAPTER 1   MODELING OUR ORDERLY WORLD   2
- 1A   Order in Our World   4
- 1B   Modeling Our World   11
- 1C   Using Mathematics for Scientific Inquiry   14

CHAPTER 2   MATTER   24
- 2A   Understanding Matter   26
- 2B   Classifying Matter   32
- 2C   States of Matter   35
- 2D   Changes in Matter   39

CHAPTER 3   THE ATOM   48
- 3A   The Atomic Model   50
- 3B   Atomic Structure   56

CHAPTER 4   THE PERIODIC TABLE   68
- 4A   Organizing the Elements   70
- 4B   Classifying the Elements   80
- 4C   Periodic Trends   86

CHAPTER 5   BONDING AND COMPOUNDS   94
- 5A   Principles of Bonding   96
- 5B   Types of Bonds   99
- 5C   Writing Chemical Formulas   108

CHAPTER 6   THE CHEMISTRY OF LIFE   122
- 6A   Organic Compounds   124
- 6B   Substituted Hydrocarbons   134
- 6C   Biochemistry   136

2 CHANGES IN MATTER   144

CHAPTER 7   CHEMICAL REACTIONS   146
- 7A   Chemical Changes   148
- 7B   Types of Chemical Reactions   156
- 7C   Energy in Chemical Reactions   159
- 7D   Reaction Rates and Equilibrium   162

CHAPTER 8   NUCLEAR CHANGES   174
- 8A   Radioactive Decay   176
- 8B   Fission and Fusion   185
- 8C   Nuclear Changes: Benefits and Risks   191

CHAPTER 9   SOLUTIONS   200
- 9A   Mixtures and Solutions   202
- 9B   Solution Concentration   212

CHAPTER 10	ACIDS, BASES, AND SALTS	222
10A	Acids and Bases	224
10B	Acidity and Alkalinity	228
10C	Salts	232

3 MATTER IN MOTION — 242

CHAPTER 11	KINEMATICS	244
11A	Describing Position	246
11B	Describing Motion	253
11C	Changing Motion	261

CHAPTER 12	DYNAMICS	268
12A	Classifying Forces	270
12B	Newton's Laws of Motion	275
12C	Types of Forces	282

CHAPTER 13	WORK AND MACHINES	294
13A	Work and Mechanical Advantage	296
13B	Lever	303
13C	Wheel and Axle	309
13D	Inclined Plane	313

CHAPTER 14	ENERGY	322
14A	Classifying Energy	324
14B	Energy Changes	332
14C	Energy Resources	337

CHAPTER 15	THERMODYNAMICS	346
15A	Temperature	348
15B	Heat	355
15C	Thermodynamics	364

CHAPTER 16	FLUIDS	370
16A	Properties of Fluids	372
16B	Gas Laws	380
16C	Fluid Mechanics	387

4 WAVES AND ENERGY — 392

CHAPTER 17	PERIODIC MOTION AND WAVES	394
17A	Periodic Motion	396
17B	Waves	403
17C	Wave Behavior	410

CHAPTER 18	SOUND	418
18A	Sound Waves	420
18B	Hearing and Music	424
18C	Using Sound Waves	430

CHAPTER 19	ELECTRICITY	440
19A	Static Electricity	442
19B	Current Electricity	450
19C	Circuits	457

CHAPTER 20	MAGNETISM	470
20A	Magnets and Magnetism	472
20B	Electromagnetism	476
20C	Generating and Using Electricity	480

CHAPTER 21	ELECTROMAGNETIC ENERGY	486
21A	Electromagnetic Waves	488
21B	The Electromagnetic Spectrum	496

CHAPTER 22	LIGHT AND OPTICS	502
22A	Light Behavior	504
22B	Color	507
22C	Reflection and Mirrors	510
22D	Refraction and Lenses	514

FEATURES

CAREERS—SERVING AS A...

Forensic Scientist: Silent Witness		8
Materials Scientist: Making an Invisibility Cloak		35
Food Chemist: Working Out of a Jam		140
Toxicologist: Pick Your Poison		153
Imagineer: Making Magic		264
Piping Engineer: Keeping the Food Flowing		389
Acoustic Engineer: Sounds Great!		434

CASE STUDIES

How Many States of Matter?	38
A World of Models	55
All Models Are Not Equal	65
Allotropes	93
Sticky Situation	121
Grain Elevators	172
Building Implosion	173
Vikings	183
Tsar Bomba	189
Road Salt	218
Maple Syrup	221
The King of Chemicals	227
Water as a Coolant	369
Galloping Gertie	414
Taser®	469
Seeing Is Believing	501

ETHICS

Reporting Scientific Data	23
Strategies and Protecting People from Radiation	66
Pseudoephedrine	118
Can Fast Food Be Nutritious?	143
Nuclear Power Generation	199
Pollution	221
Antacids	238
Radar Detectors	267
Mandatory Helmet Laws	293
Who Owns Your Photos?	521

HOW IT WORKS

Balances and Scales	18
Smoke Detectors	179
Hot and Cold Packs	208
Clocks	318
Thermostats	354
Car Suspension	401
Speakers	430
Electric Cars	466
Lasers	513

MINI LAB EXERCISES

Understanding Conversion Factors	20
Measuring Volume	31
Finding the Atomic Mass of Eggogen	63
Organizing Elements	73
Modeling Bonds in Three Dimensions	106
Modeling Hexane Isomers	133
Balanced Diet	155
Modeling Chain Reactions	190
Mass and Volume in Solutions	216
Basic Problem	234
Graphing Motion	260
A Weighty Problem	288
Law of Torques	305
Visualizing Potential Energy	331
Understanding Heating	358
Demonstration: Density Stack	379
Making Waves	402
Demonstration: Catch a Wave	436
Observing Electrostatic Charge	449
Magnetic Fields	475
Testing Sunscreen	495
Bending Light	515

WORLDVIEW SLEUTHING

Bulletproof!	44
Giving Due Credit	85
Nuclear Waste	196
Sports Drinks	204
Clean Energy	339
Urban Heat Islands	365
Wave Power Generation	405
The War of the Currents	485
Autonomous Vehicle Sensors	499

APPENDIXES

A	Understanding Scientific Terms	522
B	Math Helps	524
C	Fundamental and Derived Units of the SI	532
D	Metric Prefixes	533
E	Common Abbreviations and Symbols	534
F	Creating Graphic Organizers	536

GLOSSARY — 538
INDEX — 553
PERIODIC TABLE OF THE ELEMENTS — 570

TAKING OFF WITH PHYSICAL SCIENCE!

Have you ever thought about how pilots get all the experience they need to safely fly in all conditions? Pilots obviously need actual flight time for training, but there are many things, such as aircraft malfunctions and severe weather, for which a real airplane is not the ideal classroom! That is where *flight simulators* come in. For a simulator to be realistic, the modeling of aircraft movement has to be accurate. Programmers have to model the behavior of the aircraft as well as the aircraft's response to weather conditions and other inputs. By using the computer models in flight simulators, we can prepare pilots to handle both routine and extreme conditions. There's a lot of science and technology that goes into making these simulators realistic enough to prepare pilots for anything they might encounter.

But what exactly is science? You might think that it's discovering truth about the world around us. In reality it is about *creating models* that seek to explain and describe what we observe in the world. You're probably familiar with physical models, perhaps of ships, airplanes, or even atoms. But models can also be conceptual or mathematical—they can even be computer simulations.

Physical Science 6th Edition will provide you with the foundation for further study in the fields of chemistry and physics. These two sciences are ones that you do all the time, though you may not realize it. Chemistry impacts your life when you eat, when you use cleaners, and when you cook. Physics is instrumental in playing sports, doing art, and playing an instrument. In your study of physical science, you will learn of the models that scientists have developed and modified throughout history to better explain and describe the world around them. This study of physical science will have you answer the following questions.

1. How do scientists use models in science?
2. What are the limitations to models in science?
3. How accurately have scientific theories predicted results throughout history?
4. What does the periodic table tell me about the physical and chemical properties of elements?
5. How are energy and matter related?
6. How can I use models to describe, explain, and analyze chemical and physical systems?
7. How can I use biblical principles, outcomes, and motivations to decide what is right or wrong in the field of physical science?

So make sure your seat belts are securely fastened and prepare to launch into our study of physical science!

TAKE A PEEK INSIDE!

We've designed this textbook with you in mind. We hope it will help you appreciate the wonders of God's creation even more. Flip through the following pages to see the features that we've included to help you succeed in PHYSICAL SCIENCE. In the back of the book you'll find appendixes, a glossary, an index, and the periodic table of the elements.

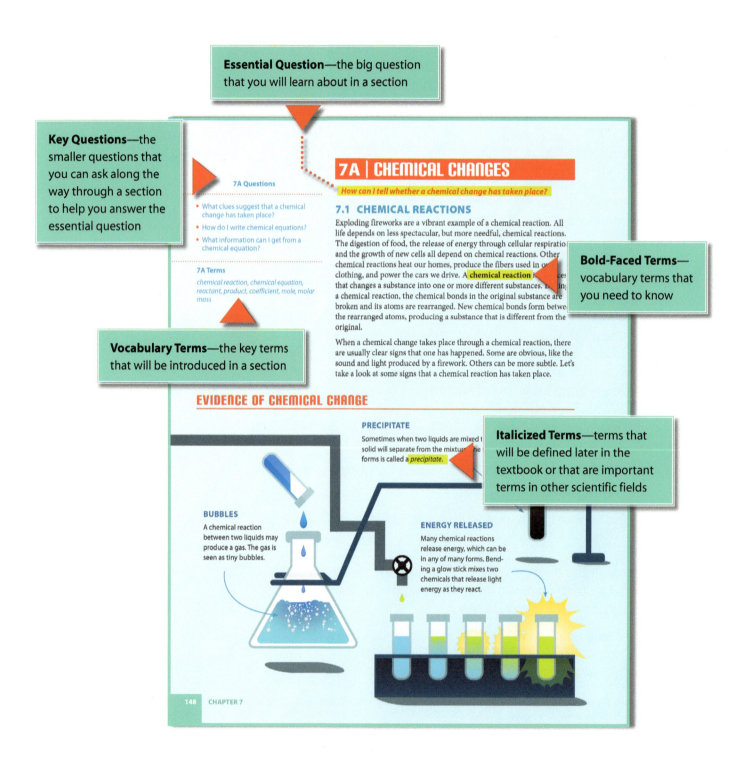

Essential Question—the big question that you will learn about in a section

Key Questions—the smaller questions that you can ask along the way through a section to help you answer the essential question

Vocabulary Terms—the key terms that will be introduced in a section

Bold-Faced Terms—vocabulary terms that you need to know

Italicized Terms—terms that will be defined later in the textbook or that are important terms in other scientific fields

Chapter Opener—a short article that highlights issues and developments in physical science that demonstrate how science intersects with your life

CHAPTER 2
Matter

IMITATING GECKOS

Geckos have a superpower: defying gravity. The ability of the gecko on the left is not due to its feet being sticky; it's because they are hairy. The tiny hairs on its feet are about 1/30th the diameter of a human hair, and there are thousands of them on every square millimeter of a gecko's feet! With each step, the gecko spreads his toes, maximizing the contact area of each foot. This action creates a tiny electrostatic attraction with the surface. The combination of these tiny forces creates a substantial attractive force. Happily for the gecko, God also designed geckos so that they can release that attraction!

Materials scientists are studying the gecko to see whether they can copy

Worldview Sleuthing Boxes—inquiry-based investigations that help you think through current topics in physical science through the lens of Scripture

WORLDVIEW SLEUTHING: BULLETPROOF!

Bulletproof technology is an extension of the ancient art of armor making. Armor proved ineffective against modern weaponry, and metal suits are too heavy for the modern battlefield. Materials scientists started to investigate cloth materials. In the late 1800s, a doctor in Arizona discovered that silk could keep a bullet from penetrating the body, but silk was also extremely expensive. Over time scientists searched for other natural and synthetic (manmade) fibers with similar properties. Today scientists are looking at other applications for these materials.

TASK
You are working as a clothing designer for the nation's Olympic sports team. You have been assigned to research advances in ballistic materials to be used in clothing that will perform well and protect speed skaters from sharp skate blades during accidents. You are to write a one-page proposal for a material to be used for these uniforms.

4. Write your proposal and show it to another person for feedback.

CONCLUSION
We are commanded in Scripture to serve others. We do this by meeting the needs of people around us. Developing materials such as these can protect the lives of athletes. Could they also help protect the lives of military and law enforcement personnel?

CHAPTER 2 REVIEW

2A UNDERSTANDING MATTER

- Matter is anything that has mass and takes up space. According to the particle model of matter, all matter is made of tiny particles (atoms and molecules) in constant random motion.
- The particle model of matter is very workable as it explains most of our observations about matter.
- Density is calculated by dividing the mass of an object by its volume.
- Mass is the amount of matter in an object, while weight is the force of gravity acting on that object.

2A Terms
matter	26
law of definite proportions	28
particle model of matter	28
atom	29
molecule	29
mass	29
volume	29
density	29
weight	30

2B CLASSIFYING MATTER

- We classify matter by its physical and chemical properties.
- Matter is classified as either a pure substance or a mixture.
- Pure substances (elements and compounds) contain only one type of substance.
- Mixtures (heterogeneous or homogeneous) are physical combinations of two or more substances in changeable proportions.

2B Terms
pure substance	33
element	33
compound	33
mixture	34
heterogeneous mixture	34
homogeneous mixture	34

2C STATES OF MATTER

- Particles in solids have low kinetic energy compared with the forces between particles. Solids have fixed shape and volume, high density, and low compressibility due to their close spacing and low energy.
- Liquid particles have enough kinetic energy to overcome some of the forces between particles. Liquids have fixed volume, high density, low compressibility due to their close spacing and low energy. Their ability to move makes them fluid and gives them a changeable shape.
- Gas particles have sufficiently high energy to overcome all the forces between them. The particles' wide spacing causes gases to have low densities and high compressibility. Their rapid motion allows them to fully occupy their container, no matter its shape or volume.

2C Terms
solid	35
liquid	35
gas	35
plasma	35

2D | REVIEW QUESTIONS

a chemical or a physi-
(III) oxide.

4. Explain what is happening to the particles in a solid as you warm it to its melting point.
5. Define *boiling point*.
6. Compare evaporation and boiling.
7. State the law of conservation of matter.
8. List the changes of state that involve adding energy.

Chapter Summary—handy statements of the big ideas of the chapter, including vocabulary lists

MATTER 45

TAKING OFF WITH PHYSICAL SCIENCE! ix

REVIEW

22. What is the percent by volume of a solution made by mixing 345 mL of ethanol with enough water to form 1325 mL of solution?
23. Create a concept map using the terms *solute, solvent, solution, solubility, concentration, concentrated, dilute, unsaturated solution, saturated solution, supersaturated solution, percent by mass, percent by volume,* and *molarity*.

Critical Thinking

24. Where would saturated, unsaturated, and supersaturated solutions be on the graph on page 220?
25. Calculate the percent by mass of 155 g of a solution that contains 142 g of solvent.
26. How many grams of salt should be added to 100 g of water to make a 14.5% salt solution?
27. Show how you could convert a percent by mass to a percent by volume. (*Hint:* Remember that mass and volume are related by density ($d = m/V$).)
28. Some classmates tell you that if they mix 100 mL of a 3.5 M sugar solution with 100 mL of a 5.0 M sugar solution, they will have 200 mL of an 8.5 M solution. Are they correct? Explain.

Use the Case Study at right to answer Questions 29–32.

29. Which material is the solvent and which is the solute in the sap?
30. What process is being used to separate the water from the sucrose in the sap?
31. What is the percent by mass of sucrose when done?
32. Why does the boiling point rise while the water is boiling off?

Use the Ethics Box below to answer Question 33.

33. Using the strategy presented in Chapter 3, write a one-page essay about how Christians should approach this issue.

Review Questions—questions that will have you recall facts, demonstrate your understanding of concepts, and cause you to use critical thinking

CASE STUDY: MAPLE SYRUP

Many people love the unique taste of maple syrup, which is a mixture made of primarily sucrose and water. The syrup is made from the sap of the sugar maple tree. The sap is collected in late winter and early spring. All of the collected sap is then boiled, removing much of the water. The process is complete when the boiling point has risen to 104.1 °C. The final syrup contains about 60 g of sucrose for every 100 g of syrup.

Case Studies—opportunities to apply what you have learned in physical science to a real-life example

Career Boxes—information about careers in physical science that can be pursued to wisely use God's world and help people

ETHICS POLLUTION

The Issue: Air Pollution

Air pollution is an issue in many areas around the world. The compounds that cause air pollution come from sources that are both natural (volcanoes, plants and animals, and forest fires) and manmade (fossil fuel power plants, motor vehicles, and aerosols). While most people think of air pollution being outdoors in major urban areas, air pollution can be indoors also. The impact of this pollution is widespread. Air pollution can damage both the natural environment and manmade structures. The health impact is enormous, increasing the number of cases of asthma and allergies as well as lung and heart disease. The World Health Organization estimates that air pollution causes about 7 million deaths each year.

Ethics Boxes—opportunities to apply a biblical worldview to ethical issues in physical science

SERVING AS A PIPING ENGINEER: KEEPING THE FOOD FLOWING

Natural gas, ketchup, ice cream—each is a fluid product that is transported through a complex system of pipes. Piping engineers are mechanical or materials engineers who specialize in designing pipe systems. Designing a facility with different-sized vats, storage tanks, and a maze of pipes connecting them is not as simple as it may sound. Many factors come into play, such as the temperature and viscosity of the product, pressure changes that occur when the diameter of a pipe changes, and choosing the most efficient route for pipes through a plant. Pipes full of liquid can be heavy, too, so the design must include sufficient support for all the equipment.

Prior to his retirement, Dave Lombard served as a piping engineer in California's agriculturally rich Central Valley. He helped design many piping systems used in the processing and packaging of foods and beverages. Dave says, "I really enjoy food and beverage work because everyone needs to eat. It's very fulfilling to be a part of that!" If you like challenging work, a career in piping engineering might be a good choice for you!

You might expect that faster-flowing fluids have higher fluid pressures, but in fact they don't. They actually have lower fluid pressures. This relationship between the increasing speed of a fluid and its decreasing pressure is known as **Bernoulli's principle**. A Swiss mathematician, Daniel Bernoulli, described the phenomenon in 1738.

Bernoulli's principle can be put to many practical uses. It partly accounts for the lift generated by aircraft wings. A curved upper wing surface is shaped so that air flows faster over the top of the wing, creating lower pressure on that side. Hose-end sprayers, used for applying fertilizers to lawns and gardens, operate on Bernoulli's principle too. Water passing over a tube inside the sprayer creates low pressure that draws the fertilizer up the tube and into the stream of water. The hourglass shape of a de Laval nozzle creates a region of low pressure in the exhaust gas of a rocket motor. The low pressure increases the speed of the vented gas, providing more thrust for a rocket.

de Laval nozzle

16C | REVIEW QUESTIONS

1. How does Pascal's principle explain the operation of a hydraulic lift?
2. What quantity remains constant within a fluid system regardless of the fluid's velocity or pressure?
3. State Bernoulli's principle.
4. The powerhead shown at right is a type of aquarium pump that circulates water. If a piece of plastic tubing is inserted into the discharge pipe on the powerhead, a stream of air can be drawn into the flowing water. Use Bernoulli's principle to explain how this is possible.

TAKING OFF WITH PHYSICAL SCIENCE!

MINI LAB
BENDING LIGHT

Rainbows are caused when white light from the sun is refracted and reflected by tiny water droplets in the atmosphere. The different colors within white light don't bend at the same rate, so they separate as they bend. You've probably seen a prism used to produce the same kind of separation. But once the colors are separated, is it possible to recombine them into white light again? Your teacher will use a light ray box and several kinds of lenses to help you think about this question.

Essential Question:
Can a rainbow be undone?

1. Do the colors of light separated by the prism follow the progression described in Subsection 22.1?
2. On the basis of what you observe, which color of light is bent the most and which the least? How can you tell?
3. Do you think it will be possible to recombine the separated colors back into white light using a lens? If so, which lens do you think will work?
4. Describe how each kind of lens affects the separated colors.

CONCLUSION
Different kinds of lenses differ in how they bend light rays. As you read further, you'll learn about these different lenses and see how they can be used to benefit people.

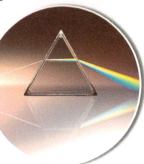

Mini Labs—short hands-on activities to get you thinking and working like a scientist

In the example on the right, you can see how light bends as it passes through different media. The ray first bends as it passes from air into water because water's index of refraction is greater than air's. The ray is bent again as it passes into glass because glass has a greater index of refraction than water. Because the boundaries in this example are parallel, the ray returns to its original direction of travel as it passes out of the glass and back into air.

Total Internal Reflection

In some situations, a ray of light *can't* pass from one medium to another. Instead, the ray reflects off the boundary between the two media and remains within the first medium. This phenomenon is called **total internal reflection** (left). It happens when a light ray's angle of incidence exceeds a certain critical value when going into a medium with a lower index of refraction.

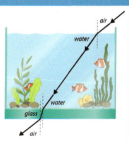

LIGHT AND OPTICS 515

18C | USING SOUND WAVES
In what other ways do we use sound waves?

18.7 SOUND TECHNOLOGIES

Acoustic Amplification
Since sound is a form of energy, it makes sense that such energy can be put to good use. And indeed, not only people but animals too use sound energy for many purposes. In this section, we'll explore some of those uses.

Acoustic amplification is the process of making a sound louder. A megaphone is a simple kind of amplifier known as an *acoustic horn*. Normally when a person speaks, the sound waves of his voice spread out over a large area. A megaphone focuses that sound energy into one specific direction, making it sound louder. If you point the large end of a megaphone at the source of a sound and listen at the smaller end, the same thing happens. This is why people sometimes cup their ears to hear better. Amplification also happens in any enclosed space with hard surfaces, where sound reflection occurs easily.

HOW IT WORKS
Speakers
Stereo speakers are described with some strange terms. Woofers, midrange, tweeters—just what do these things mean? And how does a speaker work anyway? A speaker contains a cone (1) made of stiff material connected to a voice coil (2) made of wound copper wire. The voice coil is suspended inside a powerful circular magnet (3). An amplified electrical signal enters the voice coil, creating, in essence, an electromagnet (see Chapter 21). The electrical signal causes the voice coil's magnetic field to rapidly change polarity many times per second. As this rapidly changing field interacts with the field of the magnet, the voice coil vibrates back and forth. The vibrating speaker cone pushes against the air, creating sound waves.

The size of the cone relates to the frequency of the tones produced. Large cones are needed to produce deep, low-frequency sounds—the bass end of what humans can hear. These speakers are the "woofers" in a multi-cone speaker. "Tweeters" are the small cones that produce high-pitched, high-frequency sounds. Large speakers may have three or more cones, including mid-range cones. Such speakers are better able to recreate the variety of tones carried by an electrical signal. Single-cone speakers are limited in the range of frequencies that they can produce. They are often unable to produce some low- and high-end frequencies well.

430 CHAPTER 18

How It Works—examples of physical science applications in everyday items

TAKING OFF WITH PHYSICAL SCIENCE! xi

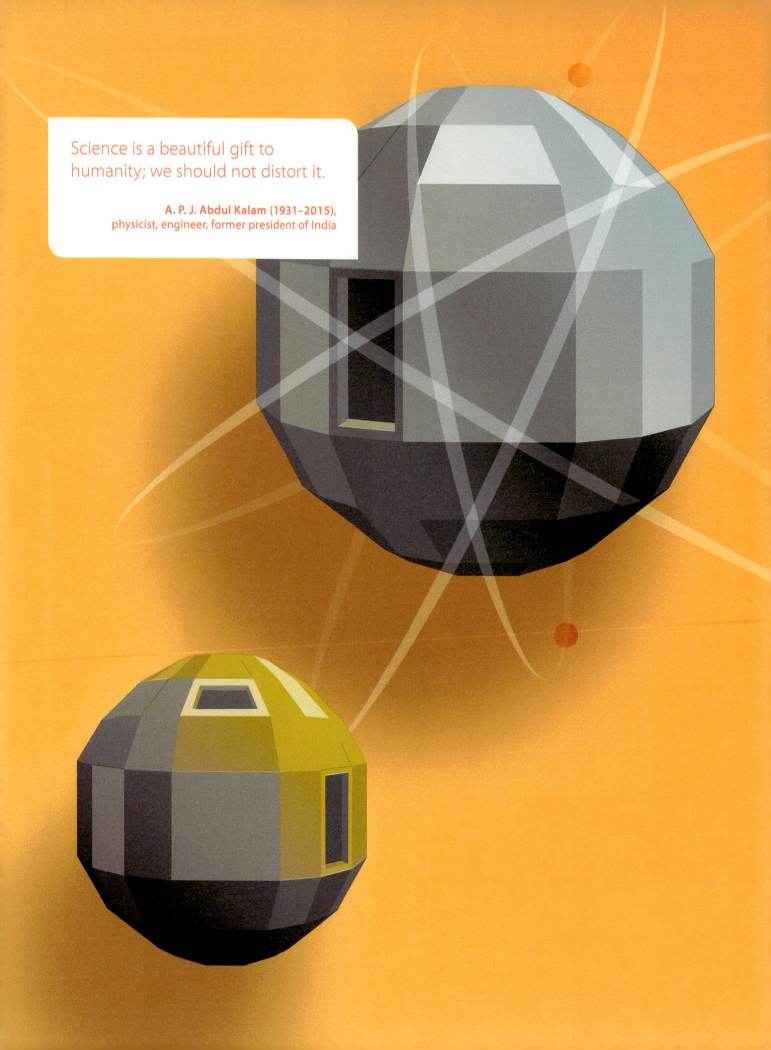

UNIT 1
THE STRUCTURE OF MATTER

CHAPTER 1: **MODELING OUR ORDERLY WORLD**

CHAPTER 2: **MATTER**

CHAPTER 3: **THE ATOM**

CHAPTER 4: **THE PERIODIC TABLE**

CHAPTER 5: **BONDING AND COMPOUNDS**

CHAPTER 6: **THE CHEMISTRY OF LIFE**

CHAPTER 1
Modeling our Orderly World

SUPER SLEUTHING SCIENTISTS

Drops of blood under the table, beige powder dusting the countertop, and a footprint in the front hall. Each of these raises multiple questions for the forensic scientist. Whose footprint is that? What is that beige powder? Whose blood is that? Is this evidence related to the crime? What other trace evidence can I find?

Forensic scientists collect evidence and conduct investigations to determine what occurred during a crime. Their task is most difficult when there are no witnesses. Forensic scientists may gather evidence at the crime scene or analyze data in a lab or on computers. Whatever their task, they must be observant so that they don't overlook evidence. They must be meticulous in handling the evidence. They conduct tests and experiments to determine the evidence's significance. Finally, they come to conclusions about evidence, helping the police arrest the perpetrator of a crime.

The culmination of the forensic scientists' work occurs in the courtroom. They present their findings and provide expert opinions about what the evidence means. They have to be both accurate and precise in their testimony. They must be well spoken and have the ability to present their findings. Every day forensic scientists use scientific inquiry to see that justice is served.

1A Order in Our World	4
1B Modeling Our World	11
1C Using Mathematics for Scientific Inquiry	14

1A | ORDER IN OUR WORLD

1A Questions

- How does physical science relate to other sciences?
- What are some evidences of order in the world around us?
- Why is there order in our world?
- How can we use science to glorify God and help others?
- How can scientists make ethical decisions?

1A Terms

science, physical science, chemistry, physics, Creation Mandate, ethics

What is the source of order in nature?

1.1 PHYSICAL SCIENCE

In the Chapter opener, you saw how forensic scientists are always asking and answering questions. This constant questioning is not something that only forensic scientists do. It is a characteristic of all scientists.

Scientists spend their careers asking questions and investigating events or facts called *phenomena* (s. phenomenon). They do this in an attempt to explain or describe the world around them. Some do this because they are naturally curious and want to learn as much as they can about the world around them. Others have a desire to use what they have learned to help other people. No matter their motivation, they all do science. But what *is* science?

Science is the systematic study of the universe to produce observations, inferences, and models. It also includes the products created through this systematic study. Science is initially divided into social science, the study of human societies and relationships, and natural science, the study of the natural world. We further divide natural science into numerous different fields. While we organize and describe these fields of study as if they were completely isolated from each other, they actually are very much interrelated.

DIVISIONS OF SCIENCE

NATURAL AND SOCIAL SCIENCE

NATURAL SCIENCE is divided into life science, earth science, and physical science.

LIFE SCIENCE
Examples: biology, botany, zoology, medicine

PHYSICAL SCIENCE
the study of nonliving matter and energy

EARTH SCIENCE
Examples: geology, oceanography, meteorology, astronomy

PHYSICS
the study of matter and energy and the interactions between them

CHEMISTRY
the study of the composition, structure, and properties of matter as well as changes in matter

Life science and physical science overlap.
Examples: biochemistry, nutrition, biophysics, pharmacology

Life science and earth science overlap.
Examples: astrobiology, ecology, climatology

Physical science and earth science overlap.
Examples: mineralogy, astronautical engineering, geophysics

Life science, earth science, and physical science overlap.
Examples: environmental science, space science

MODELING OUR ORDERLY WORLD

1.2 WHY IS ORDER IMPORTANT?

Why can we study the universe systematically? If phenomena were completely random, we would have to accept just watching events happen, but investigating them would be impossible. However, our world is orderly, and order allows us to study the events that occur.

Evidence of Order

As we wake up each morning, the periodic changes of day into night, days into weeks, weeks into months, and months into years remind us of the order in nature. The cycles that we know as seasons have been guiding farmers in the planting and harvesting of crops since Creation. We see order in the repeated patterns of the chemical and physical properties of elements.

Order in the natural world is so evident that scientists even imitate it. Chemists arrange the elements in a *periodic table* by the repeated patterns in the structure and properties of those elements. Biologists have developed a system for classifying living organisms. This classification system uses the order found in nature to categorize the different species. Order is what allows us to do science successfully.

Result of Order

Watching something random, like the movement of flames in a campfire or raindrops on a pond, may be exciting and even entertaining. But if there is no pattern to the events, then we cannot identify causes, and we certainly cannot make predictions about these events. However, since order is a built-in part of the world in which we live, we can do science. With enough observations, we can even begin to make predictions about future events on the basis of the patterns that we have observed.

One key principle that allows us to study science is the *law of cause and effect*. This law states that every effect has a specific, identifiable cause, and for every cause, there is a definite and predictable effect. We can sum up the law this way: everything happens for a reason; nothing just happens. Therefore, scientists know that they can investigate a particular phenomenon to determine its cause. Similarly, after sufficient investigation, the scientist should be able to predict the effect of a given cause.

Another important aspect of our world is the *principle of uniformity* of nature. This principle declares that nature acts the same today as it did yesterday and that we can fully expect it to act the same way tomorrow. The uniformity in nature is what allows us to make predictions in science. But even the characteristics of uniformity and predictability had to be caused by something or someone.

Source of Order

Where does this order come from? Nothing left to itself becomes more orderly than it was before. According to the law of cause and effect, the *effect* of order in the universe had to be the result of a specific, identifiable *cause*.

Genesis 1 outlines the creation of the universe. God created all things out of nothing. The universe that He created is a reflection of His very nature. When He completed His creative work on Day 6, He evaluated everything and declared that everything was good (Gen. 1:31). Some will ask, didn't *God* need a cause? Remember, however, that the law of cause and effect relates to everything *in our world*. As mentioned in Genesis 1:1, God existed before the world He created and is therefore outside of it. And God created the universe with order because He is a God of order.

1.3 WHY WE DO SCIENCE

Order in nature allows us to do science, but what makes it worthwhile? The answer to this question depends on your understanding of the world. We all view the world on the basis of assumptions we have about the world. This is called our *worldview*. As you can imagine, every decision we make is affected by our worldview.

There are many divergent worldviews, some religious and others secular. People who hold to secularism are not necessarily atheistic. Secularists simply believe that we should exclude religious beliefs from the public sphere of discourse.

Secular scientists believe that they can explain the universe and all that is in it by solely naturalistic means. There is no room for the supernatural. According to their worldview, everything began from nothing in the big bang and slowly formed over billions of years. Man is the product of evolutionary changes driven by natural selection over millions of years. Man's purpose is to understand the world around him to help him improve his life.

People with a religious worldview see the world through the lens of the teaching of a particular teacher or text. Christians develop their worldview from the teaching of the Bible. According to a biblical worldview, the world is relatively young and is the product of Creation as outlined in Genesis 1. God created the universe and all that is in it from nothing and for His good pleasure. Everything was created to glorify God. People are a special creation—made in His image.

SERVING AS A FORENSIC SCIENTIST: SILENT WITNESS

Forensic anthropologist Clea Koff has given a voice to the victims of genocide around the world when entire populations have been murdered with no witnesses. There is often no justice in cases of genocide because it is carried out during civil wars or by the dominant group in a region. Fear of the dominant group often causes these cases to go unreported. Even when reports surface, few witnesses are available or willing to testify. In these situations, teams of forensic scientists from the United Nations get involved.

Clea Koff used the remains in mass graves in Rwanda and Yugoslavia to build cases for genocide in those countries. After concluding her work there, Koff turned her attention to the United States. She used her anthropology skills to help families of missing persons. She also founded the Missing Person Identification Resource Center in Los Angeles.

Crime is a fact of life in a fallen world. Through forensic anthropology, Clea Koff has helped others find closure after loved ones had disappeared.

Clea Koff

An artist involved in forensic anthropology works on 3D facial reconstruction.

A Christian's purpose in science is the same as his purpose in life. We were created to love and glorify God (Ps. 86:9; Matt. 22:37) and to serve others (Matt. 22:39). In fulfilling our purpose in life, we can accomplish a crucial aspect of a biblical worldview—redemption. Creation was very good, but man damaged it by introducing sin into the world at the Fall. Each of us needs personal redemption, the forgiveness of our sins. But our thinking also needs to be redeemed (Rom. 12:2), as does the physical world (Rom. 8:22), because every aspect of the Creation was adversely affected by the Fall.

Our purpose in life and science stems from the Creation Mandate. The **Creation Mandate** (Gen. 1:26, 28) directs us to fill the earth and have dominion over it. God's command to humans is to exercise wise and good dominion over His creation for the glory of God and the benefit of their fellow humans. How can we accomplish this as scientists?

We do this by studying and caring for God's creation. We observe and then replicate in our models and understanding of nature the order created by God. We use science to meet the needs of other people, God's image bearers. We imitate God's creative work through inventing and engineering products to benefit people. All that we do in science should honor and glorify God while caring for His creation.

1.4 SCIENCE AND ETHICS

Scientists don't work in isolation; the science they do affects people around the world. As we do science, we must decide what is right and wrong. These decisions can become complicated very quickly. While science can tell us that a person's brain is not working, it cannot tell us whether we should disconnect the person from life support. So how do we decide whether a particular use of science is right or wrong?

What we are talking about is **ethics**—a system of moral values or a theory of proper conduct. Our worldview plays a key role in making ethical decisions. Christian ethics should be based on the three-element foundation of *biblical principles*, *biblical outcomes*, and *biblical motivations*.

Biblical Principles

God created the world and everything in it. He created man as His special creation and gave us His Word to guide our lives. While the Bible doesn't specifically address everything that we might face in life, His Word gives us enough guidance to make right decisions. God tells us what is good and evil. By studying Scripture, we can understand general principles that we can use in all situations. So the first thing that we should ask when we encounter an ethical issue is, "*What does God's Word say?*" The three biblical principles that lead to right thinking are people as God's image bearers, the Creation Mandate, and God's whole truth.

Biblical Outcomes

We also must think about what we want to achieve. God tells us in the Bible the goals that He has for us. Ethical decisions must include striving for right outcomes. A second question we must ask when faced with making an ethical decision is, "*What results are right?*" Clear goals follow from the biblical outcomes of human prospering, a thriving creation, and glorifying God.

Biblical Motivations

When God saves us, He intends for us to be changed to be more like Him. We better reflect His image as we grow. A Christian should do right because it contributes to his growth, not just because the rules tell him to or because he wants good results. A Christian should want to be more like the Lord. So the third and most important question is, "*How can I grow through this decision?*" Three biblical motivations follow from Scripture: faith in God, hope in God's promises, and love for God and others.

So how should a Christian make ethical decisions?

What does God's Word say?

What results are right?

How can I grow through this decision?

Cardiology researcher accused of falsifying data. The investigation detailed forty-five instances of the use of false data in publications.

ETHICS: CHRISTIAN ETHICS

BIBLICAL PRINCIPLES
What does God's Word say?

God's Image Bearers. Foundational to our ethical decision-making is the understanding that we all bear God's image. Therefore, we must make decisions out of respect for all people and for their protection (Gen. 1:26–28).

Creation Mandate. God's first commandment to us is to have dominion over the world that He created. Therefore, we must wisely care for God's creation. We must balance the appropriate use of the world's resources with the needs of people around the world. Nothing belongs to us; we are stewards of God's world (Gen. 1:26–28).

God's Whole Truth. God's image in people and the Creation Mandate touch on many ethical issues. But other parts of Scripture also give us helpful insights into what God wants us to do. Part of making wise ethical decisions requires that we understand what His Word teaches. We cannot live any part of our lives separated from God and His Word (2 Tim. 3:15–16).

BIBLICAL OUTCOMES
What results are right?

Human Prospering. As soon as God created mankind, He blessed him (Gen. 1:28). Throughout Scripture (Ps. 1; Matt. 5), we see that God's desire is for all people to be blessed and to prosper. Jesus came to give us life and to have us live that life abundantly (John 10:10). Our ethical decisions must align with God's will to maximize human development.

A Thriving Creation. Part of our obligation to the Creation Mandate is to ensure that creation thrives (Gen. 1:28; 2:5, 15). We wisely use and develop the earth's resources to ensure that it flourishes.

Glorifying God. Just as Jesus came to glorify God, everything we do should glorify God (Matt. 5:16; 1 Cor. 6:20; 10:31). Our decisions should show God that we love and honor Him. This obligation includes every aspect of our lives: school, work, and play. So it is not enough that our decisions help others or that creation thrives. Our decisions must always give God the honor that He is due.

BIBLICAL MOTIVATIONS
How can I grow through this decision?

Faith in God. The Bible discusses works versus faith (Jam. 2:14–26). The passage concludes that we are to live out our faith in God through our works. We are motivated to act because of our faith in God. Good works can stem only from our faith in God (Rom. 14:23; Heb. 11:6).

Hope in God's Promises. In the Bible, hope is not something that we wish for; it is something that God has promised. Biblical hope is confident expectation. The promises of God allow us to take action without fear (2 Tim. 1:7–9). Scripture teaches us that God can never lie, so we can act with the assurance that God will follow through on His promises.

Love for God and Others. As stated in 1 Corinthians 13, our greatest motivation for doing right is love. We have the love of God in us, and we do right when we are motivated by our love for God and our love for others. John 13:34–35 teaches that love is the outward sign of a transformed life.

1A | REVIEW QUESTIONS

1. Define *science* in your own words.
2. Graphically display the relationship between social science, natural science, physical science, chemistry, and physics.
3. Give an example of orderliness in nature.
4. Why does order imply a Creator?
5. What is the Creation Mandate?
6. Give two examples to show how we can fulfill the Creation Mandate.
7. Define *ethics*.
8. What are the three questions that guide Christians in making ethical decisions?

1B | MODELING OUR WORLD

How do scientists do science?

1.5 MODELING IN SCIENCE

As we mentioned, scientists seek to explain or describe the world around them. Scientists use models to help them do this. A **model** is a workable explanation or description of a phenomenon. A model may be physical, conceptual, or mathematical.

Two major categories of models in science are theories and laws. A **theory** is a model that *explains* a related set of phenomena. It can be used to predict unobserved aspects of the phenomena. On the other hand, a model, often expressed as a mathematical equation that *describes* phenomena under certain conditions, is called a **law**. It does not attempt to explain the phenomena.

So what is the benefit of using a model? Models allow scientists to focus on a particular portion of the world around them. A model helps them understand and communicate their understanding of the phenomenon being studied.

Types of Models

Because of the complexity of physical systems, scientists use various models in physical science. Physical models are quite common, including atomic models, models of waves, models for the behavior of gases, ball and stick models of molecules, and scale models of vehicles. Conceptual models can take as many forms as there are scientists. Scientists typically use conceptual models for thinking about ideal systems. Mathematical models show up everywhere in physical science. They include equations or computer modeling (virtual models) of weather, atmospheric changes, water flow, or any computer simulation.

1B Questions

- Why are models important in physical science?
- What models are important in physical science?
- How do hypotheses, theories, and laws compare?
- How do we do science?
- Why do we approach science systematically?

1B Terms

model, theory, law, workability, scientific inquiry, hypothesis

Limitations of Models

Scientific models change because scientists are always learning and improving their understanding of the world around them. People often think about scientific models as being either true or false, right or wrong. But valid models are true though incomplete all the time.

The Goal of Models: Workability

The key to good models is **workability**. What do you think this means? Scientists use two criteria to evaluate models: How well does the model explain or describe what we observe? And how accurate are predictions made with this model? If a model explains or describes the world well and makes accurate predictions, then scientists will retain it. Scientists will modify or discard models that don't do these things well.

To understand this concept of workability, we can think about how scientists have understood the relationship between the sun, earth, and the planets. The geocentric model for this relationship was developed first and stated that the earth was the center of the universe. Everything else revolved in perfect circles around the earth. The model did well at explaining the basic motion of the sun, moon, planets, and stars as seen from the earth. But it couldn't explain how some planets appeared to stop and reverse their direction (now understood as *retrograde motion*). It also couldn't explain the varying distances that the moon and planets were from the earth during their orbits. Scientists tried to adjust the geocentric model, improving its workability, but ending up making it very complicated.

In time, Nicolas Copernicus proposed the heliocentric model, with the sun at the center of our solar systems and the planets (including the earth) orbiting the sun. The heliocentric model explained the observations (retrograde motion and planetary distances) better than the geocentric model did and made better predictions. While it wasn't perfect, it was simpler and more workable than the geocentric model. So the geocentric model was discarded, and the heliocentric model replaced it. As time has passed, the heliocentric model has been modified to make it even more workable.

As we learn more about a particular branch of science, our models get better (more workable). Each successive model explains the phenomena better and makes more accurate predictions. Sometimes models are completely wrong and scientists have to start with a completely new model (a scientific revolution). As long as they remain objective, scientists aren't bothered by this. New or improved models help scientists continue to refine their understanding of the world around them.

1.6 SCIENTIFIC INQUIRY

How do scientists do science? They work through a process called **scientific inquiry**—an ongoing, orderly, cyclical approach used to investigate the world. Inquiry is similar to the scientific method that you have learned previously, but it is a continuous process that is more cyclical than linear.

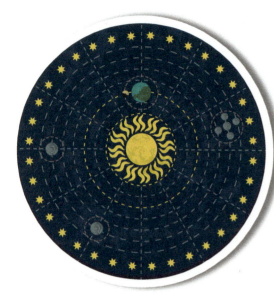

As their understanding of the solar system changed, scientists initially modified and ultimately discarded the geocentric model and replaced it with the heliocentric model.

SCIENTIFIC INQUIRY

1 *Observation.* By using his senses, a scientist collects information about the world. Observation is the central activity in science. Scientists must pay great attention to detail and be careful to collect and record data accurately.

2 *Posing Questions.* Posing questions is a fundamental skill for the scientist. The questions give him his purpose for doing science. These questions usually stem from something he observed.

3 *Research.* Scientists do lots of research. When faced with a new question, a scientist needs to determine what others already know about the topic. This research will allow him to focus his investigation on aspects of the issue that are not well understood.

4 *Forming Hypotheses.* As mentioned previously, scientists work with models all the time. One type of model is a **hypothesis**—an initial, testable explanation of a phenomenon that stimulates and guides the scientific investigation.

5 *Investigation.* Investigation is the step that most people associate with science. Often this takes the form of an experiment. A scientist designs an experiment to address the hypothesis that he formed on the basis of his question. In a controlled experiment, the scientist changes one variable (the independent variable) and observes the resulting changes in a second (dependent) variable, while holding all other variables constant.

6 *Analysis.* Once the scientist has gathered sufficient data, he has to figure out what it means. This analysis may involve graphing, reasoning, and using mathematics. Again the scientist must be careful to analyze his data accurately.

7 *Conclusion.* After analyzing his data, the scientist concludes what he has learned. This step connects all the previous activities. The conclusion indicates to what degree the experimental evidence supports or refutes the hypothesis. Hopefully it allows the scientist to answer the question.

8 *Communication.* The scientist still has work to do! Science relies on a process of *peer review*. Before publishing his findings, a scientist will have other scientists review and respond to the research. Through this process, models are strengthened, modified, or even discarded. It's part of the reason that scientists must be accurate in collecting, analyzing, and reporting their findings.

1B | REVIEW QUESTIONS

1. What is a model?
2. How do scientists use models?
3. What is a theory?
4. How does a law differ from a theory?
5. What do we call the orderly process by which we do science?
6. How does scientific inquiry differ from the scientific method?
7. What is the central activity in the orderly process of science?
8. What is a hypothesis?
9. Why do we need a process to study science?
10. Why is communication important to scientists?

1C Questions

- Why do scientists use the SI?
- How do we collect data?
- Why do I need math to do science?
- Aren't accuracy and precision the same thing?
- What can I tell about a scientist's instrument from his measurements?
- How can I change units in the SI?
- What happens when scientists are not accurate?

1C Terms

measurement, SI, accuracy, precision

There are as many scientific instruments as there are dimensions to measure.

1C | USING MATHEMATICS FOR SCIENTIFIC INQUIRY

How is math used in scientific inquiry?

1.7 MEASUREMENTS

Data

As scientists make observations, they record information about what they observed; we call it *data*. Scientists collect two types of data, qualitative and quantitative. Qualitative, or descriptive, data is observations about qualities of the object or event. These observations may be related to color, texture, or relative size, for example. Descriptive data tends to depend on the observer and is therefore less repeatable. Data can also be quantitative, meaning that it is based on numbers or quantities, in other words, **measurements**. Scientists collect quantitative data by using measuring instruments. Measured data is less dependent on the scientist, making the data more repeatable.

We take measurements of different dimensions of an object or event. Dimensions are the measurable aspects of something, such as mass, volume, length, or weight. All measurements have a numerical part and a unit, a standard of measure for comparison.

SI

You are familiar with units such as inches and feet, cups and gallons, and ounces and pounds. Most scientists don't use these units of measurement. Scientists from all around the world use a modern system of standardized metric units. We call this system **SI**, which stands for the Système International d'Unités (International System of Units). Scientists sharing data from their investigations recognized the benefit of having a common set of units. So they developed the metric system with units related to observable phenomena in the world. Over time the standards were modified to easily reproducible standards. For example, scientists defined the meter by the speed of light. This improved standardization is beneficial for all scientists. The scientific community further modified the metric system to base it on just seven fundamental units. The SI that scientists use today is this modified metric system.

SI FUNDAMENTAL UNITS

m The fundamental unit of length in the SI is the meter (m).

The kilogram (kg) is the SI fundamental unit of mass. **kg**

s The SI uses the second (s) as the fundamental unit for time.

The ampere (A) is used by scientists in the SI to measure electric current. **A**

K The SI unit for temperature is the kelvin (K).
Notice that we don't use the degree symbol with Kelvin temperatures.

Chemists use the mole (mol) as the SI unit for the amount of a substance. **mol**
A mole is a count that contains 6.022×10^{23} objects.

cd Scientists use the candela (cd) in the SI to measure the intensity of a light source.

The SI also makes use of *derived units*—mathematical combinations of two or more of the base units. For example, we measure speed in m/s, which scientists derived by dividing the unit for distance (m) by the unit for time (s).

The SI is a decimal system, which means that it is based on powers of ten. Each unit in the metric system can be multiplied or divided by a power of ten. The metric system includes prefixes that correspond to a particular power.

MODELING OUR ORDERLY WORLD

METRIC PREFIXES

 Prefix: giga- | 10^9 | Factor: 10^9 (1 000 000 000) | Example: The Three Gorges Dam produces 22.5 GW (gigawatts) of electrical power.

 Prefix: mega- | 10^6 | Factor: 10^6 (1 000 000) | Example: A local radio station transmits on a frequency of 93.5 MHz (megahertz).

 Prefix: kilo- | 10^3 | Factor: 10^3 (1000) | Example: The distance to Orlando, Florida, is 857 km (kilometers).

 Prefix: deci- | 10^{-1} | Factor: 10^{-1} (0.1) | Example: A jet fighter produces 111 dB (decibels) of sound.

 Prefix: centi- | 10^{-2} | Factor: 10^{-2} (0.01) | Example: The length of a candy bar may be 10.0 cm (centimeters).

 Prefix: mill- | 10^{-3} | Factor: 10^{-3} (0.001) | Example: A dose of cough syrup might be 10mL (milliliters).

 Prefix: micro- | 10^{-6} | Factor: 10^{-6} (0.000 001) | Example: A human hair has an average diameter of 85 µm (micrometers).

 Prefix: nano- | 10^{-9} | Factor: 10^{-9} (0.000 000 001) | Example: 680 nm (nanometers) is the wavelength of red light.

Scientists have opted for using the SI instead of the US customary system for a number of reasons. The SI is a decimal system, while the US system is based on a variety of different numbers; a base ten system makes unit conversions easier. The SI has only one unit for each dimension (e.g., meter for length), while the US system has many units. The customary system has units that are based on standards that are not repeatable, but the SI has easily reproducible standards. Finally, the SI has been adopted almost worldwide.

1.8 MEASURING

Much of science involves the collecting and recording of quantitative data. So measuring is a critically important skill for scientists. How well we measure depends on the instruments we use and how well we use them.

Look at the tape measure in the image at right. The units marked are centimeters. As we measure the length of the shelf, notice that the tens place is a 5 and the ones place is a 3. The end of the shelf does not match exactly with any of the marks on the tape. Since the smallest marks on the tape indicate tenths of a centimeter, we have to estimate measurements with this tape to the hundredths of a centimeter. You now have to estimate where between 53.8 and 53.9 the shelf ends. What do you think—should that last digit be a 5?

1.9 LIMITS OF MEASUREMENT

Uncertainty in Measurements

While quantitative data is less subjective and therefore more repeatable than qualitative data, measurements are never 100% correct. In the example above, you may read the measurement of the shelf as 53.85 cm. Your best friend might tell you that it is 53.84 cm. If you look at the book tonight while doing your homework, you may say that the measurement is 53.86 cm. So which is the correct measurement? As strange as it may seem, all three are perfectly good measurements. Every measurement has a degree of uncertainty to it. The uncertainty comes from both the instrument itself and how the scientist used it.

Accuracy and Precision

Since every measurement has some uncertainty to it, we need a way to tell how good our measurements are. Scientists use two ways to assess measurements—accuracy and precision. Accuracy and precision both compare a measurement to something else.

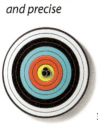

Accuracy compares a measurement to the accepted or expected value of a measurement. When using accuracy, we are looking at how much error is in the measurement. This error is the combination of the instrument uncertainty along with any errors made by the scientist.

Top: accurate but not precise
Middle: precise but not accurate
Bottom: accurate and precise

Another way to assess measurements is by using **precision**—the degree of exactness of the measurements. Precision can indicate the closeness or repeatability of measurements. Precision most often refers to the decimal place to which a scientist made a measurement. So the instrument has great influence on our precision. An instrument marked to the ones place has a precision to the tenths place because we estimate to one decimal place beyond those marked on the instrument.

MODELING OUR ORDERLY WORLD

HOW IT WORKS

Balances and Scales

Pizza! Everyone loves a pizza party. Have you ever been to one and ended up saying, "Wow! I ate way too much pizza!" If so, you may have been reluctant to get on a scale the next day.

Balances and scales basically weigh things. They measure different but related dimensions, and there is much overlap in their use and how we talk about them.

Originally, the balance worked on the principle of, well, balance. It's similar to two kids trying to balance on a seesaw. The double pan balance below was one of the first balances. It works by placing the object that you want to measure in one pan and known masses in the other. When the device is balanced, you know that the unknown mass is equal to the sum of the known masses.

A triple beam balance works on this principle. You place the unknown mass on the platform. You then slide the known masses along the three beams. Once balanced, you record the value of the known mass using the scales on the three beams. This mass is equal to the mass of the unknown object.

Mass and weight are related. On the earth, the relationship between mass and weight is constant. Scales don't measure mass but the related dimension of *weight*—force due to gravity. Within the scale are levers and a spring system. As you push down on the scale platform, the applied force transmits to the spring system. When the spring force is equal to the weight, you can read the weight of the object.

Modern electronic balances act more like scales than balances. Applying a force to the platform causes an electromagnet to create an opposing force. These opposing forces produce an electric signal, which is calibrated to indicate mass.

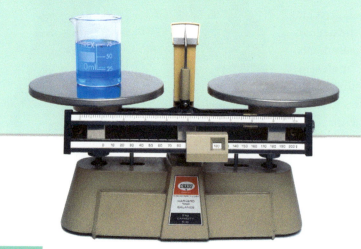

A measurement made to the thousandths place was made using an instrument marked to the hundredths place. This instrument is much more precise than an instrument marked to the ones place.

Scientists report their precision using *significant figures*. All the known digits of a measurement and the one estimated digit are significant. You will learn more about significant figures in chemistry and physics courses.

1.10 UNIT CONVERSIONS

Your best friend tells you that she did a lot of reading over the summer. She says that she spent 1,260,000 seconds reading. You may have difficulty recognizing whether this is a significant amount of time. If you knew how many hours she read, you could probably better evaluate her claim.

We can convert between any two units of measures (for the same dimension) as long as we know the *conversion factor*. A conversion factor is two quantities with different units that are equivalent to each other and written as a fraction. For example, we know that twelve inches is the same as one foot. So we could write 12 in. = 1 ft. To turn that fact into a conversion factor, divide both sides by 1 ft.

$$\frac{12 \text{ in.}}{1 \text{ ft}} = \frac{\cancel{1 \text{ ft}}}{\cancel{1 \text{ ft}}}$$

$$\frac{12 \text{ in.}}{1 \text{ ft}} = 1$$

We can express this conversion factor in two ways:

$$\frac{12 \text{ in.}}{1 \text{ ft}} \text{ or } \frac{1 \text{ ft}}{12 \text{ in.}}$$

Since the numerator and denominator are equivalent, these conversion factors equal 1, and we write them in a form that allows us to change units. When we change the unit of a measurement using a conversion factor, we really are multiplying by 1. The process is similar to how we rename fractions with different denominators. Let's say we needed to rewrite 1/3 as a fraction with 12 in the denominator. What would we do? Looking at the denominators, we know that we need to multiply the 3 by 4 to get 12. To keep the value the same as our original number, we must multiply by 4/4, which is just another form of 1.

Let's see how much your friend read this summer. This conversion will require two steps.

EXAMPLE 1-1: Unit Conversion

$$1\,260\,000 \text{ s} = ?\text{ h}$$

$$1\,260\,000 \text{ s} \left(\frac{1 \text{ min}}{60 \text{ s}}\right) = ?\text{ h}$$

Notice that at this point our units are minutes, but we are looking for hours so we need another conversion factor.

$$1\,260\,000 \text{ s} \left(\frac{1 \text{ min}}{60 \text{ s}}\right)\left(\frac{1 \text{ h}}{60 \text{ min}}\right) = \frac{1\,260\,000 \text{ h}}{3600}$$

$$= 350 \text{ h}$$

Wow! She certainly did do a lot of reading this summer.

EXAMPLE 1-2: Unit Conversion Between Metric Prefixes

How many nanoliters (nL) are in 345ML (megaliters)?

$$345 \text{ ML} = ?\text{ nL}$$

$$\frac{345 \text{ ML}}{1} \left(\frac{10^6 \text{ L}}{1 \text{ ML}}\right)\left(\frac{1 \text{ nL}}{10^{-9} \text{ L}}\right) = \frac{345 \times 10^6 \text{ nL}}{1 \times 10^{-9}}$$

$$= 345 \times 10^{15} \text{ nL}$$

As a final step, convert your answer to *scientific notation*.

3.45×10^{17} nL

EXAMPLE 1-3: Unit Conversion in Derived Units

What is the speed in km/h of a jet traveling at 153 m/s?

$$\frac{153 \text{ m}}{\text{s}} = ?\frac{\text{km}}{\text{h}}$$

$$\frac{153 \text{ m}}{\text{s}} \left(\frac{1 \text{ km}}{10^3 \text{ m}}\right)\left(\frac{60 \text{ s}}{1 \text{ min}}\right)\left(\frac{60 \text{ min}}{1 \text{ h}}\right) = \frac{550\,800 \text{ km}}{1 \times 10^3 \text{ h}}$$

$$= 551 \frac{\text{km}}{\text{h}}$$

The answer 550.8 kph was rounded to 551 kph to account for the precision of the original measurement. The original measurement had three significant figures, so our answer should have three significant figures.

Why Are Unit Conversions Important?

Scientists often learn from the mistakes that they make. Sometimes those mistakes have huge consequences. There have been some famous cases in which errors have been made in converting units. One costly conversion error resulted in the loss of the $125,000,000 Mars Climate Orbiter. Failure to properly convert between US customary units and SI resulted in the orbiter entering orbit too low, resulting in the destruction of the spacecraft.

MODELING OUR ORDERLY WORLD

MINI LAB

UNDERSTANDING CONVERSION FACTORS

Essential Question:

Where do conversion factors come from?

Equipment

metric and US customary rulers

rectangular prism

Unit conversions are simple fraction multiplication problems in which you multiply a measurement by a conversion factor to determine the value of that measurement in a different unit. The math is fairly simple, but where does the conversion factor come from?

Answer the questions and follow the lettered steps.

1. What is a conversion factor?
2. How are conversion factors derived?

Procedure

A Using the two rulers, measure the length of the rectangular prism in centimeters and inches.

3. What is the prism's length?

B Use your measurements to create the two possible conversion factors.

$$\frac{\_\_\_\_ cm}{\_\_\_\_ in.} \qquad \frac{\_\_\_\_ in.}{\_\_\_\_ cm}$$

4. Determine the conversion factors for cm to in. and in. to cm by reducing each fraction so that the denominators are 1.

C Check your values with a reference source.

Conclusion

5. What does it mean that conversion factors are equivalence statements?

Going Further

6. Give examples of other ways in which we use this equivalence concept.

1C | REVIEW QUESTIONS

1. Compare qualitative and quantitative data.
2. What does SI stand for?
3. Compare fundamental units with derived units.
4. List the seven fundamental SI units, including symbols, with the dimension that each one represents.
5. Give the symbol, factor, and exponential form associated with the metric prefix *micro-*.
6. Explain one way in which the SI is preferable to the US customary system of units.
7. Why can a measurement never be exactly correct?
8. Measure the volume of liquid in the image below.
9. Define *accuracy*.
10. Convert the following:
    a. 37.4 mL into ML
    b. 689 km/hr into m/s
    c. 34.5 m² into mm²

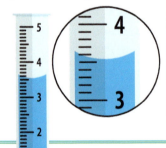

CHAPTER 1 REVIEW

1A ORDER IN OUR WORLD

- Physical science, consisting of chemistry and physics, is a field of natural science that studies matter and energy.
- All evidences of organization and design in nature are examples of order.
- God created all things both good and orderly. Disorder is a consequence of the Fall.
- God commands us to fill the earth and have dominion over it.
- Ethics is the process of deciding between right and wrong.

1A Terms

science	4
physical science	5
physics	5
chemistry	5
Creation Mandate	8
ethics	9

1B MODELING OUR WORLD

- A model is a physical, conceptual, or mathematical representation of some aspect of the world.
- Chemistry includes models of atoms, chemical reactions, the periodic table of the elements, and the behavior of gases. In physics, we model motion, the behavior of fluids, nuclear reactions, and forces.
- Hypotheses, theories, and laws are all scientific models. Hypotheses and theories explain, while laws describe.
- Scientific inquiry is a multipronged approach to doing science that includes asking questions, doing research, forming hypotheses, conducting investigations, analyzing data, and forming and reporting conclusions.
- The scientific community builds consensus through the peer review process.

1B Terms

model	11
theory	11
law	11
workability	12
scientific inquiry	12
hypothesis	13

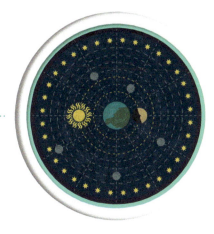

1C USING MATHEMATICS FOR SCIENTIFIC INQUIRY

- The US customary and SI systems of measurement both consist of various units for measuring different dimensions.
- The SI system is the worldwide measurement system for scientific research.
- The SI is beneficial because it is a decimal system, it has few units for each dimension, and it has easily reproduced standards.
- A good measurement is both accurate and precise.
- We do unit conversions through a series of fraction multiplications using conversion factors.

1C Terms

measurement	14
SI	14
accuracy	17
precision	17

MODELING OUR ORDERLY WORLD

CHAPTER 1

CHAPTER REVIEW QUESTIONS

Recalling Facts

1. Define *physics*.
2. Define *physical science*.
3. Describe a way that scientists imitate the order in nature.
4. What principle is the basis for expecting events to happen tomorrow the same as they happened today?
5. How do the universal principles of cause and effect and uniformity support the Christian worldview?
6. List evidences of order in nature.
7. What is the source of order in the universe?
8. Give two reasons for us to do science.
9. What is the Creation Mandate?
10. Write the three questions that lead you through ethical decision-making from a Christian perspective.
11. Summarize the three aspects of biblical motivations.
12. What is a scientific law?
13. Name two models used in physical science.
14. What activity is central to scientific inquiry?
15. What do we call the initial, testable explanation for a phenomenon used by scientists to guide their investigation?
16. Through what process is scientific inquiry assessed?
17. Color, texture, and shape are all examples of what type of data?
18. What system of units do scientists use?
19. What is the SI fundamental unit for electric current? Include its symbol.
20. What unit would be best to measure the distance from New York to San Francisco?
21. What describes the exactness of a measurement?
22. Of the two graduated cylinders on the right, which is more precise? Explain.

Understanding Concepts

23. How does physical science relate to other sciences?
24. Why is order in nature important?
25. Why is studying science a worthwhile activity for a Christian?
26. How are science and ethics related?
27. Explain how Christians should make ethical decisions.
28. Why are models important for studying science?
29. Compare hypotheses, theories, and laws.
30. What does it mean to say that a model is workable?
31. Why do we need a process to study science?

REVIEW

32. You are comparing two measurements: 35.21 cm and 49.6 cm. Which is more precise? Explain.

33. The length of a string is measured to be 22.87 mm. What do you know about the ruler that was used for this measurement?

34. If the accepted value for the length of the string in Question 33 is 24.11 mm, was the string measured accurately? Explain.

35. Convert the following:
    a. 34.5 mA into A
    b. 22.5 m/s into km/h

Critical Thinking

36. At the end of the first day of school, your science teacher shows you two candles (one taller than the other). She asks what would happen if you lit them both and then covered them with a jar. Use the scientific inquiry infographic (p. 13) to outline how you would investigate this problem.

37. Evaluate this statement: I know that the length of the board in my garage is exactly 2.74 m.

38. Why would the concept of significant figures not apply to the fact that there are twenty-three students in your class?

ETHiCS — REPORTING SCIENTIFIC DATA

No scientist can conduct all the research in any one field, so the scientific community depends on the research of others. Scientists publish papers, which are then used by other researchers to conduct future research. Much of the research in biological science is also used to make decisions related to medical treatment. We would like to be able to trust all the academic research we read. But scientists, just like all people, are sometimes tempted to lie.

In 2012, the US Office of Research Integrity concluded a ten-year investigation into a cardiology researcher accused of falsifying data. The investigation detailed forty-five instances of the use of false data in publications. Various government agencies had awarded over $8 million to this researcher on the basis of his distorted data; other researchers could have more legitimately used those funds. Additionally, other researchers cited the papers that were written on the basis of this data over one hundred times. This research may even have caused doctors to make incorrect treatment decisions for cardiac patients.

Use the Ethics box above to answer Questions 39–42.

39. What does God's Word say about how we report data?

40. What are the acceptable results?

41. What benefit does a Christian scientist gain from the acceptable results above?

42. What might motivate scientists to operate in a dishonest way?

Main image: *The amazing design of a gecko's feet enables it to climb walls and even walk on the ceiling. The tokay gecko shown in this image effortlessly walks down a wall.*

Inset: *Material scientists at Stanford University have built a stickybot that can climb walls just like a gecko.*

CHAPTER 2
Matter

IMITATING GECKOS

Geckos have a superpower: defying gravity. The ability of the gecko on the left is not due to its feet being sticky; it's because they are hairy. The tiny hairs on its feet are about 1/30th the diameter of a human hair, and there are thousands of them on every square millimeter of a gecko's feet! With each step, the gecko spreads his toes, maximizing the contact area of each foot. This action creates a tiny electrostatic attraction with the surface. The combination of these tiny forces creates a substantial attractive force. Happily for the gecko, God also designed geckos so that they can release that attraction!

Materials scientists are studying the gecko to see whether they can copy them for a variety of applications. Some scientists are working on gecko tape that acts like a gecko's feet. Others are working on robots that can climb walls, enabling them to do jobs that would be dangerous for people to do. Could scientists develop gloves and boots that could allow us to climb walls and ceilings? You'll have to *hang on* and see.

2A Understanding Matter	26
2B Classifying Matter	32
2C States of Matter	35
2D Changes in Matter	39

2A Questions

- What is matter?
- What do we think matter is made of?
- How can I measure stuff?
- Aren't mass and weight the same thing?

2A Terms

matter, law of definite proportions, particle model of matter, atom, molecule, mass, volume, density, weight

2A | UNDERSTANDING MATTER

Why is matter so important?

2.1 DEFINING MATTER

Materials scientists get inspiration from a variety of sources as they try to solve problems. They look for new ways of using existing materials. They change production methods so that the material has different properties. They also work to create new substances, such as gecko tape. But what are these materials made of?

They are all made of *matter*.

In the photo below, what do you see that you think is made of matter? Is there anything that you wouldn't consider a material? What about the lights and sounds that you would see and hear if you were at this place? While you can perceive both, neither of them are matter.

So what is matter? Matter is so fundamental that it's hard to give it a definition that is based on simpler terms. Scientists define **matter** as anything that occupies space and has mass. An easy-to-understand definition is that matter is the stuff things are made of. Notice that most things (metal, wood, even air) fit this definition. But you should also recognize that light, sound, and warmth don't. They are forms of *non-matter*.

2.2 MODELING MATTER

The definition of matter stated above doesn't help us understand what it's made of. Scientists have debated this topic for centuries.

MODELING MATTER

ANCIENT THOUGHTS

Philosophers in the sixth century BC were divided on their concept of what matter was made of. On one side were the atomists, led by Leucippus and Democritus. They believed that all matter was made of indivisible atoms and that nothing existed between them—the *void*.

Aristotle opposed this belief because he could not accept that nothing existed between atoms. He believed that matter consisted of a continuum that could be infinitely cut without changing the characteristics of that matter.

Remember that a workable model will explain most, if not all, observations, and it will make accurate predictions. Aristotle's understanding was eventually discarded. So what evidence tipped the scales in favor of a model of matter comprised of particles (atoms)? Initially, the evidence was from indirect observations, such as the behavior of gases and data from chemical reactions.

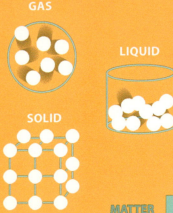

MODELING MATTER

EVIDENCE FOR PARTICLES

Law of Definite Proportions

The earliest evidence for matter as particles came from chemistry laboratories of the 1700s. Chemists were studying the masses of materials produced when they burned certain substances in air. They noticed that the ratios of the mass of the burned substances to the masses of the products were usually whole number ratios, such as 1:2 or 3:5. If matter were continuous, then scientists expected the ratios to be random decimal numbers. Also, each time a material was burned, the *reactants* always combined in the same ratio. This idea would later become known as the **law of definite proportions**.

Fe : O
2 : 3

Burning iron produces mainly iron (III) oxide with a 2:3 ratio of iron to oxygen.

Brownian Motion

In 1827, scientists had direct evidence of the random motion of matter. English botanist Robert Brown was studying tiny plant spores with a microscope. He noticed that parts from within the spores moved randomly as if something were bumping into them, but nothing was in contact with them except for the cellular fluid. Brown hypothesized that particles of the fluid were colliding with the parts from the spores, causing their random motion. Today we call this *Brownian motion*.

plant spores

Brownian motion

Diffusion

Everyone is familiar with another evidence of the constant motion of particles. For example, you might be holed up in your room studying, but you quickly know when your parents arrive with pizza for dinner! The scent of the pizza quickly spreads throughout the house due to *diffusion*, the constant random motion of the air particles in the house.

The debate between Aristotle's theories and the ideas about particles raged for 2400 years, from the sixth century BC until it was settled in the early 1900s. Finally, sufficient scientific evidence allowed scientists to develop the **particle model of matter**, which states that all physical matter exists in the form of particles (atoms or molecules) in constant motion. This model is also called the *kinetic model* of matter.

These particles are known today as atoms and molecules. An **atom** is the building block of all matter. It consists of *protons*, *electrons*, and (usually) *neutrons*. A **molecule** is a distinct group of two or more atoms *covalently bonded* together (see Chapter 5). The particle model is an excellent model because it explains most of the features of matter. Scientists accept it because it is more workable than other models. We will use the particle model as we learn more about matter.

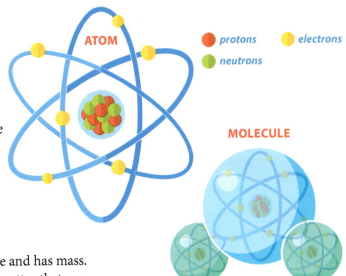

2.3 MEASURING MATTER

As defined above, matter is anything that takes up space and has mass. Mass (m) and space, or volume (V), are two aspects of matter that we can measure. **Mass** is the amount of matter in an object. The SI unit of mass is the gram, and we measure mass with a balance. **Volume** is the space enclosed or occupied by an object. Scientists can measure volume in liters or cubic meters. These are the most basic properties of matter. We can derive a third property—density—from these two. **Density** (d) is the mass of matter contained within a particular volume. The formula for density is shown below.

$$d = \frac{m}{V}$$

You can see how density is derived by dividing mass by volume. Density is an important *physical property* of matter that can help us identify substances. Let's look at some examples.

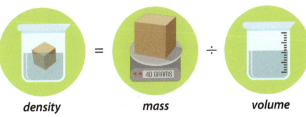

EXAMPLE 2-1: Calculating Density

A sample of aluminum has a volume of 14.8 mL and a mass of 39.96 g. What is the density of aluminum?

Write what you know.

$V = 14.8$ mL
$m = 39.96$ g
$d = ?$

Write the formula and solve for the unknown.

$$d = \frac{m}{V}$$

Plug in known values and evaluate.

$$d = 2.70 \, \frac{g}{mL}$$

Notice that the units don't cancel, so we end up with g/mL, which is the unit for density.

Can we use the formula to predict the mass from the volume and density?

EXAMPLE 2-2: Calculating Mass

What will be the mass of an iron bar with a volume of 35.74 mL? Iron's density is 7.874 g/mL.

Write what you know.

$$V = 35.74 \text{ mL}$$
$$d = 7.874 \text{ g/mL}$$
$$m = ?$$

Write the formula and solve for the unknown.

$$d = \frac{m}{V}$$
$$dV = \frac{m}{\cancel{V}}\cancel{V}$$
$$m = dV$$

Plug in known values and evaluate.

$$m = (7.874 \tfrac{g}{\cancel{mL}})(35.74 \cancel{\text{ mL}})$$
$$= 281.4 \text{ g}$$

The multiplication of the units in this problem is like any fraction multiplication. The mL in both the numerator and the denominator cancel, leaving us with grams, the unit of mass.

EXAMPLE 2-3: Calculating Volume

How much space will a sample of gold with a mass of 175.6 g occupy? The density of gold is 19.3 g/mL.

Write what you know.

$$m = 175.6 \text{ g}$$
$$d = 19.3 \text{ g/mL}$$
$$V = ?$$

Write the formula and solve for the unknown.

$$d = \frac{m}{V}$$
$$dV = \frac{m}{\cancel{V}}\cancel{V}$$
$$\frac{\cancel{d}V}{\cancel{d}} = \frac{m}{d}$$
$$V = \frac{m}{d}$$

Plug in known values and evaluate.

$$V = \frac{175.6 \cancel{g}}{19.3 \tfrac{\cancel{g}}{mL}}$$
$$= 9.10 \text{ mL}$$

In this case, the units *grams* cancel out. To understand how we end up with mL, the unit for volume, think about the division of fractions.

$$\frac{\cancel{g}}{\left(\tfrac{\cancel{g}}{mL}\right)} = \cancel{g}\left(\tfrac{mL}{\cancel{g}}\right) = mL$$

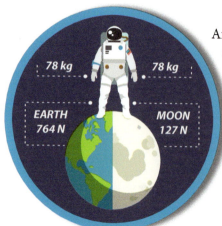

Another property of matter—weight—is closely related to mass but is different. **Weight** is the measure of gravity acting on the matter in an object. The SI unit for weight (a force) is the newton (N). The terms *weight* and *mass* are often used interchangeably because on Earth the relationship between them is constant. Mass doesn't change on the basis of location, but weight can. A 78 kg astronaut on Earth has a weight of 764 N. That same astronaut on the moon still has a mass of 78 kg but weighs only 127 N due to the much lower gravity of the moon.

MINI LAB

MEASURING VOLUME

Fish producers need to know how many trout they can safely transport in a truck. Trout are not spherical or cylindrical, so finding their volume would be difficult. But we can find their volume by water displacement. Fish producers partially fill the truck with water, then add trout until the water rises to the full level. Because of the density of trout, they know that 0.46 kg of trout displaces 0.45 kg of water. This lab activity will help you understand how to determine the volume of an irregularly shaped object.

Essential Question:

How can we find the volume of an irregularly shaped object?

Equipment
laboratory balance
metal object, irregularly shaped
plastic graduated cylinder
water

Procedure

A Measure the mass of the object and record your data on a piece of paper.

B Fill the graduated cylinder about half full with water. Measure and record the volume of water.

1. What will happen to the water as you lower the object into it?

C While tipping the graduated cylinder, slide the object into the water.

D Stand the graduated cylinder upright. Measure and record the new volume of the cylinder.

E Subtract the initial volume from the final volume and record your answer.

2. What does this calculated volume represent?

3. Calculate the density of the irregularly shaped object. Check the value with your teacher.

Conclusion

4. Why does this method allow you to find the volume of the object?

Going Further

5. Check online resources for densities of common metals. Use this information to identify the metal you think composes the object.

2A | REVIEW QUESTIONS

1. Define *matter*.
2. Give three examples of matter.
3. Describe the two competing models of matter in ancient Greece.
4. Why was Aristotle opposed to one of the models?
5. How did Robert Brown's observations support the particle model of matter?
6. State the particle model of matter.
7. Compare atoms and molecules.
8. What volume will a sample of copper with a mass of 98.2 g occupy? The density of copper is 8.96 g/mL.
9. Compare mass and weight.

MATTER 31

2B Questions

- How are mixtures different from pure substances?
- How can I recognize pure substances?
- How can I recognize mixtures?

2B Terms

pure substance, element, compound, mixture, heterogeneous mixture, homogeneous mixture

2B | CLASSIFYING MATTER

How do scientists classify matter?

2.4 WHY CLASSIFY?

Man has been classifying things since the beginning of time. Shortly after Creation, Adam named all the animal kinds that God created. Classification provides a structure within which we can conduct a scientific study. Classification systems are scientific models. Today we have the Linnaean classification system, which you learned about in life science class. Biologists classify organisms into domains, kingdoms, phyla, classes, orders, families, genera, and species. Chemists classify elements as metals, nonmetals, and metalloids, as well as into families, such as alkaline earth metals or halogens.

2.5 HOW DO WE CLASSIFY MATTER?

So how can we classify matter? We classify matter according to its properties. These properties can be physical, chemical, or nuclear. Scientists base the broadest classification of matter on whether it is a pure substance or a mixture.

CLASSIFICATION OF MATTER

A **pure substance** is a material made of only one kind of element or compound; it is not a mixture.

An **element** is a pure substance that consists of atoms with the same *atomic number* (more on this in Chapter 3). Elements are the simplest of the pure substances. Elements can exist as single atoms, molecules of one type of atom, or as a collectively bonded mass of one type of atom. All 118 known elements are listed on the periodic table of the elements. (See pages 80–81 and 570–71.)

The element chlorine is a gas that naturally occurs as a two-atom molecule.

Diamond is a crystalline solid made of the element carbon.

The element mercury is a liquid containing only atoms.

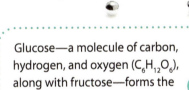

Glucose—a molecule of carbon, hydrogen, and oxygen ($C_6H_{12}O_6$), along with fructose—forms the compound sucrose (table sugar).

A **compound** is a pure substance consisting of atoms of two or more different elements that are chemically combined. The atoms may form molecules or combine as a mass of charged atoms in a repeating geometric pattern called a *crystal lattice*. Atoms of a particular compound always combine in the same proportion.

Table salt (sodium chloride, NaCl) is a crystalline compound made of sodium bonded with chlorine.

MATTER

CLASSIFICATION OF MATTER

Anything that is not a pure substance is a **mixture**—a physical combination of two or more substances (elements, compounds, or other mixtures) in a changeable ratio. The two key aspects of mixtures are (1) that they are only physically combined, not chemically, and (2) that the proportion of substances within the mixture is variable. Scientists further classify mixtures by their appearance.

A **heterogeneous mixture** does not have a uniform appearance since the substances are unevenly distributed. Sometimes the different substances can be seen with the unaided eye or a low-powered microscope.

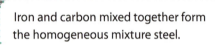

Iron and carbon mixed together form the homogeneous mixture steel.

A chocolate chip cookie is a heterogeneous mixture because the chocolate is not evenly distributed throughout the cookie.

When different substances are mixed so that they have a uniform appearance throughout, a **homogeneous mixture** results. Another term for homogeneous mixture is *solution*.

Tea is another homogeneous mixture.

Oil and water is another heterogeneous mixture because the oil and water remain separated with a nonuniform appearance.

2B | REVIEW QUESTIONS

1. Why do we classify things?
2. What is a pure substance?
3. Define *element*.
4. What characteristics make a mixture *not* a pure substance?
5. Compare heterogeneous and homogeneous mixtures.
6. Give two examples of pure substances and two examples of mixtures not included in this section.
7. How does a compound differ from a mixture?
8. Classify each of the following as an element, compound, or mixture.
    a. magnesium chloride ($MgCl_2$)
    b. copper
    c. trail mix

2C | STATES OF MATTER

How can particles in a solid be moving?

You can see matter in different physical forms, or *states*. The four most common states of matter are **solid**, **liquid**, **gas** (or vapor), and **plasma**. At times we can find a substance existing in more than one state at the same time. Scientists will often refer to these different states as *phases*. Notice the three distinct phases of water (right). Solid ice, liquid water, and water vapor are all present at the same time. The water vapor is in the bubbles within the water and in the beaker between the liquid water and the glass cover. The cloud above the beaker is not water vapor, but rather tiny droplets of liquid water, which is why you can see it.

2C Questions

- How do solids, liquids, and gases compare?
- How do the particles in different states of matter move?
- How can the particles of a solid be moving if the solid stays still?

2C Terms

solid, liquid, gas, plasma

SERVING AS A MATERIALS SCIENTIST: MAKING AN INVISIBILITY CLOAK

Imagine a fabric that could bend light around you when you put it on, making you invisible. Magic, right? No, science! Graduate student Joseph Choi and professor John Howell of Rochester University developed the Rochester cloak (right) that bends light around small objects, making them invisible.

Materials science is an interdisciplinary field in which scientists work to develop new materials and improve existing ones. Materials scientists use many approaches to produce an invisibility cloak. Some approaches use retro-reflective projection technology in which an image of the background behind an object is projected onto the surface of that object. Other approaches use carbon nanotubes or other materials to bend light around an object. Some scientists are using optics to achieve invisibility. Materials scientists envision using this technology for aviation, cars, medicine, and many other applications.

The creativity of materials scientists imitates God's creative work. God has given them the skills and understanding to make these advances. The work of these scientists directly affects the lives of those around them.

2.6 STATES AND THE PARTICLE MODEL

Recall that the particle model states that all matter consists of tiny particles in constant random motion. The states of matter are determined by the relationship between the *kinetic energy* (energy of motion) of the particles and the attractive forces between them. In a solid substance, for example, the particles vibrate in place, so their movements are very small. Their kinetic energy is insufficient for overcoming the attractive forces within the material. If the substance is warmed, its particles move faster and it becomes a liquid once its kinetic energy is high enough to overcome some of the attractive forces between the substance's particles. If the material continues to be warmed, it eventually gains enough energy to overcome all the attractive forces and become a gas.

Different materials may be in different states of matter even if their kinetic energies (temperatures) are the same. At room temperature, for example, oxygen is a gas, water is a liquid, and aluminum is a solid. This occurs because they have different degrees of attractive forces between their particles. So the oxygen is a gas because it has low attractions between its particles, the water is a liquid because it has medium attractions, but the aluminum is a solid due to its high attractive forces. As you can see, the particle model explains why we observe different states of matter at different temperatures.

The relative position of the particles and how they move determine the properties of solids, liquids, and gases. The spacing between particles controls density and compressibility, the degree to which particles in the material can be pushed together. The particles' ability to move determines how well the material holds its shape and volume and whether it can flow.

STATES OF MATTER

SOLID

Particle spacing: *close*
Particle motion: *vibrating in place*
Volume: *fixed*
Shape: *fixed*
Compressibility: *low*
Density: *high*
Fluid? *no*

Crystalline Solids—solids with particles arranged in regular repeating patterns, or lattices

Amorphous Solids—solids that consist of a mass of particles with no discernible pattern

LIQUID

Particle spacing: *close*
Particle motion: *able to slide past each other*
Volume: *fixed*
Shape: *changes to fill a container from the bottom*
Compressibility: *low*
Density: *between that of a solid and that of a gas*
Fluid? *yes*
Viscosity: *The attractive forces between the liquid particles determine the viscosity of a liquid.*

Viscosity is a measure of a fluid's resistance to flowing.

GAS (VAPOR)

Particle spacing: *widely spaced*
Particle motion: *high speed*
Volume: *changes to fill the container*
Shape: *changes to fill the container*
Compressibility: *high*
Density: *low*
Fluid? *yes*
Pressure: *due to collisions with container surface*

PLASMA

The most common state of matter in the universe is not one that we encounter often in our daily lives. Plasma is a gas-like state of matter formed at very high temperatures that consists of high-energy ions and free electrons. Plasma is the state of matter of our sun and other stars.

CASE STUDY: HOW MANY STATES OF MATTER?

If someone were to ask you how many states of matter there are, you would probably answer four: solid, liquid, gas, and plasma. But particle physicists today would say that there are thirty-three actual or theoretical states!

What are these twenty-nine other states of matter? They are states of matter that exist only under very precisely controlled conditions. Even when these conditions exist, some of these states are very unstable and decay into some other state quickly. All this matter is made of particles, though the particles aren't always "normal" protons, neutrons, and electrons.

Some of these are high-energy states of matter, such as quark-gluon plasma (QGP) and color-glass condensate (CGC). QGP exists at either extremely high temperatures or very high densities. CGC is a theoretical state of matter that occurs when *nuclei* move near the speed of light (extremely high kinetic energy). Scientists at the Large Hadron Collider have observed behavior in matter consistent with the predicted behavior in a CGC.

Even more of these states of matter are considered low-energy states, such as the Bose-Einstein condensate (BEC) and superfluids. BEC forms in samples with very low densities when scientists cool them to almost *absolute zero* (–273 °C or 0 K). This is the lowest temperature possible. Superfluids are extremely low-temperature fluids that also have extremely low viscosities.

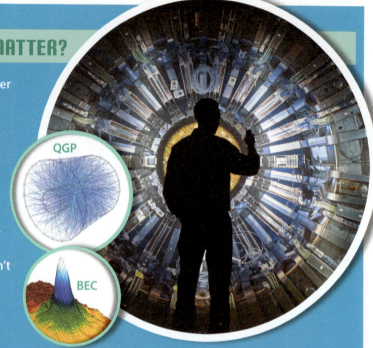

There are other states of matter that don't necessarily fall into the high-energy or low-energy categories. These include states like liquid crystals, quantum spin liquids, and exotic matter. Each different state has properties that make it interesting to study and potentially useful in a variety of applications.

1. How are the four common states of matter similar to the other states?
2. What are two major divisions in the uncommon states of matter?
3. What do scientists use to classify all states of matter, even the unusual states mentioned here?
4. What are some ways that scientists study these other types of matter?

2C | REVIEW QUESTIONS

1. List the four most common states of matter.
2. What factors determine the state of a substance?
3. Describe a solid.
4. Compare crystalline and amorphous solids.
5. What property allows you to describe liquids and gases as fluids?
6. What is viscosity?
7. Compare the degree of motion of the particles in solids, liquids, and gases.
8. What state of matter has changeable shape, low compressibility, and fixed volume?
9. What is a plasma?

2D | CHANGES IN MATTER

How does matter change?

If you were going to make a pillow, would you make it out of river rocks or duck feathers? You would use feathers, of course! Materials scientists, engineers, and even pillow makers choose materials on the basis of the properties that best suit them for the intended application. These properties depend on the particle nature of the materials. Once again, the particle model of matter demonstrates that it is a good model because it is workable. Scientists categorize the properties of materials as physical, chemical, or nuclear (see Chapter 8 for nuclear properties).

2.7 PHYSICAL PROPERTIES

A **physical property** is anything about a substance that can be observed or measured without altering the substance's chemical composition. Physical properties are helpful in identifying a particular substance and include color, texture, physical state, and so on. Physical properties can be explained according to the particle model of matter.

2D Questions

- What are physical properties?
- What are chemical properties?
- How can I tell whether a change in matter is physical or chemical?
- What does it mean to conserve matter?
- How does energy affect states of matter?

2D Terms

physical property, physical change, chemical change, chemical property, melting, melting point, freezing, vaporization, evaporation, boiling, boiling point, law of conservation of matter, condensation, sublimation, deposition

Ductility is the ability of a solid material to be pulled into a wire. The particles in solids are rigidly held in place, but external forces may rearrange the particles while still allowing them to be strongly attracted.

The ability of some solid materials to be hammered or pressed into sheets is called *malleability*. The particles in metals are rigidly held in place but can be rearranged in response to a force, in this case, a *compression force*.

Conductivity is the ease with which a material allows the transfer of either electrical or thermal energy. If the substance contains charged particles that are free to move, it can transfer thermal or electrical energy easily.

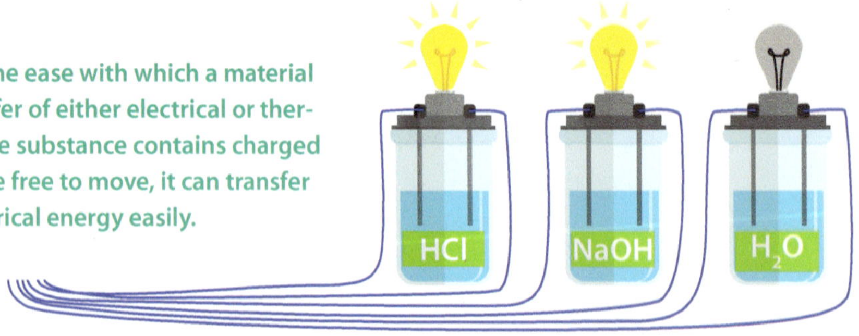

Luster is the quality of how a material reflects light. Materials that have free electrons are lustrous because the free electrons absorb and reemit all the colors of light, making an object shiny.

Any change in matter that does not alter the composition of a substance is called a **physical change**. Physical changes will change the appearance of a substance. We could change the shape, texture, color, dimensions, state, and so on. The key to recognizing that a change is physical is knowing that the substance is chemically the same after the change has occurred. Some examples include melting butter, making ice cubes, and crumpling paper.

2.8 CHEMICAL PROPERTIES

Some changes alter the chemical composition of substances. Scientists call these **chemical changes**, or *chemical reactions*. You will learn more about chemical changes in Chapter 7.

Chemical changes occur according to the chemical properties of a substance. A **chemical property** describes how a substance changes in the presence of another substance or under certain conditions. There are many chemical properties, but two are particularly important for this course.

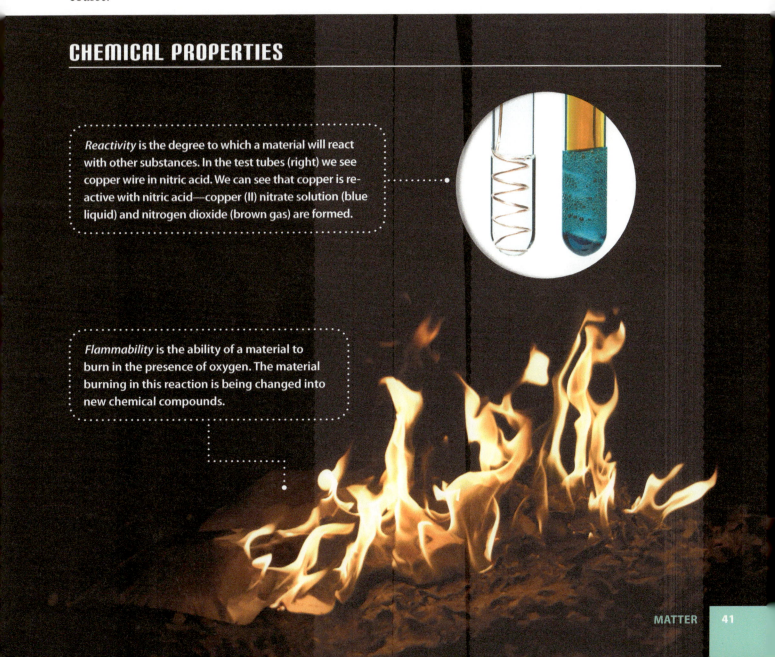

CHEMICAL PROPERTIES

Reactivity is the degree to which a material will react with other substances. In the test tubes (right) we see copper wire in nitric acid. We can see that copper is reactive with nitric acid—copper (II) nitrate solution (blue liquid) and nitrogen dioxide (brown gas) are formed.

Flammability is the ability of a material to burn in the presence of oxygen. The material burning in this reaction is being changed into new chemical compounds.

2.9 CONSERVATION OF MATTER

While matter changes all the time, the *amount* of matter never changes. This statement may seem to contradict your observations. You have seen water in a glass seem to disappear over time. You may have watched wood burn until it was all gone. You eat food and hours later you're

CHANGES OF STATE

a Materials in the solid state have the lowest amount of energy. As energy enters a solid, the particles vibrate faster and faster. Once the particles have enough energy, the vibrations are large enough to overcome the forces holding them rigidly in place. The solid reaches the melting point and starts to melt. **Melting** is the change of state from a solid to a liquid. The **melting point** is the temperature at which a solid turns to a liquid.

b The change of state that is the opposite of melting is freezing. As energy leaves a liquid, the particles move slower. When the particles lose enough energy, attractive forces will lock them into a rigid arrangement. The change of state from a liquid to a solid, which occurs when a substance's temperature decreases to its freezing point, is called **freezing**. The freezing point is the same temperature as the melting point.

c **Vaporization** is the change of state from a liquid to a vapor (gas). In a liquid, the particles remain in contact with each other but have enough energy to be mobile, or to move from place to place, within the liquid. As energy moves into the material, particles can gain enough energy to overcome all the attractive forces. Vaporization occurs in two different ways: through evaporation and through boiling.

- **Evaporation** is the relatively slow form of vaporization in which particles on the surface of the liquid state obtain sufficient energy to change to the gaseous state through the random collisions of particles. Evaporation occurs at any temperature between the freezing and boiling points of the liquid but can happen only at the surface of the liquid.

- **Boiling** is the relatively fast form of vaporization. As the substance is heated, the particles' energy creates a pressure equal to the air pressure outside liquid. Gas bubbles of the substance form within the liquid and rise to the surface. Boiling can occur anywhere in the liquid, but can occur only at the **boiling point**, the temperature at which the liquid starts to boil. This temperature changes as air pressure changes.

hungry again. It seems as if matter is disappearing in all of these instances. But what is happening is that the matter is changing into a different state or substance. Matter can't disappear—it's against the law!—the **law of conservation of matter**. It is a fundamental natural law that states that matter can neither be created nor destroyed, but can only change forms. This happens when a substance changes state.

d **Condensation** is the change of state from a vapor (gas) to a liquid and is the opposite of vaporization. As particles in a gas lose energy to their surroundings, the particles slow down. When they are moving slowly enough, the particles can get trapped in the liquid state. Dew forming on a cool summer night is an example of condensation.

e While not as common, it is possible for a solid to transition directly to a vapor without melting first. This change is called **sublimation** and occurs when solid particles get enough energy to change to vapor. It occurs only at the surface of the solid and is the opposite of deposition. If you have ever noticed that ice cubes seem to shrink in the freezer, then you have noticed sublimation. Dry ice—solid carbon dioxide— and moth crystals are two examples of substances that sublime at room temperature.

f On a cold winter morning, as you scrape frost from the windows of the family car, you are not scraping frozen water droplets. The solid frost formed directly from a vapor (gas), bypassing the liquid state, in a process called **deposition**. This process is the opposite of sublimation.

When scientists are working with changes of state, they will often use a *phase diagram*. The diagram at right shows the pressures and temperatures at which each state, or phase, exists. The lines between the phases each represent changes in state. The red circle on the graph is a unique point called the *triple point*—the temperature and pressure at which solid, liquid, and gas exist in equilibrium.

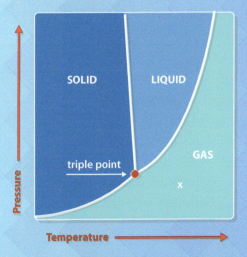

WORLDVIEW SLEUTHING: BULLETPROOF!

Bulletproof technology is an extension of the ancient art of armor making. Armor proved ineffective against modern weaponry, and metal suits are too heavy for the modern battlefield. Materials scientists started to investigate cloth materials. In the late 1800s, a doctor in Arizona discovered that silk could keep a bullet from penetrating the body, but silk was also extremely expensive. Over time scientists searched for other natural and synthetic (manmade) fibers with similar properties. Today scientists are looking at other applications for these materials.

TASK

You are working as a clothing designer for the nation's Olympic sports team. You have been assigned to research advances in ballistic materials to be used in clothing that will perform well and protect speed skaters from sharp skate blades during accidents. You are to write a one-page proposal for a material to be used for these uniforms.

PROCEDURE

1. Research the history and current information about bulletproof technology by doing keyword searches for "bulletproof," "ballistic materials," "Kevlar™," and "Twaron™."
2. Research other uses of similar materials by doing keyword searches for "aramid" and "para-aramid synthetic fiber."
3. Plan your proposal and collect any required photographs. Remember to cite your sources.
4. Write your proposal and show it to another person for feedback.

CONCLUSION

We are commanded in Scripture to serve others. We do this by meeting the needs of people around us. Developing materials such as these can protect the lives of athletes. Could they also help protect the lives of military and law enforcement personnel?

2D | REVIEW QUESTIONS

1. What is a physical property?
2. What is a chemical change?
3. Identify each of the following as a chemical or a physical change.
    a. Iron rusts and becomes iron (III) oxide.
    b. Sugar dissolves in iced tea.
    c. Paper burns.
4. Explain what is happening to the particles in a solid as you warm it to its melting point.
5. Define *boiling point*.
6. Compare evaporation and boiling.
7. State the law of conservation of matter.
8. List the changes of state that involve adding energy.

CHAPTER 2 REVIEW

2A UNDERSTANDING MATTER

- Matter is anything that has mass and takes up space. According to the particle model of matter, all matter is made of tiny particles (atoms and molecules) in constant random motion.
- The particle model of matter is very workable as it explains most of our observations about matter.
- Density is calculated by dividing the mass of an object by its volume.
- Mass is the amount of matter in an object, while weight is the force of gravity acting on that object.

2A Terms

matter	26
law of definite proportions	28
particle model of matter	28
atom	29
molecule	29
mass	29
volume	29
density	29
weight	30

2B CLASSIFYING MATTER

- We classify matter by its physical and chemical properties.
- Matter is classified as either a pure substance or a mixture.
- Pure substances (elements and compounds) contain only one type of substance.
- Mixtures (heterogeneous or homogeneous) are physical combinations of two or more substances in changeable proportions.

2B Terms

pure substance	33
element	33
compound	33
mixture	34
heterogeneous mixture	34
homogeneous mixture	34

2C STATES OF MATTER

- Particles in solids have low kinetic energy compared with the forces between particles. Solids have fixed shape and volume, high density, and low compressibility due to their close spacing and low energy.
- Liquid particles have enough kinetic energy to overcome some of the forces between particles. Liquids have fixed volume, high density, low compressibility due to their close spacing and low energy. Their ability to move makes them fluid and gives them a changeable shape.
- Gas particles have sufficiently high energy to overcome all the forces between them. The particles' wide spacing causes gases to have low densities and high compressibility. Their rapid motion allows them to fully occupy their container, no matter its shape or volume.

2C Terms

solid	35
liquid	35
gas	35
plasma	35

CHAPTER 2

2D CHANGES IN MATTER

- Physical properties are characteristics that can be observed without changing the chemical composition of a substance.
- Chemical properties of matter are characteristics that change the chemical composition of a substance when observed.
- The law of conservation of matter states that matter can't be created or destroyed but only changed in form.
- As a material warms, its particles move faster as it transitions from a solid to a liquid by melting. Liquids vaporize to gases as energy is added. Solids can change directly to gases by sublimation.
- Evaporation and boiling are different forms of vaporization that occur at different temperatures, different locations, and through different mechanisms.
- As a material cools, it transitions from a gas to a liquid by condensing. Liquids can change to a solid through freezing as energy is lost. Gases can change directly to a solid by deposition.
- Melting occurs at the melting point of a substance; boiling occurs at its boiling point.

2D Terms

physical property	39
physical change	41
chemical change	41
chemical property	41
melting	42
melting point	42
freezing	42
vaporization	42
evaporation	42
boiling	42
boiling point	42
law of conservation of matter	43
condensation	43
sublimation	43
deposition	43

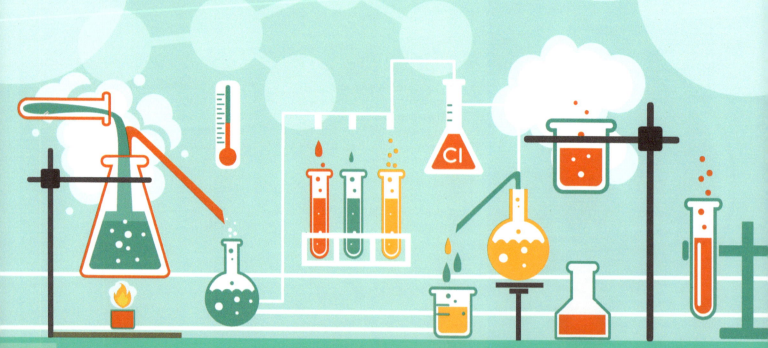

CHAPTER REVIEW QUESTIONS

Recalling Facts

1. What has mass and takes up space?
2. Give three examples of non-matter.
3. Give two pieces of evidence that led to the acceptance of the particle model of matter.
4. Why was the particle model of matter accepted over competing models?
5. Define *weight*.
6. Define *mixture*.
7. What do we call a mixture that appears the same throughout?
8. Give an example of an element, a compound, a heterogeneous mixture, and a homogeneous mixture.
9. The state of matter of a substance depends on what two properties?
10. Copy the following table onto your paper and fill it in to compare the different states of matter. (See Appendix F for creating graphic organizers.)

	Solid	Liquid	Gas
Volume			
Shape			
Compressibility			
Particle Spacing			
Density			
Particle Motion			
Fluid?			

11. A substance with low density and changeable shape and volume that is highly compressible is in what state of matter?
12. What term is used to describe the ability of liquids and gases to flow?
13. What is a physical change?
14. Define *chemical property*.
15. Give the opposite change of state for the following processes.
    a. freezing
    b. condensation
    c. deposition
16. Give an example of deposition.

Understanding Concepts

17. How do Aristotle's model of matter and the particle model of matter represent matter?
18. Considering the formula for density, how will the volume of a piece of lead with a density of 11.34 g/mL change if you double the mass of the sample?
19. How can the weight of an object change even when its mass remains constant?
20. What mass of silver will occupy a volume of 87.75 mL? The density of silver is 10.49 g/mL.
21. Compare a pure substance with a mixture.
22. Draw a hierarchy chart using the terms *matter*, *pure substance*, *element*, *compound*, *mixture*, *heterogeneous mixture*, and *homogeneous mixture*. (See Appendix F.)
23. Describe a typical liquid.
24. Relate the spacing of particles in solids, liquids, and gases to their motion according to the particle model.
25. If particles in any substance are in constant motion, why can't solids flow?
26. What causes different liquids to vary in their viscosity?
27. Compare physical and chemical changes.
28. The mass of ash remaining after a log burns is much less than the original mass of the log. If the law of conservation of matter is true, how can you explain this?
29. Describe what happens to the particles in a material as it is warmed from a solid to a liquid and ultimately to a gas.
30. Explain how matter changes state as energy is removed.

Critical Thinking

31. A classmate tells you that air is not actually matter because you cannot see it. Do you agree? Explain.
32. Do we know for sure that the particle model is correct? Explain.
33. Water is unusual in that it is less dense in its solid form than in its liquid form. Hypothesize as to why this may be. Why might this be beneficial?
34. Given that solids have high density, liquids medium density, and gases low density, why do most types of wood float in water?
35. A gel is a combination of a solid network structure distributed throughout a liquid. It acts like a solid in most cases because the network structure keeps the liquid from flowing. An aerogel is a solid that is made by removing the liquid from a gel, leaving the solid network with air where the liquid had been. What properties would you expect an aerogel to have?
36. In the phase diagram on page 43, at the position marked by the x, the material is a gas. If you move vertically, increasing pressure at a constant temperature, the substance will change from a gas to a liquid. Why do you think this will happen?

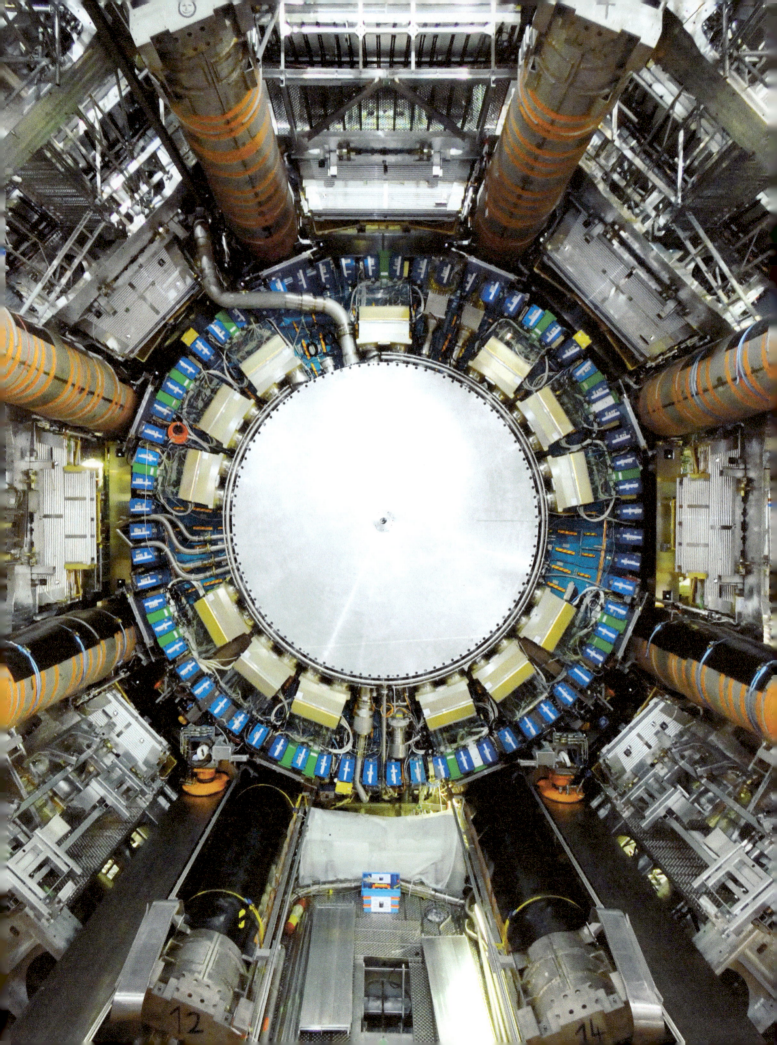

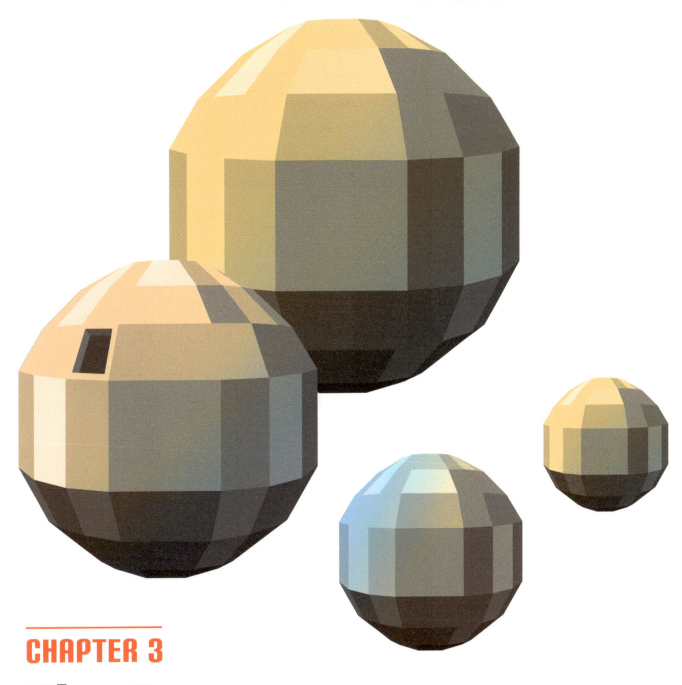

CHAPTER 3
The Atom

SIMPLY SMASHING!

What do you think is shown in the image on the left? It looks like something out of a sci-fi movie, right? Except the Large Hadron Collider isn't science fiction—it's science reality. The collider, built along the border between France and Switzerland, is located in a circular, underground tunnel whose circumference is a whopping *27 kilometers*! Incredibly, this world's-largest machine is used to explore the universe's tiniest particles, particles that are far smaller than atoms. By smashing these particles together at speeds near the speed of light, scientists hope to unlock the mysteries surrounding the basic properties of matter.

3A The Atomic Model	50
3B Atomic Structure	56

3A Questions

- Who were some of the key figures that contributed to our understanding of atoms?
- What were the significant discoveries that shaped atomic theory?
- If the Bohr model has been replaced, why do we still use it?
- How does the most current atomic model depict atoms?
- How does workability drive the development of models?

3A Terms

law of electrostatic charges, plum pudding model, nuclear model, quantum mechanics, Bohr model, energy level, quantum-mechanical model

3A | THE ATOMIC MODEL

How have people thought about matter?

3.1 THE HISTORY OF THE ATOMIC MODEL

Many of the particles that are hurled together by the Large Hadron Collider are *subatomic particles*—particles that are smaller than atoms, which themselves were once thought to be the smallest possible bits of matter. But until very recently, people couldn't see atoms. Even now they are visible only when using the world's most powerful—and expensive—microscopes. And even with such advanced technology, atoms still appear only as blurry, indistinct spheres.

You can't see any of their internal structure. So how did we come to know so much about atoms? That story is a fascinating one. It shows both how scientific modeling works and how the work of scientists builds on the earlier work of others.

The Ancient Greeks

Scientists had been thinking about atoms long before they were certain that atoms existed. You might be surprised to learn that ancient philosophers thought about the nature of matter. One of them, Democritus, a Greek, believed that matter existed in the form of very small particles, invisible to the human eye, which could not be further divided. He called these particles *atomos*, from a Greek word that means "indivisible." It's from that word that we get our modern English word *atom*.

It seems silly to us today, but many of the scholars of his time dismissed Democritus's notion of indivisible atoms. They believed instead that matter was continuous (i.e., it could be subdivided an unlimited number of times), an idea championed by Aristotle. No one bothered to test the two models—that way of thinking was still many centuries off.

C. 400 BC

Democritus suggested that atoms were tiny, indivisible particles. He also believed that their shapes and textures were related to the type of matter in which they were found: atoms in solids had hooks to hold them together, water atoms were slippery, and so on.

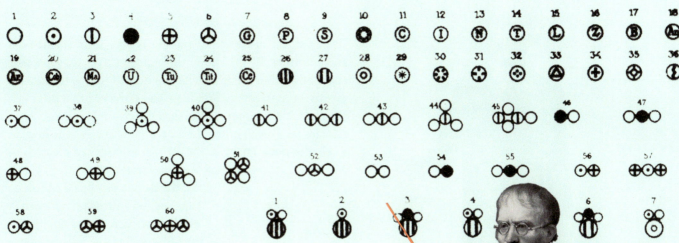

From Democritus to Dalton

By the turn of the nineteenth century, mounting evidence from the study of chemistry hinted that Democritus may have been right. The law of definite proportions, for instance, is what one would expect if matter were made of distinct particles, but not if it were continuous as Aristotle believed. It was time for some additions to the particle model of atoms.

An English schoolteacher, John Dalton, proposed such additions in 1803. Drawing in part on his own studies of the chemical reactions of gases, Dalton suggested several properties of atoms:

1. Elements are made of atoms.
2. Atoms are indivisible and cannot be destroyed.
3. The atoms of an element are all alike.
4. The atoms of one element are different from the atoms of all other elements, especially their masses.
5. Atoms combine chemically in small, whole number ratios.

Dalton's atomic model held up well for nearly a century. But scientists continued to delve into the mystery of atoms and developed better instruments for doing so. Along the way they discovered that some parts of Dalton's model were incorrect. Some changes needed to be made.

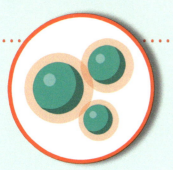

1803

Dalton imagined atoms as hard spheres of different sizes and weights, surrounded by heat envelopes.

THE ATOM

cathode ray

The Discovery of the Electron

The existence of electrical charge has been known since ancient times. By the end of the eighteenth century it was also known that electrical charges existed in two forms, positive and negative. Scientists had observed that opposite electrical charges attract each other, while same charges repel each other. This became known as the **law of electrostatic charges**. By 1875, physicists had created a device that operated on this principle. Called a *Crookes tube*, the device produced a visible beam called a *cathode ray* when an electric current was applied to it. Scientists later learned that cathode rays could be deflected by both magnetic and electric fields.

On the basis of his research with Crookes tubes, university professor and physicist Joseph John Thomson theorized in 1897 that cathode rays were made of negatively charged particles. Using certain measurements, he was able to estimate the mass of these particles. This mass turned out to be extremely small—about 1/1800 of the mass of a hydrogen atom. Thomson called these tiny particles *corpuscles*, but scientists soon began calling them *electrons*. Thomson hypothesized that the electrons were coming from within the atom, which meant that the atom was *not* indivisible! Recognizing that the current atomic model wasn't workable, Thomson suggested a new model for atoms in 1904. Because atoms are electrically neutral (having neither a positive nor negative charge), Thomson proposed that the negatively charged electrons churned rapidly within a uniform positively charged matrix. This image of an atom reminded some scientists of a traditional English Christmas pudding, so Thomson's model became known as the **plum pudding model**.

Christmas pudding

1904

Thomson's plum pudding model suggested negatively charged electrons embedded in a positive substance.

The Nuclear Model

But not everyone accepted Thomson's plum pudding model. A Japanese physicist, Hantaro Nagaoka, found fault with Thomson's notion that opposite electrical charges could mix in the manner that Thomson suggested. In that same year of 1904, Nagaoka proposed a new model. In his *Saturnian model*, the electrons orbited around a massive, positively charged center in a flat ring, much like the rings of Saturn orbit their planet.

New Zealander Ernest Rutherford had been one of Thomson's star pupils. By 1908 he had earned a Nobel Prize in Chemistry for his work on radioactive decay, but his greatest contribution to science was yet to come. Rutherford knew of the differences between the plum pudding and Saturnian models. Now a professor himself, he directed two of his students in a series of experiments to explore the nature of the inner atom. These experiments were performed between 1908 and 1913. In Thomson's model, an atom's electric charges were very spread out; Rutherford believed that a heavy and fast-moving type of particle, called an *alpha particle*, directed at an atom should be able to easily pass through it with little or no deflection.

To the surprise of all, a few of the alpha particles deflected away from the gold foil at very sharp angles. The Thomson model did not predict this. Such a result could be possible only if most of the atom's mass and positive charge were densely packed into a very small space. Rutherford called this dense, central portion of an atom the *nucleus*. Fascinatingly, Rutherford's discovery also showed that most of an atom consists of empty space. Again, the current model didn't explain the observations; it was not workable. These experiments later gave rise to a new model of the atom—the **nuclear model**—in which the atom was made up of a dense, positively charged central nucleus surrounded by negatively charged electrons.

1904
Nagaoka's atomic model had a massive and positively charged center surrounded by a flat ring of electrons.

1911
Rutherford's model of the atom had a tiny, but massive, positively charged nucleus surrounded by electrons.

THE ATOM 53

3.2 THE MODERN ATOMIC MODEL

As science began to unlock the secrets of the atom at the turn of the twentieth century, they saw things that classical physics—the work of Isaac Newton and others—could not explain. Classical physics was, and still is, fine for explaining the motion and energy of things like cars, rockets, and planets, but electrons were a whole new ball game. A new branch of physics was born, one that explored the behavior of matter and energy at the atomic and subatomic levels. The new branch became known as **quantum mechanics**.

The Bohr Model

Many of Rutherford's students became Nobel Prize winners in their own right. One of them, Niels Bohr from Denmark, turned his attention to learning more about how electrons move within atoms. In Rutherford's nuclear model, the electrons simply moved randomly about in the space surrounding the nucleus. If this were really happening, then atoms would constantly emit light and all matter would glow. So what was really happening? Bohr took a clue from the behavior of heated elements. When heated to high temperatures, each element produces an *emission spectrum* (p. spectra; see left) that shows only certain wavelengths of light and no others.

Bohr adjusted the nuclear model to explain this phenomenon. In the **Bohr model** of the atom, electrons do not travel randomly about the nucleus. Instead, electrons can move only in distinct **energy levels**, spherical regions located at fixed distances from the nucleus. If an electron absorbs energy, such as when being heated, it can "jump" to a higher energy level—one that is farther away from the nucleus. Almost instantly, though, it "falls" back to its original energy level. In the process, it loses its extra energy in the form of a *photon*, a packet of light energy that produces a single wavelength of color. An element absorbs and emits energy in very specific amounts only, called *quanta* (s. quantum).

Like the models before it, the Bohr model would soon need further refining. Even so, it still remains in widespread use today. Compared with later models, the Bohr model is fairly simple and easy to grasp. Though it is not completely correct in its depiction of atomic structure, it remains useful for showing how certain processes work, such as chemical bonding. It also represents energy levels in a way that is easier to visualize than the later improved model.

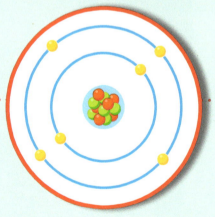

1913

Bohr's atomic model placed electrons in orbits at specific distances from the nucleus.

54 CHAPTER 3

The Quantum-Mechanical Model

As their understanding of quantum mechanics grew, physicists soon realized that electrons didn't travel in neat, spherical orbits as predicted by the Bohr model. Nor could scientists even precisely pinpoint where an electron would be at any point in time. In the resulting **quantum-mechanical model** of the atom, the spherical orbits of the Bohr model were replaced by *orbitals*. Rather than being clearly traveled pathways, orbitals are indistinct and often oddly shaped regions where electrons are *likely* to be found. The quantum-mechanical model is still the primary model being used by physicists and chemists today.

The key factor that drove the development of new atomic models is *workability*. Could the model in use at a given time explain what scientists were seeing in their laboratories? As long as the answer was *yes*, the model remained useful. But if the answer ever became *no*, then the model would be revised or replaced in order to explain the new observations.

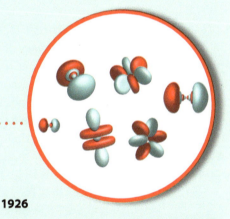

1926

The oddly shaped, overlapping orbitals of the quantum-mechanical model indicate areas where electrons in different energy levels are *likely* to be found.

CASE STUDY: A WORLD OF MODELS

What do you think of when you hear the word *model*? Perhaps you think of a plastic model car or airplane that is built from a kit. That's a kind of physical model, and scientists and engineers do make use of models like that. But there are lots of other kinds of models too. Mathematical formulas can model shapes or processes. A map models geographical features. Computer models analyze data and make predictions that are based on patterns discovered within the data.

What kind of model is being used in each of the following scenarios? Explain your reasoning for each choice.

1. Meteorologists predict the path of a developing hurricane.
2. Your class builds DNA molecules made of toothpicks and colored marshmallows.
3. A fish hatchery manager calculates how much food to feed to fish in a pond on the basis of the size of the pond and the amount of fish in it.
4. A student reads a book about mountain formation.

3A | REVIEW QUESTIONS

1. Name the scientist that is usually associated with each of the following:
    a. discovery of electrons
    b. electrons found in distinct energy levels
    c. first modern atomic model
    d. discovery of the nucleus
2. Why was Thomson's plum pudding model a significant departure from previous atomic models?
3. Why is the Bohr model still used, even though it is not completely accurate in its description of an atom's structure?
4. Following Ernest Rutherford's discovery of the atomic nucleus, what aspect of an atom's structure were scientists primarily trying to discern?
5. How is the quantum-mechanical model different from the Bohr model?
6. Why are newer models said to be more workable than older models?

3B | ATOMIC STRUCTURE

3B Questions

- How are protons, neutrons, and electrons similar and different?
- What are some of the basic properties of atoms?
- How are the isotopes of an element similar and different?
- How can we identify specific isotopes?
- How do ions form?
- How can I tell which isotope of an element is most abundant?
- How can we protect people from radiation?

3B Terms

electron, proton, neutron, atomic number, isotope, mass number, ion, anion, cation, atomic mass

What is it like inside an atom?

3.3 SUBATOMIC PARTICLES

The development of the modern atomic model revealed that atoms are made of three basic particles—protons, neutrons, and electrons. Each affects the structure and properties of the atom.

Electrons

Electrons (indicated by the symbol e⁻) are the smallest of the main subatomic particles, having a mass of 9.1094×10^{-31} kg. As you learned in the previous section, they were also the first subatomic particle to be discovered. They exist at a considerable distance outside the nucleus. Each electron carries a single negative electrical charge (–1e). Electrons are primarily responsible for all the chemical properties of atoms, including how bonds form between atoms. Many other properties of atoms are also determined by the number and arrangement of electrons around an atom's nucleus.

Protons

A **proton** (p⁺) is a subatomic particle located in the atom's nucleus. The discovery of the proton was not as straightforward as that of the electron. After Rutherford's famous experiments, scientists knew that the positive charge of an atom was concentrated in the nucleus. But the nature and identity of the particles that carried the positive charge revealed themselves only gradually. The name *proton* was first used in print in 1920. Protons have a mass of 1.6726×10^{-27} kg, which is about 1836 times the mass of an electron. They carry a single positive electrical charge (+1e).

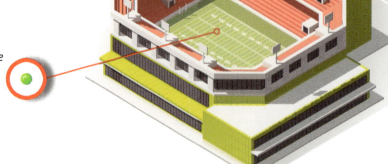

If the nucleus of an atom were the size of a pea, the average atom would be the size of a football stadium!

56 CHAPTER 3

Neutrons

A second kind of subatomic particle found in the nucleus, the **neutron** (n), carries no electrical charge. It was therefore difficult to detect and was not discovered until 1932, though its existence had been predicted earlier. One of Ernest Rutherford's former students, James Chadwick, first isolated neutrons. A neutron has slightly more mass than a proton, about 1838 times that of an electron.

3.4 PROPERTIES OF ATOMS

Atomic Number

What makes silver unlike sulfur? Why is carbon so different from copper? The answer to these questions lies in the structure of the atoms that make up each element. Since each element is composed of its own kind of atom, something must make each kind of atom unique. That something is the distinctive number of protons in an atom's nucleus. For example, every gold atom has seventy-nine protons. If a proton could be removed from a gold atom, the remaining atom would no longer be a gold atom—it would be an atom of platinum because all atoms with seventy-eight protons are platinum atoms. As in gold and platinum, all atoms of a given element have the same number of protons. The unique number of protons in the atoms of each element is that element's **atomic number** and identifies that element. Atomic numbers are always whole numbers because atoms don't contain any partial protons.

The atomic number also tells us the number of electrons surrounding the nucleus in a *neutral atom*, an atom with balanced electric charges. Since protons carry a single positive charge and electrons each carry a single negative charge, an atom is neutral only if it has equal numbers of protons and electrons. For example, the atomic number of carbon is 6, so a neutral carbon atom has six protons and six electrons. The atomic numbers for all the elements are shown on a periodic table (pp. 80–81 and 570–71).

Properties of Subatomic Particles

Particle: electron
Relative Mass (proton = 1): 1/1836 u
Actual Mass (kg): 9.1094×10^{-31} kg
Symbol: **e⁻**
Charge: –1*e*

Symbol: **n**
Charge: 0

Symbol: **p⁺**
Charge: +1*e*

Particle: neutron
Relative Mass (proton = 1): 1 u
Actual Mass (kg): 1.6749×10^{-27} kg

Particle: proton
Relative Mass (proton = 1): 1 u
Actual Mass (kg): 1.6726×10^{-27} kg

Mass Number

On the basis of what you just read, what do you think makes every atom of carbon a carbon atom? If you said that it's the six protons in the nucleus of every atom of carbon, then you're correct! But there's more than just protons in the nuclei of carbon atoms—there are also neutrons. Does every carbon atom have the same number of neutrons? It turns out that the answer to that question is *no*. Most naturally occurring carbon atoms have six neutrons, but some carbon atoms have seven or even eight neutrons. These atoms of the same element with differing numbers of neutrons are called **isotopes**.

Each isotope of carbon is given a unique identifier consisting of its name followed by a number, called its *isotope name*. The isotope with six neutrons is called carbon-12. The number *12* is this isotope's **mass number**, the total number of particles in its nucleus. It indicates that there are twelve particles in each carbon-12 nucleus—six protons and six neutrons. The carbon isotope with seven neutrons is carbon-13, and the isotope with eight neutrons is carbon-14. Like atomic numbers, mass numbers are always whole numbers.

If you know the mass number for any isotope of an element, it is a simple matter to figure out how many neutrons that isotope has. To do so, one subtracts the element's atomic number from the isotope's mass number. The result is the number of neutrons present in the isotope. For example, how would one find the number of neutrons in each atom of sodium-23? By consulting a source such as a periodic table, we learn that sodium's atomic number is 11. Subtracting 11 from 23 shows us that each atom of sodium-23 has 12 neutrons.

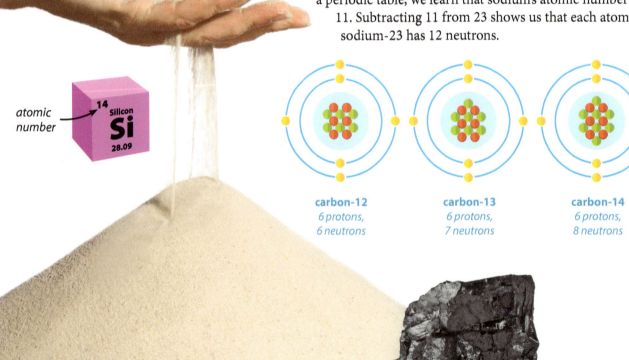

carbon-12
6 protons,
6 neutrons

carbon-13
6 protons,
7 neutrons

carbon-14
6 protons,
8 neutrons

Isotope Notation

An isotope can also be indicated by using *isotope notation*, which shows both the atomic and mass numbers of the element along with its chemical symbol. Starting with the element's symbol, which can be found in a periodic table (see 80–81 and pp. 570–71), the isotope's mass number is placed to the upper left and the element's atomic number is placed to the lower left. The general form is

$$^{A}_{Z}X,$$

where X is the chemical symbol of the element, A is the mass number, and Z is the atomic number. As a memory aid, always put the larger number (mass number) above the lower number (atomic number). Thus the isotope of sodium mentioned on the previous page can be written as

$$^{23}_{11}Na.$$

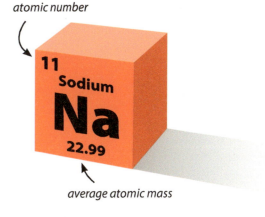

EXAMPLE 3-1: Isotope Notation

Chlorine-35 is the most common isotope of chlorine. In its pure form it is a toxic, greenish-yellow gas. Using the periodic table, write its isotope notation. How many protons, neutrons, and electrons are in one neutral atom of chlorine-35?

Answer: The symbol for chlorine is Cl and its atomic number is 17, making its isotope notation $^{35}_{17}Cl$. One neutral atom of chlorine-35 has seventeen protons, eighteen neutrons, and seventeen electrons.

Isotopes and Physical Properties

Isotopes of the same element can sometimes have quite different physical properties. Carbon-12, for example, is not only extremely common—it's also a prime building block of living things. As you might imagine, carbon-12 is a very stable sort of atom. Its cousin carbon-14 is not quite so stable. In fact, carbon-14 is radioactive (see Section 8A), a fact you might recall from studying life science. Scientists have used this property to try to determine the ages of some kinds of artifacts in a process known as *radiocarbon dating*.

THE ATOM 59

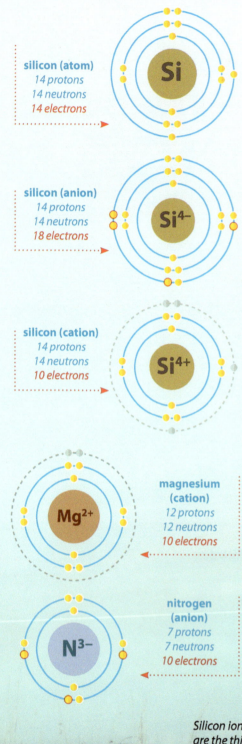

silicon (atom)
14 protons
14 neutrons
14 electrons

silicon (anion)
14 protons
14 neutrons
18 electrons

silicon (cation)
14 protons
14 neutrons
10 electrons

magnesium (cation)
12 protons
12 neutrons
10 electrons

nitrogen (anion)
7 protons
7 neutrons
10 electrons

3.5 IONS

So far we've seen that all the atoms of one element have the same number of protons but can have different numbers of neutrons. But as we hinted at earlier, atoms of an element can *also* have different numbers of electrons. Because of their unequal numbers of protons and electrons, such charged atoms are known as **ions** to distinguish them from electrically balanced atoms. Ions can form by either gaining or losing electrons.

Under certain conditions, atoms can become ions by gaining extra electrons. In such cases, the atom then has more negatively charged electrons than it has positively charged protons. Its electric charges are no longer balanced. Ions that have more electrons than protons have an excess of *negative* charge and are called **anions**. Similarly, some atoms can lose electrons. Again, this makes them electrically imbalanced, but in this instance they have more protons than electrons, because they have *lost* electrons. Ions that have more protons than electrons have an excess of *positive* charge and are known as **cations**. Ions play a critical role in the way that many chemical compounds form. You'll see how that works in Chapter 5. As a memory aid you can think of anion as a negative ion and think of ca+ion, with the t as a plus symbol.

The amount of excess positive or negative charge in an ion can be indicated by using a number to show the *amount* of charge followed by a positive or negative sign to show the *type* of charge. The number and charge are shown in superscript next to the element's symbol. The symbol 2+, for example, indicates that an ion has an excess of two positive charges. Magnesium easily forms ions by losing two electrons. A magnesium cation is indicated by the symbol below.

$$Mg^{2+}$$

In the same manner, 3– would show that an ion has an excess of three negative charges. When an ion has only a single positive or negative charge, only the type of charge (+ or –) is shown.

$$K^+$$

Silicon ions are found in sand, while magnesium cations are the third-most common ion in seawater.

3.6 ATOMIC MASS

You've learned that both the atomic number and the mass number for any element are whole numbers. But you might notice something odd if you look at a periodic table. Take carbon, for example (see right). On a periodic table you'll see the letter C, the symbol for carbon, along with carbon's name and usually several numbers. There's a number 6, and we've already discussed that carbon has six protons, so that 6 in the box must be carbon's atomic number. But what about the number 12.01? That's not a whole number, so it can't be a mass number. So what's going on?

The decimal numbers that you see on a periodic table are the atomic masses of the elements. The **atomic mass** of an element is the weighted average of the masses of all the naturally occurring isotopes of that element. But before we consider how those weighted averages are determined, we first need to address the SI unit for atomic mass. Atomic masses are measured in *unified atomic mass units*, and the unit's symbol is the lowercase letter u. One u is not an exact quantity—it is defined as one-twelfth of the mass of a carbon-12 atom. Since a carbon-12 atom has six protons and six neutrons in its nucleus, 1 u approximates the mass of either one proton or neutron. As we saw earlier, protons and neutrons have slightly different masses, but the difference is very small. For most purposes, the difference in mass can be ignored. So 1 u is used to represent the mass of either particle.

But why is atomic mass called a "weighted average"? You already know how to calculate normal averages—you add up all the values in a set and then divide the total by the number of values. But how could you possibly add up *all the masses* of every carbon atom in existence? It's an absurdly impossible task! But what scientists *can* do is make measurements. They have made many measurements of the relative amounts of carbon's two stable isotopes, carbon-12 and carbon-13. These measurements have shown that about 98.93% of all the stable carbon on the earth is carbon-12. The remaining 1.07% is carbon-13. So most of the carbon on the earth is carbon-12, but not all of it.

If all Earth's carbon were carbon-12, then finding its average mass would be easy. Every carbon atom would have a mass of 12 u, so the average would also be 12 u. But since a small amount of carbon is carbon-13, the average mass should be more than 12 u—but how much more? That's where finding the weighted average comes in. To find the weighted average, we will multiply the mass of each isotope by its relative abundance and then add the resulting figures. The calculation is shown below.

$$12\,u(0.9893) + 13\,u(0.0107) = 12.01\,u$$

If you think about this process in reverse, you can see that the atomic masses listed on a periodic table give hints about which isotopes are most common. For example, lithium has two stable (nonradioactive) isotopes—lithium-6 and lithium-7—and its atomic mass is 6.94 u. Simply ask yourself, is the weighted average of 6.94 u closer to the mass of lithium-6 or to lithium-7? Since the value 6.94 is much closer to 7 than to 6, then most of the lithium on the earth must exist in the form of the lithium-7 isotope, and in fact it does. This determination gets a little trickier for elements that have more than two stable isotopes. Mercury, for instance, has seven stable isotopes and its average atomic mass value of 200.59 u is close to the atomic mass of mercury-201. But mercury-201 is *not* the most common isotope of mercury. Both mercury-202 (the most common isotope) and mercury-200 are more common than mercury-201.

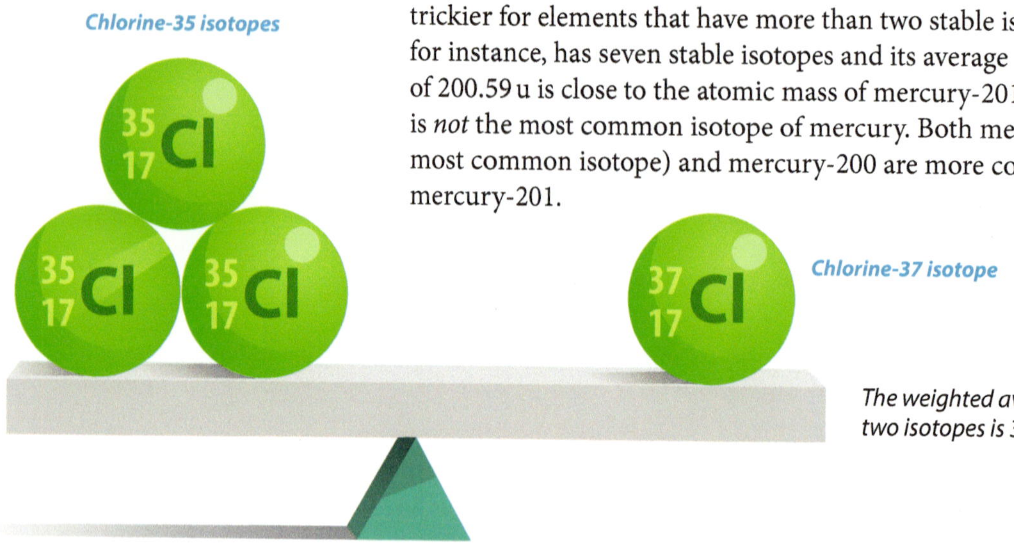

Chlorine-35 isotopes

Chlorine-37 isotope

The weighted average of chlorine's two isotopes is 35.45 u.

3B | REVIEW QUESTIONS

1. Which type of subatomic particle is *not* found inside the nucleus?
2. Which two subatomic particles have approximately equal masses?
3. State the type of electric charge associated with each of the three main subatomic particles.
4. Lithium's atomic number is 3. How many electrons does a neutral lithium atom have? Explain.
5. Why does carbon-14 have a mass number of 14 if it has only six protons?
6. One isotope of tellurium, tellurium-123, has seventy-one neutrons. What must tellurium's atomic number be?
7. One isotope of sulfur has sixteen protons and eighteen neutrons. What is the name of this isotope?
8. What is the isotope notation for phosphorus-31?
9. How does an ion differ from an atom?
10. Fluorine's atomic number is 9. Indicate the type and amount of excess charge on a fluorine atom that has gained one electron.
11. The charge on a certain ion is indicated as 2+. Has the atom gained or lost electrons? Explain.
12. Why doesn't an atom's atomic mass necessarily have a whole-number value?
13. Bromine has two stable isotopes: bromine-79 and bromine-81. Bromine's atomic mass is 79.90 u. Which isotope of bromine is more common?

MINI LAB

FINDING THE ATOMIC MASS OF EGGOGEN

Scientists have discovered a new element—eggogen! Your task in this lab activity is to help those scientists determine eggogen's atomic mass.

Essential Question:

How do scientists determine the atomic mass of an element?

Procedure

A You and your team need to devise a plan for determining the atomic mass of eggogen. You will have only five minutes to come up with your plan. During those five minutes, you may examine the eggogen "atoms" in any manner you choose. But you may not open the atoms, and you may not make any mass measurements at this time.

B Your plan may include up to a *maximum of three* mass measurements.

1. How will your group determine the atomic mass of eggogen?

C Now carry out your plan. You will have five minutes to complete this task.

Equipment

eggogen atoms (10)
laboratory balance

Conclusion

2. What is the atomic mass of eggogen?

3. How similar is your answer to that of other groups?

4. Describe at least one way in which you could improve the certainty of the value for eggogen's atomic mass.

5. You were not allowed to open the eggogen atoms. How is this similar to the limitations that scientists must work around when examining real atoms?

Going Further

6. Did your plan allow you to determine whether any "isotopes" of eggogen exist? If so, how many were there, and what is their ratio within your sample?

7. Suggest a plan for determining which eggogen isotope, if any, is most common and what the ratio of eggogen isotopes is in "nature" (i.e., all the samples in your class).

CHAPTER 3

3A THE ATOMIC MODEL

- The Greek philosopher Democritus used the word *atomos* to describe indivisible particles of matter.
- John Dalton proposed a comprehensive theory of atoms in 1803.
- J. J. Thomson discovered electrons in 1897 and suggested a new model of the atom that came to be known as the plum pudding model.
- Ernest Rutherford and his students confirmed the existence of the nucleus, a small and dense central portion of an atom.
- Niels Bohr established that electrons move in distinct orbits with particular energy levels at fixed distances from the nucleus.
- Despite not being the most up-to-date model, the Bohr model remains useful for its ability to illustrate some chemical processes.
- The quantum-mechanical model of atoms replaced orbits with orbitals that showed where electrons were likely to be found.
- Workability was a key factor in the development of new atomic models.

3A Terms

law of electrostatic charges	52
plum pudding model	52
nuclear model	53
quantum mechanics	54
Bohr model	54
energy level	54
quantum-mechanical model	55

3B ATOMIC STRUCTURE

- Basic subatomic particles include positively charged protons and electrically neutral neutrons, both found in the nucleus, and negatively charged electrons found outside the nucleus.
- A neutron is slightly more massive than a proton, and both are roughly 1800 times more massive than an electron.
- Atoms of the same element always have the same number of protons, but the isotopes of an element have different numbers of neutrons.
- Isotope notation indicates the name of an element along with its atomic number and mass number.
- Ions exist as either positively charged cations that have lost electrons or as negatively charged anions that have gained electrons.
- The atomic mass of an element is the weighted average of the masses of the element's stable isotopes.

3B Terms

electron	56
proton	56
neutron	57
atomic number	57
isotope	58
mass number	58
ion	60
anion	60
cation	60
atomic mass	61

CHAPTER REVIEW QUESTIONS

Recalling Facts

For Questions 1–5, match each atomic model with the scientist most closely associated with it.

1.
2.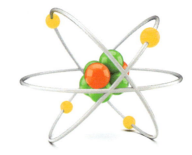
3.
4.
5.

a. John Dalton
b. J. J. Thomson
c. Hantaro Nagaoka
d. Ernest Rutherford
e. Niels Bohr

6. What key discoveries shaped the development of the atomic model?
7. Where are protons, neutrons, and electrons located in an atom?
8. How does the mass of a proton or neutron compare with that of an electron?
9. What property of an atom determines what element it is?
10. How is the number of electrons in a neutral atom related to the number of protons?

REVIEW

Understanding Concepts

11. Explain how each new atomic model was more workable than the last.
12. What parts of Dalton's atomic theory were disproven by later discoveries?
13. Why is the Bohr model still used today even though it is no longer believed to accurately represent the structure of an atom?
14. If an atomic model cannot explain a scientific observation, what should happen to it?
15. Summarize our current understanding of the locations of the basic subatomic particles within an atom.
16. Why were neutrons discovered later than either protons or electrons?

An atom of copper-63 contains 29 protons. Use this information to answer Questions 17–20.

17. What is copper's atomic number?
18. How many neutrons does copper-63 have?
19. What is copper-63's mass number?
20. How many electrons does a neutral copper-63 atom have?
21. Two of selenium's stable isotopes are selenium-78 and selenium-79. Which subatomic particle must each of these isotopes have the same number of?
22. Describe two ways that atoms of an element can be different from each other.
23. Explain the difference between *mass number* and *atomic mass*.
24. Write the isotope notation for the following isotopes.
    a. lithium-7
    b. lead-207
    c. iron-56
    d. calcium-42

A neutral atom of potassium (atomic number 19) loses one electron. Use this information to answer Questions 25–27.

25. Will the resulting ion have an excess positive or negative charge?
26. What kind of ion is formed by the loss of the electron?
27. Write the symbol for this ion using potassium's atomic symbol (K).
28. Hydrogen has three stable isotopes: hydrogen-1, hydrogen-2, and hydrogen-3. Hydrogen's atomic mass is 1.008. Which isotope do you think is most common? Explain.

Critical Thinking

29. Regarding his discovery of the nucleus, Ernest Rutherford said, "It was almost as incredible as if you had fired a 15-inch shell [an artillery shell] at a piece of tissue paper and it came back and hit you." Why was that so?
30. Why is it more proper to describe the quantum-mechanical model of atoms as *workable* rather than *correct*?
31. Could a magnesium cation, Mg^{2+}, form by the addition of two protons to the nucleus of a magnesium atom? Explain.
32. While helping a friend with his homework, you notice that he has written the isotope notation for bromine-81 as $^{35}_{81}Br$. What is wrong with your friend's notation, and why would the isotope that it suggests be physically impossible?

CASE STUDY:
ALL MODELS ARE NOT EQUAL

You may recall from a study of life science that scientists also model the beginnings of life on Earth. Until the mid-nineteenth century, most models of origins included God as the Creator of life, and some of these models also did not allow for change within created kinds of organisms. But around the turn of the nineteenth century, some scientists began to consider that living things might change, or *evolve*, over time. This idea became predominant after Charles Darwin introduced his theory of natural selection. Since then, most scientists have abandoned the Creation model despite the fact that it remains very workable.

Use the Case Study above to answer Questions 33–35.

33. How are the models of origins similar to the atomic models that you read about in this chapter?
34. One of the major differences between models of origins and models of atoms concerns the idea of testability. Explain how the two models are different in this regard.
35. Why do you think that most scientists have abandoned the Creation model of origins despite its workability?

CHAPTER 3

ETHiCS: STRATEGIES AND PROTECTING PEOPLE FROM RADIATION

Do you remember the discussion of ethics in Chapter 1? If you are a little rusty, you should review pages 9–10 because we are going to study a strategy for evaluating the ethics of issues in physical science. Then we will apply this strategy to evaluate responses to radon gas exposure.

The Strategy

Strategies are helpful because they provide a way to deal with the complexities of ethics and help people organize their thoughts. This strategy is organized into six steps.

❶ What information can I get about this issue?

Often you may need to do some research on a topic before evaluating whether it is right or wrong. For this exercise, you can find the needed information in the example that follows.

❷ What does the Bible say about this issue?

Study the Bible to see what it says about this issue. Sometimes God's Word directly addresses an issue. Many times you have to study the Scriptures to learn how its principles apply to the issue. Consider the biblical principles of the image of God in man, the Creation Mandate, and God's whole truth.

❸ What are the acceptable and unacceptable options?

When trying to solve an ethical dilemma, you must look at all the options. Some of these potential solutions may not be obvious, while some may be talked about by almost anyone discussing the issue. Using the biblical teaching and principles that you considered in Step ❷, decide which of the options are acceptable from a biblical point of view. Discard those that are biblically unacceptable.

❹ What may be the consequences of the acceptable options?

Remember the biblical outcomes of human flourishing, a thriving creation, and God's glory? Some of the acceptable options may have consequences that fit with the biblical outcomes. Others may not. Reject the options that have consequences that are inconsistent with human flourishing, a thriving creation, and God's glory.

❺ What are the motivations of the acceptable options?

Remember the biblical motivations of faith in God, hope in God's promises, and love for God and others? Besides considering principles that inform our decisions and consequences biblically, we also need to be asking how this option will help me and others grow. We should also reject options that don't help us grow biblically in faith, hope, and love.

❻ What action should I take?

Suggest a course of action or form an opinion. Be prepared to justify the action that you suggest on the basis of the analysis in the previous steps.

The Issue: Radon Gas Exposure

❶ What information can I get about this issue?

Several times in this chapter, you've seen the word *stable* used to describe some isotopes. Unstable isotopes attempt to become stable by giving off excess particles or energy or both. These emissions are forms of *radiation* (see Chapter 8). Many people have some notion that radiation is a bad thing produced by nuclear power plants or atomic explosions, but did you know that there are naturally occurring radioactive isotopes? One source of these isotopes is the radioactive element uranium. Uranium is found throughout the earth's crust in small amounts. As uranium breaks down, it forms another radioactive element, radon, which is a gas. When radon gas seeps out of the earth's crust, it can be inhaled. This is not normally a problem because of the very small amounts produced. But radon gas can collect in places such as mines or in the basements of houses. In either of those places it has the potential to accumulate in amounts sufficient to cause health problems in humans.

❷ What does the Bible say about this issue?

Radioactive elements were unknown at the time the Bible was written, so exposure to radiation is not directly addressed in the Bible. But the Bible does teach us to have concern for our fellow image bearers, so dealing with radiation exposure is something we should certainly consider.

❸ What are the acceptable and unacceptable options?

An unacceptable option is to do nothing. Worse yet would be to allow people to be potentially harmed by radon gas exposure if you knew in advance that the danger existed. Exodus 22:5–7 contains some laws that show the importance that God places on the accidental or inadvertent damage of another person's property due to carelessness. How much more valuable is a human life compared with an item of property! Some acceptable options include mapping of areas at risk for radon exposure (see image on the facing page), requiring homeowners and businesses to disclose the presence of radon gas, and developing affordable ways to reduce radon gas exposure.

REVIEW

❹ What may be the consequences of the acceptable options?

One consequence would be the flourishing of human life. Knowing both where radon gas risks are highest and how best to reduce exposure to it when present would allow people to make informed decisions about where to live and how to build safe residences and businesses. Besides lowering the risks for exposure and illness to individuals, protection against radon gas exposure would lower the costs to society for treating and possibly compensating victims of exposure.

❺ What are the motivations of the acceptable options?

Alerting people to the possible hazards associated with exposure to radon gas in places where they live or work is an example of loving my neighbor as myself. Helping people take action against radon gas exposure can also demonstrate God's love and concern for His image bearers.

❻ What action should I take?

Now that I know about the risks of radon gas, I can make wise decisions for my future. For example, if I move to a different part of the country, I can consult resources that indicate where the risk of radon gas exposure is greatest and act accordingly. I can ask my employer whether my place of work has been tested for exposure. If I buy a house, I can request that the property be tested as a condition of the sale. This will help protect both me and my family.

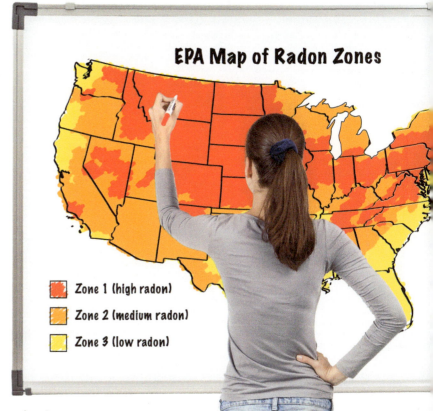

Use the Ethics box above to answer Questions 36–40.

36. Which step(s) in this process explore(s) biblical principles?
37. Which step(s) in this process explore(s) biblical outcomes?
38. Which step(s) in this process explore(s) biblical motivation?
39. What might happen in Step ❷ if Step ❶ is not carried out thoroughly?
40. Why is Step ❻ necessary?

DIAMOND ANVIL CELL

FORCE

← particles

FORCE

CHAPTER 4
The Periodic Table

DIAMONDS DELIVERING DATA

Diamonds are beautiful! Surprisingly, they are made of the same element as charcoal for your grill: carbon. We all know that diamonds are valued as gemstones, but they are valued for other reasons too. They are the hardest natural substance and have the highest thermal conductivity. These two characteristics make them great for industrial uses, such as cutting and grinding.

But did you know that diamonds are used for research also? A device called the *diamond anvil cell* enables scientists to create extreme pressures and temperatures. The design of the device (inset image left) is such that a sample is squeezed between the tips of two diamonds. Because of diamond's hardness and high thermal conductivity, scientists can apply immense pressures and high temperatures to a sample during testing. Diamonds are also transparent, which allows illumination of samples.

Diamond anvil cells are tiny laboratories in which scientists reproduce the most intense conditions in our universe—such as pressures and temperatures found in Earth's core. They "push" living cells to extremes to see whether they can survive. Diamonds may be a researcher's best friend.

4A Organizing the Elements	70
4B Classifying the Elements	80
4C Periodic Trends	86

4A | ORGANIZING THE ELEMENTS

How does the periodic table relate to elements in the real world?

4.1 THE NEED FOR A TABLE

Chemical Discoveries

Today we know of 118 elements. Half of these were discovered only in the last 160 years, but some have been known since ancient times. The Bible refers to seven materials—gold (atomic number 79), silver (47), tin (50), copper (29), lead (82), iron (26), and brimstone, that is, sulfur (16)—that we know as elements today. While the Bible doesn't refer to these substances as elements, the concept that elements are the basis for matter is an ancient one.

Ancient Greeks thought that all matter was made of five elements, each with its own unique shape and characteristics. The original four elements were earth, air, fire, and water. Aristotle later added a fifth—aether. All other matter was thought to be made of combinations of these basic elements.

Interest in elements and other materials led to the rise of a group of scientists called *alchemists*. Much of their early work was aimed at turning low-value materials, such as lead, into high-value substances like gold. But some alchemists were interested in studying elements for purely scientific reasons.

4A Questions

- Why do we need a periodic table?
- Who developed the periodic table?
- How has the periodic table changed over time?
- Can the periodic table predict new elements?
- Why is the periodic table arranged the way it is?

4A Terms

periodic law, periodic table, family, group, valence electron, period

Finding Elements on a Periodic Table

Throughout this chapter, the first reference to an element will include a number in parentheses. This number is the element's atomic number. Use atomic numbers to familiarize yourself with the location of elements on the table.

Robert Boyle

In the seventeenth and eighteenth centuries, Robert Boyle and Antoine Lavoisier began a transition from alchemy to chemistry. Robert Boyle was an alchemist who believed that it was possible to change one element into another. His work solidified the position that matter consisted of indivisible particles of elements, rejecting the Greek concept of matter as mixtures of the five basic elements. In 1661, he published *The Skeptical Chemyst*, in which he called for experimentation to be the basis of science.

Antoine Lavoisier

Likewise, Antoine Lavoisier played a crucial role in our understanding of the elements. In 1779, while researching combustion reactions, he demonstrated that combustion required oxygen (8). Therefore fire formed as oxygen and other materials—the fuel—reacted. Fire couldn't be an element if it needed something else—oxygen—to form it. By this time, chemists had discovered many of the naturally occurring elements, and Lavoisier made one of the first lists of these elements. His list included the thirty-three elements known in his day.

Jacob Berzelius

As the list of known elements grew, scientists quickly recognized the need to organize all this data. Many scientists came up with their own symbols to represent each element. These different symbols made the sharing of information difficult. In 1814, Swedish chemist Jacob Berzelius came up with a system that would become the accepted standard. As Carolus Linnaeus did with animals, Berzelius gave each element a Latin name. He then represented each element with the capitalized first letter of its name. If needed, he used the first two letters of the name, with the second letter in lowercase. Berzelius represented oxygen with an O, and he used N to represent nitrogen (7).

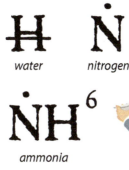

Berzelius also combined these symbols to represent compounds. He would use the letters to represent the elements and *superscripts*, or other symbols, to represent how many atoms of each element were present in the compound. Berzelius came up with symbols for the forty-seven elements known to him.

THE PERIODIC TABLE

Modern Element Symbols

As the number of known elements has grown, the need for a standard way to represent them has become even greater than in the past. Our modern system is much like Berzelius's system, though it uses *subscripts* to indicate the number of atoms. The symbol for each element still consists of either one or two letters. In our modern system, twelve of the elements have a symbol that consists of a single capital letter, the first letter of the name. For instance, we use U to represent uranium (92). The rest of the elements use a capital letter followed by a lowercase letter. For example, Chlorine (17) has the symbol Cl, and Zn is the symbol for zinc (30). Remember that in two-letter symbols the first letter is always capitalized and the second letter is always lowercase. This helps us figure out whether we are dealing with an element or a compound. For example, are we using cobalt (27) with the symbol Co or carbon monoxide, a compound made of carbon (C) and oxygen (O) and whose *chemical formula* is CO?

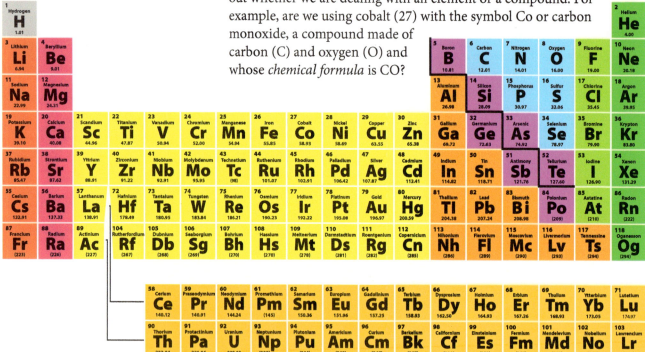

Tungsten (74) has the symbol W, which is the first letter in *wolfram*, tungsten's name in German.

At first glance, not all the symbols make sense, such as Na for sodium (11), Fe for iron, and Ag for silver. But remember that the elements were given names in Latin. Some of the symbols are based on the element's name in this original language. Na (sodium) came from *natrium*, Fe (iron) from *ferrum*, and Ag (silver) from *argentum*—all words from Latin.

4.2 DEVELOPMENT OF THE PERIODIC TABLE

Can you imagine trying to memorize the names, symbols, and characteristics of all known elements? In the seventeenth and eighteenth centuries, this wouldn't have been too difficult. But as the number of elements grew, many scientists looked for ways to arrange the information known about them.

MINI LAB

ORGANIZING ELEMENTS

As scientists gathered more information, the need to organize their data increased. This became evident as chemists discovered more elements. How would you organize elements? In this activity, you will arrange fifteen "known" elements.

Essential Question:

How can we use properties to organize elements?

Procedure

A Study the data on each of the element cards.

1. What data do you think will be most helpful for sorting the elements?

B Choose the property or properties by which you are going to sort.

C Arrange cards in the form of a table, starting with the first row. Start new rows when needed.

2. Which property did you use for sorting the elements? Why did you make that choice? If you used more than one property for sorting, was one considered of first importance compared with the other?

3. How did you decide to start a new row?

4. Were any of the elements difficult to sort?

5. Are there any gaps in your table of elements?

Conclusion

6. What trends can you see in the properties?

Going Further

7. What would a missing element imply?

8. If you have an element missing, predict the characteristics of that element.

Equipment
Set of fifteen element cards

Periodicity

German chemist Johann Döbereiner made a key discovery while studying properties of the elements in the early 1800s. He noticed groups of elements, like chlorine, bromine (35), and iodine (53), with similar properties. Since these groups always seemed to have three elements, he called them *triads*. While most chemists didn't accept Döbereiner's work at the time, his efforts prompted a few to look for patterns in the properties of elements. These scientists started to notice repeating patterns when they arranged the elements in order by increasing atomic mass. This repeated pattern is called *periodicity*.

THE PERIODIC TABLE   73

In the 1860s, John Newlands expanded on the concept of periodicity. When he arranged the elements in order by atomic mass, he found pairs of elements with similar properties. These pairs of similar elements differed in their position by a multiple of eight. This led Newlands, in 1865, to develop his *law of octaves*, which states that when elements are arranged by atomic mass, the properties of every eighth element are similar. Newlands is now credited with the earliest form of the *periodic law*, which stated that the properties of the elements vary with their atomic masses in a regularly repeating way.

	No.		No.		No.		No.		No.		No.		No.		No.
H	1	F	8	Cl	15	Co & Ni	22	Br	29	Pd	36	I	42	Pt & Ir	50
Li	2	Na	9	K	16	Cu	23	Rb	30	Ag	37	Cs	44	Os	51
G	3	Mg	10	Ca	17	Zn	24	Sr	31	Cd	38	Ba & V	45	Hg	52
Bo	4	Al	11	Cr	18	Y	25	Ce & La	33	U	40	Ta	46	Tl	53
C	5	Si	12	Ti	19	In	26	Zr	32	Sn	39	W	47	Pb	54
N	6	P	13	Mn	20	As	27	Di & Mo	34	Sb	41	Nb	48	Bi	55
O	7	S	14	Fe	21	Se	28	Ro & Ru	35	Te	43	Au	49	Th	56

Mendeleev and Periodic Law

In 1869, Russian chemist Dmitri Mendeleev organized the elements in a table much like our modern periodic table. He arranged the elements in columns by increasing atomic masses. This arrangement displayed the periodic nature of the properties across the rows. Over time Mendeleev switched the rows and columns of the table. This change placed elements with similar characteristics in columns like our modern table.

Period	I	II	III	IV	V	VI	VII	VIII	O
1	H 1								He 4
2	Li 7	Be 9.4	B 11	C 12	N 14	O 16	F 19		Ne 20
3	Na 23	Mg 24	Al 27.8	Si 28	P 31	S 32	Cl 35.5		Ar 39
4	K 39	Ca 40	— 44	Ti 48	V 51	Cr 52	Mn 55	Fe 56 Co 59 Ni 59 Cu 63	
5	Cu 63	Zn 65	— 68	— 72	As 75	Se 78	Br 80		Kr 83
6	Rb 86	Sr 87	Y 88	Zr 90	Nb 94	Mo 96	— 100	Ru 104 Rh 104 Pd 106 Ag 108	
7	Ag 108	Cd 112	In 113	Sn 118	Sb 122	Te 125	I 127		Xe 131
8	Cs 133	Ba 137	Di 138	Ce 140	—				
9	—	—	—	—	—				
10	—	—	Er 178	La 180	Ta 182	W 184	—	Os 195 Ir 197 Pt 198 Au 199	Rn (222)
11	Au 199	Hg 200	Tl 204	Pb 207	Bi 208	—	—		
12				Th 231	—	U 240	—		

The most striking features of Mendeleev's periodic table (bottom of previous page) were the blank spaces (unshaded cells) that he left between some elements. These spaces were for unknown elements that he predicted would be found in the future. He based his predictions on the periodic law. When Mendeleev ordered the elements on the basis of atomic mass, he recognized that in some cases there weren't elements with the expected properties. Therefore he concluded that there must be elements that would later fill those spots. At first, other scientists ignored these predictions, but the discoveries of gallium (31) and germanium (32) proved that Mendeleev was correct.

4.3 THE MODERN PERIODIC TABLE

Reordering the Table

English physicist Henry Moseley made an important discovery in 1914. He recognized that he was able to use a new technique called *X-ray spectroscopy* to determine the atomic number for each element. As a result, chemists rearranged the periodic table on the basis of increasing atomic number. This change corrected some misplaced elements from Mendeleev's table. Moseley's discovery also led to an update of the **periodic law**, which now states that the properties of the elements vary in a periodic way with their atomic numbers. This allowed scientists to produce the periodic table of the elements that we use today. The arrangement of the periodic table relates to the structure of the atoms themselves. Pages 80–81 show the modern **periodic table** of the elements—a table of the chemical elements arranged in a way that displays their periodic properties in relationship to their atomic numbers.

The most updated versions of the periodic table include all the elements through atomic number 118. Four of these elements (113, 115, 117, and 118) were only recently added after a lengthy naming process by the International Union of Pure and Applied Chemistry (IUPAC), the group that is responsible for standardization in chemistry.

ORGANIZATION OF THE PERIODIC TABLE

FAMILY

You will notice that a periodic table is arranged in rows and columns. The first row starts with hydrogen (1), then helium (2), but then there is a new row. This is due to the periodic law. Lithium (3), needed to be placed below hydrogen because lithium and hydrogen have similar properties.

A **family**, or **group**, is a set of elements in the same column on the periodic table. Elements in the same group have similar structures, which result in their acting like each other. Just like members of your family have similarities, the elements within a group have similar physical and chemical properties.

Most importantly, each element in the group has the same number of **valence electrons**, which are the electrons in the outermost energy level of a neutral atom. These electrons take part in chemical bonding, as you will see in Chapter 5.

On the periodic table, columns are numbered 1–18. These *group numbers* identify all the elements in that group. Thus, hydrogen and lithium are both Group 1 elements, and oxygen and sulfur are in Group 16.

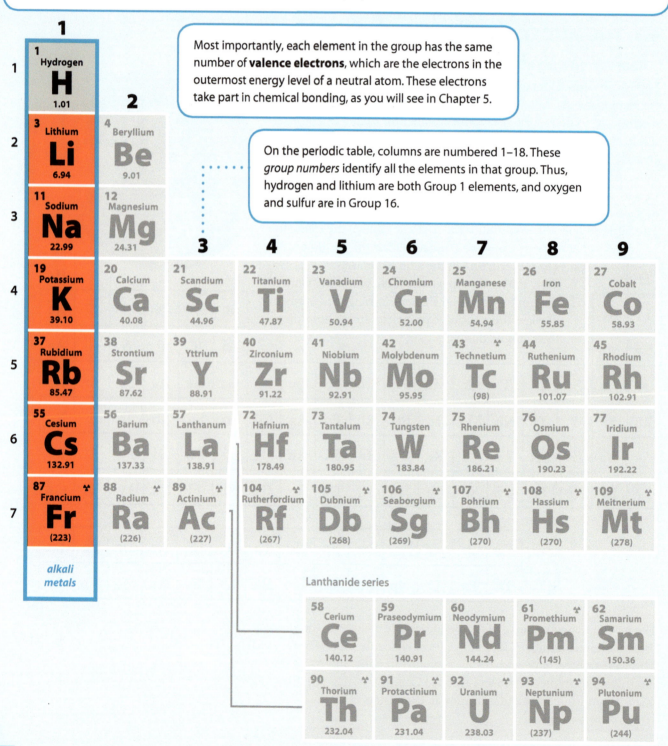

Lanthanide series

Actinide series

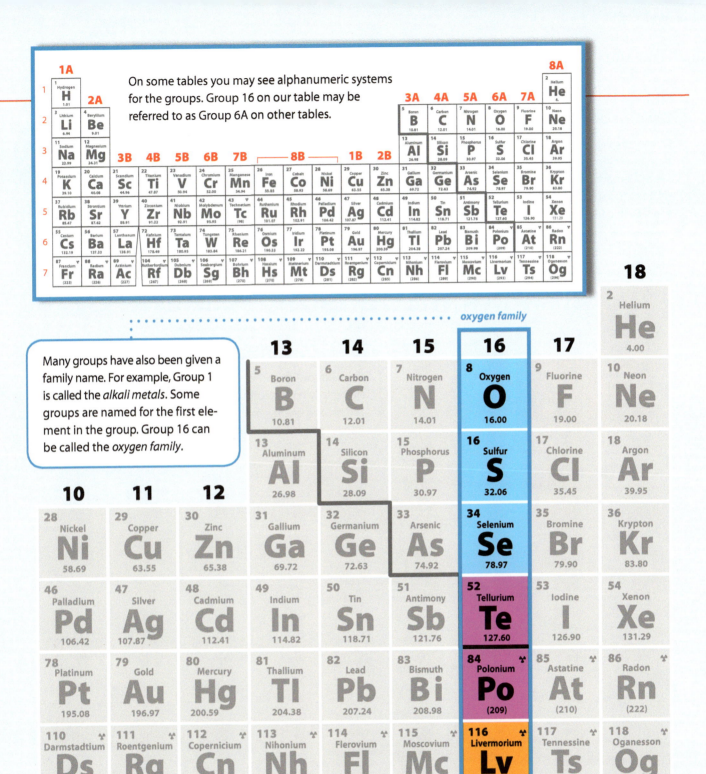

On some tables you may see alphanumeric systems for the groups. Group 16 on our table may be referred to as Group 6A on other tables.

Many groups have also been given a family name. For example, Group 1 is called the *alkali metals*. Some groups are named for the first element in the group. Group 16 can be called the *oxygen family*.

THE PERIODIC TABLE 77

ORGANIZATION OF THE PERIODIC TABLE

PERIODS

Scientists call a row on the periodic table a **period**. Both the columns (representing the different families of elements) and the rows on the periodic table relate to the structure of the atoms, but in different ways. Each period represents an energy level.

The rows on the periodic table are labeled 1–7. The row that a certain element is located in also designates the energy level of its valence electrons. Thus calcium (20), in Period 4, has valence electrons in its fourth energy level, and radon (86) in Period 6 has its valence electrons in its sixth energy level.

Hydrogen and helium, the two elements in the first period, have only one energy level with electrons in it.

The next row down, Period 2, has eight elements whose electrons are in the first and second energy levels, and so on.

In theory, an atom could have an infinite number of energy levels, but so far we know only of elements with up to seven energy levels.

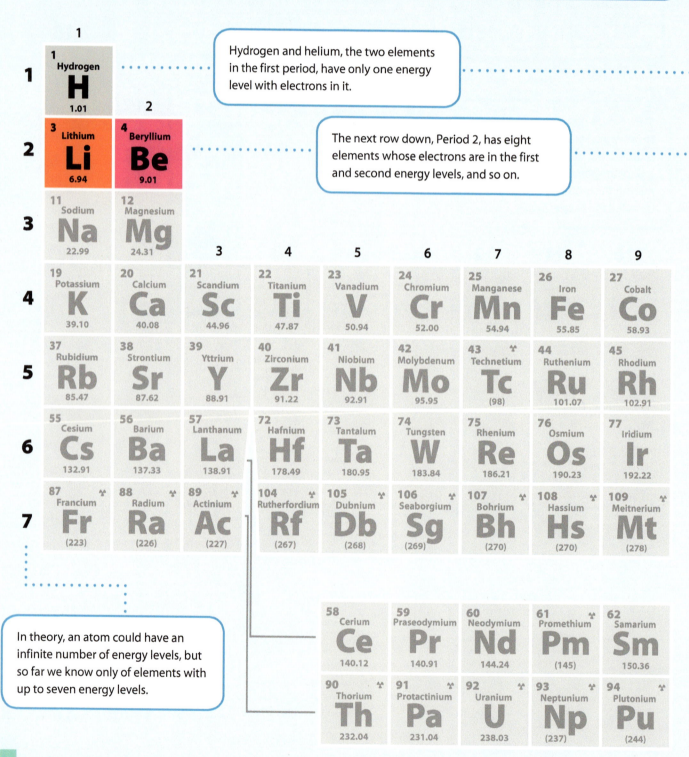

CHAPTER 4

4A | REVIEW QUESTIONS

1. When did people first know about substances that we call elements?
2. Why were people driven to develop a way to organize the elements?
3. How did Lavoisier's study of combustion prove that fire was not an element?
4. Why do you think that potassium (19) has the symbol K?
5. What were Döbereiner's triads?
6. What was the most remarkable feature of Mendeleev's periodic table?
7. Give a definition for the periodic table.
8. What is a family or group of elements and how is it represented on the periodic table?
9. What is a period of the periodic table? What does it tell us about an atom?

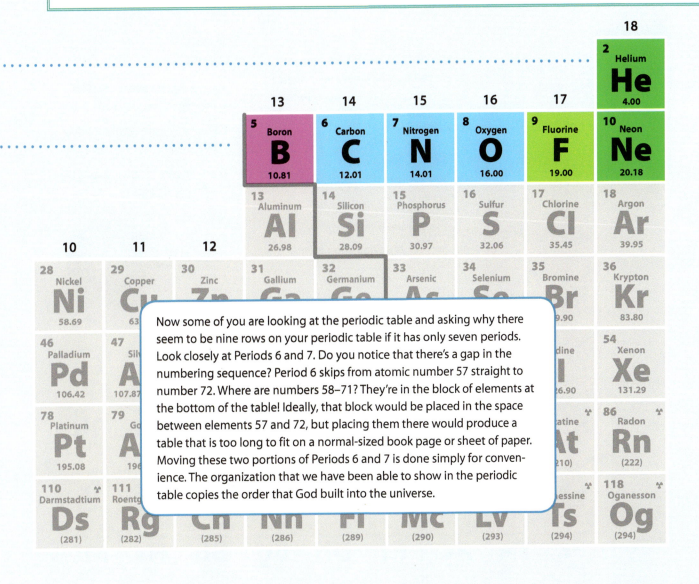

Now some of you are looking at the periodic table and asking why there seem to be nine rows on your periodic table if it has only seven periods. Look closely at Periods 6 and 7. Do you notice that there's a gap in the numbering sequence? Period 6 skips from atomic number 57 straight to number 72. Where are numbers 58–71? They're in the block of elements at the bottom of the table! Ideally, that block would be placed in the space between elements 57 and 72, but placing them there would produce a table that is too long to fit on a normal-sized book page or sheet of paper. Moving these two portions of Periods 6 and 7 is done simply for convenience. The organization that we have been able to show in the periodic table copies the order that God built into the universe.

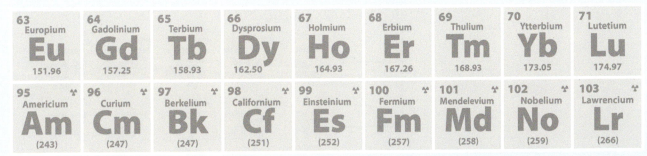

THE PERIODIC TABLE

4B Questions

- Where are metals and nonmetals on the periodic table?
- How do metals, metalloids, and nonmetals compare?
- How can we know the chemical family of an element?
- How many valence electrons does each element have?

4B Terms

metal, metalloid, nonmetal, alkali metal, alkaline-earth metal, transition metal, inner transition metal, mixed group, halogen, noble gas

4B | CLASSIFYING THE ELEMENTS

How is the periodic table useful?

4.4 TYPES OF ELEMENTS

Recall that the main point of the periodic table is to organize the elements to help us understand them better. Scientists arranged the table to group elements by their properties. One key characteristic of an element is how metallic it is. Elements range from highly metallic to nonmetallic as we move across the periodic table from left to right.

Metals—Almost 80% of elements are metals, which have few valence electrons. Metals are found to the left of the heavy stairstep line on the periodic table.

- alkali metals
- alkaline-earth metals
- transition metals
- inner transition metals
- post-transition metals

Typical Properties:

- state: exist as a dense, ductile, malleable, lustrous solid
- conductivity: are highly conductive, electrically and thermally
- reactivity: are reactive, especially with nonmetals

copper

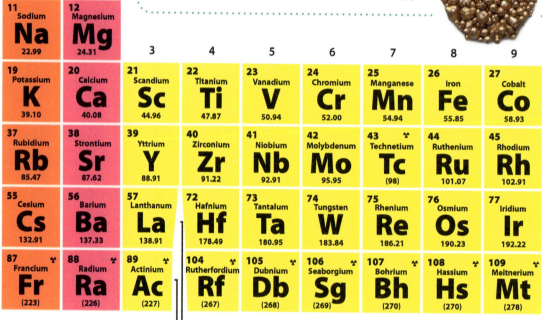

80 CHAPTER 4

Metalloids—These elements have characteristics between those of metals and nonmetals. Metalloids are located along the stairstep line and are also called *semiconductors*.

Typical Properties:

- state: exist as a brittle solid with metallic luster
- conductivity: are fairly conductive, increasingly so as temperature rises
- reactivity: varies

▨ *metalloids*

Nonmetals—These elements typically have four or more valence electrons and do not exhibit the general properties of metals. Nonmetals are to the right of, but not touching, the heavy stairstep line on the periodic table.

Typical Properties:

- state: exist as a gas, a liquid, or a dull, brittle solid
- conductivity: are poorly conductive, electrically and thermally
- reactivity: varies

▨ nonmetals
▨ halogens (also nonmetals)
▨ noble gases (also nonmetals)

iodine

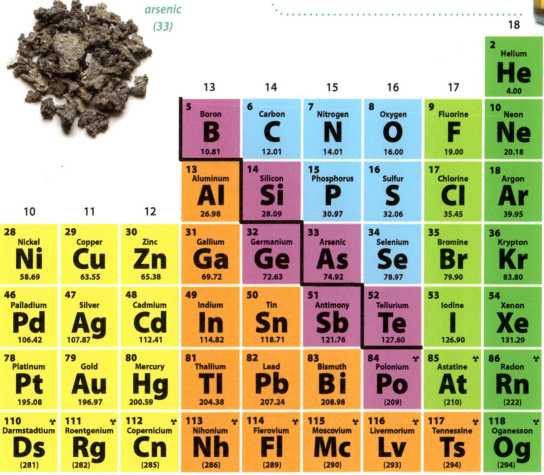
arsenic (33)

THE PERIODIC TABLE

4.5 ELEMENT FAMILIES

The metal-to-nonmetal spectrum is a broad classification of elements, but remember that each column represents a family of elements. Just as in your family, there are distinctive traits in each of these families. Each family contains elements whose properties are similar. These properties arise because all the elements in the family have atomic structures, particularly the arrangements of their electrons, which are alike. Elements in each family have the same number of valence electrons. The descriptions of the families on this page and the next two pages specify how many valence electrons the elements have. Is there a pattern?

Alkali metals are elements in Group 1. They have one valence electron, which they can easily lose to form a 1+ cation (see page 60). The fact that they easily lose this electron makes them very reactive; in fact, these are the most reactive of all the metals. They are so reactive that these elements are never found in their pure form in nature. They are always found as parts of compounds.

Sodium bursts into flames as it contacts water.

Elements in Group 2 are called **alkaline-earth metals**. They each have two valence electrons, which they tend to lose, making them 2+ cations. Alkaline-earth metals are only slightly less reactive than alkali metals. These elements are often added to alloys, homogeneous mixtures that contain at least one metal, to make them stronger. Beryllium (4) is added to copper to make it stronger. Magnesium (12) alloys are used to build light but strong aircraft structures.

sterling silver

	Alkali metals
	Alkaline-earth metals
	Transition metals
	Inner transition metals
	Post-transition metals
	Metalloids
	Nonmetals
	Halogens (also nonmetals)
	Noble gases
☢	Radioactive isotopes

Groups 3–12 are called the **transition metals**. Most of these elements have one or two valence electrons. These electrons are easily lost, making cations with charges of 1+ or 2+. We often use these elements in alloys, such as sterling silver (a silver and copper alloy) or steel, an iron and carbon (6) alloy.

The two rows of elements shown below the main body on most periodic tables are the **inner transition metals**. They typically have two valence electrons, and so they form 2+ cations. The elements in the first row are rare and are used in lighting, lasers, superconductors, and magnets. The elements from the second row are all radioactive, and some are used as nuclear fuels.

This sample of holmium (67) shows the characteristic metal properties of inner transition metals.

Inner transition metals get their name from their proper placement—between the transition metals in the periodic table.

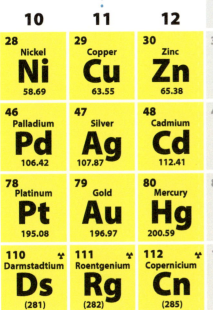

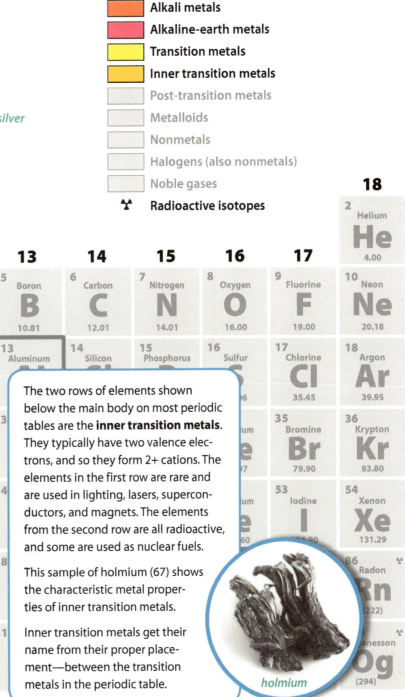
holmium

THE PERIODIC TABLE   83

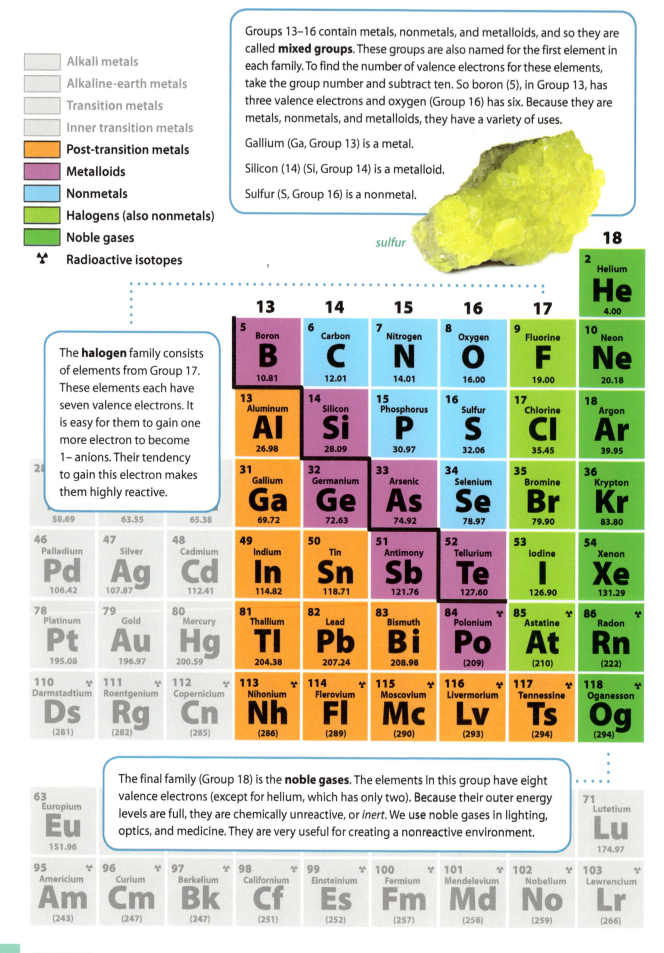

WORLDVIEW SLEUTHING: GIVING DUE CREDIT

Discoveries in science are very important. They are important because of the new data gained and the potential for new applications. They can also advance a scientist's career and standing within the scientific community. Being credited with a discovery may mean great wealth for a scientist in addition to historical recognition. While some scientists are not interested in fame and fortune, recognition for years or decades of work is always appropriate.

TASK
You are working for IUPAC and are representing one of two teams of scientists involved in discovering element 113. You will prepare and take part in a debate to determine which scientific team deserves credit for the discovery and will therefore get to name the element.

PROCEDURE
1. Research the process of naming new elements by doing a keyword search for "naming new elements."
2. Research the history of how element 113 was discovered by doing a keyword search for "newest elements" and "discovering nihonium."
3. Plan your debate and collect any required materials. Remember to cite your sources.
4. Participate in the debate.

CONCLUSION
The discovery of a new element is a significant event in chemistry and results in great recognition for the scientist that accomplishes it. We should always want to recognize the achievements of others. As Christians, we should keep in mind that the goal of science is to learn about God's creation, giving glory to God and helping others.

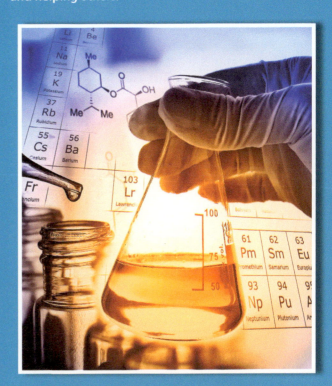

4B | REVIEW QUESTIONS

1. What is a metal?
2. Using the periodic table, would you expect oxygen or selenium (34) to act more like a nonmetal? Explain.
3. You have a sample of an element. At room temperature, it is a lightweight, brittle solid with a metallic luster. It has fair conductivity that increases with increased temperature. Do you think the material is a metal, nonmetal, or metalloid? Explain.
4. What do we call the family in Group 1?
5. Describe an alkaline-earth metal.
6. Why do most periodic tables show the inner transition metals below the main body of the table?
7. Describe a halogen.
8. How many valence electrons do the following elements have?
    a. magnesium
    b. antimony (51)
    c. chlorine

4C Questions

- Why is the periodic table shaped the way it is?
- Is there an easy way to show elements and their electrons?
- How does an element's position on the table relate to its properties?

4C Terms

electron dot notation, atomic radius, electronegativity

4C | PERIODIC TRENDS

What can an element's position on the periodic table tell us about the element?

4.6 ATOMIC STRUCTURE & THE PERIODIC TABLE

People often look at the periodic table and wonder why it is not a rectangle or some other regular shape. The shape of the periodic table is related to the structures of the atoms themselves. The periods, or rows, on the periodic table are related to the energy levels of the atoms. The groups are elements with similar electron arrangements, which is the reason for their similar properties. The arrangement of the periodic table tells us a lot about the composition and structure of each element's atoms.

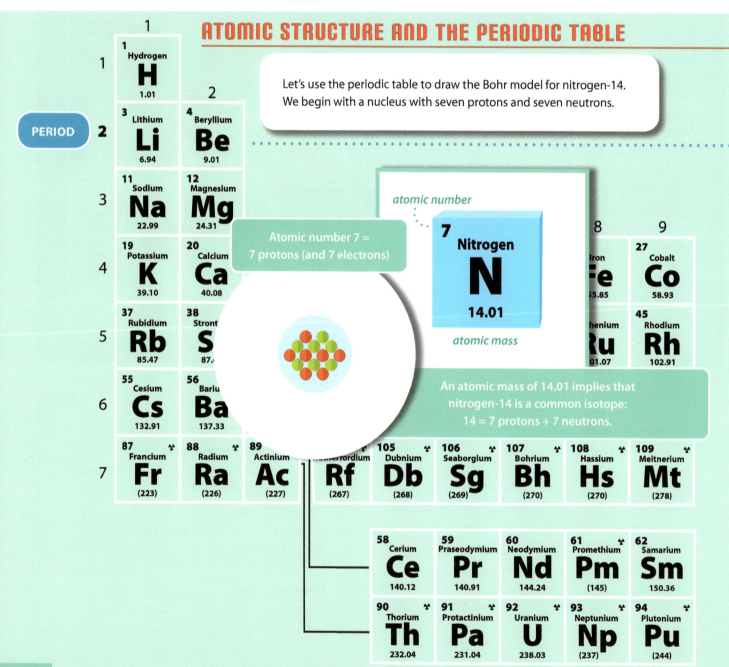

86   CHAPTER 4

Let's look at oxygen on the periodic table. Oxygen is element number 8. It is to the right of the stairstep, has an average atomic mass of 16.00, is in Group 16, and is in the second period. What does this tell us about oxygen atoms?

Oxygen's being to the right of the stairstep tells us that it is a nonmetal. Its atomic number tells us that oxygen has eight protons, which means that as a neutral atom it must also have eight electrons. The average atomic mass implies that oxygen-16 is likely a common isotope with eight neutrons. The fact that oxygen is in Group 16 tells us that it has six valence electrons, which must be in the second energy level because oxygen is in the second period. And we can know all of this just by looking at where oxygen is located on the table! The periodic table is like a snapshot of the atomic structures of all the known elements.

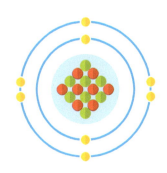

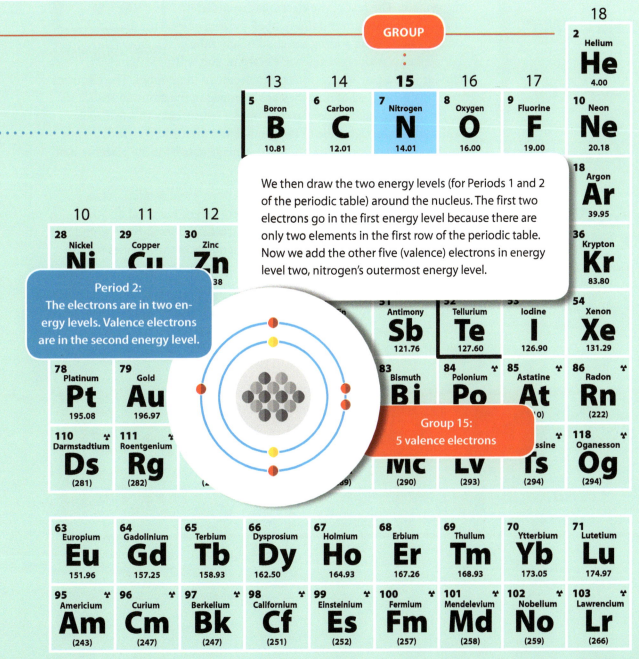

We then draw the two energy levels (for Periods 1 and 2 of the periodic table) around the nucleus. The first two electrons go in the first energy level because there are only two elements in the first row of the periodic table. Now we add the other five (valence) electrons in energy level two, nitrogen's outermost energy level.

Period 2: The electrons are in two energy levels. Valence electrons are in the second energy level.

Group 15: 5 valence electrons

So what happens to the atomic structure as we move from left to right across a period on the periodic table? Let's look at Period 2 on the table. As we move across the row, each element adds a proton, an electron, and typically some neutrons.

In many cases, the Bohr model shows more information than is needed. Therefore we can use a simpler drawing called an electron dot notation to represent atoms. **Electron dot notation** consists of the element's chemical symbol surrounded by the atom's valence electrons. The chemical symbol represents the nucleus and all of the *non-valence electrons*.

To write the electron dot notation, start with the chemical symbol. Then begin adding valence electrons. Add the first valence electron to the right side of the symbol. Add the second valence electron to form a pair on the right side of the symbol. The rest of the electrons are added counterclockwise around the symbol, placing one on each side before adding a second to any side. For elements with eight valence electrons, you will end up with a pair of electrons on each side of the symbol. The electrons are in pairs because of how they are arranged in the atom. In reality, it doesn't matter which side you start on; for consistency this textbook will start on the right side of the symbol and then add electrons counterclockwise around the atom. See the image above for the placement of valence electrons.

electron dot notation

Look at the electron dot notations for Groups 1 and 2 at left. As we move down each column, we add an energy level, but what do you notice about the number of valence electrons? It doesn't change, and this consistent structure is what causes the elements in the families to have similar properties.

Notice below that as we move across the rows (with the transition metals being the exceptions), we continue to add valence electrons. But also notice what happens as we move down a column (family). Even though we add an energy level, the number of valence electrons remains constant.

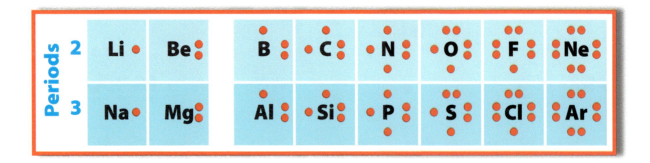

4.7 PERIODIC TRENDS

Atomic Radius and Electronegativity

Because of the changes in structure throughout the elements on the periodic table, there are trends in the properties of those elements as well. Two key trends are atomic radius and electronegativity.

TRENDS IN ATOMIC RADIUS

Atomic radius is simply the distance from the center of an atom's nucleus to the electrons in its outermost energy level. The size of an atom affects things such as how it will arrange itself with other atoms in molecules. It also affects how each atom will react and bond, since reactions and bonding depend on how close the valence electrons are to the nucleus. As we move down a column, the addition of energy levels, like layers on an onion, makes an atom bigger. As we move to the right across a period, adding protons and electrons causes the atoms to get smaller. This happens because opposite charges attract and more protons (positive) and electrons (negative) create a stronger force, pulling the electrons closer to the nucleus.

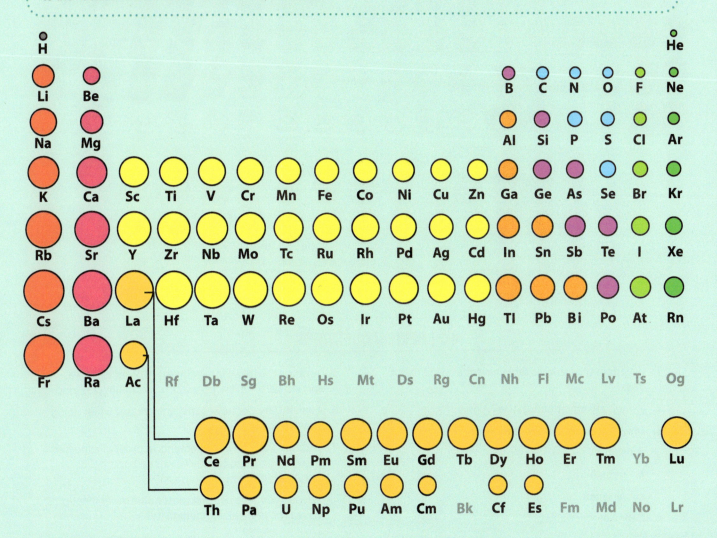

THE PERIODIC TABLE 89

TRENDS IN ELECTRONEGATIVITY

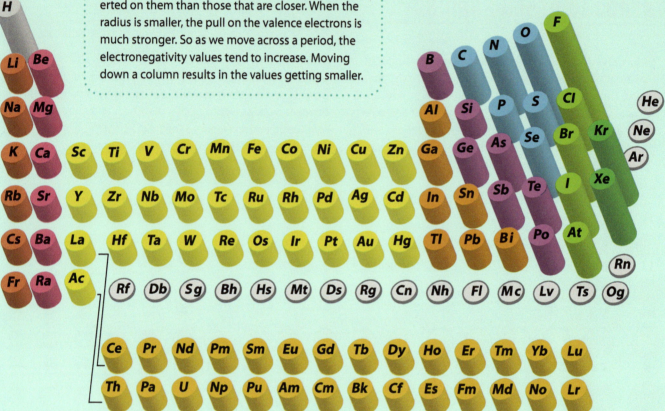

The measure of an element's ability to attract and hold electrons when bonded to other atoms is called **electronegativity**. The trend in electronegativity is related to the atomic radius. Electrons that are more distant from the nucleus have less pull exerted on them than those that are closer. When the radius is smaller, the pull on the valence electrons is much stronger. So as we move across a period, the electronegativity values tend to increase. Moving down a column results in the values getting smaller.

4C | REVIEW QUESTIONS

1. Why does the periodic table have the shape that it does?
2. What is electron dot notation?
3. Draw the electron dot notation for
   a. sulfur
   b. potassium
   c. nitrogen
4. Define *atomic radius*.
5. Arrange the following elements in order of increasing radius: bromine, potassium, vanadium (23), zinc.
6. Explain why electronegativity decreases when going down a column.
7. Arrange the following elements in order of expected increasing electronegativity values: barium (56), beryllium, calcium, magnesium.
8. Why do you suppose that most noble gases (Group 18) have little or no electronegativity?

CHAPTER 4 REVIEW

4A ORGANIZING THE ELEMENTS

- Scientists developed the periodic table to organize the data about the elements.
- Scientists noticed repeated patterns in the properties of elements, leading others to organize the periodic table.
- At first, chemists ordered the table on the basis of atomic mass, but this misplaced a few elements. Changing the table by using atomic numbers to order it made the model more workable.
- Dmitri Mendeleev predicted the discovery of elements that seemed to be missing from his periodic table.
- The periodic table's arrangement corresponds to atomic structure. The periods (rows) represent energy levels and the families (columns) represent similar electron arrangements.

4A Terms

periodic law	75
periodic table	75
family	76
group	76
valence electron	76
period	78

4B CLASSIFYING THE ELEMENTS

- The periodic table places metals on the left side and nonmetals on the right side, separated by a heavy stairstep line. Metalloids along the stairstep have both metallic and nonmetallic properties.
- Metals tend to be lustrous, malleable, ductile, and reactive solids with high conductivity.
- Nonmetals can be solids, liquids, or gases, with varying reactivity and poor conductivity.
- The columns on the periodic table represent families of elements.
- Members of a family have the same number of valence electrons and properties.

4B Terms

metal	80
metalloid	81
nonmetal	81
alkali metal	82
alkaline-earth metal	82
transition metal	83
inner transition metal	83
mixed group	84
halogen	84
noble gas	84

THE PERIODIC TABLE 91

CHAPTER 4

4C PERIODIC TRENDS

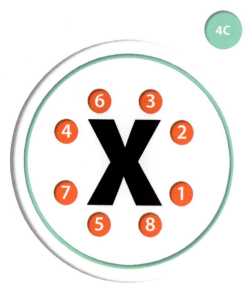

- Both the number of valence electrons that an element has and the energy level in which the valence electrons are found can be determined by the element's location on the periodic table.
- Electron dot notation is a simple way to represent atoms of elements and their valence electrons.
- The properties of elements change as we move across rows or down columns because of changes in atomic structure from one element to the next.
- Atomic radius *decreases across a row* as the attractive electric force between the nucleus and electrons increases. Atomic radius *increases down a column* due to the addition of energy levels.
- Electronegativity *increases across a row* as the valence electrons are nearer to the nucleus and *decreases down a column* as the valence electrons are farther from the nucleus.

4C Terms

electron dot notation	88
atomic radius	89
electronegativity	90

CHAPTER REVIEW QUESTIONS

Recalling Facts

1. Of what elements did the ancient Greeks believe all matter consisted?
2. Name a key contribution that Robert Boyle made to the study of chemistry.
3. What was Berzelius's contribution to the study of elements?
4. Define *periodicity*.
5. Name two changes made to Mendeleev's original periodic table to produce the one that we use today.
6. State the periodic law.
7. Define *valence electron*.
8. Using the periodic table, list two pairs of elements that would have been out of order on Mendeleev's periodic table (according to atomic mass) compared with our current table. Assume that scientists knew about all the elements up to and including lead.
9. Describe a nonmetal and list its properties.
10. What trait of alkali metals results from the fact that they easily lose their one valence electron?
11. How many valence electrons do the transition metals typically have?
12. Why is the oxygen group considered one of the mixed groups?
13. Which family's atoms generally have eight valence electrons? What property do these electrons produce?
14. Name the family for each of the following elements.
    a. lead
    b. calcium
    c. promethium (61)
15. Why are electrons paired in electron dot notation?

Understanding Concepts

16. Why did Berzelius establish his system of chemical symbols?
17. Does CN represent the element copernicium (112) or a compound of carbon and nitrogen? Explain.
18. Why do we need a periodic table?
19. Place in chronological order the following people who contributed to the development of the periodic table: Berzelius, Döbereiner, Mendeleev, Mosley, Newlands.
20. List the group number for each element.
    a. cadmium (48)
    b. cesium (55)
    c. carbon
21. How does the shape of the periodic table reflect the order that we see in atoms and in the universe?
22. Using the periodic table, would you expect copper or gallium to exhibit more metallic properties? Explain.
23. Using the periodic table, classify the following elements as metal, nonmetal, or metalloid.
    a. antimony
    b. strontium (38)
    c. bromine

REVIEW

24. What is similar about elements in a particular family?
25. What type of ion do the alkali metals form and why?
26. All of the transition metals have at least how many energy levels? Explain.
27. On the basis of its position on the periodic table alone, what do we know about sulfur?
28. What information about a neutral atom of silicon does the periodic table tell us?
29. Write the electron dot notation for the following elements.
    a. magnesium
    b. fluorine (9)
    c. indium (49)
30. Explain why atomic radius changes as we move to the right across a period.
31. Arrange the following elements in order of increasing radius: oxygen, polonium (84), selenium, sulfur.
32. Explain why electronegativity changes as it does across a period.

Critical Thinking

33. Why did people keep working on developing the periodic table? How are they continuing to work on the table now?
34. You want an element that is conductive, malleable, and ductile but not very reactive. Where on the periodic table would you look for your element? Explain.
35. What is wrong with the Bohr model of carbon shown below?

36. Ionization energy is the measure of how much work needs to be done to remove a valence electron from a neutral atom. Thinking about the trends in atomic radii and electronegativity, hypothesize as to how you think ionization energy would change across a row. Explain.

Use the Case Study at right to answer Questions 37–41.

37. What is an allotrope?
38. Of what element(s) are diamond and graphite made?
39. Considering your answer to Question 38, how can you explain the significant differences in properties between diamond and graphite?
40. Would a change from one allotrope to another be a chemical change or a physical change?
41. Why can we model the characteristics of elements in such an orderly way?

CASE STUDY: ALLOTROPES

As mentioned in the Chapter opener, diamonds are made of pure carbon, but so is the graphite in your pencil. Diamond and graphite are *allotropes* of carbon. Allotropes are different forms of the same element in the same state. Allotropes' internal structures differ, resulting in remarkably different characteristics.

Many elements exhibit allotropy. Metals such as polonium, iron, and cobalt each have allotropes. Some metalloids—antimony, arsenic, and silicon—exhibit allotropy. In addition to carbon, other nonmetals like oxygen (ozone and O_2) and phosphorus (15) can be found in different forms. Tin transforms from metallic tin into an allotrope that is a metalloid below 13.2 °C. Materials can change allotrope forms in response to the effects of temperature, light, and pressure.

While graphite and diamond are probably the two most well-known allotropes of carbon, there are a total of nine allotropes of carbon. Each allotrope has distinct properties, all stemming from varied internal structures. Graphite consists of thin sheets of carbon. The sheets easily separate, making graphite useful as pencil lead. Diamonds form from very strong crystal structures, giving them their extreme hardness. Carbon can even form exotic, soccer-ball shaped structures sometimes referred to as *buckyballs*.

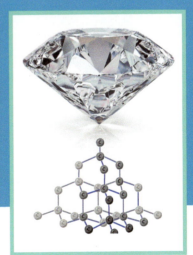

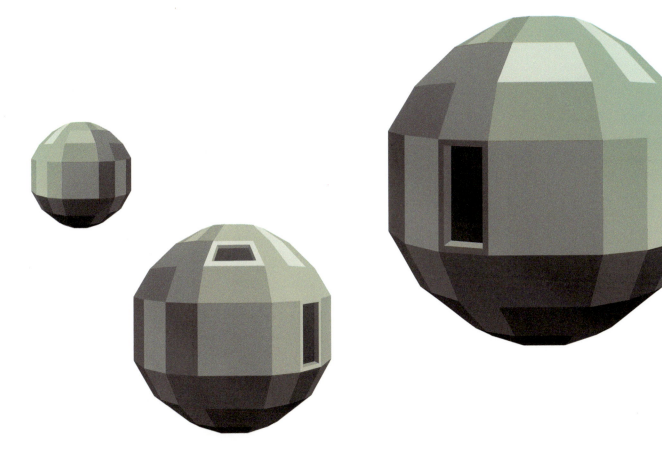

CHAPTER 5
Bonding and Compounds

TREE TRUNK TREASURE

What do you think is the link between these trees and the family car? Need a hint? The substance being harvested from these trees, once processed, can be used for a wide variety of products including garden hoses, waterproof rainwear, pencil erasers—and car tires. Have you guessed the connection yet? The trees in the picture are rubber trees, and the sticky, white latex from their bark is used to make natural rubber. Latex is just one of many naturally occurring chemical compounds that people depend on to meet everyday needs.

5A Principles of Bonding	96
5B Types of Bonds	99
5C Writing Chemical Formulas	108

5A Questions

- What is a chemical bond?
- Why do chemical bonds form?
- Are molecules always made from more than one element?
- What determines how atoms bond?
- Are the properties of compounds the same as those of the elements of which they are made?

5A Terms

chemical bond, octet rule

5A | PRINCIPLES OF BONDING

How do compounds form?

5.1 CHEMICAL BONDS

Rubber, gasoline, starch, aspirin—these very different substances all have something in common. They are each made of compounds that are held together by chemical bonds. A **chemical bond** is an electrostatic attraction that forms between atoms when they share or transfer valence electrons. Chemical bonding is what makes it possible for the rather small number of naturally occurring elements to combine and form the great variety of compounds that exist in the world around us.

Chemical bonds store useful energy within molecules. The energy stored in the chemical bonds in gasoline powers your family's car. When chemical bonds in molecules are broken and new compounds are formed, some of that stored energy may become available to do work, like moving a car. Humans don't run on gasoline, but our food contains chemical bonds, just like gasoline. The energy stored in those bonds is what powers us too.

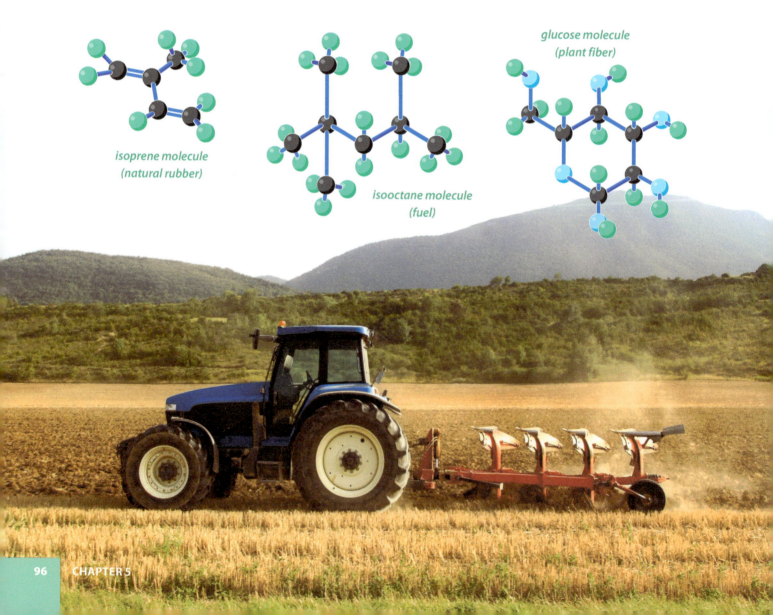

Why Compounds?

Why do atoms form chemical bonds in the first place? It turns out that most atoms are chemically unstable when they are not bonded to other atoms. This is why few elements are found in their pure form in nature. Instead, they are usually found combined with other elements in the form of compounds. Groups of atoms bonded together as compounds are usually more stable than those individual atoms would be by themselves.

Element or Compound?

An atom of one kind of element doesn't always have to combine with an atom of a different element in order to gain stability. Some pure elements in nature occur as chemically bonded atoms of that element. Both oxygen and nitrogen in Earth's atmosphere exist mostly in the form of molecules that are made of two atoms of the same element bonded together. Sulfur is commonly found as ring-shaped molecules made of eight sulfur atoms.

5.2 THE OCTET RULE

The instability in atoms and their tendency to form chemical bonds is mainly due to incomplete valence energy levels (see Chapter 4). Atoms generally are most stable when they have a full eight electrons in their valence energy level. This principle is called the **octet rule** of bonding. There are a few exceptions to the octet rule. Hydrogen and helium, for example, have electrons only in their first energy level, which can have at most two valence electrons. The next three elements—lithium, beryllium, and boron—usually lose their valence electrons when bonding, again leaving electrons only in their first energy levels. You'll see later how the octet rule is applied in chemistry to determine how atoms will bond to other atoms. Extensive experimental evidence shows that an atom can fill an octet in its outer level by bonding with other atoms to form molecules and compounds.

Atoms achieve greater stability through bonding in one of two ways. First, atoms can share electrons. The shared electrons can fill the valence levels of all the atoms bonded together at the same time. Second, atoms can gain or lose electrons through a process of electron transfer to gain a full octet. Atoms with few valence electrons, which are loosely held, can easily lose them, exposing the full octet of the next lower energy level. On the other hand, if an atom needs a few electrons to complete an octet, it can acquire them from other atoms or its surroundings. Ions formed by losing or gaining electrons are pulled together by their opposite charges. No matter the type of bonding, the end result is the same—atoms are more stable with completely filled outer energy levels.

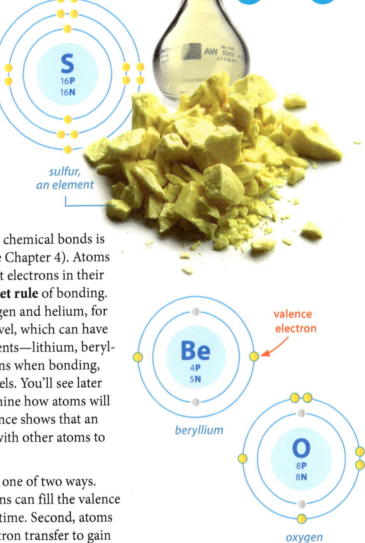

5.3 CHANGES IN PROPERTIES

Have you ever tried to set water on fire? Silly question, right? Water, as you probably know, is a compound made of the elements hydrogen and oxygen. At room temperature, water is a liquid that is useful for drinking and bathing. It's essential for life as we know it. It's also not flammable, so it's terrible for starting fires but great for putting them out. Pure hydrogen and oxygen are another story. Both are colorless, odorless, and extremely reactive gases that easily, and sometimes explosively, combine with other elements. Hydrogen and oxygen are chemically reactive precisely because their atoms are unstable. When they combine with each other, the resulting compound—water—is much more stable than its parent elements. Chemical bonding changes the physical and chemical properties of all the substances involved.

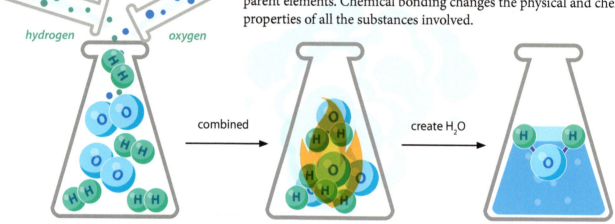

What's true for water is usually true for other compounds as well. They tend to be quite different from the elements of which they are made. Think about pencil lead for a moment. We call it "lead," but it's actually a mixture of clay and *graphite*, a form of pure carbon (see page 93). Would you consider stirring some powdered graphite into your coffee or sprinkling it on a bowl of oatmeal? Of course not! But if those same carbon atoms found in graphite are combined and bonded in a certain way with hydrogen and oxygen atoms, a surprising change takes place. They produce a white, crystalline, and sweet-tasting solid—table sugar. Sugar tastes great, but it's useless for writing. Its physical and chemical properties are just far too different from those of its parent element, carbon.

5A | REVIEW QUESTIONS

1. A chemical bond forms when atoms _____ electrons.
   a. share
   b. transfer
   c. either share or transfer
   d. both share and transfer
2. Why do atoms form chemical bonds?
3. Are molecules always made of atoms of different elements? Explain.
4. What is the octet rule?
5. What are two ways that an atom can meet the octet rule requirement?
6. Describe two properties of water that are different from the properties of the elements from which it is formed.

5B | TYPES OF BONDS

Why do atoms bond in different ways?

5.4 TYPES OF BONDS

Remember, most atoms are seeking to have eight valence electrons, and they can accomplish this by gaining, losing, or sharing electrons. As you saw in Chapter 4, not all atoms exert the same amount of tug on their electrons. Atoms with high electronegativities hold on to their valence electrons tightly and are eager to acquire more. Those with low electronegativities, on the other hand, pull weakly on their valence electrons and easily give them up. The difference (or lack thereof) between the electronegativities of two atoms governs what sort of bond they will make.

5B Questions

- How does the role of electrons differ in covalent, ionic, and metallic bonding?
- What's an easy way to illustrate simple compounds?
- Can two atoms form more than one bond between them?
- How does the polarity of a bond compare with that of a molecule?

5B Terms

covalent bond, ionic bond, formula unit, metallic bond, diatomic molecule, Lewis structure, polarity

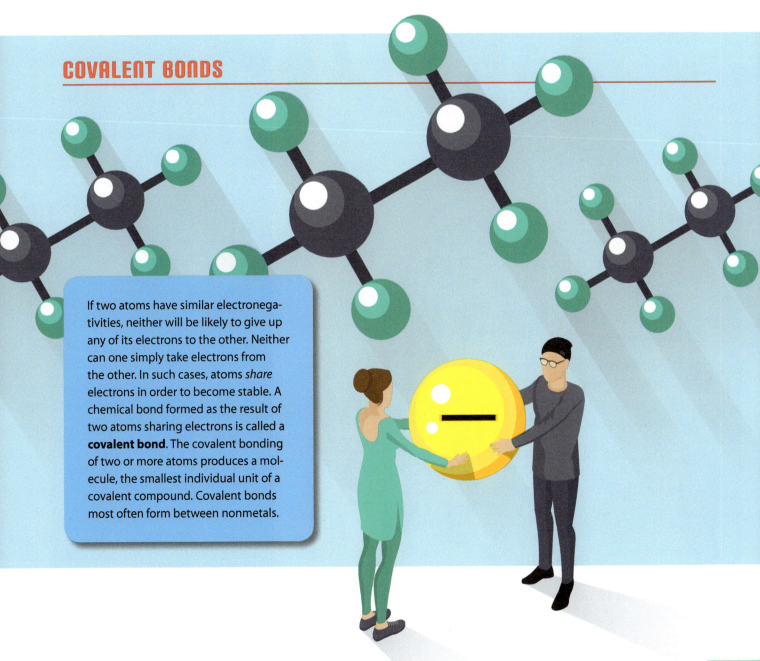

COVALENT BONDS

If two atoms have similar electronegativities, neither will be likely to give up any of its electrons to the other. Neither can one simply take electrons from the other. In such cases, atoms *share* electrons in order to become stable. A chemical bond formed as the result of two atoms sharing electrons is called a **covalent bond**. The covalent bonding of two or more atoms produces a molecule, the smallest individual unit of a covalent compound. Covalent bonds most often form between nonmetals.

IONIC BONDS

When atoms with very different electronegativities bond, one typically loses one or more valence electrons and the other gains the lost electrons. The atom with low electronegativity, which loses electrons, becomes a cation. The atom with high electronegativity, which gains electrons, becomes an anion. The opposite electrical charges on these ions attract each other to form an **ionic bond**. Ionic bonds normally form between a metal and a nonmetal.

Ionic substances form crystals rather than molecules (see Chapter 2). Each ion is equally attracted to a number of oppositely charged ions within a lattice. Crystals of a particular ionic solid always form with the same ratio of cations to anions. Thus, the chemical formula for an ionic substance represents a **formula unit**, the smallest ratio of the ions within the compound.

METALLIC BONDS

Many metals, especially transition metals, exhibit a third type of chemical bond. These elements have similar low electronegativities. This causes atoms of those metals to hold their valence electrons very loosely. As a result, the electrons are able to move about easily from one atom to another. They are not shared or exchanged between any two atoms in particular in the manner of either a covalent bond or an ionic bond. The result of this metallic bonding is a "sea" of freely moving electrons within the lattice structure of the metal. A **metallic bond** is the attraction between metal atoms and their sea of shared electrons. Most of the typical properties of metals are due to these metallic bonds.

5.5 HOW COVALENT BONDING WORKS

Diatomic Molecules

The simplest example of two atoms with similar electronegativities sharing electrons in order to satisfy the octet rule occurs when the atoms are of the same element. Molecules made of two atoms, whether of the same element or not, are called **diatomic molecules**. Let's look at how a molecule of chlorine (Cl_2) forms. Both chlorine atoms have seven valence electrons and need one more to complete an octet (see right, top). The electronegativity of chlorine is high, so it is not willing to give up an electron. And since each atom has the same electronegativity, neither can remove an electron from the other. But if each chlorine atom shares one of its electrons with the other, both atoms can have an octet. The shared electrons count toward meeting the octet rule requirement for both atoms. The two shared electrons, called a bonding pair, form a *single covalent bond* (right, bottom) between the two chlorine atoms.

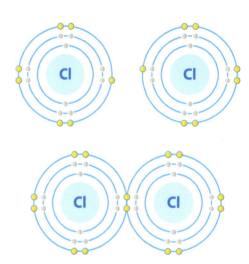

Chlorine forms a covalent bond by sharing a single electron with another chlorine atom. When counting valence electrons, the shared electrons count for both atoms.

The example just given uses Bohr models to show how electrons are shared. But there's a simpler way to model a covalent bond using **Lewis structures**. Chemist Gilbert Lewis introduced this system in 1916. By rotating the unpaired dots in the electron dot notation for two chlorine atoms, the single dots can be shown forming a bonding pair. The pair of dots between the symbols identifies a covalent bond. The Lewis structure for the bonding of chlorine is shown below.

$$:\ddot{Cl}\cdot + \cdot\ddot{Cl}: \rightarrow :\ddot{Cl}:\ddot{Cl}: \text{ or } Cl-Cl$$

Covalent bonds may be represented by a pair of dots, but more often a single dash between the bonded atoms is used to distinguish between the electrons involved in bonding and those that are not. When dashes are used in Lewis structures, the nonbonding electrons may or may not be shown.

Seven elements are found in nature as diatomic molecules. You'll need to know these for Chapter 7. The seven diatomic elements are hydrogen, nitrogen, oxygen, fluorine, chlorine, bromine, and iodine. A memory aid for these is the *Hydrogen Seven*. Notice on the periodic table below that the diatomic elements have been shaded. Hydrogen is in the upper left corner, but the other six form the numeral seven starting at nitrogen, element number 7.

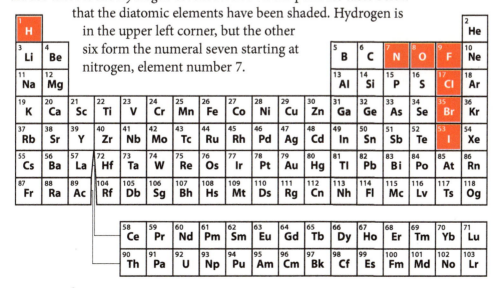

diatomic oxygen

diatomic nitrogen

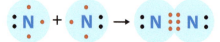

Multiple Bonds

Like chlorine, pure oxygen occurs in the form of diatomic molecules. But there's a difference. An oxygen atom has six valence electrons, so it lacks two electrons to make a complete valence octet. To address the shortage, an oxygen atom needs to share not just one, but two, of its electrons with another oxygen atom. Sharing two pairs of electrons forms a *double covalent bond*. Double bonds are stronger than single bonds. In Lewis structures they are represented by two dashes instead of two pairs of dots. Even stronger *triple covalent bonds* form when atoms must share three electrons to complete their octets. Diatomic nitrogen is one example.

Polyatomic Molecules

Of course, matter in our world consists of much more than just diatomic elements. Most substances are made of polyatomic molecules (Gk. *poly*, meaning "many") that contain different kinds of atoms. Such molecules are also held together by covalent bonds. Let's look at water as one example. As you probably already know, water consists of one part oxygen for every two parts hydrogen. We've just seen that each oxygen atom needs two valence electrons to complete an octet. That gives us a hint as to why a molecule of water contains two hydrogen atoms.

If you locate hydrogen on the periodic table, you'll see that its atomic number is 1 and that it is in Group 1 and in the first period. That means that a neutral hydrogen atom has one valence electron (Group 1) and only one energy level (Period 1). You'll recall from Chapter 4 that the first energy level of every atom can hold only two electrons. This means that there are a few elements, including hydrogen, that are exceptions to the octet rule. They can be stable if their first energy level is at full capacity—just two electrons.

Now we can see how forming a molecule of water satisfies the electron needs of each atom in the molecule. Each hydrogen atom shares its one electron with the oxygen atom, and the oxygen atom shares two electrons, one with each hydrogen atom. This forms two single covalent bonds—one between each hydrogen atom and the oxygen atom.

Covalent bonds within other polyatomic molecules are formed in similar fashion. Take a look at the Lewis structures for two common substances and see whether you can identify the kinds of atoms and bonds found in them.

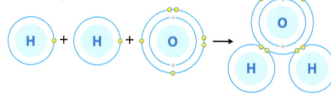

Polarity

Do atoms share their electrons "nicely" with each other? It turns out they don't! It's more like a tug of war. Remember that elements vary in their electronegativity—their pull on electrons. Unequal pulling results in unequal sharing, which results in an unequal distribution of electric charge. **Polarity** is the name we use for this unequal distribution of electric charge. Remember the quantum mechanical model of atoms? Recall that electrons inhabit a region that looks like a cloud. The cloud surrounding a bonded pair of atoms is shifted toward the more electronegative atom of the pair.

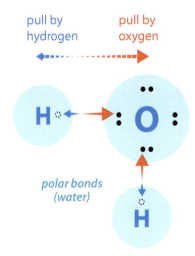

polar bonds (water)

Let's see how this works in a molecule of water. Oxygen is the second-most electronegative element in existence. Hydrogen's electronegativity is much lower. As a result, the oxygen atom in a water molecule (top right) pulls on its shared electrons much more than the hydrogen atoms do. This means that the shared electrons in a water molecule spend more of their time near the oxygen atom than they do near the hydrogen atoms. We call this unequal sharing of electrons a *polar covalent bond*. Not all covalent bonds are polar. If the bond exists between two atoms of the same element, such as in oxygen or nitrogen gas, then each atom exerts an equal pull on the shared electrons. Equal sharing like this produces a *nonpolar bond* (middle right).

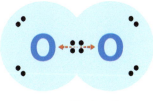

nonpolar bond (oxygen)

The unequal sharing of electrons can produce more than just polar bonds. It can produce *polar molecules* as well. This results when the unequal sharing of electrons causes some portion of a molecule to have a negative charge while another portion is positive. Again, let's see how this happens in a water molecule (bottom right). When the shared electrons in a water molecule are near the oxygen atom, they create a temporary region of excess negative electrical charge around the oxygen atom. At the same time, their absence from the hydrogen atoms creates a region of temporary excess positive charge around those atoms. Overall, this causes the oxygen side of the molecule to have a slight negative charge and the hydrogen side to have a slight positive charge. The polarity of water molecules causes them to be slightly "sticky." The positive side of one water molecule is attracted to the negative sides of other water molecules.

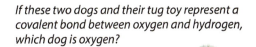

polar molecule (water)

If these two dogs and their tug toy represent a covalent bond between oxygen and hydrogen, which dog is oxygen?

BONDINGS AND COMPOUNDS

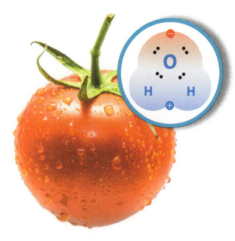

The cohesion and surface tension that allow spilled water to bead on a tomato are both caused by the polarity of water molecules.

The polarity of water accounts for many of its physical and chemical properties, including its being a liquid at room temperature. If water were not polar, life as we know it could not exist!

5.6 HOW IONIC BONDING WORKS

In the last section, we saw how atoms with similar electronegativities can share electrons to fill their valence energy levels. But if two atoms have very different electronegativities, they will form a different kind of bond. In ionic bonding, atoms don't share valence electrons. Instead, the atoms transfer electrons. One of them, the donor, will give away one or more of its electrons. The other, the receptor, will accept the electrons. Another way to think about this is as very unequal sharing—so much so that the nonmetal takes the metal's valence electrons. We'll see how this works by looking closely at another very common substance—table salt.

Table salt, or sodium chloride, is an ionic compound made of equal parts, or a one-to-one ratio, of sodium and chlorine. Chlorine's electronegativity is very high, the third highest on the periodic table in fact. Chlorine holds on to its valence electrons very tightly. Sodium, on the

The weakly electronegative metals in Groups 1 and 2 easily form salts with the strongly electronegative nonmetals in Groups 13–17.

other hand, has only one valence electron. It's in the third energy level, so it's rather far from the nucleus. Therefore, the positive charge in sodium's nucleus attracts its one valence electron only weakly. Chlorine's strong pull on electrons, coupled with sodium's weak hold on its valence electron, allows an atom of chlorine to remove the valence electron from an atom of sodium. In so doing, chlorine completes an octet of valence electrons. When sodium loses its one valence electron, its third energy level is emptied. Since its second energy level already has an octet of electrons, the exchange of electrons fulfills the octet rule for both atoms.

Transferring sodium's one valence electron to chlorine exposes the eight electrons in sodium's already-full second energy level.

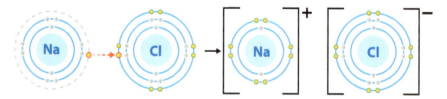

But more than just a simple exchange of an electron has taken place in this scenario. By losing an electron, the sodium atom has become a sodium cation. The chlorine atom, by accepting the electron, has become a chloride anion (the switch to the *-ide* suffix is a naming convention—an agreed upon way to name substances). The attraction between oppositely charged particles is what holds the pair together.

As we did with covalent bonds, we can use Lewis structures to show ionic bonds. There are a couple of differences though. The Lewis structures for ionic compounds do not show all of the valence electrons that you might expect. Take a look at the Lewis structure for sodium chloride.

$$[Na]^+ \; [:\!\ddot{C}\!l\!:]^-$$

Can you spot the differences? Recall from Chapter 4 that electron dot notation, and thus Lewis structures, show only valence electrons. Where is sodium's one valence electron? It has been transferred over to chlorine, which is now shown with eight electrons instead of its original seven. Also note that the resulting electric charge is shown with each ion produced by the electron exchange.

Ionic bonding commonly occurs between highly electronegative nonmetals on the right-hand side of the periodic table and the weakly electronegative metals in Groups 1 and 2. The compounds formed in this manner are referred to as salts, and table salt is just one example. Other salts that you may be familiar with include Epsom salt (magnesium sulfate), found in bath salts, and baking soda (sodium bicarbonate). As you see, chemical bonds are very important for giving us some of the materials and energy that we need to live.

5B | REVIEW QUESTIONS

1. Describe the role of electrons in covalent, ionic, and metallic bonds.

2. What do the red electrons below represent?

   $$:\!\ddot{O}\!::\!\ddot{O}\!:$$

3. Under what circumstances will a triple bond form?

Use the Lewis structure for chloromethane and the table of electronegativites below to answer Questions 4–5.

Element	Electronegativity
Carbon	2.6
Chlorine	3.2
Hydrogen	2.2

4. How many of the bonds shown in the Lewis structure for chloromethane are polar? Explain.

5. Is chloromethane a polar molecule? Defend your answer.

6. Do ionic bonds tend to form between atoms with similar or dissimilar electronegativities?

7. If the atoms in an ionic bond are not sharing electrons, what keeps the atoms together?

MINI LAB

MODELING BONDS IN THREE DIMENSIONS

Methane is a colorless, odorless gas at normal temperatures. One source of methane is the decomposition of dead plant matter in wetlands; there it is an important component of *swamp gas*. Methane is also considered a greenhouse gas.

The Lewis structure for methane (below left) is, of course, a model. Carbon and hydrogen don't look like the letters *C* and *H* in real life. Real bonds don't look like straight lines either. But there's another big difference between this model and real-life methane. The Lewis diagram is a flat, two-dimensional model, whereas methane is three-dimensional. In this lab activity, you'll try to figure out the shape of a methane molecule.

Essential Question:

What does a methane molecule actually look like?

Equipment

foam ball, 5 cm.

foam balls, 2.5 cm. (4)

toothpicks (4)

protractor

1. Describe what you think the three-dimensional shape of methane might be.

PROCEDURE

A Take a moment to think about the Lewis structure for methane. Each line that connects two atoms represents a shared electron pair. Electrons all have the same negative charge, and negative charges repel each other. The slight positive charge on each hydrogen atom, a result of its polar bond with carbon, also repels its hydrogen neighbors. The result of all this is that each bonding pair in a methane molecule tries to get as far away as possible from every other bonding pair.

B With the 5 cm. foam ball as the central carbon atom, use toothpicks to connect the four 2.5 cm. balls (representing hydrogen) to the carbon atom.

methane

2. What do the toothpicks represent in this model?

C Move the hydrogen atoms around until you think you have them as far apart as possible.

ANALYSIS

D Use the protractor to measure the *bond angle* between two adjacent hydrogen atoms. The bond angle is the angle formed by two bonds with the central carbon atom at the vertex of the angle.

3. If the hydrogen atoms are as far apart as possible, what should be true about all the bond angles within the molecule?

E Using your answer in Question 3, make adjustments to your model, if needed.

CONCLUSION

4. What bond angle measurement produces the desired spacing of the hydrogen atoms?

5. How does your final version of methane compare to your prediction in Question 1?

GOING FURTHER

6. Replacing the four hydrogen atoms in methane with fluorine atoms creates a molecule of tetrafluoromethane. Unlike hydrogen, fluorine is more electronegative than carbon, so a fluorine atom bonded with carbon has a negative charge rather than a positive one. How do you think this negative charge affects the three-dimensional shape of tetrafluoromethane compared with that of methane?

5C Questions

- How can I predict the chemical formula for a compound?
- How do I write out the chemical formula for a compound?
- What are polyatomic ions?

5C Terms

chemical formula, binary compound, polyatomic ion, oxidation state

5C | WRITING CHEMICAL FORMULAS

How do you know how many atoms are in a chemical formula?

5.7 CHEMICAL FORMULAS FOR BINARY IONIC COMPOUNDS

Chemical Formulas: A Quick Review

You're probably already somewhat acquainted with chemical formulas. A **chemical formula** is a shorthand way of identifying a chemical compound. Chemical formulas also give us information about the composition of the compounds they identify. The chemical formula for water, H_2O, is familiar to most people. *H* is the symbol for hydrogen and *O* is the symbol for oxygen. The subscript *2* after the *H* tells us that each water molecule has two hydrogen atoms. There is no subscript after the *O*, so it's understood that a water molecule has only a single oxygen atom. Formulas for other compounds are written in a similar way. Each formula includes the element symbols that identify the kinds of atoms in the compound, along with subscripts to indicate the number of atoms of each element.

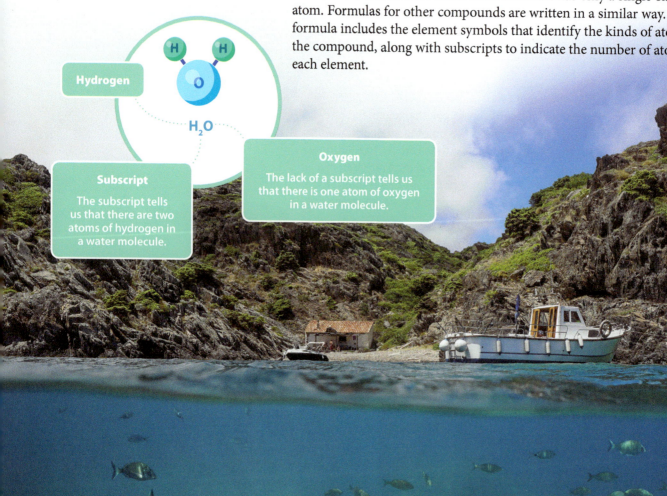

Hydrogen

Subscript
The subscript tells us that there are two atoms of hydrogen in a water molecule.

Oxygen
The lack of a subscript tells us that there is one atom of oxygen in a water molecule.

Predicting Formulas for Binary Ionic Compounds

It's one thing to already know the formula for a chemical compound, but can we also predict how different atoms will bond together to form them? How do we know, for instance, that sodium will bond with fluorine in a one-to-one ratio to make sodium fluoride, the stuff in your toothpaste? For many simple **binary compounds**—those made from only two elements—formulas can be predicted on the basis of where those elements are found on the periodic table. This is especially true for ionic compounds—those made from oppositely charged ions. The number of electrons that each atom can donate or accept determines the ratio of elements in the compound. We'll soon see how this works, but it helps to first understand how ionic compounds are named.

TOOTHPASTE INGREDIENTS

ACTIVE INGREDIENTS:
SODIUM FLUORIDE EP 0.32%, TRICLOSAN 0.3%
OTHER INGREDIENTS:
COPOLYMER, HYDRATED SILICA, GLYCERINE, SORBITOL, SODIUM LAURYL SULPHATE, FLAVOR, CELLULOSE GUM, TITANIUM DIOXIDE, SODIUM SACCHARIN, CARRAGEENAN, SODIUM HYDROXIDE, WATER, COLORS: E104, E133.

Naming Ionic Compounds

How would you name an ionic compound, such as that formed by bromine and potassium? A hint is found in some of the names that you've already seen in this chapter, such as sodium chloride. Sodium, the cation, is named first, followed by the name of the anion, chlorine. Other ionic compounds are named in like manner. But notice that there is a slight twist. The names of the anions below are modified to include the suffix *–ide*. Thus, chlor*ine* is changed to chlor*ide*. Similarly, when bromine forms an anion, it becomes brom*ide*. So by giving the name of the metal cation first followed by the nonmetal anion with its modified ending, we can see that the compound of bromine and potassium is called potassium bromide.

COMMON ANIONS

Name	Symbol	Name	Symbol
fluoride	F^-	oxide	O^{2-}
chloride	Cl^-	sulfide	S^{2-}
bromide	Br^-	nitride	N^{3-}
iodide	I^-	phosphide	P^{3-}

Got that? Okay, then let's get back to predicting formulas.

BONDINGS AND COMPOUNDS

PREDICTING AND WRITING THE FORMULA FOR AN IONIC COMPOUND

Na· — First, locate the cation (the metal) on the periodic table. Sodium is a Group 1 alkali metal with one valence electron. It needs to lose the electron to fulfill the octet rule.

 Next, find the anion (the nonmetal). Fluorine is a Group 17 halogen, so we know that it is one valence electron short of fulfilling the octet rule.

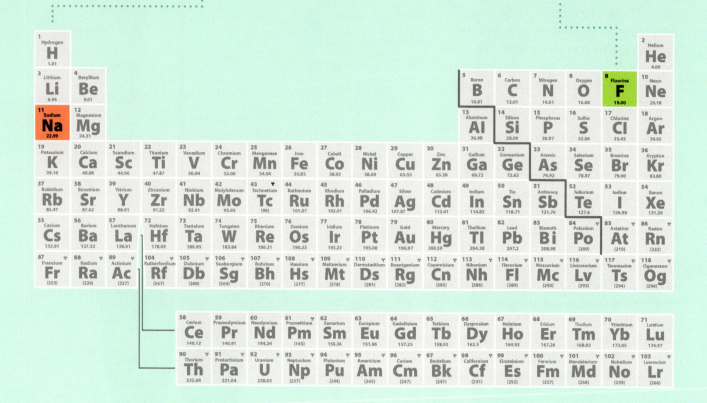

 Because sodium is a metal and fluorine is a nonmetal, we know that the two will form an ionic bond. And since each needs to lose or gain only one electron to satisfy the octet rule, we know that sodium and fluorine will bond in a one-to-one ratio.

 In the formula for sodium fluoride, the symbol for the cation is written first, followed by the symbol for the anion. There is only one of each ion, so no subscripts are needed. The formula is therefore NaF.

Sodium can form many kinds of chemical salts, and not all of them are made in a one-to-one ratio with a nonmetal. For example, sodium can also combine with oxygen to form sodium oxide. In what ratio do you think sodium and oxygen will combine? Use the technique described above to find out.

EXAMPLE 5-1: Predicting and Writing the Formula for a Compound of Sodium and Oxygen

In which group is sodium? How many electrons can it donate?

Sodium is in Group 1. It has one electron to donate.

In which group is oxygen? How many electrons can it accept?

Oxygen is in Group 16. It can accept two electrons.

How many sodium atoms are needed to fulfill the electron requirements for oxygen?

Since each sodium atom has only one electron to donate, two are necessary to fulfill oxygen's need for two electrons.

What is the ratio of sodium to oxygen in sodium oxide?

The ratio of sodium to oxygen in sodium oxide must be two atoms of sodium for every atom of oxygen, or 2:1.

What is the formula for this compound?

Since there are two sodium cations for every oxide anion, the formula is Na_2O.

Predicting chemical formulas isn't always as straightforward as this example suggests. For instance, magnesium can combine with nitrogen to form magnesium nitride. We can use the same procedure demonstrated in Example 5-1, but we'll need to add an extra step.

EXAMPLE 5-2: Predicting and Writing the Formula for Magnesium Nitride

In which group is magnesium? How many electrons can it donate?

Magnesium is in Group 2. It has two electrons to donate.

In which group is nitrogen? How many electrons can it accept?

Nitrogen is in Group 15. It can accept three electrons.

How many magnesium atoms are needed to fulfill the electron requirements for nitrogen?

Each nitrogen atom needs three electrons, but each magnesium atom has only two to donate. It will take at least two magnesium atoms, with four total electrons to donate, to meet nitrogen's need for three electrons. But that will leave one electron left over, requiring at least one more nitrogen atom to accept it. Atoms can bond only in whole number ratios, so we need to find the smallest number of each atom to combine so that no electrons are left over. To do this, we'll borrow a skill from mathematics—finding a *least common multiple*.

What is the least common multiple of two and three (the numbers of electrons donated and received by magnesium and nitrogen, respectively)?

The least common multiple of two and three is six. Two will divide into six three times, and three will divide into six two times. Therefore, every three magnesium atoms (with two electrons to donate each) can fulfill the electron needs of two nitrogen atoms (three electrons to receive each).

What is the ratio of magnesium to nitrogen in magnesium nitride?

Since three magnesium atoms are required to fulfill the electron needs of two nitrogen atoms, the ratio of magnesium to nitrogen in magnesium nitride must be 3:2.

What is the formula for this compound?

Since there are three magnesium cations for every two nitride anions, the formula is Mg_3N_2.

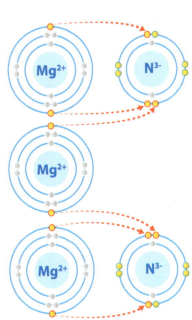

BONDINGS AND COMPOUNDS

Predicting and Writing Formulas for Ionic Compounds Containing Polyatomic Ions

So far, all of the ionic compounds that we have looked at have been made of monatomic ions, such as Mg^{2+} or Cl^-. But some ions consist of more than one atom. A **polyatomic ion** is a group of covalently bonded atoms that *together* have gained or lost electrons. The bonded atoms act as a single ionized particle. One such polyatomic ion is known as phosphate. It consists of a single atom of phosphorus covalently bonded to four atoms of oxygen. Together they act as a single ion that has accepted three electrons, giving it a charge of 3–. One phosphate ion can bond with three sodium atoms, which have each lost one electron and have a charge of 1+. The compound formed is called *trisodium phosphate*, a common cleaning agent. A table showing the names and chemical formulas of common polyatomic ions is shown on the left. Note that you'll see both subscript *and* superscript numbers in their symbols. Remember, subscripts indicate the number of atoms and superscripts indicate charge. The amount of charge is equal to the number of electrons that each polyatomic ion has donated or accepted. A positive charge indicates that the ion has lost electrons, while a negative charge shows that electrons have been gained.

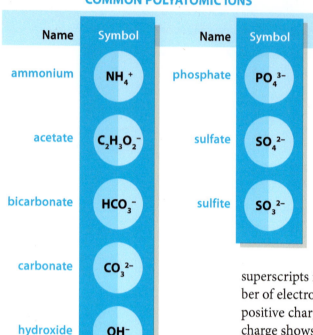

COMMON POLYATOMIC IONS

Name	Symbol	Name	Symbol
ammonium	NH_4^+	phosphate	PO_4^{3-}
acetate	$C_2H_3O_2^-$	sulfate	SO_4^{2-}
bicarbonate	HCO_3^-	sulfite	SO_3^{2-}
carbonate	CO_3^{2-}		
hydroxide	OH^-		
nitrate	NO_3^-		

When predicting the formulas for compounds containing polyatomic ions, the polyatomic ion is treated as a single particle. The same goes for writing formulas for compounds with polyatomic ions. The only difference occurs when more than one polyatomic ion is in the formula. In such instances, the polyatomic ion is shown in parentheses with a subscript number outside of them.

EXAMPLE 5-3:
Predicting and Writing the Formula for Magnesium Hydroxide

In which group is magnesium? What is its charge when it forms ions?

Magnesium is in Group 2. It has two electrons to donate; therefore ionized magnesium will have a charge of 2+.

What is the charge on a hydroxide ion?

1–

How many of each ion are needed to balance their positive and negative charges?

Since magnesium's charge is 2+ and hydroxide's is 1–, each magnesium cation can balance the electrical charge of two hydroxide anions.

What is the ratio of magnesium to hydroxide in magnesium hydroxide?

1:2

What is the formula for this compound?

Since there are two hydroxide anions for every magnesium cation, the formula is $Mg(OH)_2$.

Predicting Formulas Using Oxidation States

The method that we just looked at works well for predicting the formulas of some ionic binary compounds but not for all of them. For instance, copper and oxygen can form ionic bonds. Copper has one valence electron to donate. Oxygen needs two. It seems easy enough to predict that two copper atoms will donate their electrons to one oxygen atom, forming Cu_2O. That combination meets the electron needs of all three atoms. And that is in fact one way that the two elements can bond—but it's not the only way. Copper can also bond with oxygen in a one-to-one ratio, forming CuO. This formula isn't suggested by a straightforward analysis of the location of each element on the periodic table.

To reflect these varied ways that two elements can bond with each other, chemists have developed another way to predict chemical formulas. On the basis of many years of observing the ways that elements bond with each other, each element has been assigned one or more oxidation states. An element's **oxidation state** shows the electric charge gained or lost by that element when it forms a compound. This definition treats all chemical bonds as though they are ionic in nature, even though they may be covalent in reality. Oxidation states are whole numbers whose values can be positive, negative, or zero. Usually one or two oxidation states for an element, called primary oxida-

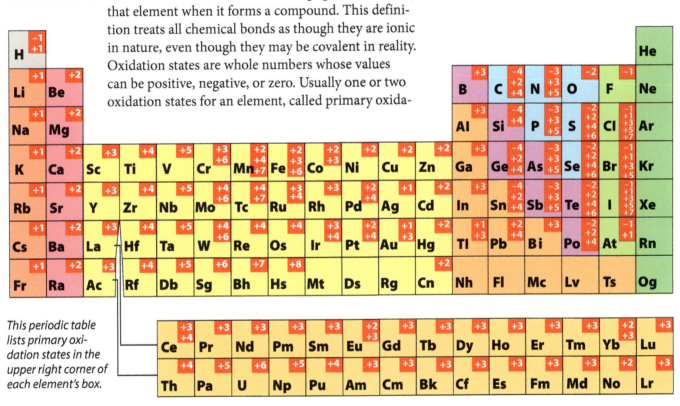

This periodic table lists primary oxidation states in the upper right corner of each element's box.

tion states, are more common than the others. For main block elements (those in Groups 1–2 and 13–18), one of the common oxidation states is usually the same as the number of electrons predicted to be gained or lost on the basis of an element's location on the periodic table. If shown on the table, common oxidation states are normally printed in bold type. When predicting a chemical formula using oxidation states, the oxidation states for all the elements involved must add up to zero. Let's look at some examples.

EXAMPLE 5-4:
Predicting the Ratio of Cesium to Oxygen in Cesium Oxide Using Oxidation States

What is cesium's most common oxidation state?

+1

What is oxygen's most common oxidation state?

−2

On the basis of its most common oxidation state, how many cesium atoms will bond with one oxygen atom?

Since the oxidation states of all the bonded atoms must add up to zero, two cesium atoms at +1 each are required to balance the −2 oxidation state of one oxygen atom.

What ratio of cesium to oxygen do these oxidation states indicate in cesium oxide?

On the basis of these oxidation states, the ratio of cesium to oxygen in cesium oxide will be two atoms of cesium for every atom of oxygen, or 2:1.

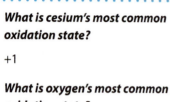

The familiar green patina on copper statues is copper (II) carbonate.

Notice that in Example 5-4, the predicted ratio of atoms is the same as what we would predict according to the location of both elements on the periodic table. The compound produced in this case is ionic and is named accordingly. The cation, *cesium*, is written first, followed by the anion, *oxide*. But the copper compounds we mentioned earlier require an extra step in the naming process. The names of compounds containing transition metals, like copper, add a Roman numeral in parentheses between the names of the ions. The Roman numeral indicates the oxidation state of the metal cation. Thus, Cu_2O is copper (I) oxide because copper's oxidation state in this compound is +1. CuO is copper (II) oxide.

Let's look at another example of using oxidation states to predict the formula for a compound that includes a transition metal.

EXAMPLE 5-5: Predicting the Formula and Name for a Compound of Iron and Oxygen If Iron's Oxidation State is +3

Which oxidation state for oxygen should I use?

Oxygen doesn't have a −3 oxidation state to balance iron's +3; its most common oxidation state is −2.

On the basis of oxygen's most common oxidation state, what is the most likely ratio of iron atoms to oxygen atoms?

Since 3 is not a whole number multiple of 2, we'll need to use least common multiples again. The least common multiple of 3 and 2 is 6; therefore the ratio of iron atoms (+3 oxidation state) to oxygen atoms (−2 oxidation state) is 2:3.

What is the formula for this compound?

Again, start with the cation, iron, followed by the anion, oxide. This yields the formula Fe_2O_3.

What is the name of this combination of iron and oxygen?

Using the naming process for ionic compounds, the name of the metal cation is given first, followed by the name of the anion. Since iron is a transition metal, its oxidation state must be included in parentheses. The full name of the compound is thus iron (III) oxide.

5.8 CHEMICAL FORMULAS FOR BINARY COVALENT COMPOUNDS

carbon monoxide

carbon dioxide

To this point we have considered only ionic compounds. Now let's look at some examples that show how oxidation states can be used to predict the formulas for covalent compounds. We'll look at a group of compounds called *carboxides*—compounds made of carbon and oxygen.

EXAMPLE 5-6:
Predicting the Ratio of Carbon to Oxygen in Carboxides

What are the most common oxidation states for carbon? What is the most common oxidation state for oxygen?

Carbon's most common oxidation states are –4 and +4. We've seen that oxygen's most common oxidation state is –2. Since the oxidation states in a compound must add up to zero, we'll need to use the +4 oxidation state for carbon.

What is the ratio of carbon to oxygen according to these oxidation states?

To balance the +4 oxidation state of one carbon atom, two oxygen atoms are needed. The ratio of carbon to oxygen is therefore 1:2. The formula for this compound is CO_2.

Does carbon have any other oxidation states that are positive multiples of 2?

Yes! Carbon also has a +2 oxidation state.

What would be the ratio of carbon to oxygen using a +2 oxidation state for carbon?

Since the +2 oxidation state of a carbon atom can be balanced by a single –2 oxygen atom, the ratio of carbon to oxygen for this compound would be 1:1. Its formula is simply CO.

Naming Covalent Compounds

As you just saw, carbon can covalently bond with oxygen in multiple ways. Neither element is a metal, so how do chemists know which element to list first in the names of these compounds? How can chemists distinguish between the 1:2 carbon-to-oxygen compound and the 1:1 version? It turns out that chemists have a system for doing this. In some ways it is similar to the system for naming binary ionic compounds, but there are a few key differences.

Carbon monoxide (a key component of smog) or carbon dioxide? The difference is kind of important!

The names of binary covalent compounds indicate both the elements involved and the number of atoms of each element. To name a binary covalent compound, follow the few simple rules below.

❶ The element with the *lower group number* is written first.

❷ If both elements are in the same group, then the element with the *higher period number* is written first.

❸ The second element to be named is modified using the *-ide* suffix, just as you saw when naming ionic compounds (e.g., ox*ide*, sulf*ide*, nitr*ide*).

❹ A system of *Greek prefixes* (see below) is used to indicate the number of atoms of each element in the compound. The exception to this rule is when there is only one atom of the first element; in such cases, no prefix is used.

Let's apply these rules to carboxides. For both of them, carbon has the lower group number (14 compared with 16 for oxygen), so carbon will be named first ❶. In each case, there is only one carbon atom, so carbon, being named first, requires no prefix ❹. Since oxygen will be named second, it requires the *-ide* suffix and will be written as *oxide* ❸. The first example had two oxygen atoms, which calls for the prefix *di-* ❹. This compound is thus *carbon dioxide*, the stuff in your carbonated drinks. The second example had only one oxygen, which will be indicated with the prefix *mon-*. This compound is *carbon monoxide*. Recall that carbon monoxide is a toxic gas that is a component of smog (page 115).

As you can see from the examples just shown, the names of chemical compounds are not random. A lot of thought went into giving them names that not only serve as unique identifiers, but also provide information about the number and kinds of atoms in each compound as well.

BINARY COVALENT COMPOUND PREFIXES

Prefix	Number of atoms	Example
mon–, mono–	●	nitrogen monoxide (NO)
di–	●●	silicon dioxide (SiO_2)
tri–	●●●	boron trifluoride (BF_3)
tetr–, tetra–	●●●●	carbon tetrachloride (CCl_4)
pent–, penta–	●●●●●	diphosphorus pentoxide (P_2O_5)
hex–, hexa–	●●●●●●	sulfur hexafluoride (SF_6)

5.9 USING COMPOUNDS WISELY

Millions of chemical compounds can be made from the less than 120 elements found on the periodic table. The many different ways that atoms can bond make this great variety possible. Part of exercising wise dominion over God's earth is seeking to learn as much as we can about how the world works, including its compounds. The knowledge that we gain from studying compounds and bonding can change lives for the better. From finding new materials to discovering new medicines, our efforts to understand compounds have helped God's image bearers to thrive.

Of course, the study of chemical compounds has had a dark side too. Chemists have produced many less noble compounds—for example, addictive drugs, weapons of war, and toxins that pollute our world. These chemicals are often the very same ones that can be put to more beneficial uses. Godly wisdom and compassion for others are needed to decide how best to use the compounds that chemistry makes possible.

5C | REVIEW QUESTIONS

1. Using tellurium's location on the periodic table, predict one of its common oxidation states.
2. Your friend writes down the formula for a compound of oxygen and lithium as OLi_2. Is this correct? Explain.
3. What is the name of the compound in Question 2?
4. Predict the formula and give the name for a compound of calcium and fluorine.
5. When naming binary compounds, how is determining which element is named first in an ionic compound different than doing so for a covalent compound?
6. What is a polyatomic ion?
7. Predict the formula and give the name for a compound of magnesium and phosphate.
8. What is the difference between manganese (II) oxide and manganese (III) oxide? How will this difference be reflected in the chemical formulas of the two compounds?
9. How is the naming of the second element in a binary ionic compound similar to that for a covalent compound? How is it different?
10. For the compound whose formula is NCl_3, is the correct name *nitrogen trichloride*, *nitrogen chloride*, or *trichlorine mononitride*? Explain.
11. (See the Ethics box on the next page.)

ETHICS: PSEUDOEPHEDRINE

If you've ever taken Sudafed to reduce nasal congestion, you've probably taken the controversial drug pseudoephedrine. Since 2006, federal guidelines have limited how much of the drug a person can buy in a thirty-day period. This obviously requires that records be kept on purchases and purchasers alike. Why does the government track such data? Is pseudoephedrine dangerous? Your task is to use the skills that were modeled for you in Chapter 3 (see pages 66–67) to formulate a position on the sale of pseudoephedrine and other drugs.

Recall that the model for thinking about ethical issues included six questions. The questions are repeated for you here along with some tips on how to go about answering them. Included at the end of the exercise is a self-assessment rubric. Use it to measure your performance on this task.

❶ What information can I get about this issue?

Do an internet search using the keyword "pseudoephedrine." Find the information you need to answer the following questions.

a. What is pseudoephedrine?
b. What is pseudoephedrine's intended medical use?
c. How is pseudoephedrine used illegally?
d. Are there effective substitutes for pseudoephedrine?

You may think of other questions that are relevant to the topic as well.

❷ What does the Bible say about this issue?

You won't find the word "pseudoephedrine" in the Bible. But think about the reasons *why* people use pseudoephedrine. Does the Bible have anything to say about that?

❸ What are the acceptable and unacceptable options?

Pseudoephedrine could simply be banned, or it could be completely legal and easy to obtain. Naturally, there's an entire range of options in between those two extremes. What might some of those options be?

❹ What may be the consequences of the acceptable and unacceptable options?

How would restricting or not restricting access to pseudoephedrine affect the people that use it?

❺ What are the motivations of the acceptable options?

How do the acceptable options that you've identified demonstrate things like respect for God's laws and growing in Christ-like character?

❻ What action should I take?

Using all the information that you have found and your answers to the previous questions, think about what sorts of actions you should take with regard to the sale and use of pseudoephedrine. You might even consider what stance you should take regarding the sale and use of medications in general.

11. Write a three-paragraph response on the ethical use of pseudoephedrine. In the first paragraph, discuss principles from Scripture that may guide Christians in this area. In the second paragraph, discuss the possible outcomes of heeding or ignoring the guidelines of Scripture and the motivations for choosing each option. In the third paragraph, state what actions you should take regarding the issue.

	Task completed	Task partially completed	Task not completed
I gathered and used additional information for my response.	two or more sources	only one source	no sources
I reviewed Scripture for relevant principles on using drugs.	two or more passages	only one passage	no passages
I stated both the acceptable and unacceptable options.	both	only one	neither
I stated the consequences of the acceptable and unacceptable options.	yes	—	no
I considered the biblical outcomes of human prospering, a thriving creation, and glorifying God.	all three	only one or two	no outcomes
I considered the biblical motivations of faith in God, hope in God's promises, and love for God and others.	all three	only one or two	no motivations
I stated what action should be taken.	yes	—	no

CHAPTER 5 REVIEW

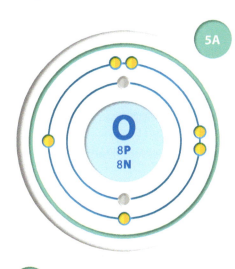

5A PRINCIPLES OF BONDING

- A chemical bond is an electrostatic attraction that exists between atoms due to the sharing or exchanging of valence electrons.
- Molecules form when atoms bond with other atoms in an attempt to gain stability.
- With few exceptions, atoms are most stable when they have eight valence electrons; this is known as the octet rule.
- The properties of compounds are usually very different from the properties of the elements from which they are formed.
- Atoms of some elements will bond with each other to form molecules, such as the diatomic elements.

5A Terms
chemical bond	96
octet rule	97

5B TYPES OF BONDS

- Covalent bonds are produced when atoms must share electrons due to having similar electronegativities. A group of two or more covalently bonded atoms is a molecule.
- When atoms have very different electronegativities, the more weakly electronegative atom donates electrons to the stronger. The resulting ions are held together in a crystal structure in a ratio represented as a formula unit.
- The atoms of metals bond by mutually sharing all of their valence electrons, forming an electron sea.
- A covalent bond consists of a bonding pair of electrons. Atoms can share up to three pairs of electrons, forming either single, double, or triple bonds.
- Atoms having different electronegativities share electrons unequally, resulting in polar bonds.
- An uneven distribution of electric charge on a molecule produces a polar molecule.

5B Terms
covalent bond	99
ionic bond	100
formula unit	100
metallic bond	100
diatomic molecule	101
Lewis structure	101
polarity	103

5C WRITING CHEMICAL FORMULAS

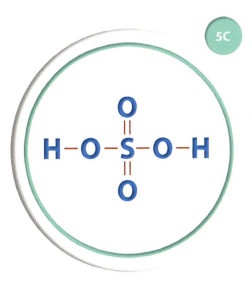

- The chemical formulas for some ionic compounds can be predicted by the location of the component elements on the periodic table.
- Oxidation states are used to predict chemical formulas for both ionic and covalent compounds.
- In naming and writing binary ionic compounds, the cation is identified first, followed by the anion.
- Ionic compounds may contain polyatomic ions, which are groups of covalently bonded atoms that act as single ionized particles.
- The naming and writing of binary covalent compounds take into account the locations of the elements on the periodic table as well as the number of atoms of each element.

5C Terms
chemical formula	108
binary compound	109
polyatomic ion	112
oxidation state	113

BONDINGS AND COMPOUNDS

CHAPTER 5

CHAPTER REVIEW QUESTIONS

Recalling Facts

1. What is a chemical bond?
2. (True or False) Compounds often have physical and chemical properties that are similar to those of the elements of which they are made.
3. What is a covalent bond?
4. What is used to represent the ratio of ions in a group of ionically bonded atoms?
5. Describe a metallic bond.
6. Define the term *multiple bond*.
7. (True or False) A triple bond holds three atoms together.
8. What is a polar covalent bond?
9. If two atoms have very different electronegativities, will they form an ionic bond or a covalent bond?
10. What does a chemical formula tell about a compound?
11. What does an oxidation state indicate about an element?

Understanding Concepts

12. Must atoms of one element always bond with atoms of a different element in order to become stable? Explain.
13. Describe an example of an element that does not need to meet the octet rule requirement in order to become stable.
14. Identify whether each of the following elements would be more likely to gain or lose electrons in order to form an ionic bond.
    a. bromine
    b. rubidium
    c. strontium
15. Will two iodine atoms form an ionic or covalent bond? Explain.

Refer to the Lewis structure for sulfur dioxide below to answer Questions 16–18.

16. Copy the Lewis structure for sulfur dioxide onto your paper, then circle all of the shared pairs of electrons.
17. Knowing that the electronegativity of oxygen and sulfur are 3.4 and 2.6 respectively, do you think that sulfur dioxide contains polar or nonpolar bonds? Explain.
18. Actual sulfur dioxide molecules are bent, just as the Lewis structure shows. Are molecules of sulfur dioxide polar or nonpolar? Explain.
19. If two sulfur atoms bond with each other, will they form a single bond or a double bond? Explain.
20. Will the sulfur molecule in Question 19 be polar or nonpolar? Explain.
21. In a compound of strontium and fluorine, which element forms an anion? Which forms a cation?
22. Which of the following compounds are salts? Explain.
    a. selenium dioxide
    b. barium iodide
    c. sulfur dibromide
    d. triphosphorus pentanitride
23. Name the elements and the number of atoms of each element in the chemical formula $C_5H_{11}NO_3$.
24. Which of the following Lewis structures is the correct one for beryllium chloride? Explain your choice.
25. Predict the names and formulas for compounds containing the following elements or polyatomic ions.
    a. sodium and bromine
    b. calcium and chlorine
    c. barium and nitrate
    d. nickel and fluorine
26. What is (are) the primary oxidation state(s) for each of the following elements?
    a. scandium
    b. sulfur
    c. zirconium
    d. iodine
27. Using primary oxidation states (see table on page 113), predict the names and formulas for two different compounds composed of nitrogen and oxygen.
28. Are the compounds you predicted in Question 27 the only possible compounds of nitrogen and oxygen? Explain.

REVIEW

Critical Thinking

29. Magnesium is a soft, silvery metal. Chlorine is a poisonous, greenish gas that irritates lung tissue. Both elements do not naturally occur in their pure forms. They can react together to form a compound. Find out what this compound is and contrast its properties with those of magnesium and chlorine.

30. Create a hierarchy chart using the following terms: *chemical bonds, covalent bond, ionic bond, metallic bond, shared electrons, transferred electrons, metal-to-metal bond, metal-to-nonmetal bond, nonmetal-to-nonmetal bond, table salt, sugar, copper* wire.

31. Food oils consist of nonpolar molecules. How does this explain their usefulness in nonstick cooking sprays?

32. Name the elements, and the number of atoms of each, in the chemical formula $Al_2(SO_4)_3$.

33. How would life on Earth be affected if every element had exactly the same electronegativity?

34. The chemical formula for vinyl acetate, the compound in PVA glue, is $C_4H_6O_2$. The chemical formula for methyl acrylate is also $C_4H_6O_2$, but it's used for making carpet fibers, not glue. Do you like butter? One of the chemicals that gives butter its distinct flavor is butanedione, whose chemical formula again is—you guessed it—$C_4H_6O_2$. How do you suppose it is possible for three compounds to have the same chemical formula and yet have such very different properties?

Use the Case Study on the right to answer Questions 35–36.

35. How are the molecules in CA and PVA similar?

36. How are the molecules in CA and PVA different?

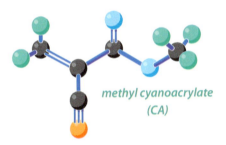

methyl cyanoacrylate (CA)

Key:
black = carbon
blue = oxygen
green = hydrogen
orange = nitrogen

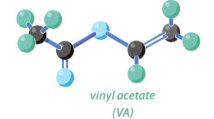

vinyl acetate (VA)

CASE STUDY: STICKY SITUATION

What do sticky notes, furniture, diving wetsuits, running shoes, and paperback books all have in common? They're all things that are made with adhesives, or glues. When you think of the word "glue," you probably think of white craft glue. But there are many other kinds of glues, and they are used in the making of thousands of everyday products. You might be surprised to learn that two very different glues—CA (sometimes called superglue) and PVA (white glue)—are made from only four elements: carbon, hydrogen, oxygen, and nitrogen. Models of the two compounds from which CA and PVA are made are shown below left.

BONDINGS AND COMPOUNDS 121

One of the most distinctive smells is that of a skunk. A collection of organic chemicals produces its pungent odor.

CHAPTER 6
The Chemistry of Life

SKUNK-SAVING SCENT

The phrase "lifesaving chemical" may bring many things to mind. Some of you may know people who have undergone chemotherapy. Most of you have taken medicine when you were sick, and everyone needs the compounds in foods to stay healthy. Some people need to carry lifesaving medicines, such as insulin (for diabetes) or epinephrine (for allergic reactions), with them at all times.

Skunks carry a supply of lifesaving chemicals with them too. God designed skunks with a highly effective defense system—skunk scent. A skunk warns a threat by hissing, stomping its feet, and doing a little dance. If the warning is not heeded, a skunk will spray from two glands at the base of its tail, aiming at the face of an assailant. In addition to its telltale odor, skunk scent is also an eye irritant that can temporarily blind an attacker. The mixture is so effective that most predators avoid skunks at all costs. The great horned owl is the only predator that routinely preys on skunks. While almost everyone seeks to sidestep its stench, skunks certainly savor the sweet scent of this lifesaving chemical.

6A Organic Compounds	124
6B Substituted Hydrocarbons	134
6C Biochemistry	136

6A Questions

- Why is carbon able to make so many compounds?
- What are hydrocarbons?
- How can we change the properties of hydrocarbons?

6A Terms

organic compound, hydrocarbon, saturated hydrocarbon, unsaturated hydrocarbon, aromatic hydrocarbon, benzene ring, isomer

6A | ORGANIC COMPOUNDS

Why is carbon so important?

6.1 DEVELOPMENT OF ORGANIC CHEMISTRY

Most of us have seen the term *organic* in the grocery store. Up and down the aisles we are surrounded by signs for organic carrots, lettuce, tomatoes, and much more. In this context, the term is meant to make us think that these are healthier options. This may be true because often these products are grown through organic farming methods, such as using only pesticides and fertilizers that come from natural, living sources—other plants and animals.

In chemistry, the term *organic* began with a similar meaning because all the then-known organic compounds came from living things. Over time we have changed our definition, but we are getting ahead of ourselves. Today organic chemistry touches every aspect of our lives. All around your house you can find many products that are the work of organic chemists. Plastics, fuels, dyes, drugs, and many other goods are made using modern organic chemistry. But this is not a new field of science.

Since ancient times, people have used many compounds from plants and animals. They used these as medicines, poisons, and dyes. Because

they believed that these products came from organisms, scientists called them *organic compounds*. The study of these compounds became known as *organic chemistry*, a term coined by Jacob Berzelius in the early 1800s.

Scientists thought that these compounds contained the "vital force," and that only living organisms could make them. While many scientists studied these compounds, no one had been able to produce them. In 1828, German chemist Friedrich Wöhler inadvertently made urea, an organic compound, in the laboratory. Through this accidental discovery, Wöhler refuted the vital force theory and started a period of synthesizing many organic compounds.

Today we still divide chemistry into organic and inorganic chemistry, but our definitions have changed. Chemists originally defined an organic compound as any compound produced by a living organism that therefore contained the vital force. Today, scientists define an **organic compound** as one that contains carbon. While some compounds (e.g., carbon dioxide and carbonates) have carbon in them but are not considered organic, all organic compounds contain carbon. So today, organic chemistry is the study of carbon-containing compounds.

6.2 VERSATILE CARBON

Bonding

There are millions of organic compounds. Just as a small number of types of Lego™ blocks is the basis for many structures, so the atoms of one element—carbon—provide the framework for all organic matter. This is because the structure of carbon atoms makes them extremely versatile, allowing for great variety in the compounds that are possible.

Carbon is a Group 14 element, which means that it has four valence electrons. These electrons allow carbon atoms to bond with up to four other atoms. Most organic compounds form on a framework of carbon atoms bonded to each other, like a carbon backbone, with bonds that are quite strong. The structure of carbon, with its four valence electrons, also allows it to form single, double, and triple bonds. These three properties of carbon atoms—four valence electrons that can bond with

carbon

MOLECULAR ARRANGEMENT

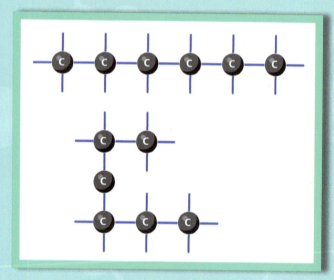

STRAIGHT CHAIN

The most basic form of an organic molecule is a *straight chain*.

Straight chains consist of a single continuous series of any number of carbon atoms bonded to each other. The term "straight" doesn't imply a straight line, just a single path of carbon atoms from one end of the molecule to the other. Straight chains will often have angles and bends in them.

While the chain at top left is clearly a straight chain, the one at bottom left is also a straight chain, even though it has two turns in it. In both cases there is a single continuous path of six carbon atoms.

BRANCHED CHAIN

Even if the types and number of atoms remain the same, changing how the atoms are arranged will result in molecules with different properties.

Organic molecules can be made more complex by including branches. A *branched chain* has carbon atoms that connect to other carbon atoms that are not on the ends of a straight chain. So there is more than one path that you could follow when counting the carbon atoms.

Like the straight chains, these branched chains (right) still contain six carbon atoms. But one of the carbon atoms has been removed from the continuous chain and now branches off from the main chain. Because the branched arrangement is different from a straight chain, it behaves differently than the straight chain from which it derives.

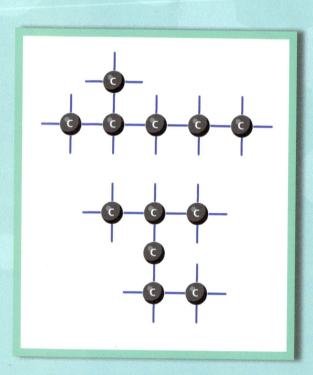

four other atoms, strong carbon–carbon bonds, and the potential for multiple bonds—enable it to combine in many ways with other atoms to form many substances. While other members of Group 14 have similar characteristics, carbon's smaller size allows it to form stronger C–C bonds compared with Si–Si bonds, for example. But silicon's closeness to carbon makes it a great candidate for building blocks of life in science fiction stories!

Arrangement

Most organic compounds have a structure that is based on multiple carbon atoms. In Chapter 4 we learned that the structure of an atom determines the properties of the elements. Similarly, the atoms that make up an organic molecule and the arrangement of those atoms dictate the properties of the molecule. The framework of carbon atoms can take one of three main forms—straight chains, branched chains, or rings.

RINGS

The third main form for organic molecules is a *ring*, which is made by connecting the two ends of a continuous chain.

Notice at right that a straight chain of six carbon atoms has been turned into a ring of six carbon atoms by connecting the two carbon atoms at the ends of the chain. Rings can also have branches (bottom left), or they can be a branch attached to a chain (bottom right).

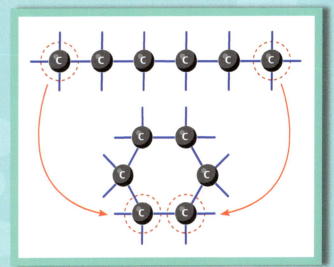

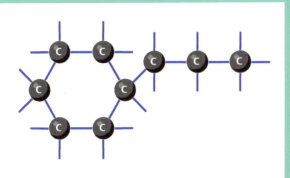

Above is a three-carbon branch attached to one of the carbon atoms in a six-carbon ring.

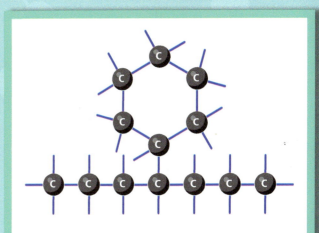

In the image above, a six-carbon ring is a branch off the fourth carbon atom in a seven-carbon chain.

NUMERICAL PREFIXES	
Number of carbon atoms	Prefix
1 ●	meth-
2 ●●	eth-
3 ●●●	prop-
4 ●●●●	but-
5 ●●●●●	pent-
6 ●●●●●●	hex-
7 ●●●●●●●	hept-
8 ●●●●●●●●	oct-
9 ●●●●●●●●●	non-
10 ●●●●●●●●●●	dec-

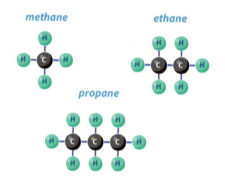

6.3 HYDROCARBONS

Organic compounds always contain carbon. Notice that carbon atoms, with their C–C bonds, are the framework for organic molecules. In addition to carbon, hydrogen is the most common element in organic compounds. Compounds made of only carbon and hydrogen atoms are called **hydrocarbons**.

We can refer to hydrocarbons by their names, molecular formulas, or structural formulas. A hydrocarbon's name starts with a prefix for the number of carbon atoms in the longest chain or ring. Some of these prefixes will be familiar to you; in the table at left, you likely know *pent-*, *hex-*, *oct-*, and *dec-*. Do you recognize *eth-*, *prop-*, or *but-* (BYOOT)? The ending of a hydrocarbon's name indicates the types of bonds between the carbon atoms; we'll discuss that a little later.

Molecular formulas are the same chemical formulas that you learned in Chapter 5. A *structural formula* is a drawing that shows not only the atoms but also the bonds and their arrangement in the molecule. Though more complex, the structural formulas are similar to the Lewis structures that you learned about in Chapter 5. The examples shown on pages 126–27 for straight chains, branched chains, and rings were just the carbon-atom frames for structural formulas; to emphasize the carbon backbones and basic shapes, they didn't include any other atoms bonded to the carbon atoms.

CLASSIFYING HYDROCARBONS

One way to classify hydrocarbons is by the types of bonds between the carbon atoms.

SATURATED HYDROCARBONS (ALKANES)

If only single bonds exist between the carbon atoms in the molecule, we call it a **saturated hydrocarbon**. Because the carbon atoms have only single bonds, these hydrocarbons have the greatest number of hydrogen atoms possible. So a saturated hydrocarbon is saturated with hydrogen.

Alkanes are saturated hydrocarbons. The names of alkanes end with *-ane*. The simplest alkane, methane, consists of one carbon atom bonded to four hydrogen atoms. Its name comes from *meth-* (one carbon) and *-ane* (only single bonds).

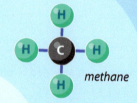
methane

Alkanes have low reactivity. Low-number alkanes are natural gases and liquid fuels. We use high-number alkanes to produce plastics, waxes, and asphalt.

Propane's name tells us that it has three carbon atoms (*prop-*) and only single bonds (*-ane*). Propane is a natural gas commonly used for home heating and gas grills.

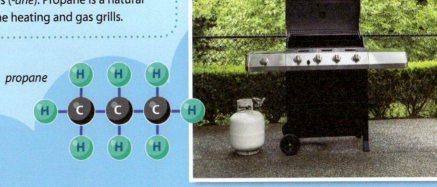

propane

Regular gas has an octane rating of 87, while fuel with an octane rating of 91 or higher is known as premium gas. Octane has eight single-bonded carbon atoms.

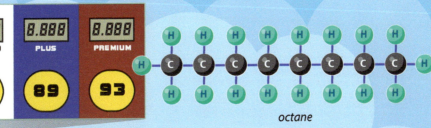

gas octane ratings

octane

UNSATURATED HYDROCARBONS (ALKENES AND ALKYNES)

Hydrocarbons that contain any double or triple bonds between carbon atoms are called **unsaturated hydrocarbons**. Because of the presence of multiple bonds, these hydrocarbons have fewer hydrogen atoms than alkanes with the same number of carbons.

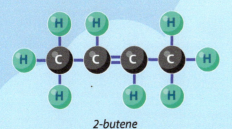

2-butene

Alkenes are unsaturated hydrocarbons with at least one double bond between carbon atoms. The names for alkenes end in *-ene*. Alkenes undergo more types of reactions than alkanes do because they are unsaturated and other atoms can be added to them. This four-carbon (*but-*) molecule with one double bond (*-ene*) is 2-butene.

THE CHEMISTRY OF LIFE 129

CLASSIFYING HYDROCARBONS

Ethene's name tells us that it contains two carbon atoms (*eth-*) with one double bond (*-ene*) between them. Ethene, also known as ethylene, is used to help fruit ripen.

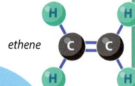
ethene

polyethylene bag

Many ethene molecules can be joined in long chains to form polyethylene, which we use in both routine and unusual ways, from plastic bags (left) to replacement hip sockets.

An unsaturated hydrocarbon that contains at least one triple bond between carbon atoms is called an *alkyne*. The names for these hydrocarbons end with *-yne*. For example, pentyne contains five (*pent-*) carbon atoms with one triple bond (*-yne*). The addition of the triple bond makes alkynes the most reactive class of hydrocarbons.

ethyne

Ethyne, also known as acetylene, has two (*eth-*) carbon atoms and a triple bond (*-yne*). It is highly combustible; its 3300 °C flame makes it perfect for welding.

The model below shows calicheamicin, a powerful antitumor drug. Chemists form it by bonding an alkene between two alkynes.

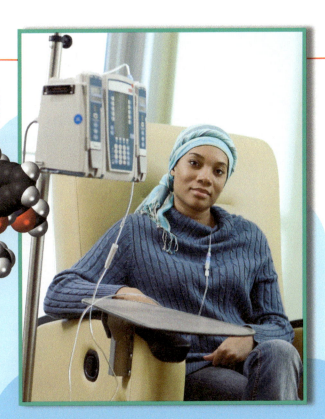

AROMATIC HYDROCARBONS

Early in the study of organic chemistry, scientists discovered some compounds with strong, often sweet, aromas. Because of this feature, scientists called them aromatic hydrocarbons. Further study revealed that **aromatic hydrocarbons** all contain at least one benzene ring and that some are not very aromatic. English chemist Michael Faraday is given credit for discovering benzene, the simplest aromatic hydrocarbon, in 1825. Forty years later, Friedrich Kekulé, a German chemist, suggested benzene's unique structure.

A **benzene ring** is an unsaturated hydrocarbon containing six carbon atoms, each with an attached hydrogen atom. In the Chapter opener, you read about skunk scent, which includes an aromatic hydrocarbon. So much for sweet-smelling!

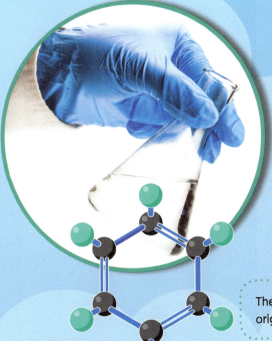

The image at left shows how the C–C bonds in a benzene ring were originally thought to alternate between double and single bonds.

Now our understanding of a benzene ring's structure is that the electrons involved in the C–C bonding are free to move around the ring. The standard symbol for a benzene ring (right) uses a hexagon to represent the six carbon atoms and an inner circle to indicate the freely shared electrons. So there really are no double bonds in a benzene ring, and it does not react in the ways that typical unsaturated hydrocarbons do.

THE CHEMISTRY OF LIFE   131

6.4 ISOMERS

Do you recall what determines the properties of molecules? If you thought of the atoms that make up the molecule and the arrangement of those atoms, then you're correct. If we rearrange the atoms in a hydrocarbon, thus changing its form, we change the way it behaves. Two molecules with the same molecular formula but whose structures differ are called **isomers**.

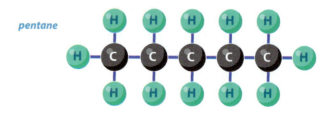
pentane

Both pentane and isopentane have the same molecular formula, C_5H_{12}—they each contain five carbon atoms and twelve hydrogen atoms. By moving one of the carbon atoms from an end of a pentane molecule and connecting it to form a branch, we form isopentane. This small change in structure produces significant changes in physical and chemical properties. Isopentane is more stable than pentane and has lower melting and boiling points.

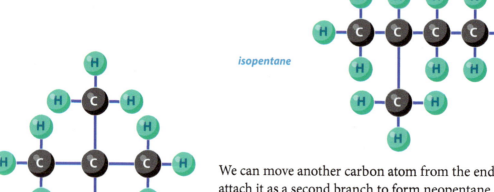
isopentane

We can move another carbon atom from the end of isopentane and attach it as a second branch to form neopentane. This makes an isomer that is even more stable than isopentane. The varied properties of isomers allow scientists to select a particular isomer for a particular function.

neopentane

6A | REVIEW QUESTIONS

1. Why did scientists originally think that man could not synthesize organic compounds?
2. How did the definition of *organic compound* change over time?
3. How do you recognize a branched chain molecule?
4. What are the three basic structures that organic molecules can have?
5. Compare saturated and unsaturated hydrocarbons.
6. Why does *benzene* have the *-ene* ending?
7. What is an aromatic hydrocarbon?

MINI LAB

MODELING HEXANE ISOMERS

Many of us have used rubbing alcohol (isopropyl alcohol, bottom right) as a disinfectant or hand sanitizer. It also has a distinct smell. Isopropyl alcohol is an isomer of propanol. Isomers have the same molecular formula, but their structures differ. In this lab activity, you will build models of the isomers of hexane.

1. What does the name hexane tell us?

Procedure

A Draw the structural formula for hexane.

2. What is the molecular formula for hexane?

B Using the large foam balls to represent carbon atoms, the small foam balls to represent hydrogen atoms, and the toothpicks as bonds, create a model of hexane. Remember that each carbon atom can have only four bonds and each hydrogen atom can have only one.

C Rearrange your atoms to try to form an isomer of hexane.

D Draw the structural formula for the isomer.

3. How many carbon atoms and hydrogen atoms does this isomer have?

E Rearrange your atoms again to try to form as many isomers of hexane as you can.

Conclusion

4. How many isomers did you form?

Going Further

F Rearrange your atoms again to form a hexane ring. Remember that each carbon atom can have only four bonds.

5. Is this an isomer of hexane? Explain.

6. Why are isomers important?

Essential Question:

How can we rearrange the atoms in a molecule?

Equipment

foam balls, 5 cm (6)

foam balls, 2.5 cm (14)

toothpicks (19)

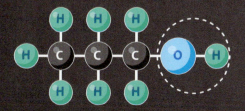

The OH at the end of this propane molecule makes this propanol.

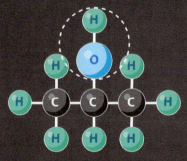

By moving the hydroxyl group (OH) to the middle of the molecule, we turn propanol into isopropyl alcohol.

THE CHEMISTRY OF LIFE

6B | SUBSTITUTED HYDROCARBONS

6B Questions

- Can organic compounds contain other elements?
- What makes various substituted hydrocarbons different from each other?
- What are substituted hydrocarbons used for?

6B Terms

substituted hydrocarbon, functional group, alcohol, aldehyde, ketone

What other elements can be found in organic compounds?

6.5 SUBSTITUTION

Modern food processors are similar to blenders, but they can also do the work of five or six other machines. How can they do this? These machines have attachments that allow them to do different tasks. They have blades for slicing, shredding, chopping, grating, and mixing. One can attach other tools for juicing or even kneading bread dough. Just substitute one blade for another and you have a machine with vastly different properties.

Chemists can do the same thing with hydrocarbons. By replacing, or *substituting*, one of the hydrogen atoms with a different atom or group of atoms, they can create a **substituted hydrocarbon**. This new substance has properties that differ from those of the original hydrocarbon. The atom or group of atoms that scientists insert to form the substituted hydrocarbon is called a **functional group**, or a *substituent*. The same compounds that chemists make may also be found occurring naturally, however, and are not always man-made. Notice the common functional groups in the table on the left.

COMMON FUNCTIONAL GROUPS

Name	hydroxyl	carbonyl	halogen
Structure	O–H	C=O	Cl, F, I, Br
Hydrocarbon	alcohol	ketone or aldehyde	alkyl halide
Example	propan-2-ol	methanal	chloroethane

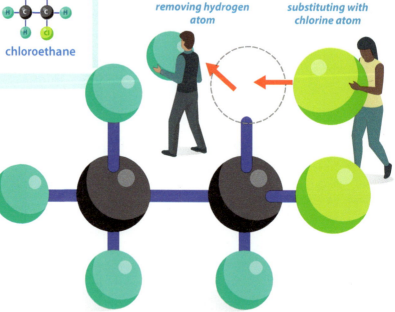

6.6 COMMON SUBSTITUTED HYDROCARBONS

Substituted hydrocarbons form when a hydrogen atom in a hydrocarbon is replaced with a functional group. While many functional groups contain oxygen, others include nitrogen, sulfur, or one of the halogens.

SUBSTITUTED HYDROCARBONS

ALCOHOLS

Alcohols are hydrocarbons in which a *hydroxyl group* (OH) replaces at least one of the hydrogen atoms.

Typical Properties: higher boiling points and greater solubility in water than similar hydrocarbons

Name: alcohols identified by the *-ol* ending

ethanol

Common Uses: as solvents, in detergents, and as fuel additives

ALDEHYDES

Aldehydes are hydrocarbons in which a double-bonded oxygen atom replaces two hydrogen atoms at the end of a chain. The carbon atom double bonded to an oxygen atom is called a *carbonyl group* (C=O).

Typical Properties: vary depending on the hydrocarbon to which the carbonyl group is attached

Name: aldehydes identified by the *-al* ending

methanal (formaldehyde)

Common Use: industrial applications, especially making plastics

KETONES

Ketones are hydrocarbons in which a carbonyl group is present on a carbon atom that is not at the end of a chain.

Typical Properties: greater solubility in water than their unsubstituted counterparts

Name: ketones identified by the *-one* ending

hexan-2-one

Common Uses: as solvents, in medicine, and in making polymers
Many plastic-model cements, nail polishes, and nail polish removers contain ketones.

These are just three of the substituted hydrocarbons. Each has distinct properties that are based on which functional group the hydrocarbon has and where the group is placed within the molecule.

6B | REVIEW QUESTIONS

1. What is a functional group?
2. What type of substituted hydrocarbon might have an iodine atom in place of one of the hydrogen atoms?
3. What substituted hydrocarbon(s) contain(s) a carbonyl group?
4. How do you know that pentanol is an alcohol?
5. How do ketones and aldehydes differ?
6. Name a substituted hydrocarbon that may be used as a fuel additive.
7. Why do different substituted hydrocarbons have different properties?

THE CHEMISTRY OF LIFE   135

6C | BIOCHEMISTRY

What molecules are needed for life?

6C Questions

- What are polymers?
- How do our bodies use polymers?
- Can food be fast *and* healthy?

6C Terms

polymer, monomer, carbohydrate, protein, amino acid, lipid, nucleic acid, nucleotide, DNA

6.7 POLYMERS

Have you ever watched as a long train passes a railroad crossing? Car after car rumbles by. Most of the world's products move around on freight trains. The conductor at the stockyard created the train by connecting smaller units, or boxcars.

Chemistry has its own version of these trains—gigantic molecules formed by linking many smaller molecules together. These *macromolecules* are called **polymers**. The smaller molecules that are put together to form a polymer are called **monomers**.

Polymers are made either by linking identical monomers or by connecting different ones. The properties of polymers are determined by the monomers that make them and how those smaller units are arranged. While many polymers form in nature, others are synthetic.

Glutamine (left) is one of the amino acids that form proteins, which are critical macronutrients for our bodies. Many foods, especially from animals, are good sources of proteins.

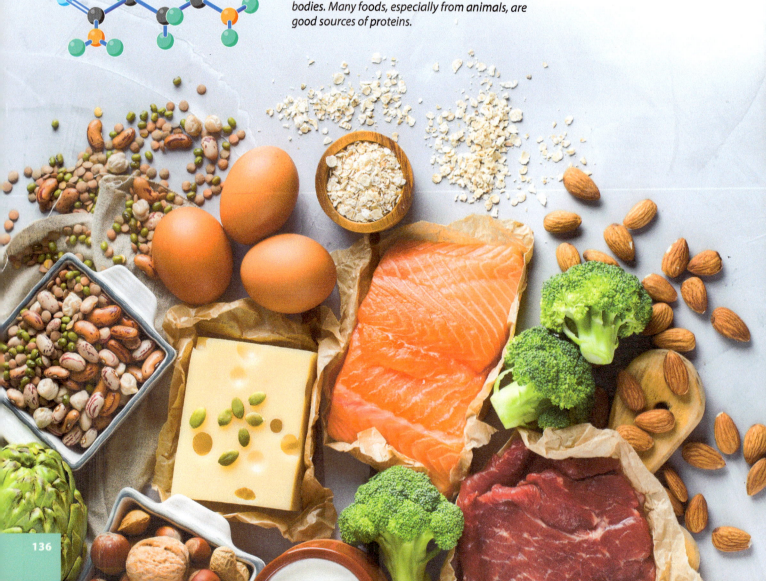

136

6.8 SYNTHETIC POLYMERS

As scientists studied hydrocarbons, they realized that they could link many small hydrocarbons to form polymers. Many of the products that we use daily are made from these manmade polymers. While working with a methane derivative, scientists accidentally made a long chain of ethane molecules. Today many food containers and plastic bottles are made from this material, called polyethylene (see page 130). We use polyethylene because of its flexibility and high-impact strength.

Other scientists worked to create manmade fibers to replace natural fibers that are costly or difficult to obtain. The desire to replace silk and cotton led to the discovery of nylon. By linking a couple of different substituted hydrocarbons, scientists made the first nylon. Today there are many forms of nylon, each with distinct properties. Scientists also created other fibers such as Nomex® for flame-resistant clothing, bulletproof Twaron LFT® for ballistic vests, and neoprene for water-resistant cloth.

We don't always think about the amazing properties of synthetic polymers such as nylon. Do you think the skydiver above is thinking about the nylon that made his parachute?

In the image above you can see nylon forming at the boundary between two liquids.

6.9 BIOLOGICAL MACROMOLECULES

Organic chemistry was originally thought to be the chemistry of living things. Today this branch of chemistry is defined as the chemistry of carbon-containing compounds. But the chemistry of living things is still a major field of study and one that we call *biochemistry*.

MACRONUTRIENTS

CARBOHYDRATES

Carbohydrates are compounds made of carbon, hydrogen, and oxygen atoms, with the hydrogen and oxygen usually in a 2:1 ratio. Carbohydrates are often polymers of sugars in our food, and the most common sources are fruits, vegetables, and grains.

Starches are more complex carbohydrates and are formed by linking glucose molecules to form large polymers, some of which contain branches of smaller glucose chains off the main chain. Plants make starch for energy storage. Starches are a key source of energy for humans.

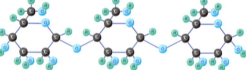

starch, a chain of glucose molecules

Sugars are simple carbohydrates that provide quick energy to people and animals. Common sugars include glucose and fructose. Plants make glucose during photosynthesis—to produce energy during cellular respiration, to convert to starches for storage, or to change into cellulose for plant structures. Glucose circulates in the blood of animals and humans and is called *blood sugar*. It is the primary form of energy for humans and animals. Animals and people convert excess glucose into a polymer called *glycogen* for short-term storage. The sugar we are probably most familiar with is sucrose, or table sugar, which consists of a glucose and a fructose molecule bonded together.

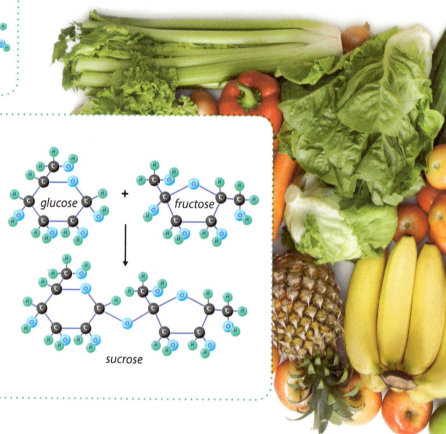

PROTEINS

Another major group on a nutrition label is proteins—the building blocks of muscles, blood, skin, tendons, and hair. **Proteins** are polymers formed by linking amino acids. **Amino acids** are organic molecules with both an amine (NH_2) and a carboxyl group (C=OOH). Proteins contain hundreds of amino acids, formed by the amine group of one monomer bonded to the carboxyl group of the next. About twenty amino acids are needed to form all of the proteins that are essential for our bodies. Our bodies can produce some of these amino acids, but others must come directly from our food. Even in cases where we can produce the amino acids, it is more efficient to get them from food sources. Common sources of proteins are meats, dairy products, eggs, beans, and nuts.

LIPIDS

The third important group of compounds in biochemistry is lipids. **Lipids** are organic compounds that include fats, oils, waxes, and cholesterol. Fats get a lot of negative attention, but we need fats to maintain good health. They help with many processes within our cells as well as with brain function. Like hydrocarbons, fats can be saturated or unsaturated.

Saturated fats have only single bonds between carbon atoms within a portion of the molecule called a *fatty acid chain*. They are typically solid at room temperature, come from animal sources, and are simply called *fats*.

Unsaturated fats contain one or more double bonds in their fatty acids. They typically come from plants, remain liquid at room temperature, and are called *oils*. To increase the shelf life of foods, manufacturers will add hydrogen atoms to unsaturated fats—breaking the double bonds and changing them into saturated fats. There is concern about the long-term effects of these modified fats.

THE CHEMISTRY OF LIFE 139

SERVING AS A FOOD CHEMIST: WORKING OUT OF A JAM

Diabetes, a disease in which a person cannot process sugar properly, affects over eight percent of American adults. These people must limit the amount of sugar they eat. The food industry attempts to provide sugar-free foods, including jams and jellies, but often these products just don't taste as good!

Geoff Dubrow, a food science PhD candidate, is working to address this problem. He and others are studying the composition and structure of the compounds in both standard and sugar-free fruit spreads. As they find new compounds, they test to see whether they affect how people perceive the flavor of the spreads. With this data, food companies hope to be able to provide delicious jams and jellies even to people with restricted diets.

Meeting the needs of people around us is part of the Creation Mandate. While jams and jellies may not be a vital need, products such as these allow people with diabetes to enjoy foods they once had to avoid. Even food scientists can glorify God and help other people.

Nucleic Acids

As important as the nutritional molecules discussed above are, another group of biochemical polymers is found in every living cell in every organism on the earth. These are nucleic acids and are central to the working of cells. **Nucleic acids** are polymers that contain the instructional code for the reproduction, growth, and all other processes of cells. Without nucleic acids, cells would cease to function.

Nucleic acids form by linking **nucleotides**—monomers that consist of a sugar, a phosphate group, and a nitrogen-containing base. The order of the nucleotides in nucleic acids encodes the information for the cell.

There are two forms of nucleic acids, depending on what sugar is in the nucleotide. If the sugar is ribose, then the nucleic acid is RNA—ribonucleic acid. RNA is critical for protein synthesis in cells and for communication within a cell. The other nucleic acid, DNA (deoxyribonucleic acid), contains a sugar derived from ribose. **DNA** is the nucleic acid that directs the reproduction and growth of cells in all living organisms.

NUCLEOTIDES

phosphate · sugar · nitrogen base

6C | REVIEW QUESTIONS

1. What are polymers?
2. Define *carbohydrate*.
3. How do plants, animals, and humans use sugars?
4. What are proteins?
5. What are the parts of a nucleotide?
6. What is the role of DNA in cells?

CHAPTER 6 REVIEW

6A ORGANIC COMPOUNDS

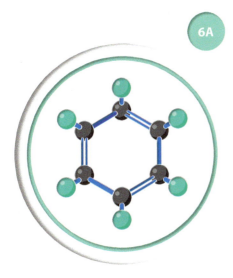

- Organic chemistry is the study of the composition, structure, and properties of carbon-containing compounds.
- Organic compounds form as straight chains, branched chains, and rings.
- Hydrocarbons are organic compounds consisting of only carbon and hydrogen atoms.
- Saturated hydrocarbons contain only single bonds between carbon atoms; unsaturated hydrocarbons have double or triple bonds.
- Aromatic hydrocarbons contain six-carbon benzene rings.
- Molecules with the same chemical formula but whose atoms are arranged differently are called isomers.
- Differences in the structures of molecules result in differences in the properties of compounds.

6A Terms

organic compound	125
hydrocarbon	128
saturated hydrocarbon	128
unsaturated hydrocarbon	129
aromatic hydrocarbon	131
benzene ring	131
isomer	132

6B SUBSTITUTED HYDROCARBONS

- Substituted hydrocarbons are organic compounds where one or more hydrogen atoms have been replaced with other atoms or groups of atoms called functional groups.
- Functional groups substituted into hydrocarbons change the composition and therefore the properties of the original molecule.

6B Terms

substituted hydrocarbon	134
functional group	134
alcohol	135
aldehyde	135
ketone	135

6C BIOCHEMISTRY

- Polymers are huge molecules formed by linking monomers together.
- Many manmade products are synthetic polymers, such as plastics, nylon, and synthetic rubber.
- Carbohydrates, proteins, lipids, and nucleic acids are organic compounds involved in biochemical processes.
- Carbohydrates, sugars, and starches provide most energy for living organisms. They are also building blocks for other structures.
- Examples of lipids are the fats, oils, and waxes in our foods.
- Proteins, which are made from amino acids, are the building blocks for many tissues in our bodies.
- Nucleic acids encode and store information for cells. They direct functions such as reproduction, growth, and development in the cell.

6C Terms

polymer	136
monomer	136
carbohydrate	138
protein	139
amino acid	139
lipid	139
nucleic acid	140
nucleotide	140
DNA	140

THE CHEMISTRY OF LIFE

CHAPTER 6

CHAPTER REVIEW QUESTIONS

Recalling Facts

1. What event changed scientists' thinking about organic compounds and chemistry?
2. Name three characteristics of carbon atoms that allow them to form so many different organic compounds.
3. What are hydrocarbons?
4. What are isomers?
5. How do substituted hydrocarbons differ from hydrocarbons?
6. What is a hydroxyl group, and what is its relationship to alcohols?
7. What is a carbonyl group? In which group(s) of substituted hydrocarbons would you find one?
8. Name two uses of alcohols.
9. Give two examples of synthetic polymers and how we use them.
10. What biochemical macromolecules do our bodies use?
11. What monomers link to form proteins?
12. What are good food sources of proteins?
13. What are lipids?
14. What are nucleic acids?

Understanding Concepts

15. Why is the term *straight chain* somewhat misleading?
16. Compare alkanes, alkenes, and alkynes.
17. Identify each of the following as an alkane, alkene, alcohol, aldehyde, or ketone.

    a.   b. 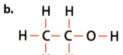 c.

18. Create a concept map using the following terms: *hydrocarbon, saturated hydrocarbon, unsaturated hydrocarbon, single bond, double bond, triple bond, alkanes, alkenes,* and *alkynes*.
19. Why is butyne much more reactive than butane?
20. What changes the properties of hydrocarbons?
21. Explain how monomers and polymers are related.
22. Explain how starches and sugars are related.
23. How do plants, animals, and humans use starches?
24. Evaluate the statement, "For a healthy diet, we must avoid eating fats."
25. Create a concept map using the following terms: *molecule, polymer, monomer, carbohydrate, starch, sugar, lipid, protein, energy, cellulose,* and *glycogen*.
26. Compare the structure and function of RNA and DNA.

Critical Thinking

27. What is the mathematical relationship between the number of hydrogen atoms and the number of carbon atoms in alkanes?
28. Why does methene not exist?
29. Are the two molecules shown below isomers of each other? Explain.

    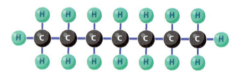

    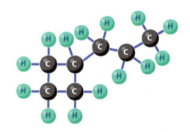

30. Change the propane molecule shown below into
    a. a ketone.
    b. an alcohol.
    c. an aldehyde.

    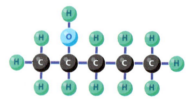

31. The IUPAC name for the isomer shown below is pentan-2-ol. Why do you think it is so named?

32. Compare carbohydrates, proteins, and lipids by copying the following table on your paper and filling in the cells.

	Carbohydrates	Lipids	Proteins
Made of chains of			
Uses in the body			
Examples in food			

REVIEW

ETHiCS — CAN FAST FOOD BE NUTRITIOUS?

We live in a fast-paced world. People run from event to event without time for a breath, much less time to think about what they are eating. Twenty-five percent of Americans eat at least one fast-food meal each week, and Americans eat twenty percent of their meals in their cars. Drive through many commercial districts and you may see five, ten, even fifteen fast-food restaurants. Watch an hour of television, and you will see at least one fast-food commercial. The food looks delicious, and it will be ready . . . well, fast, so that you can keep moving. And it is food, right?

But is all food the same when we consider its macronutrients—proteins, lipids, and carbohydrates? What about when we start to look at its micronutrients, such as vitamins and minerals?

THE ISSUE: EATING WELL

33. What information can you get about this issue? Do a keyword search for "fast food nutrition" and "fast food's impact on health." Write a summary of what you have learned.

34. What does the Bible say about this issue? Does the Bible specifically address what people eat? Can you find general principles in Scripture about what we eat?

35. What are the acceptable and unacceptable options of fast food? Should we never eat any fast food? Should you eat and drink whatever you feel like? And there are many options between those two. What might some of those options be?

36. What might be the consequences of adopting the acceptable or unacceptable options? What might be the result of never eating any fast food? What could result from eating and drinking whatever you feel like?

37. What are the motivations for the acceptable options? How do the acceptable options that you've identified demonstrate things such as respect for God's laws and growing in Christlike character?

38. What action should you take? Using all the information that you have found and your answers to the previous questions, think about what sorts of actions you should take with regard to fast food.

	Task completed	Task partially completed	Task not completed
Additional Information			
Biblical Principles			
Acceptable and Unacceptable Options			
Consequences			
Biblical Outcomes			
Biblical Motivations			
Opinion Statement			

THE CHEMISTRY OF LIFE

It is a profound and necessary truth that the deep things in science are not found because they are useful; they are found because it was possible to find them.

J. Robert Oppenheimer (1904–67), American physicist

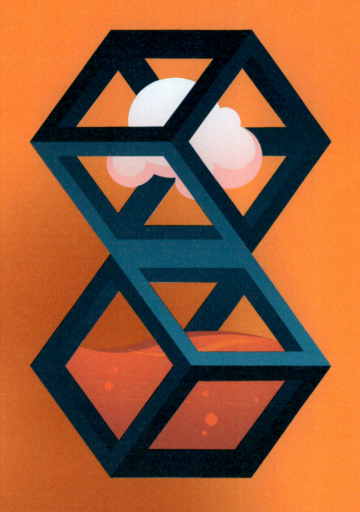

UNIT 2
CHANGES IN MATTER

CHAPTER 7: **CHEMICAL REACTIONS**

CHAPTER 8: **NUCLEAR CHANGES**

CHAPTER 9: **SOLUTIONS**

CHAPTER 10: **ACIDS, BASES, AND SALTS**

Main photo: *Explosions are rapid combustion reactions and can be devastating when uncontrolled.*

Inset: *Residents comb through the wreckage after the Mont Blanc explosion destroyed most of Halifax.*

CHAPTER 7
Chemical Reactions

LITTLE CHANGES, BIG EFFECTS

In Halifax, Nova Scotia, the tranquility of Sunday morning, December 6, 1917, was suddenly shattered by an enormous blast. A French cargo ship, the SS *Mont Blanc*, fully laden with explosives for the Allied war effort in Europe, collided with another ship, caught fire, and exploded. The energy released was the equivalent of 2900 metric tons of TNT being detonated—the largest manmade explosion in history before the advent of atomic weapons. About 2000 people lost their lives.

What makes the Halifax tragedy even more astonishing is that all that energy was released when unimaginably tiny atoms within the explosive compounds swiftly rearranged themselves. Happily for us, not all chemical reactions are as powerful or as deadly as this example. God designed chemical reactions to sustain life on Earth.

7A	Chemical Changes	148
7B	Types of Chemical Reactions	156
7C	Energy in Chemical Reactions	159
7D	Reaction Rates and Equilibrium	162

7A | CHEMICAL CHANGES

How can I tell whether a chemical change has taken place?

7A Questions

- What clues suggest that a chemical change has taken place?
- How do I write chemical equations?
- What information can I get from a chemical equation?

7A Terms

chemical reaction, chemical equation, reactant, product, coefficient, mole, molar mass

7.1 CHEMICAL REACTIONS

Exploding fireworks are a vibrant example of a chemical reaction. All life depends on less spectacular, but more needful, chemical reactions. The digestion of food, the release of energy through cellular respiration, and the growth of new cells all depend on chemical reactions. Other chemical reactions heat our homes, produce the fibers used in our clothing, and power the cars we drive. A **chemical reaction** is a process that changes a substance into one or more different substances. During a chemical reaction, the chemical bonds in the original substance are broken and its atoms are rearranged. New chemical bonds form between the rearranged atoms, producing a substance that is different from the original.

When a chemical change takes place through a chemical reaction, there are usually clear signs that one has happened. Some are obvious, like the sound and light produced by a firework. Others can be more subtle. Let's take a look at some signs that a chemical reaction has taken place.

EVIDENCE OF CHEMICAL CHANGE

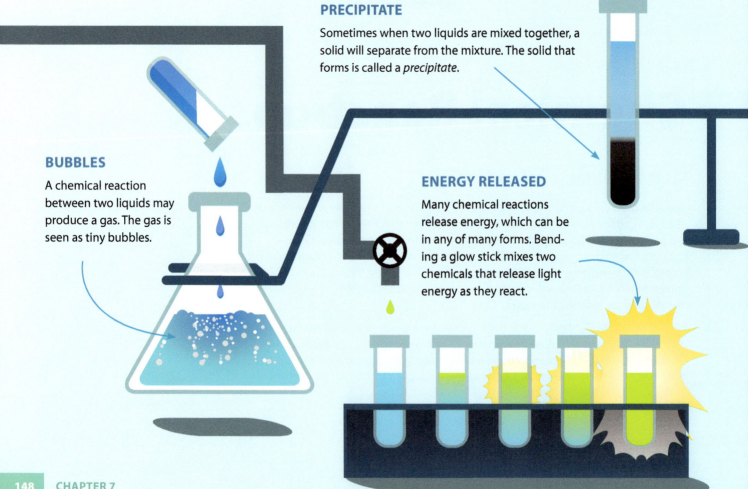

BUBBLES
A chemical reaction between two liquids may produce a gas. The gas is seen as tiny bubbles.

PRECIPITATE
Sometimes when two liquids are mixed together, a solid will separate from the mixture. The solid that forms is called a *precipitate*.

ENERGY RELEASED
Many chemical reactions release energy, which can be in any of many forms. Bending a glow stick mixes two chemicals that release light energy as they react.

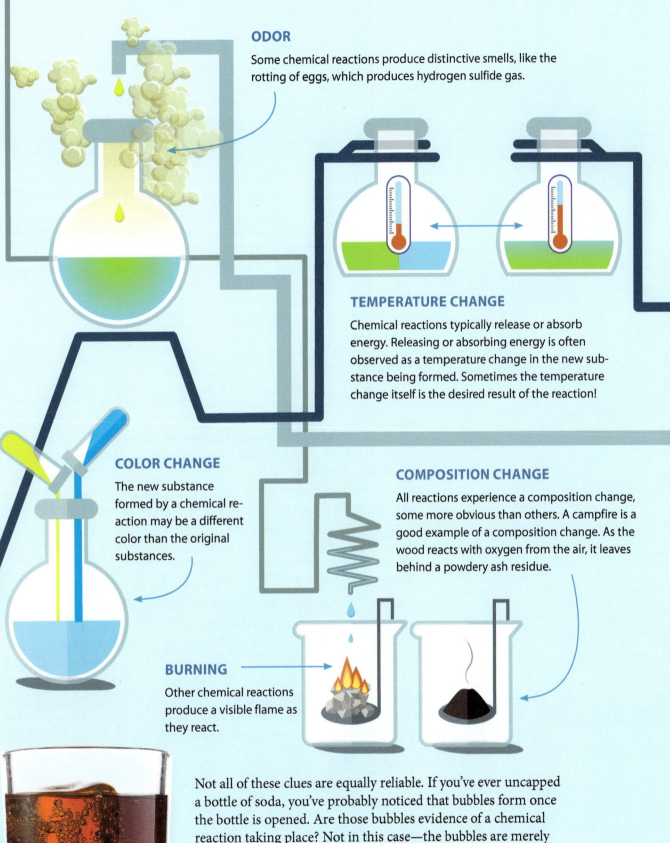

ODOR
Some chemical reactions produce distinctive smells, like the rotting of eggs, which produces hydrogen sulfide gas.

TEMPERATURE CHANGE
Chemical reactions typically release or absorb energy. Releasing or absorbing energy is often observed as a temperature change in the new substance being formed. Sometimes the temperature change itself is the desired result of the reaction!

COLOR CHANGE
The new substance formed by a chemical reaction may be a different color than the original substances.

COMPOSITION CHANGE
All reactions experience a composition change, some more obvious than others. A campfire is a good example of a composition change. As the wood reacts with oxygen from the air, it leaves behind a powdery ash residue.

BURNING
Other chemical reactions produce a visible flame as they react.

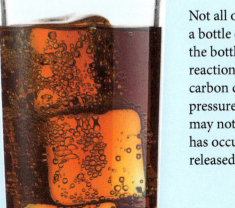

Not all of these clues are equally reliable. If you've ever uncapped a bottle of soda, you've probably noticed that bubbles form once the bottle is opened. Are those bubbles evidence of a chemical reaction taking place? Not in this case—the bubbles are merely carbon dioxide that has come out of the soda solution once the pressure inside the bottle is released. So bubbles by themselves may not be enough to say with certainty that a chemical reaction has occurred. Now, bubbles with a color change *and* heat energy released? That's a different story!

CHEMICAL REACTIONS

7.2 WRITING CHEMICAL EQUATIONS

Chemists—and chemistry students—do more than just look for signs of chemical reactions. They also study, create, describe, and communicate about chemical reactions. But describing a chemical reaction with words alone is not a wholly satisfactory method. Take for instance a chemical reaction between hydrogen and oxygen to form water. We could create a word equation such as, "Two units of hydrogen gas react with one unit of oxygen gas to yield two units of water." That's a lot of words to describe a fairly simple reaction—most reactions are far more complex. Chemists need a simpler way to model chemical reactions. For this purpose they use a **chemical equation**, a combination of chemical formulas and symbols that models a chemical reaction. Let's look at what the hydrogen and oxygen reaction looks like when it's written as an equation.

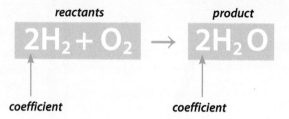

A chemical equation is much like a mathematical equation. The substances on the left side of the equation, hydrogen and oxygen, are indicated by their chemical formulas, H_2 and O_2 (don't forget your diatomic elements from Chapter 5). The plus sign shows that hydrogen and oxygen will be combined in this reaction. The formula for water on the right of the equation shows that water is what is formed by the reaction. The substances that enter into a chemical reaction are called the **reactants**. The substances that they combine to produce are called the **products**. The arrow symbol, like the equal sign in math, means "yields."

What about the number "2" in front of the formulas for hydrogen and water? Go back to the original description of this reaction. Notice the words "two units" and "one unit." The number placed in front of a chemical formula within a chemical equation is called a **coefficient**, similar to coefficients in math class. It shows how many units (e.g., atoms, molecules, or formula units) of each reactant and product are in the chemical equation. Hydrogen and oxygen always combine in a ratio of two units of hydrogen to one unit of oxygen to form two units of water. A coefficient is not shown if only one unit of a substance is indicated in an equation.

7.3 BALANCING CHEMICAL EQUATIONS

There's another reason why coefficients are needed for chemical equations. On the basis of many observations, chemists know that chemical reactions do not create any new matter, nor do they destroy any matter. The number and kinds of atoms in the products of a reaction are always the same as the number and kind found in the reactants. This relates back to the law of conservation of matter concept that you learned in

Chapter 2. To see how this works, let's look at our equation again, but this time without the coefficients.

$$H_2 + O_2 \rightarrow H_2O$$

This new equation isn't completely wrong. After all, it does show that hydrogen combines with oxygen to yield water, and that is true. But let's count up the number of each kind of atom shown in both the reactants and the product. On the reactants side, we can see two atoms of hydrogen and two atoms of oxygen. But on the product side there are two hydrogen atoms and only one oxygen atom. We call this an *unbalanced* chemical equation. We know from the law of conservation of matter that the missing oxygen atom couldn't have been destroyed during the reaction. We need to account for that second oxygen atom, and to do so we need to balance the equation. Let's see how that's done.

EXAMPLE 7-1: Writing Balanced Chemical Equations

Step 1: *Write the word equation for the reaction.*

Hydrogen plus oxygen yields water.

Step 2: *Write the chemical equation for the reaction.*

$$H_2 + O_2 \rightarrow H_2O$$

Step 3: *Balance the chemical equation with coefficients, if necessary.* Here we see right away that the number of oxygen atoms on both sides of the equation is not balanced. We need to double the number of oxygen atoms on the right side (product side) of the equation. We'll do that by placing a 2 in front of the formula for water. That coefficient doubles the number of units of water, which are molecules in this case.

$$H_2 + O_2 \rightarrow 2H_2O$$

Notice that we did not place a subscript 2 after the symbol for oxygen in H_2O in order to double the number of oxygen atoms! The reason for this is simple and important: adding subscripts changes the identity of a chemical formula. H_2O is water; H_2O_2 is not!

Step 4: *Check to see whether the addition of the coefficient balances the equation.* In this case, adding the coefficient for water corrected the imbalance for oxygen but unbalanced the number of hydrogen atoms on each side.

Step 5: *Repeat Steps 3 and 4, if necessary.* Adding the coefficient 2 for water doubled the number of hydrogen atoms on the product side (note that it is 2 units of water × 2 hydrogen atoms in each unit = 4 total hydrogen atoms). We need to double the number of hydrogen atoms on the reactant side to balance the new number on the product side.

$$2H_2 + O_2 \rightarrow 2H_2O$$

A quick check of this equation shows that it is now balanced. It has the same number of hydrogen and oxygen atoms on each side of the equation.

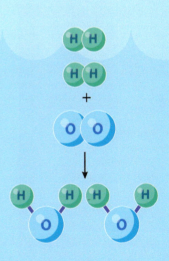

EXAMPLE 7-2: Writing Balanced Chemical Equations

Rust is a combination of iron and **oxygen** that we first saw back in Chapter 5 as iron (III) oxide. Let's see how the chemical equation for this reaction is balanced.

Step 1: Write the word equation for the reaction.

Iron plus oxygen yields iron (III) oxide (rust).

Step 2: Write the chemical equation for the reaction.

$$Fe + O_2 \rightarrow Fe_2O_3$$

Step 3: Balance the chemical equation with coefficients, if necessary.

The presence of two oxygen atoms on the left and three on the right shows that least common multiples will be necessary. The least common multiple of 2 and 3 is 6, so a coefficient of 3 will need to be added to the reactant oxygen, and a 2 will be necessary for the product.

$$Fe + 3O_2 \rightarrow 2Fe_2O_3$$

This balances the number of oxygen atoms on each side but leaves the iron atoms unbalanced. Since there are four iron atoms in the product, the addition of a 4 as the coefficient for the iron reactant easily fixes the problem.

$$4Fe + 3O_2 \rightarrow 2Fe_2O_3$$

A quick check shows that the equation is now balanced. The balanced equation reveals that four atoms of iron react with three molecules of oxygen to produce two formula units of iron (III) oxide.

7.4 WORKING WITH CHEMICAL EQUATIONS

Why do we bother with balancing chemical equations? First of all, we balance equations so that they will conform to the law of conservation of matter. But we also need to think about how chemists work. Do chemists work with individual atoms and molecules? Of course not! Even very small masses of atoms contain enormous numbers of particles. One nanogram (a billionth of a gram) of hydrogen gas contains over 597 *trillion* hydrogen atoms! To make working with these gigantic numbers easier, scientists use a quantity called a *mole*. Just as we can talk about a *pair* (2) of shoes, a *dozen* (12) doughnuts, or a *case* of copier paper (5000 sheets), a mole is a defined quantity of items. One **mole** (mol) is 6.02×10^{23} particles, either atoms, ions, or molecules.

Balanced chemical equations don't just tell us how many individual atoms, ions, or molecules take part in a reaction. They also tell us how many moles of those particles will react together. We've seen that carbon and oxygen can combine in a one-to-one ratio to form carbon monoxide. So if a chemist wanted to combine one mole of carbon atoms with oxygen to form carbon monoxide, how many oxygen atoms would be needed? The answer is one mole. Similarly, the chemical equation we saw in the previous section, $4Fe + 3O_2 \longrightarrow 2Fe_2O_3$, can be read as, "Four moles of iron atoms react with three moles of oxygen molecules to produce two moles of iron (III) oxide formula units."

Mole-Mass Conversions

Still, chemists don't have time to count out moles of atoms. No one does! But there is a way to *measure* out moles of atoms! You saw in Chapter 5 that the atomic mass of an element as shown on the periodic table represents the average mass of an atom of that element. For example, an average carbon atom's mass is 12.01u. But 12.01 also happens to be the mass in grams of one mole of carbon atoms. So 12.01g of carbon contain 6.02×10^{23} atoms of carbon. This relationship is true for every other element on the periodic table as well. The mass of one mole of a substance is its **molar mass**. The molar mass of oxygen atoms is 16.00 g, of calcium atoms is 40.08 g, and of tin atoms is 118.71g.

> ### SERVING AS A TOXICOLOGIST:
> ### PICK YOUR POISON
>
> The wildlife personnel shown below are not just splashing and having fun. They're applying a chemical called *rotenone* to a pond. The rotenone will kill all the fish in the pond. That sounds like a terrible thing—why would they do it? It's usually done to eliminate invasive species. Rotenone degrades rapidly once applied, so native fish can be stocked back into the pond after several days. Rotenone is derived from plants. Indigenous peoples discovered that crushing these plants and adding them into a body of water would cause fish to come to the surface for air, where they could easily be caught.
>
> The science of identifying and studying toxic substances, such as rotenone, is called *toxicology*. Toxicologists decode the chemical structure of toxins and figure out how they work. Many of these poisonous substances are found in nature. Like rotenone, some of them can be put to beneficial uses, such as helping to manage the environment or treat disease. There are still lots of toxins out there waiting to be discovered—maybe you'll be the first to identify one of them.

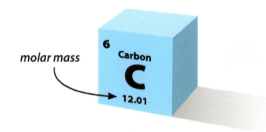

Suppose a chemist wished to carry out the iron (III) oxide reaction described in the last section. To produce 2 mol of iron (III) oxide, he needs 4 mol of iron and 3 mol of oxygen. With the help of conversion factors, he can easily calculate how many grams of each substance he needs. The relationship between moles of iron and grams of iron is shown in the conversion factors below.

$$\frac{1 \text{ mol Fe}}{55.85 \text{ g Fe}} \text{ and } \frac{55.85 \text{ g Fe}}{1 \text{ mol Fe}}$$

Likewise, the conversion factors for moles and grams of oxygen are as follows.

$$\frac{1 \text{ mol O}_2}{32.00 \text{ g O}_2} \text{ and } \frac{32.00 \text{ g O}_2}{1 \text{ mol O}_2}$$

Note that these conversion factors show oxygen as O_2 because pure oxygen exists as diatomic molecules. Therefore 1 mol of oxygen molecules consists of 2 mol of oxygen atoms. Now let's see how these conversion factors are used to solve the chemist's problem.

EXAMPLE 7-3

How many grams of iron and oxygen are needed to make 2 mol of iron (III) oxide?

Use conversion factors to convert 4 mol of iron and 3 mol of oxygen to grams.

$$4 \text{ mol Fe} \times \frac{55.85 \text{ g Fe}}{1 \text{ mol Fe}} = 223.4 \text{ g Fe}$$

$$3 \text{ mol O}_2 \times \frac{32.00 \text{ g O}_2}{1 \text{ mol O}_2} = 96.00 \text{ g O}_2$$

To make 2 mol of iron (III) oxide, the chemist needs 223.4 g of iron and 96.00 g of oxygen.

In the same manner as you saw for oxygen, the molar masses of compounds can be found by adding together the molar masses of the compound's elements. Remember, the formula Fe_2O_3 tells us that each formula unit of iron (III) oxide contains two atoms of iron and three atoms of oxygen. So 1 mol of Fe_2O_3 will contain 2 mol of iron and 3 mol of oxygen. So the calculation for the molar mass of iron (III) oxide will be as shown below.

$$\left[\left(\frac{2 \text{ mol Fe}}{1 \text{ mol Fe}_2\text{O}_3}\right)\left(\frac{55.85 \text{ g Fe}}{1 \text{ mol Fe}}\right)\right] + \left[\left(\frac{3 \text{ mol O}}{1 \text{ mol Fe}_2\text{O}_3}\right)\left(\frac{16.00 \text{ g O}}{1 \text{ mol O}}\right)\right] = \frac{159.7 \text{ g Fe}_2\text{O}_3}{1 \text{ mol Fe}_2\text{O}_3}$$

7A | REVIEW QUESTIONS

1. Why is melting ice not an example of a chemical change?
2. List five evidences that suggest a chemical change may have occurred.
3. How does a chemical equation model a chemical reaction?
4. Write a chemical equation that is based on the following word equation: "Carbon combines with oxygen to form carbon dioxide." You do not need to balance the equation.
5. Why do chemical equations need to be balanced?
6. Determine whether each of the following equations is balanced. Write "balanced" if the equation is balanced. If it is not, then add the coefficients needed to balance it.
    a. $N_2 + 3H_2 \rightarrow 2NH_3$
    b. $C + S_8 \rightarrow CS_2$
    c. $K + O_2 \rightarrow K_2O$

7. Write the balanced chemical equation for the decomposition of hydrogen peroxide (H_2O_2) into water and oxygen.

Use the equation below to answer Questions 8–9.

$$CH_4 + 4Cl_2 \rightarrow CCl_4 + 4HCl$$

8. A chemist wants to react 1 mol of methane (CH_4) with chlorine gas (Cl_2) to produce carbon tetrachloride (CCl_4). How many moles of chlorine gas will he need for this reaction?
9. How many grams of chlorine gas will the chemist need to react with 1 mol of methane?

MINI LAB

BALANCED DIET

Want to make balancing chemical equations more appetizing? Then give this visual method a try!

Procedure

A Choose one color of candy to represent atoms of carbon, another for hydrogen, a third for oxygen, and a fourth to represent any other element.

B Set the ruler on the desk in front of you, pointing away. Set two cups to the left of the ruler and one to the right.

C Then have a look at the following chemical equation.

$$C_2H_2 + H_2 \rightarrow C_2H_6$$

D Use your colored candies to fill the cups, representing the reactants and product of the reaction. It will help if you place your "models" before you in the same order as they occur in the reaction.

1. What do the cups represent?
2. What does the ruler represent?

E Carefully count up the number and kind of each atom.

3. Does your model show the same number and kind of each atom on both the reactant and product sides of the equation?

F If necessary, add more cups of candies representing additional molecules of one or more of the substances in the equation until you have the same number and kind of atoms on each side of the equation. Remember, you can't add only part of a molecule! Molecules enter into reactions only as whole units.

4. What did you have to add in order to balance the equation?
5. What is the balanced form of the equation?

Going Further

G Use this edible model technique to balance the following equations. After finishing each model, write the balanced equation.

6. $NH_3 + O_2 \rightarrow N_2O + H_2O$
7. $C_4H_8O_4 + O_2 \rightarrow CO_2 + H_2O$

> **Essential Question:**
> How can I tell whether a chemical equation is balanced?
>
> **Equipment**
> colored candies, such as gumdrops
> small cups
> ruler

CHEMICAL REACTIONS

7B Questions

- Are all chemical reactions the same?
- How can I tell one kind of chemical reaction from another?
- What is combustion?

7B Terms

synthesis reaction, decomposition reaction, single-replacement reaction, double-replacement reaction, combustion, oxidation, reduction

7B | TYPES OF CHEMICAL REACTIONS

Are there different kinds of chemical reactions?

7.5 CLASSIFYING CHEMICAL REACTIONS

So far, the reactions that we have looked at in this chapter have been ones in which two reactants have combined to form a single product. Are other kinds of reactions possible? Yes! Let's take a look at different kinds of reactions.

REACTIONS

SYNTHESIS REACTIONS

Synthesis reactions combine two or more reactants into a single, more complex product. These reactions generally take the form X + Y → XY, where X, Y, and XY represent different substances. The reactants can be elements, such as when zinc and sulfur combine to form zinc sulfide. They could also be compounds, such as when carbon dioxide and water react to form carbonic acid.

$$CO_2 + H_2O \rightarrow H_2CO_3$$

Key feature: A single, more complex product results from two or more reactants.

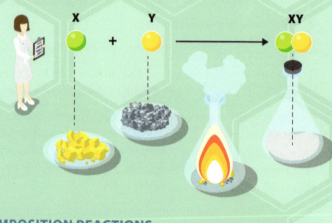

DECOMPOSITION REACTIONS

The opposite of a synthesis reaction, a **decomposition reaction**, is one in which a single reactant breaks down into two or more products. Such reactions have the form XY → X + Y. Decomposition reactions usually require an input of energy, such as when electrolysis (breaking down by electricity) is used to separate water into hydrogen and oxygen: $2H_2O \rightarrow 2H_2 + O_2$. Oil refineries use decomposition reactions to break the large molecules in some petroleum products into smaller, more useful molecules such as those found in gasoline.

Key feature: A single reactant becomes two or more products.

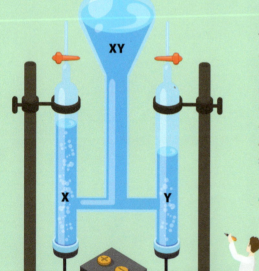

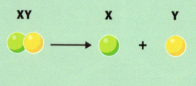

SINGLE-REPLACEMENT REACTIONS

In a **single-replacement reaction**, one element in a compound is replaced by another element. The general form for this type of reaction is XY + Z → ZY + X. When copper metal is submerged into a solution of silver nitrate, an interesting transformation takes place. The clear solution turns blue and the copper takes on a silvery coating. The equation for this reaction is

$$2AgNO_3 + Cu \rightarrow Cu(NO_3)_2 + 2Ag.$$

Notice how the copper atoms have replaced the silver atoms in $AgNO_3$, forming $Cu(NO_3)_2$.

Key feature: Both reactants and products include an element and a compound.

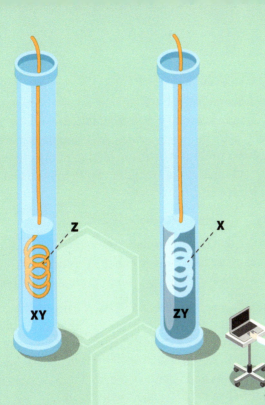

DOUBLE-REPLACEMENT REACTIONS

In a **double-replacement reaction**, two compounds swap cations or anions with each other. The general form for these reactions is WX + YZ → WZ + YX. In one such reaction, sodium chloride and silver nitrate swap ions to form sodium nitrate and silver chloride. The silver chloride forms a white precipitate. The equation for this reaction is

$$NaCl + AgNO_3 \rightarrow NaNO_3 + AgCl.$$

Key feature: Two compounds swap ions, producing two new compounds as products.

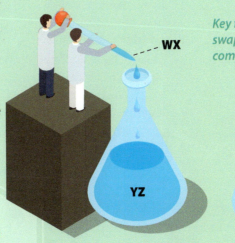

COMBUSTION REACTIONS

In a **combustion** reaction, a substance, usually called a *fuel*, reacts with oxygen. We often refer to this as *burning*. Your home may be warmed in winter by an unspectacular, but vital, example of combustion. Natural gas, used to heat many homes, contains methane (see Chapter 6). It reacts with oxygen to form carbon dioxide and water vapor, releasing heat energy in the process.

$$CH_4 + 2O_2 \rightarrow CO_2 + 2H_2O$$

Key feature: A reactant (the fuel) combines with oxygen, often producing large amounts of heat and light.

CHEMICAL REACTIONS

7.6 OXIDATION AND REDUCTION

Chemists have long known that many elements, especially metals, will react with oxygen. Such reactions are called *oxidations*, and the compounds that are formed are known as *oxides*. An example is mercury (II) oxide, an orange powder produced by the oxidation of mercury. Notice that this oxidation reaction is also a synthesis reaction since two reactants combine to form a single product.

$$2Hg + O_2 \longrightarrow 2HgO$$

Chemists later realized that during oxidations, metals lose electrons, so the term **oxidation** came to mean any instance in which electrons are lost. The term is still used even if oxygen is not present in the reaction.

But what happens to the electrons that are "lost" during oxidation? They're not truly lost—they are actually gained by another reactant. This gaining of electrons during a chemical reaction is called, perhaps confusingly, **reduction**. In the case of mercury (II) oxide, oxygen is the element that gains electrons and is thus reduced. But as is true for oxidation, oxygen is not necessary for reduction to occur.

It should be apparent that oxidation and reduction go hand in hand. If an element is oxidized during a reaction, another element must be reduced. Because of this, such reactions are often called *redox* reactions, from the words *red*uction and *ox*idation. Redox reactions do not always involve the complete transfer of electrons between elements. As you learned in Chapter 5, water molecules contain polar bonds that result when electrons are not equally shared. The synthesis of water from hydrogen and oxygen is considered a partial redox reaction because hydrogen's electrons are not completely transferred to oxygen.

7B | REVIEW QUESTIONS

1. Compare decomposition and synthesis reactions.
2. Compare single- and double-replacement reactions.
3. How can you identify a combustion reaction?
4. What kind of reaction is shown in the following equation? Explain.

   $$2Na + Cl_2 \longrightarrow 2NaCl$$

5. Compare oxidation and reduction.
6. The following reaction is a redox reaction. Which element goes through oxidation? Which goes through reduction? Besides being a redox reaction, how else could this reaction be classified?

   $$Mg + S \longrightarrow MgS$$

7C | ENERGY IN CHEMICAL REACTIONS

Do chemical reactions always give off energy?

7.7 UNDERSTANDING ENERGY EXCHANGES

Sure, something like a rocket launch is an exciting chemical reaction to see, and it obviously releases lots of energy too. Smoke, noise, heat, flames—there's plenty of evidence that a dramatic reaction is happening. But what about this old, rusting truck? It's experiencing a chemical reaction, too, only it takes years to happen instead of a few minutes. Does rusting release energy? Yes, it does, but it sure doesn't seem like it!

As you learned in Chapter 5, energy is stored in the chemical bonds that form between atoms. To start a chemical reaction, additional energy is required to break the chemical bonds within reactants, but when bonds are reformed afterward in products, energy is released. This relationship between absorbing and releasing energy determines whether the overall reaction itself will give off or absorb energy.

7C Questions

- How does energy change during a chemical reaction?
- Can energy be created or destroyed?

7C Terms

exothermic reaction, endothermic reaction, activation energy, law of conservation of energy

ENERGY OUT—ENERGY IN

EXOTHERMIC REACTIONS: ENERGY OUT

Some reactions release more energy than they absorb. Reactions that release this energy as heat are called **exothermic reactions**. Combustion is a type of exothermic reaction. The graph below shows that the products of an exothermic reaction contain less stored energy than the reactants.

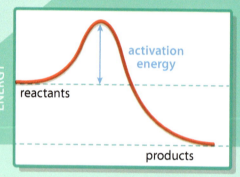

ENDOTHERMIC REACTIONS: ENERGY IN

The graph below shows a reaction whose products contain *more* stored energy than its reactants. A reaction like this that absorbs more energy than it releases is an **endothermic reaction**. The photosynthesis that makes much of life on Earth possible is an endothermic reaction.

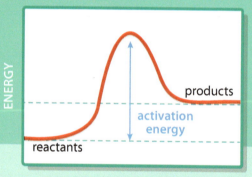

Activation Energy

Have you ever wondered why the first hill on a roller coaster is the tallest one? The energy gained from that first hill is needed by the roller coaster train to complete its journey around the track. Without that initial energy, the ride isn't much fun! Chemical reactions also need added energy to get started. The energy needed to get a chemical reaction started is its **activation energy**. You may have noticed that the graphs above each show a sort of "hill" in the middle of the reaction. That hill represents the extra energy needed by the reactants in order to get a reaction started.

7.8 CONSERVATION OF ENERGY

A campfire almost completely consumes the wood that is burned, while photosynthesis creates sugars in plants. It's tempting to think that exothermic reactions destroy energy while endothermic reactions create it. Such thinking would be a serious error, though, for it is contrary to one of the fundamental laws of science. The **law of conservation of energy** states that during a chemical reaction no energy is created or destroyed, but only changed from one form to another. The diagrams at right show how this works.

As the stored chemical energy in a piece of firewood is changed into thermal energy, where does it go? It is radiated out into and absorbed by the environment (including you, if you are standing near enough). This energy that seems to be "lost" is actually shown on the graph (above, right) as the difference in the y-axis values for the energy found in the reactants and products. Where does the stored chemical energy in a plant come from? It started out as light energy produced by the sun. Similar to the exothermic example, the difference in energy between reactants and products can also be seen in the graph (far right) of an endothermic reaction. In this case the difference between the y-axis values represents the energy that is gained during the reaction. So even though some reactions may *look* like violations of the law of conservation of energy, they clearly are not.

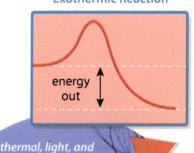

Exothermic Reaction

energy out

thermal, light, and sound energy

stored chemical energy (firewood)

ENERGY OUT

light energy (sun)

stored chemical energy (plant)

ENERGY IN

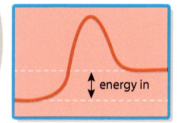

Endothermic Reaction

energy in

7C | REVIEW QUESTIONS

1. A reaction whose products have more energy than its reactants is a(n) _____ reaction.

2. Is combustion a type of exothermic or endothermic reaction?

3. What is activation energy?

Use the information in the graph at right to answer Questions 4–5.

4. Label parts A, B, and C of the graph.

5. Does the graph represent an exothermic or endothermic reaction?

6. What must be true about the energy given off by an exothermic reaction and the energy remaining in the products according to the law of conservation of energy?

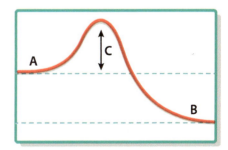

CHEMICAL REACTIONS

7D Questions

- Do reactants always react together?
- How can we change the speed of a reaction?
- What is equilibrium?
- How can a system in equilibrium change?

7D Terms

collision model, reaction rate, catalyst, inhibitor, reversible reaction, chemical equilibrium, Le Châtelier's principle

7D | REACTION RATES AND EQUILIBRIUM

Why do some things burn slowly while others explode?

7.9 REACTION RATES

We see reactions happen all the time. Cars rust, wood burns, dead plants decay, and fireworks explode. Some reactions happen lightning fast, while others occur with almost glacial slowness. Some things explode and others do not. Why is this true?

Collision Model

As we think about why some reactions happen quickly while others are slow, we must consider the conditions needed for a reaction to occur at all. Our current model, the **collision model**, states that we need three things to happen before reactants will react. First, the reactant particles must collide with each other. Then, when they collide, they must be

FACTORS THAT AFFECT REACTION RATES

SURFACE AREA

Reactions happen only where the reactants are in direct contact. If the surface area of the reactants is increased, then more reactant particles will be in contact with each other and more reactions will take place. For a given mass of reactant, a large number of small pieces have a greater combined surface area than a single large piece.

- more reactants in contact

STIRRING

Stirring the reactants moves more reactants into contact with each other, causing more reactions to occur.

- more reactants in contact
- slightly more energy

TEMPERATURE

Increased temperature means the particles move faster, resulting in more reactions.

- more collisions
- collisions have more energy

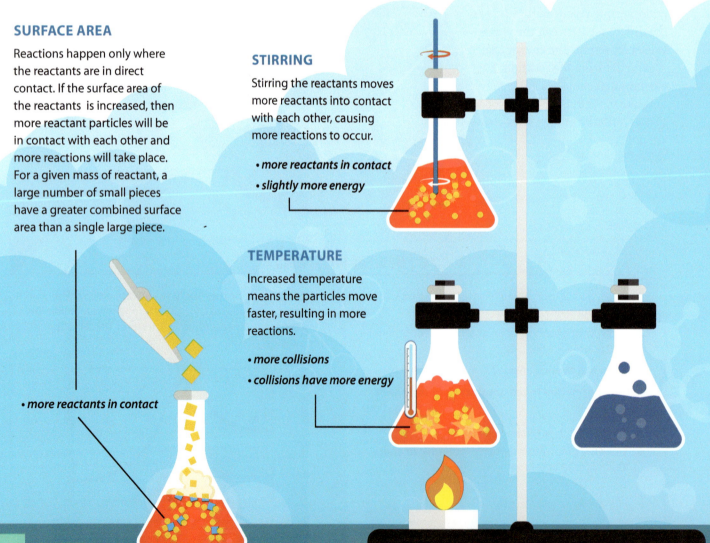

aligned properly in order to rearrange their bonds. Finally, there must be enough energy. But what does "enough energy" mean? Let's look again at the energy diagrams for endothermic and exothermic reactions.

To move from the reactants to the products on this graph, we have to get all the way to the top of the activation energy "hill." The activation energy is the "enough energy." If the collision occurs, is in the right alignment, and there is the proper amount of energy, then the reaction occurs. How quickly this all happens depends on the likelihood that the reactant particles will collide with the proper alignment and energy.

The speed of a reaction is called its **reaction rate**. This is an indication of how quickly reactants change into products. Many things can cause a chemical reaction to speed up or slow down, that is, change the reaction's rate. These things include temperature, concentration, surface area, and stirring. Let's look at how these things affect the collisions between particles in a reaction.

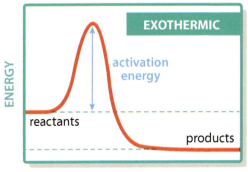

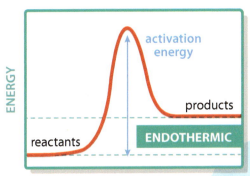

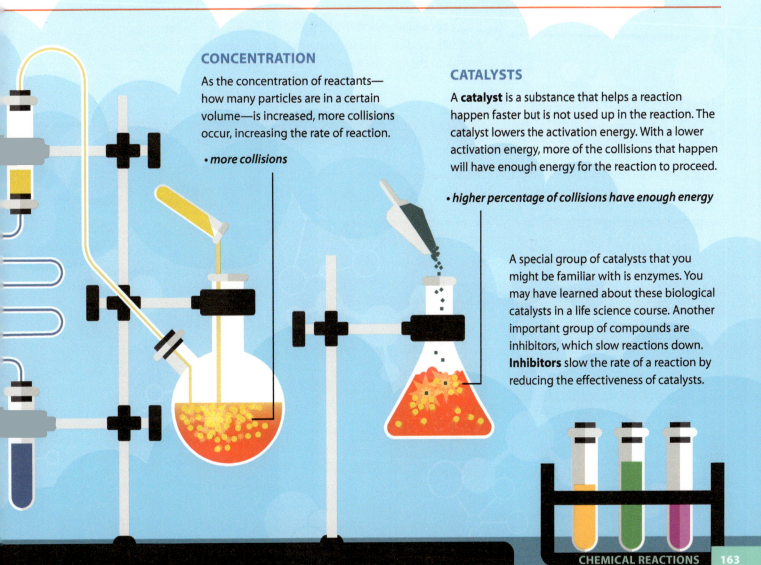

CONCENTRATION

As the concentration of reactants—how many particles are in a certain volume—is increased, more collisions occur, increasing the rate of reaction.

• more collisions

CATALYSTS

A **catalyst** is a substance that helps a reaction happen faster but is not used up in the reaction. The catalyst lowers the activation energy. With a lower activation energy, more of the collisions that happen will have enough energy for the reaction to proceed.

• higher percentage of collisions have enough energy

A special group of catalysts that you might be familiar with is enzymes. You may have learned about these biological catalysts in a life science course. Another important group of compounds are inhibitors, which slow reactions down. **Inhibitors** slow the rate of a reaction by reducing the effectiveness of catalysts.

CHEMICAL REACTIONS

7.10 REVERSIBLE REACTIONS AND EQUILIBRIUM

Reversible Reactions

You might think that a reaction continues until all the reactants are turned into products. Think about a burning candle, which is a combustion reaction. If we were to light the candle and leave it, we would expect it to burn until all the fuel (the candle wax) was used. This reaction is an irreversible reaction because it progresses only in one direction. The candle and oxygen form soot and some gases. The soot and gases can't recombine to form the candle again. But many reactions are **reversible reactions**. Not only do the reactants form products, but the products can also react together to re-form the original reactants. The equations for these reactions differ from those you have already seen in that there are yield arrows in both directions. The reaction between hydrogen and iodine to form hydrogen iodide gas is reversible.

$$H_2 + I_2 \rightleftharpoons 2HI$$

For the reverse reaction to occur, the product(s) must remain contained. If this is done for the hydrogen iodide reaction, some of the hydrogen iodide will decompose back into hydrogen and iodine gases.

Equilibrium

Every reversible reaction starts at a particular rate depending on the original conditions in its container. As the amounts of reactants and products change, the reaction rate changes too.

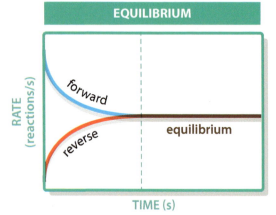

UNDERSTANDING EQUILIBRIUM

Equilibrium represents a state in which multiple influences cancel each other out. The system is said to be balanced.

Forces in Equilibrium

Equilibrium can refer to forces that balance each other. Think of balancing with a friend on a seesaw. The forces that you each apply to the seesaw cancel each other out and the seesaw is balanced, that is, it's in equilibrium. There are still forces acting on the seesaw, but its motion remains unchanged because the forces are balanced.

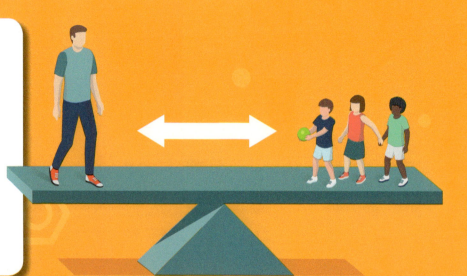

Physical Equilibrium

Physical processes, such as evaporation and condensation, can be in equilibrium too. If you fill a jar with warm water and then close the lid, some of the water will evaporate. But it will reach a point at which no more water appears to evaporate. When this point is reached, the water continues to evaporate, but it does so at the same rate that vapor is condensing. The system is in physical equilibrium.

2 molecules evaporate

2 molecules condense

The number of vapor and liquid molecules remains constant.

Chemical Equilibrium

As the amounts of substances in a reaction vessel change, the rates of the forward and reverse reactions change. **Chemical equilibrium** occurs when the forward and reverse reactions each happen at the same rate. When the system reaches this equilibrium, we say that the reaction has stopped.

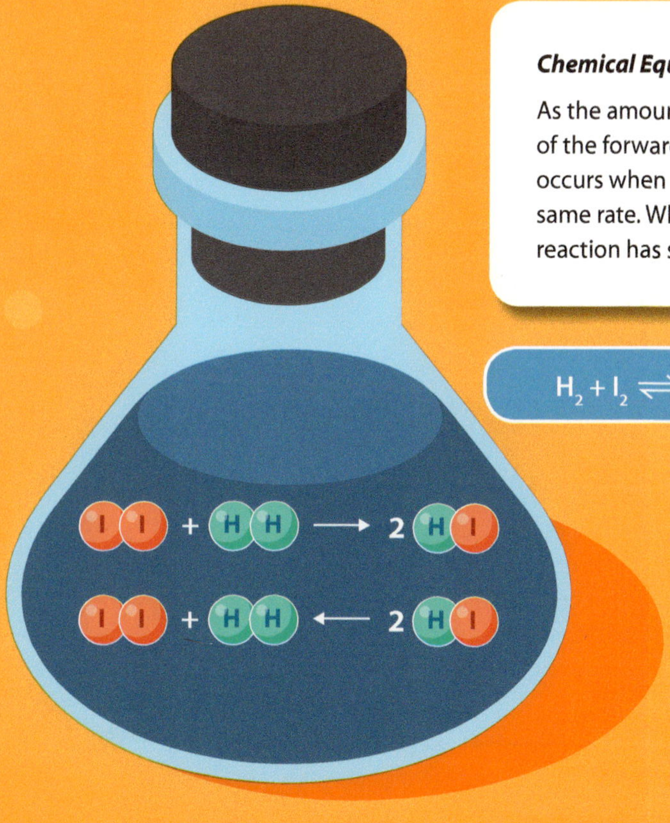

$$H_2 + I_2 \rightleftharpoons 2\,HI$$

However, understand that both the forward and reverse reactions continue to occur, but at the same rate. There appears to be no change in the system—it is in chemical equilibrium. The number of each type of particle remains constant, a state called the *equilibrium state*.

CHEMICAL REACTIONS 165

7.11 CHANGES TO EQUILIBRIUM

Once a system is in equilibrium, changes to the system can alter the equilibrium state. This happens as either the forward or reverse reaction rate increases. Over time the system will reach a new equilibrium state. Changes to temperature, pressure, or concentration can cause these changes. Let's use the following equation for the exothermic reaction of nitrogen and hydrogen gases forming gaseous ammonia to see how changing conditions can affect the equilibrium state.

CHANGING CHEMICAL EQUILIBRIUM

FACTORS THAT AFFECT EQUILIBRIUM
(numbers shown indicate number of molecules)

$N_2 + 3H_2 \rightleftharpoons 2NH_3$

EQUILIBRIUM STATE

N_2: 100, H_2: 25, NH_3: 150

Concentration (Adding or Removing Substances)

Adding any of these substances will shift the equilibrium state toward the other side of the reaction. Adding ninety more H_2 molecules will result in the forward reaction occurring, producing ammonia. *This will use up some of the added hydrogen molecules.*

+90 H_2 → N_2: 73, H_2: 34, NH_3: 204

$N_2 + 3H_2 \rightleftharpoons 2NH_3$
shifts right

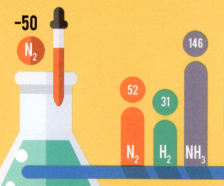

Removing any substance will shift the equilibrium state toward the side of the reaction from which we removed the substance. If fifty nitrogen molecules are removed, some of the ammonia will decompose into nitrogen and hydrogen, *replacing some of the nitrogen that we removed.*

−50 N_2 → N_2: 52, H_2: 31, NH_3: 146

$N_2 + 3H_2 \rightleftharpoons 2NH_3$
shifts left

Temperature (Adding or Removing Energy)

Increasing the temperature will shift the reaction toward the side of the equation that absorbs energy. Lowering the temperature will shift the system toward the side that releases energy. The formation of ammonia is exothermic and its decomposition is endothermic. Increasing the temperature (right) will cause more reactants to be made, *absorbing some of the added energy.*

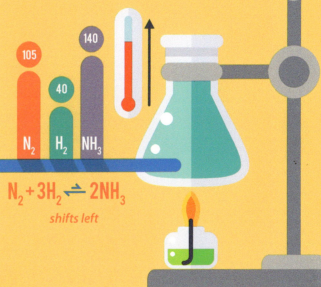

$N_2 + 3H_2 \rightleftharpoons 2NH_3$
shifts left

$N_2 + 3H_2 \rightleftharpoons 2NH_3$
shifts right

Lowering the temperature (left) will cause more ammonia to be produced because the forward reaction results in the release of energy, *warming the system after it was cooled.*

Pressure

The pressure inside a closed container is related to the number of particles it contains—the greater the number of particles, the greater the pressure. If the pressure is increased, a response that results in fewer gas particles is needed. This will lower the pressure within the container. As you can see in the equation, there are more gas particles in the reactants, nitrogen (1) and hydrogen (3), than in the product, ammonia (2); this synthesis reaction results in a net decrease in the number of particles. If the pressure acting on this reaction is increased, the need to reduce the number of particles in response will shift the equilibrium state toward the side of the equation that has fewer particles. In this case, that shift would be toward the product, thus *reducing some of the added pressure.*

Lowering the pressure on this reaction (below) would shift the reaction toward the reactants, producing more particles and *restoring some of the removed pressure.*

$N_2 + 3H_2 \rightleftharpoons 2NH_3$
shifts left

CHEMICAL REACTIONS

LE CHÂTELIER'S PRINCIPLE

Notice the results of the changes that were made to the ammonia synthesis reaction. In each case, the system tries to undo the change. French Chemist Henri Le Châtelier studied the effects of changes to reactions that were in equilibrium. Today **Le Châtelier's principle** states that any chemical system in equilibrium will adjust its equilibrium state in such a way as to reduce the effect of any changes made to the system.

EXAMPLE 7-4:
Predicting the Effect of Changes on Chemical Equilibrium

Consider the exothermic reaction of gaseous sulfur dioxide and oxygen to form the gas sulfur trioxide.

$$2SO_2 + O_2 \rightleftharpoons 2SO_3$$

What will happen if SO_3 is added?

If we add SO_3, the system will try to correct the change, so the reverse reaction will increase, forming additional reactants and decomposing some of the added SO_3.

What will be the effect if the temperature of the system is increased?

Energy is added when the temperature is raised. This will cause the reaction to proceed in the endothermic direction, shifting the equilibrium state toward the left and using up some of the added energy. SO_3 will decompose into SO_2 and O_2.

What will occur if the pressure within the system is increased?

If the pressure is increased, the system will shift to reduce the pressure. This is done by decreasing the number of gas particles. Since the synthesis of SO_3 decreases the number of particles within the system, an increase in pressure will shift the equilibrium to the right, using SO_2 and O_2 to form more SO_3.

7D | REVIEW QUESTIONS

1. According to the collision model, what are the three requirements for a reaction to occur?
2. Predict how the following factors affect reaction rate (if they do at all).
   a. raising the temperature
   b. decreasing the concentration of the reactants
   c. decreasing the surface area of the reactants
   d. stirring
   e. applying a magnetic field
3. Why are there no enzymes that slow down the rates of chemical reactions in living things?
4. Define *chemical equilibrium*.
5. What is the effect of adding a chemical substance to a reversible reaction that is in equilibrium? Give an example.
6. Explain Le Châtelier's principle.
7. Predict how the following factors will shift equilibrium for the following reaction (if they do at all) using Le Châtelier's principle. The forward reaction is exothermic.

   $$4Li + O_2 \rightleftharpoons 2Li_2O$$

   a. raising the temperature
   b. decreasing the concentration of lithium
   c. increasing the pressure
   d. increasing the concentration of lithium oxide
   e. applying a magnetic field

CHAPTER 7 REVIEW

7A CHEMICAL CHANGES

- Evidences that a chemical reaction may have occurred include formation of a precipitate or bubbles, the release of energy, temperature or color change, odor, burning, or a change in composition.
- A single evidence by itself may not be sufficient to prove that a chemical reaction has occurred since some physical changes also produce such evidence.
- Chemical equations use chemical formulas and coefficients to show the number and kinds of reactants and products in a chemical reaction.
- The number and kind of atoms shown in chemical equations must be balanced in order to fulfill the requirements of the law of conservation of matter.
- Balanced chemical equations show not only the number of individual particles that enter into a chemical reaction but also the number of moles of each substance.
- The number of moles of a substance can be converted to the mass of that substance using mole-mass conversion factors.

7A Terms

chemical reaction	148
chemical equation	150
reactant	150
product	150
coefficient	150
mole	152
molar mass	153

7B TYPES OF CHEMICAL REACTIONS

- The five main kinds of chemical reactions are synthesis, decomposition, single-replacement, double-replacement, and combustion reactions.
- Combustion occurs when a substance reacts with oxygen.
- In a redox reaction, one reactant is oxidized (loses electrons) while another is reduced (gains electrons).

7B Terms

synthesis reaction	156
decomposition reaction	156
single-replacement reaction	157
double-replacement reaction	157
combustion	157
oxidation	158
reduction	158

CHEMICAL REACTIONS 169

CHAPTER 7

7C ENERGY IN CHEMICAL REACTIONS

- Exothermic reactions release thermal energy, while endothermic reactions absorb it.
- Chemical reactions require an initial input of energy—the activation energy—in order to proceed.
- The law of conservation of energy states that during a chemical reaction no energy is created or destroyed, but only changed from one form to another.

7C Terms

exothermic reaction	160
endothermic reaction	160
activation energy	160
law of conservation of energy	161

7D REACTION RATES AND EQUILIBRIUM

- Reactions occur only when reactants collide with proper alignment and enough energy.
- A reaction rate can be changed by changing the temperature of the reaction, by stirring the reactants, or by changing the concentration or surface area of the reactants.
- Catalysts increase reaction rate by lowering the activation energy. Inhibitors slow the rate by interfering with catalysts. Enzymes are biological catalysts.

- A reversible chemical reaction has reached equilibrium when the forward and reverse reaction rates are equal. The reaction appears to have stopped.
- Le Châtelier's principle states that a system in equilibrium will adjust to reduce the effect of any changes made to the system.

7D Terms

collision model	162
reaction rate	163
catalyst	163
inhibitor	163
reversible reaction	164
chemical equilibrium	165
Le Châtelier's principle	168

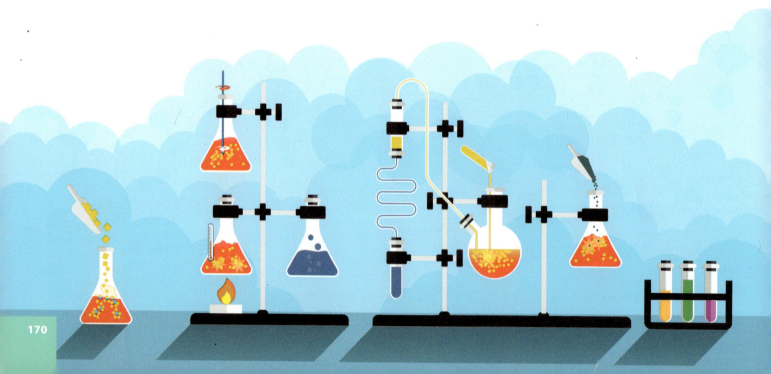

REVIEW

CHAPTER REVIEW QUESTIONS

Recalling Facts

1. Which of the following is never an evidence of chemical change?
   a. formation of bubbles
   b. change in phase from solid to liquid
   c. change in temperature
   d. production of an odor

2. What is a combustion reaction?

3. Why does a reaction in which one substance is oxidized happen only in the presence of another substance that is reduced?

4. Which type of reaction gives off thermal energy? Which kind absorbs thermal energy?

5. How much energy is enough to cause a reaction to occur?

6. List the conditions given in the textbook that can cause a change in the equilibrium state.

7. Why does changing the temperature change the equilibrium state?

Understanding Concepts

In Questions 8–10, write a balanced chemical equation for each reaction.

8. Magnesium and oxygen react to form magnesium oxide.

9. Sodium sulfide decomposes into sodium and sulfur (consider sulfur to exist as individual atoms).

10. Zinc and hydrogen chloride react to form zinc chloride and hydrogen gas (assume +2 oxidation state for zinc).

11. For each of the equations that you balanced in Questions 8–10, how many moles of each reactant and product were involved in the reaction?

Calculate the molar mass for each of the following compounds.

12. K_2O
13. $AgNO_3$
14. C_2H_6
15. MgS

16. If 4 mol of hydrogen molecules are reacted with 4 mol of iodine molecules, how many grams of hydrogen iodide will be produced?

Classify each of the following as either a synthesis, decomposition, single-replacement, or double-replacement reaction.

17. $2AgNO_3 + Zn \rightarrow 2Ag + Zn(NO_3)_2$
18. $2KClO_3 \rightarrow 2KCl + 3O_2$
19. $BaCl_2 + Na_2SO_4 \rightarrow BaSO_4 + 2NaCl$
20. $2K + Cl_2 \rightarrow 2KCl$

21. Compare combustion reactions with redox reactions.

22. Why is it incorrect to say that all oxidation reactions involve oxygen?

CHAPTER 7

Use the graph below to answer Questions 23–25.

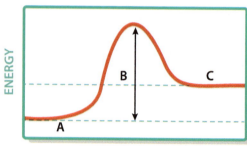

23. Which portion of the graph shows the energy of the products?
24. Which portion of the graph shows the activation energy?
25. Does this graph show an exothermic or endothermic reaction?
26. What will happen if the necessary substances for a reaction are placed together but the activation energy is not available?
27. Can a chemical reaction ever give off more energy than was originally contained in its reactants? Explain.
28. Why does increasing the temperature increase a reaction rate?
29. Do reactants that collide always react together? Explain.
30. Compare catalysts, inhibitors, and enzymes.

For Questions 31–32, consider the following equation for a system in equilibrium. All substances in the reaction are gases and the forward reaction is exothermic.

$$2C_2H_6 + 7O_2 \rightleftharpoons 4CO_2 + 6H_2O$$

31. What would be the effect of adding CO_2 to the system?
32. What would be the effect of cooling the system?

For Questions 33–34, consider the following equation for a system in equilibrium. All substances in the reaction are gases and the forward reaction is exothermic.

$$2CO + O_2 \rightleftharpoons 2CO_2$$

33. How could you cause more CO_2 to be produced?
34. Hypothesize about how changes in volume would affect the equilibrium state.

Critical Thinking

Use the Case Study on the left to answer Questions 35–36.

35. Review the information in Section 7.9 on reaction rates. What conditions found in grain storage silos do you think contribute to the danger of possible explosion?
36. What might be done to reduce the risk of explosion in a grain silo?

CASE STUDY: GRAIN ELEVATORS

Grain elevators are large facilities used for storing harvested grain and loading it onto ships or railway cars. On occasion such elevators have exploded, often causing considerable damage and even loss of life. The culprit is the very fine grain dust that circulates within the storage silos at the elevator site.

REVIEW

37. When sugar ($C_{12}H_{22}O_{11}$) and sulfuric acid (H_2SO_4) are mixed, they form a steaming black solid that grows out of the mixture. Using this information, evaluate whether a chemical change has taken place. Support your claim with evidence.

38. Given the following equation, what could you have changed to produce the results shown in the table? The forward reaction is exothermic and all substances are gases.

$$CO + 2H_2 \rightleftarrows CH_3OH$$

Equilibrium State	CO	H_2	CH_3OH
Initial	21	18	374
Final	24	24	371

CASE STUDY: BUILDING IMPLOSION

Some pro sports teams play in iconic stadiums, such as Chicago's Wrigley Field. But the old Seattle Kingdome, once home to baseball's Mariners and football's Seahawks, was *not* one of those venues. The Kingdome's vast interior and small baseball crowds once earned it the nickname *The Tomb*. In addition, hit baseballs that crazily bounced off structures like loudspeakers were often still in play, which made for some interesting game outcomes. Not too many tears were shed when the decision was made to demolish the old stadium and build a new one.

Tearing down such a massive structure can be very expensive and time-consuming. One solution: bring the old structure down with explosives, a process called *implosion*.

Use this Case Study to answer Questions 39–40.

39. Think about what a demolition needs to accomplish. Would that best be done by an exothermic or endothermic chemical reaction?

40. Implosions are done with many smaller demolition charges rather than a single large one. Why do you think this is so?

CHEMICAL REACTIONS 173

CHAPTER 8
Nuclear Changes

RED, RIPE, AND IRRADIATED?

The world population is quickly heading toward 8 billion people. How can we ensure that all these people have enough food and that the food they have is safe? The food we eat comes from all over the world. But before it reaches the market, some will decay or be eaten by insects. And of the food that does reach us, some is tainted by disease. Scientists are always working to solve these issues.

Scientists have learned that by treating food with *ionizing radiation*, a form of *nuclear radiation*, they can slow decay, kill insects, and destroy the bacteria that cause disease. But is irradiated food safe to eat? Over thirty years of testing has shown us that treated food is safe for humans. Furthermore, the radiation does not affect the nutritional value of food or change its taste. People in the United States eat over 100 million kilograms of treated food each year.

In this chapter, we will learn about nuclear radiation, including the benefits and risks of this amazing aspect of chemistry. So grab a red, ripe, and possibly irradiated apple as we learn about radiation.

Main image: *The apples shown at left are being treated with ionizing radiation to kill insects and microbes.*

Inset: *The green emblem indicates irradiated food.*

8A	Radioactive Decay	176
8B	Fission and Fusion	185
8C	Nuclear Changes: Benefits and Risks	191

8A Questions

- How was radioactivity discovered?
- Why don't all atoms decay?
- What types of radioactive decay exist?
- Can we know when an atom will decay?
- How can we use decay to determine the age of an object?

8A Terms

strong force, radioactive decay, alpha decay, beta decay, gamma decay, half-life

8A | RADIOACTIVE DECAY

Why do only some isotopes decay?

8.1 RIDDLES IN THE DARK

In Chapter 2 you learned about physical changes—changes that alter the appearance of something but not its makeup. Chapter 2 also mentioned chemical changes, which you learned more about in Chapter 7. A chemical change alters the composition of a material by breaking and forming bonds to rearrange the atoms into new substances. In this chapter, you will look at *nuclear changes*, which change the particles that are in the nucleus, usually resulting in a different element.

DISCOVERING RADIOACTIVITY

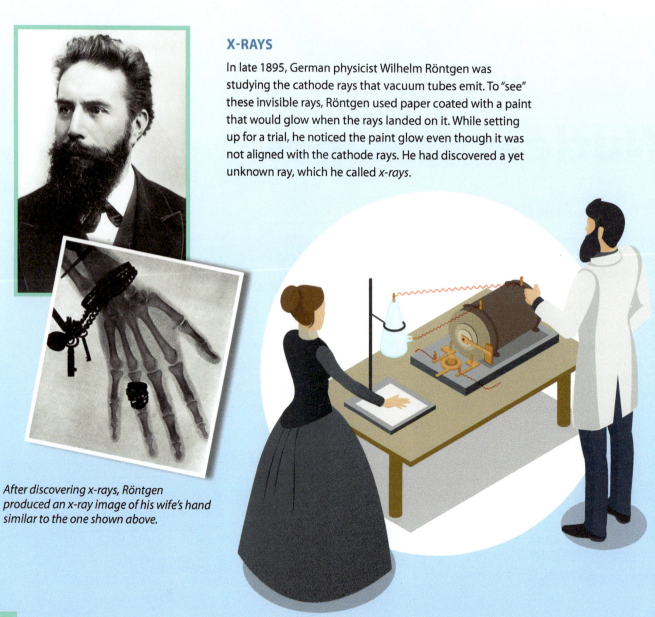

X-RAYS

In late 1895, German physicist Wilhelm Röntgen was studying the cathode rays that vacuum tubes emit. To "see" these invisible rays, Röntgen used paper coated with a paint that would glow when the rays landed on it. While setting up for a trial, he noticed the paint glow even though it was not aligned with the cathode rays. He had discovered a yet unknown ray, which he called *x-rays*.

After discovering x-rays, Röntgen produced an x-ray image of his wife's hand similar to the one shown above.

RADIOACTIVE DECAY

In 1896, French physicist Henri Becquerel was studying phosphorescence—the way that some materials emitted light of a certain color when exposed to sunlight. Becquerel wanted to know whether these materials would also emit x-rays when exposed to sunlight. To test his hypothesis, Becquerel exposed uranium salts to sunlight, wrapped them in black paper, and placed them on photographic paper. As expected, the paper showed signs that it had been exposed to x-rays. While waiting for a sunny day to do a second trial, he left the unexposed wrapped salts in a drawer with the photographic paper. Becquerel developed the paper even though he had not exposed the salts to sunlight. The paper showed the same x-ray exposure because the salts emitted x-rays without being exposed to the sun. Becquerel had discovered *radioactivity*—the spontaneous emission of particles and energy from an atom's nucleus. This is also known as *radioactive decay*.

Rutherford and Villard discovered the charges and relative penetrating abilities of alpha and beta particles and gamma rays.

Rutherford

Villard

TYPES OF DECAY

In 1899, Ernest Rutherford and Paul Villard continued the study of radioactivity. They identified three types of radioactive decay, which they named *alpha*, *beta*, and *gamma* for the first three letters of the Greek alphabet. They learned that alpha particles have a positive charge and beta particles negative. They also learned that alpha particles are much more massive than beta particles. Gamma rays were discovered to be emitted energy, having neither mass nor charge. They found that each type of radiation also has a unique ability to penetrate materials. Today we know of many other types of decay.

NUCLEAR CHANGES

8.2 NUCLEAR STABILITY

So why do some isotopes emit radiation while others don't? It's an issue of nuclear stability. In Chapter 5 you learned about chemical bonding—how atoms bond by sharing or transferring valence electrons to become chemically stable. But radioactivity is related to the stability of the nucleus.

Do you recall the particles found in the nucleus? All nuclei contain protons, and most contain neutrons. Remember that protons have a positive charge and that neutrons are neutral. Since like charges repel, we might expect that nuclei with more than one proton would break apart as the protons push each other away. But there seems to be an attractive force that scientists call the **strong force** that holds protons and neutrons together in nuclei. At very short distances, it is the strongest of four fundamental forces known to science (the others being gravity, the electromagnetic force, and the weak force). The strong force can hold a nucleus together if there is a proper mix of protons and neutrons. In unstable nuclei, the strong force isn't powerful enough to hold the nucleus together. The unstable nucleus changes or is said to decay, resulting in a nucleus with a lower, more stable, energy state.

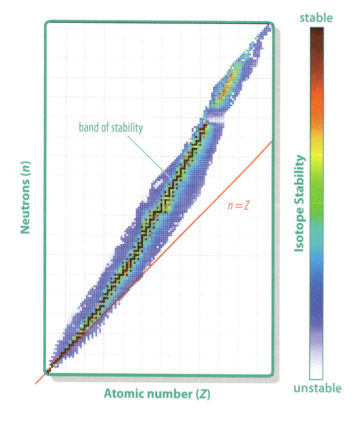

The graph at left shows all the known isotopes for the elements plotted by number of neutrons (n) versus number of protons (Z). Stable isotopes are shown with black data points and form the belt of stability—isotopes with the proper ratio of neutrons to protons that results in stable nuclei. Any isotope outside this belt will experience **radioactive decay**—the naturally occurring change of an unstable isotope to a more stable one as its nucleus emits particles, energy, or both. The red and green data points represent isotopes that are fairly stable, decaying very slowly. The blue data points show isotopes that are very unstable and that decay rapidly. The belt of stability eventually ends because all elements with atomic number 84 and above are unstable. The process of decaying moves isotopes closer to the belt of stability. Some need only one decay event to make them stable, while others need a series of decay events.

The stable isotopes are in black. All the other colors indicate unstable isotopes that undergo some form of radioactive decay. Isotopes farthest from the black isotopes (blues) are the most unstable. While there are many isotopes that are unstable, the majority of atoms exist as stable isotopes.

HOW IT WORKS

Smoke Detectors

Have you ever burned something in the oven? You turn the oven off, turn on a fan, and wave your arms around. You are trying to disperse the smoke before it sets off the smoke detector. Too late!

Working smoke detectors don't just notify us that the pizza is burnt. They lower the death rate in house fires by more than 50%. But how do these devices detect fire and sound the alarm? There are two basic designs—photoelectric and ionization devices. Some detectors combine these two technologies.

A photoelectric device contains a light source and sensor, which are set up so that the light is aimed away from the sensor. If smoke enters the device, the light reflects off the smoke particles. When enough light is reflected onto the sensor, the warning tone sounds. This design is most effective for smoky, smoldering fires.

An ionization device contains a small sample of americium-241, which undergoes alpha decay. The sample is in an ionization chamber that has a positive end and a negative end. As the isotope decays, emitted alpha particles collide with oxygen and nitrogen atoms in the chamber, knocking electrons from those atoms. These free electrons are attracted to the positive end of the chamber and the cations to the negative end. These charges produce a current in the detector circuit. If smoke enters the chamber, it blocks some of this current. The alarm sounds when the current drops below a set limit. This design is best at detecting flaming fires.

Smoke detectors save lives, but only if they are installed and working. Keep those batteries fresh!

NUCLEAR CHANGES

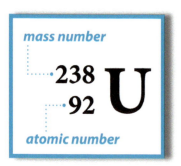

8.3 RADIOACTIVE DECAY

Chemists use equations similar to chemical equations to show the changes that occur during nuclear decay. The only way that these equations differ is that they use isotope notation instead of chemical symbols. Noting the isotope notation for uranium-238 (left) recall that the subscript shows the atomic number, while the superscript shows the mass number. As in chemical equations, the law of conservation of matter still applies to equations for nuclear reactions. Therefore the atomic numbers and mass numbers on each side of the equation must be equal to each other.

TYPES OF RADIOACTIVE DECAY

ALPHA DECAY

Alpha decay results in the emission of an *alpha particle*—a helium-4 nucleus—and some energy. An alpha particle is made of two protons and two neutrons, leaving it with a 2+ charge. The symbol for an alpha particle is ^4_2He, or α. This decay event lowers the mass number of the remaining nucleus by 4 and the atomic number by 2.

Uranium-238 emits an alpha particle and some energy as it moves toward a more stable element. Uranium-238 has to go through fourteen decay events to become stable lead-206. The first decay event in this *decay chain* is uranium-238 emitting an alpha particle to become thorium-234.

$$^{238}_{92}U \rightarrow {}^{234}_{90}Th + {}^4_2He$$

Notice how the mass numbers, 238 = 234 + 4, and the atomic numbers, 92 = 90 + 2, on each side are equal to each other.

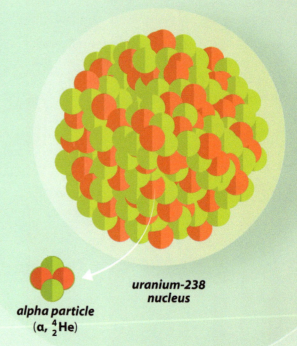

uranium-238 nucleus

alpha particle (α, ^4_2He)

electron (ß, $^0_{-1}e$)

BETA DECAY

Beta decay emits a high-energy electron—a *beta particle*—and some energy from the nucleus. It's produced when a neutron splits into a proton and an electron. The resulting nucleus has the same mass number as before, but its atomic number is greater by 1. The symbol for a beta particle is $^0_{-1}e$, or β.

In the decay chain of uranium-238, the second step is the emission of a beta particle from thorium-234 to form protactinium-234.

$$^{234}_{90}Th \rightarrow {}^{234}_{91}Pa + {}^0_{-1}e$$

Again, the sums of the mass numbers and atomic numbers on each side are equal to each other, 234 = 234 + 0 and 90 = 91 + (−1).

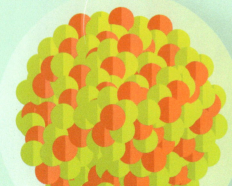

thorium-234 nucleus

EXAMPLE 8-1: Balancing Nuclear Decay Equations

Radon-222 goes through alpha decay to become more stable. Use the law of conservation of matter to write the balanced equation for this decay event.

Write the isotope notation for radon-222.

$$^{222}_{86}Rn$$

Since this is alpha decay, we will have an alpha particle along with our unknown element on the product side of our equation.

$$^{222}_{86}Rn \rightarrow {}^{4}_{2}He + {}^{A}_{Z}X$$

According to the law of conservation of matter, our mass numbers on each side of the equation must be equal. Therefore,

$$222 = 4 + A$$
$$222 - 4 = A$$
$$A = 218$$

Now we can enter the mass number of our unknown.

$$^{222}_{86}Rn \rightarrow {}^{4}_{2}He + {}^{218}_{Z}X$$

The law of conservation of matter also requires the atomic numbers on each side of the equation to be equal. Therefore,

$$86 = 2 + Z$$
$$86 - 2 = Z$$
$$Z = 84$$

Now we can enter the atomic number of our unknown.

$$^{222}_{86}Rn \rightarrow {}^{4}_{2}He + {}^{218}_{84}X$$

Knowing the atomic number of our element, we look up the symbol on the periodic table. Element number 84 is polonium (Po).

$$^{222}_{86}Rn \rightarrow {}^{4}_{2}He + {}^{218}_{84}Po$$

GAMMA DECAY

Gamma decay involves the emission of a *gamma ray*, a high-energy light wave, called a *photon*. There is no change in the composition of the nucleus, but the decrease in total energy results in a more stable nucleus.

$$^{60}_{28}Ni \rightarrow {}^{60}_{28}Ni + {}^{0}_{0}\gamma$$

nickel-60 nucleus

NUCLEAR CHANGES

Table 8-1

Half-Life	Fraction Remaining	Fraction Decayed
0	1	0
1	1/2	1/2
2	1/4	3/4
3	1/8	7/8
4	1/16	15/16
5	1/32	31/32
6	1/64	63/64
7	1/128	127/128

8.4 HALF-LIFE

Look back at the graph (page 178) showing the belt of stability. Recall that the isotopes farther from the stable isotopes are the most unstable. Each unstable isotope decays at its own consistent rate. The rate at which an isotope decays is called its **half-life**—the length of time needed for half of the atoms in a sample to decay. After one half-life, 50% of the atoms in a sample will have decayed (changed) to some other element, and the other 50% will remain unchanged. What do you think will happen after a second half-life? Will the rest of the atoms have decayed? Let's look at an example to see what happens.

UNDERSTANDING HALF-LIFE

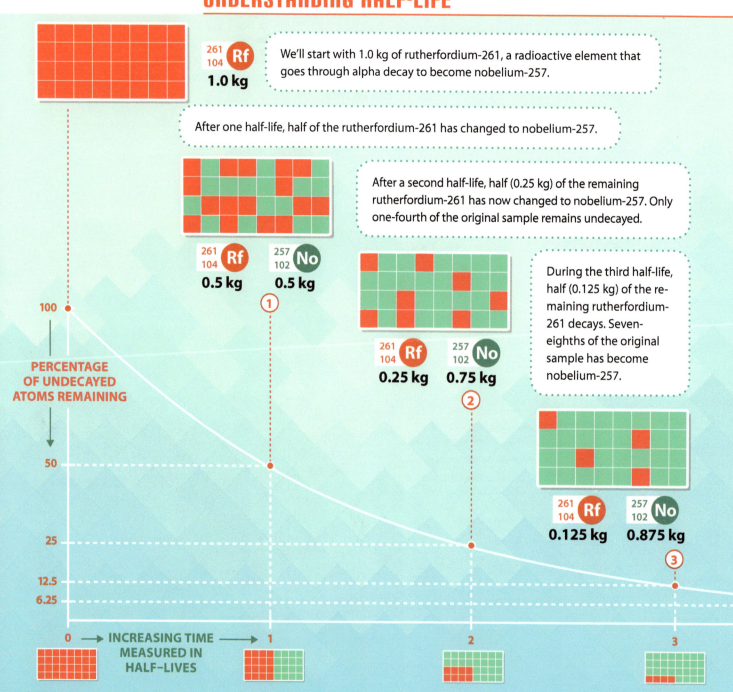

We'll start with 1.0 kg of rutherfordium-261, a radioactive element that goes through alpha decay to become nobelium-257.

After one half-life, half of the rutherfordium-261 has changed to nobelium-257.

After a second half-life, half (0.25 kg) of the remaining rutherfordium-261 has now changed to nobelium-257. Only one-fourth of the original sample remains undecayed.

During the third half-life, half (0.125 kg) of the remaining rutherfordium-261 decays. Seven-eighths of the original sample has become nobelium-257.

182 CHAPTER 8

CASE STUDY: VIKINGS

The Viking Age ranged from 793 to 1066, during which the Vikings explored, raided, traded, and even settled regions from the Mediterranean Sea to North America. They traveled great distances in their wooden longboats. A process called radiometric dating allows scientists to estimate the age of a longboat by comparing the current amount of a radioactive isotope with the amount originally in a sample.

Since the ships were made of wood, scientists use carbon-14 to estimate when the ships were built. Carbon-14 is a radioactive isotope found in the environment. Trees take in carbon-14 from the air. As long as a tree is alive, it keeps taking in carbon-14, maintaining its carbon-14 level equal to that in the environment. When a tree is cut down, it stops taking in carbon-14. A tree's carbon-14 content decreases as the carbon-14 decays into nitrogen-14. By measuring the amounts of carbon-14 and nitrogen-14, a scientist can determine the percent change from the original amount. Then using the half-life for carbon-14 (5730 y), the date when the ship was built can be estimated.

1. What is radiometric dating?
2. Why does the amount of carbon-14 decrease after a tree is cut down?
3. Radiometric dating is a controversial topic. Why do you think that is?

By the time a fourth half-life has passed, 15/16 of the rutherfordium-261 has decayed.

$^{261}_{104}$ **Rf** $^{257}_{102}$ **No**
0.0625 kg   0.9375 kg

During the fifth half-life, half of the remaining rutherfordium-261 decays, leaving only 1/32 of the rutherfordium along with 31/32 kg of nobelium-257.

$^{261}_{104}$ **Rf** $^{257}_{102}$ **No**
0.03125 kg   0.96875 kg

Notice that the number of red boxes representing rutherfordium-261 gets cut in half during each half-life. This reduction could go on for a long time. Eventually, you would get down to one atom. What do you think would happen during the half-life after that?

NUCLEAR CHANGES 183

The Understanding Half-Life infographic on the previous page spread shows an ideal example. In reality, the half-life is the time for half of the sample to *probably* decay. When the samples are large, the statistics predict fairly well what will happen for the entire sample. But we can never predict what will happen to any one particular atom. When you get down to a handful of atoms, about half of the sample will decay in a half-life, but it's not a sure thing. Each atom has a 50% probability of decaying during one half-life, sort of like flipping a coin. So when you get down to a single atom, you have a 50% chance that the atom will decay in one half-life.

EXAMPLE 8-2: Half-Life Problem

Rutherfordium-261 is a radioactive isotope that turns into nobelium-257 through alpha decay. It has a half-life of 78 s. Start with a 3.2 g sample.

$$^{261}_{104}Rf \rightarrow {}^{4}_{2}He + {}^{257}_{102}No$$

How long will it take until 0.40 g of rutherfordium-261 remains?

We start by determining what fraction of rutherfordium-261 remains.

We can see that $\frac{0.4 \text{ g}}{3.2 \text{ g}} = \frac{1}{8}$ of the original rutherfordium-261 remains. According to Table 8-1, 1/8 of the original substance will remain after three half-lives. Since each half-life is 78 s long, in 234 s (3 × 78 s) there would be 0.40 g of rutherfordium-261 remaining.

How much rutherfordium-261 will remain after 312 s?

First determine how many half-lives 312 s represents.

$$\frac{312 \text{ s}}{78 \text{ s}} = 4 \text{ half-lives}$$

Table 8-1 shows that after four half-lives, 1/16 of the sample remains.

$$3.2 \text{ g} \times \frac{1}{16} = 0.20 \text{ g}$$

Therefore 0.20 g of rutherfordium-261 remains after 312 s.

How much of the sample will have changed into nobelium-257 after 390 s?

$$\frac{390 \text{ s}}{78 \text{ s}} = 5 \text{ half-lives}$$

After five half-lives, 31/32 of the sample will be changed to the new isotope.

$$3.2 \text{ g} \times \frac{31}{32} = 3.1 \text{ g}$$

Therefore 3.1 g of rutherfordium-261 has changed in 390 s.

8A | REVIEW QUESTIONS

1. Define *radioactive decay*.
2. Identify three types of radioactive decay.
3. Complete the following equation. What type of decay is this?

    $$^{227}_{90}Th \rightarrow {}^{A}_{Z}X + {}^{4}_{2}He$$

4. Where does the beta particle (electron) come from in the decay event shown below?

    $$^{137}_{55}Cs \rightarrow {}^{137}_{56}Ba + {}^{0}_{-1}e$$

5. Define *half-life*.
6. How many grams of an 8.4 g sample of carbon-14 would remain after two half-lives?

8B | FISSION AND FUSION

Why is the sun so hot?

In Section 8A, you learned about radioactive decay, which happens when an unstable nucleus spontaneously emits particles, energy, or both to become more stable. As scientists studied these decays, they wondered whether it were possible to trigger such nuclear changes. They learned that by crashing a high-energy particle into the nucleus of an atom, they could cause the nucleus of one element to change into the nucleus of a new one, a process called *artificial transmutation*. Apparently, the alchemists had been on to something (see page 70), though they were trying to accomplish these changes through chemical reactions.

How do nuclear reactions compare with the chemical reactions that you learned about in Chapter 7? Chemical reactions involve only the valence electrons of the atoms. In those reactions, bonds are broken and formed to rearrange the atoms as new substances. Nuclear reactions involve the nucleus, including both the protons and neutrons. Nuclear changes also produce more energy than chemical reactions produce. There are two types of nuclear reactions—fission and fusion.

8B Questions

- How do fission and fusion compare?
- Where does the missing mass go in a nuclear reaction?

8B Terms

fission, chain reaction, critical mass, fusion

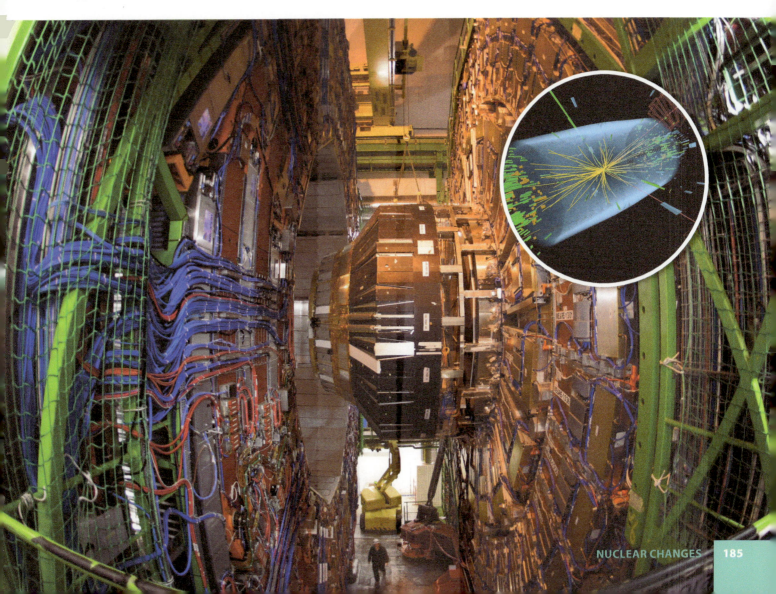

The image below shows a section of the Large Hadron Collider (LHC). Notice the man walking through the collider.
Inset: This image shows the results of a particle collision like those done by the LHC.

NUCLEAR CHANGES 185

8.5 FISSION

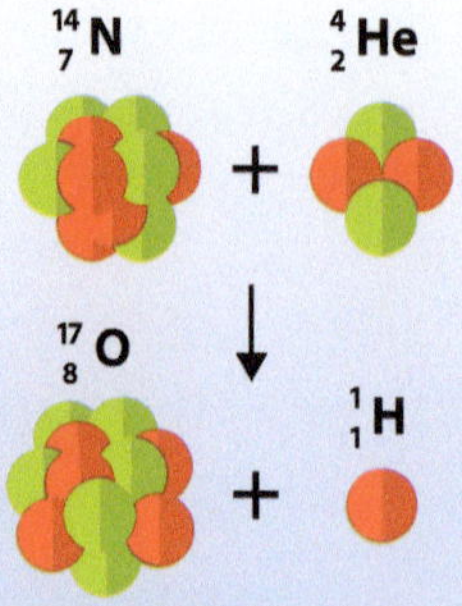

In 1925, Patrick Blackett caused the first artificial transmutation. He shot alpha particles into nitrogen-14 nuclei. Each effective collision produced an oxygen-17 nucleus and a hydrogen atom. In the 1930s, German scientists Otto Hahn and Fritz Strassmann used this process in an attempt to make new, heavier-than-uranium elements. By crashing high-energy neutrons into uranium atoms, they expected to create atoms with atomic numbers greater than 92. When they instead made barium (atomic number 56), they were shocked. They asked others to confirm their results. After Otto Frisch and Lise Meitner did just that, Hahn and Strassman realized that they had discovered a nuclear reaction, which they named *fission*. They chose this name because this process was similar to the fission of cells. Nuclear **fission** is a nuclear reaction in which a large nucleus is split into smaller nuclei by bombardment with a high-energy particle. Fission reactions release huge amounts of energy, which is why we use them in power plants and nuclear weapons.

$$^{14}_{7}N + {}^{4}_{2}He \rightarrow {}^{17}_{8}O + {}^{1}_{1}H$$

UNDERSTANDING FISSION

REACTIONS

The most common fuels for fission reactions are uranium-235 and plutonium-239. When uranium-235 goes through a fission reaction, it can produce different products. Below are two of the possible equations for the fission of uranium-235. The $^{1}_{0}n$ symbol represents the neutron that scientists use to start the fission reaction.

$$^{235}_{92}U + {}^{1}_{0}n \rightarrow {}^{137}_{52}Te + {}^{97}_{40}Zr + 2{}^{1}_{0}n$$

$$^{235}_{92}U + {}^{1}_{0}n \rightarrow {}^{142}_{56}Ba + {}^{91}_{36}Kr + 3{}^{1}_{0}n$$

CHAIN REACTION

The reactions shown at right are two that can occur when an atom of uranium-235 undergoes fission. Notice that in both cases more neutrons result than went into the reaction. The neutrons produced can start other fission reactions.

Fission processes, such as these, in which the neutrons produced trigger more fission events are called **chain reactions** (see below).

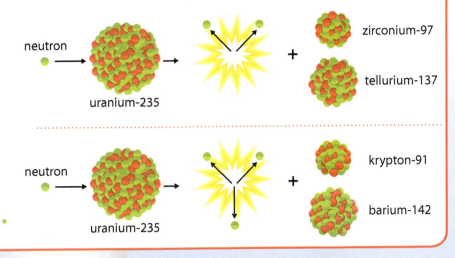

Water-cooling towers like these at the Grohnde nuclear power plant in Germany are the feature that most people identify with nuclear power plants. But it is the domed building on the left of the complex that houses the reactor.

CRITICAL MASS

To maintain a chain reaction, there must be the right amount of fuel. If there is not enough fuel, called *subcritical mass*, or if the fuel is spread out too much, then the neutrons can escape without starting other reactions. The smallest mass of fissionable material that can sustain a chain reaction is called the **critical mass**. Having more fissionable material than the critical mass—a *supercritical mass*—results in too many reactions being started and control is lost, as occurs in a nuclear weapon.

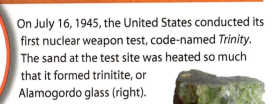

On July 16, 1945, the United States conducted its first nuclear weapon test, code-named *Trinity*. The sand at the test site was heated so much that it formed trinitite, or Alamogordo glass (right).

NUCLEAR CHANGES 187

8.6 FUSION

While fission reactions can be helpful for producing energy, life would not exist on Earth without nuclear fusion. We depend on fusion because our sun produces energy by fusion reactions. Every second, our sun emits 3.8×10^{26} J of energy—enough energy to power the United States for 3.7 million years! While only a small portion of this energy reaches us, it provides almost all the energy ultimately used by living things.

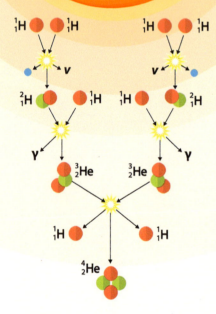

Fusion is a nuclear reaction in which small nuclei combine to form a more massive nucleus. These reactions occur only under extreme pressures and temperatures. Fusion in the sun occurs within the core, where the temperature is about 15.7 million K. Four hydrogen nuclei in our sun fuse together in a three-step process to become a helium nucleus. The equation for this is $4\,^1_1H \rightarrow\,^4_2He + 2\,^0_1e$ and takes place as a series of three steps:

$$^1_1H + ^1_1H \rightarrow\ ^2_1H + ^0_1e \text{ (occurs twice)}$$

$$^1_1H + ^2_1H \rightarrow\ ^3_2He \text{ (occurs twice)}$$

$$^3_2He + ^3_2He \rightarrow\ ^4_2He + 2\,^1_1H \text{ (occurs once)}$$

In the above fusion equations, the ^0_1e symbols look like beta particles, but they are actually positrons, which is why the atomic number is positive and not negative. Scientists would love to be able to produce energy through nuclear fusion, and many organizations are working to make it a reality.

8.7 NUCLEAR ENERGY

Every nuclear change involves the release of energy. As isotopes decay, they emit particles and energy to become more stable. As a nucleus splits in a fission reaction, it yields two smaller nuclei and a large amount of energy. Fission of uranium-235 produces 2.7 million times more energy than an equal mass of anthracite coal (the most energy-dense coal). Recall from our discussion about the sun that the fusing of small nuclei releases even larger amounts of energy. Fusion produces three to four times the energy that fission produces. But where does this energy come from?

In each of these cases, the total mass of the products is less than the mass of what went into the reaction. At first glance, this looks like a violation of the law of conservation of matter. Where did the missing mass go? Albert Einstein answered this question with his famous equation, $E = mc^2$. This equation states that mass and energy are interchangeable forms of the same thing. Remember that the law of conservation of matter states that matter cannot be created or destroyed but only transformed. In nuclear decays and reactions, the "missing mass" changes into energy.

$$E = mc^2$$

energy = mass × speed of light²

CASE STUDY: TSAR BOMBA

In August of 1945, many people were celebrating around the world. The United States had dropped two fission bombs and brought World War II to an end. The world was at peace—or was it? As soon as the war ended, the Soviet Union decided that it also wanted nuclear weapons. In 1949, the Soviet Union tested its first nuclear weapon and entered into a cold war with its former ally. Each country worked to build larger weapons and stay ahead of the other.

The United States tested many devices on islands in the South Pacific Ocean. On March 1, 1954, the United States tested its largest nuclear weapon in a test code-named *Castle Bravo*. Instead of the planned 5 Mt (five-megaton) yield, the explosive equivalent of 5 million tons of TNT, the blast had an actual yield of 15 Mt. This error convinced scientists in the United States that they didn't need to build any larger weapons.

Castle Bravo inspired the Soviets. The Soviet leader, Nikita Khrushchev, challenged his scientists to show the Americans what the Soviets could do. The result, a fusion weapon code-named *Vanya* but nicknamed *Tsar Bomba* in the West, produced the largest man-made explosion ever. The Soviets designed it as a three-stage, 100 Mt device. Stage 1 was a fission reaction using uranium-238. This reaction would start Stage 2, a fusion reaction of hydrogen. This second stage then started a larger fusion reaction, Stage 3. The test used only the first two stages, which made it a 50 Mt weapon. Even this scaled-down version yielded more than 1500 times the explosive force of either of the two bombs that ended World War II.

1. Why do you think that this period of history was called the *Cold War*?
2. The bombs dropped that ended World War II were fission devices. What does this mean?
3. What enabled the Tsar Bomba to have so much more power than the fission bombs?

MINI LAB

MODELING CHAIN REACTIONS

Essential Question:

How can we model fission reactions with dominoes?

Equipment
dominoes (15)
stopwatch
centimeter ruler

We don't usually think about how our electricity gets to us or how it is produced. Did you know that about 20% of our power is generated by nuclear power plants? These plants produce energy by fission. The process begins by smashing high-energy neutrons into the fuel, usually uranium-235 or plutonium-239. Each reaction produces two or more high-energy neutrons that can start more fission events. The key to keeping control of the reactions in the power plant is to control the number of new high-energy neutrons that are available.

1. What is a chain reaction?
2. What is the term for the smallest amount of matter needed to sustain a fission chain reaction?

Procedure

A Standing the dominoes on their narrow end, arrange them in a line spaced 3.0 cm apart.

B Time how long it takes for all 15 dominoes to fall when you push over the first domino.

3. What does a domino represent?
4. What does the domino's falling represent?

C Repeat Steps **A** and **B** but with the dominoes set 4.0 cm apart.

5. Did all 15 dominoes fall? How did the time change?

D Repeat Steps **A** and **B** but with the dominoes arranged so that each domino will fall against two other dominoes.

6. Did all 15 dominoes fall? How did the time change?

Conclusion

7. How does this exercise model a chain reaction?

Going Further

E Rearrange your "atoms" once again like Step **D**. Using the ruler, can you interrupt part of the chain reaction?

8. What does Step **E** have to do with nuclear reactors?

8B | REVIEW QUESTIONS

1. Define *fission*.
2. Explain what will happen if there is a subcritical mass for a fission reaction.
3. Define *fusion*.
4. What is the nuclear fuel for the fusion within the sun?
5. Complete the equation for the fusion of helium-4 nuclei, shown below.

$$^{4}_{2}He + ^{4}_{2}He \rightarrow ^{A}_{Z}X$$

6. What conditions are required for fusion to occur?
7. Where does the missing mass go in a nuclear reaction?

8C | NUCLEAR CHANGES: BENEFITS AND RISKS

What are the risks and benefits of nuclear changes?

You may not know it, but radiation, both natural and manmade, surrounds us all the time. Sources in nature include soil, rocks, food, water, the atmosphere, and even space. Most of this radiation is low level and safe. We are exposed to manmade radiation from nuclear medicine, medical procedures, and electronic devices. These manmade sources produce radiation that is typically more energetic and often more harmful. We need to balance the helpful aspects of using nuclear processes against the harmful effects of being exposed to too much radiation.

8C Questions
- How does radioactivity affect us?
- How can we protect ourselves from radiation?
- What are some benefits of radiation?

8C Terms
radiotracer, somatic damage, genetic damage

8.8 USES OF RADIATION

We've already seen many uses for radiation, such as protecting food supplies, producing power, detecting fires, and estimating the age of historical objects. Radiation also has other uses, such as in medicine and security.

USING RADIATION

IMAGING

In the past, when doctors treated certain diseases, the only way to look inside our bodies was through surgery, and at times this was more dangerous than the disease itself. With the discovery of radiation, scientists devised new ways for doctors to "look" inside the body. X-rays pass easily through muscle and fat, but more of the rays are absorbed by bones. This allows doctors to see through the skin, muscle, fat, and tendons to see the bones within. Within a year of their discovery, x-rays were being used by doctors.

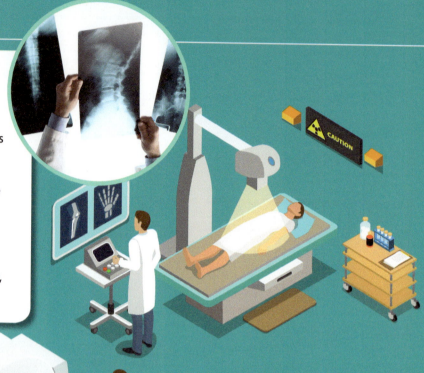

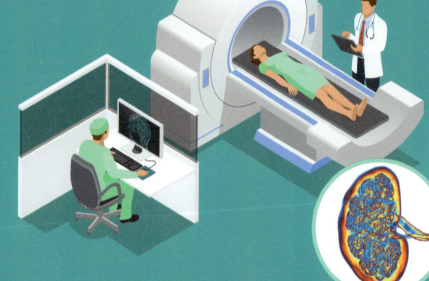

Computed tomography (CT) machines improve on the x-ray concept. CT machines also use x-rays but take millions of images, which a computer turns into cross-sectional images of the body. Shown (left) is a CT scan of a kidney.

NUCLEAR MEDICINE

Nuclear medicine also uses radioactive materials to treat disease, especially cancer. Again, the doctor chooses an isotope that will concentrate in the organ or system that he is treating. These isotopes are often chosen to produce damaging radiation in the hopes of killing the diseased (cancer) cells in the organ.

RADIOTRACERS

Radioactive isotopes help doctors diagnose many illnesses. Doctors inject these isotopes, called **radiotracers**, to see how they move through or collect in a certain organ or system. The PET scan (right) shows how fluorine-18 has collected in a brain. The red shows large amounts of the fluorine. When selecting radiotracers, doctors must consider the part of the body they are studying. They also need to select isotopes with fairly short half-lives, long enough to do the testing but without exposing the patient to more radiation than needed. The doctor's goal is to treat the patient while minimizing the harmful effects.

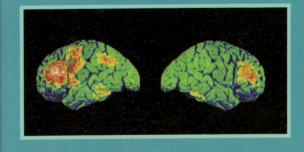

Medical Tracers

Isotope	Half-Life	Target
carbon-11	20 min	brain, thyroid
carbon-14	5730 y	pancreas
cobalt-57	272 d	intestines
fluorine-18	110 min	bone
iodine-123	13 h	thyroid
krypton-81m	13 s	lungs
oxygen-15	122 s	brain, heart
technetium-99m	6.0 h	numerous

SECURITY

Before air passengers enter secure areas of an airport, security personnel use x-ray machines to scan them and their luggage. They are able to detect weapons, bombs, and other dangerous objects. Advanced systems use the same CT technology that is used in hospitals. Law enforcement agencies also use different x-ray systems to scan for weapons, drugs, and even people being smuggled into the country. Below, backscatter x-rays show thirty-seven people hidden inside a truck full of bananas.

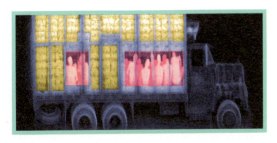

NUCLEAR CHANGES

Other ways that we benefit from radiation include promoting food safety, as noted in the Chapter opener, and in scientific and industrial applications, which you can study in the worldview sleuthing activity on page 196. These benefits arose because scientists did the pure, investigative science needed to learn new things about God's creation.

8.9 EFFECTS OF RADIATION

The discovery of x-rays in November 1895 had doctors excited about the possibilities. Everyone was experimenting to see what could be done with them. Then in February 1896, Vanderbilt University researchers reported the first *injuries* from the new rays. Other reports flooded in, and doctors realized that while radiation can be useful, many forms of radiation were also harmful.

The degree to which radiation can damage cells depends on how it affects them. As some radiation enters the cells, it can knock electrons out of the atoms and molecules, creating ions. This type of radiation is called *ionizing radiation*. Other forms of radiation, such as radio waves, microwaves, and visible light, do not have enough energy to turn atoms into ions and are less dangerous. We must be careful not to expose people to too much ionizing radiation while using it to help improve their lives.

The amount of damage that ionizing radiation does depends on certain factors. How deep does radiation go into the person? How massive is the particle? How much energy does the radiation have? How much radiation is the person exposed to? How long was he exposed?

Radiation can do two types of damage. Damage to cells that are not involved in reproduction is called **somatic damage**. This type of damage harms the organism but can't be passed on to offspring. Somatic damage causes injury and illness, such as cancer. Damage to the DNA is called **genetic damage**. This damage will affect the reproduction and growth of new cells. It may also be passed to offspring if the damaged DNA is in a reproductive cell. To limit the extent of damage, we must be able to detect radiation and then take steps to protect people from exposure.

8.10 DETECTING RADIATION

Detection of radiation allows us to know when we need to protect ourselves from that radiation. There are a number of devices for detecting radiation. Some tell us only whether there is radiation present, while others tell us whether there is much or little present. Still other devices specifically indicate the amount of radiation present.

Somatic damage is radiation damage to any non-reproductive cell. It may be visible, as in a sunburn, or it may be completely unseen.

While alpha particles can be very damaging, they are also stopped by thin layers like the skin. But they are very dangerous if they are swallowed or inhaled.

DETECTING RADIATION

GEIGER COUNTER

One device used for detecting radiation is the *Geiger counter*. This device consists of a closed tube containing inert gas molecules that connects to an indicating circuit and a counter. As ionizing radiation enters the tube, it knocks electrons from the atoms of gas in the tube. The electrons are attracted to a positively charged wire that runs the length of the tube. As these electrons enter the wire, they produce a current, which produces a click that can be heard and is displayed on the counter. The current produced indicates the amount of radiation present. Geiger counters are limited—they can't identify the type of radiation, and they also work only for low levels of radiation.

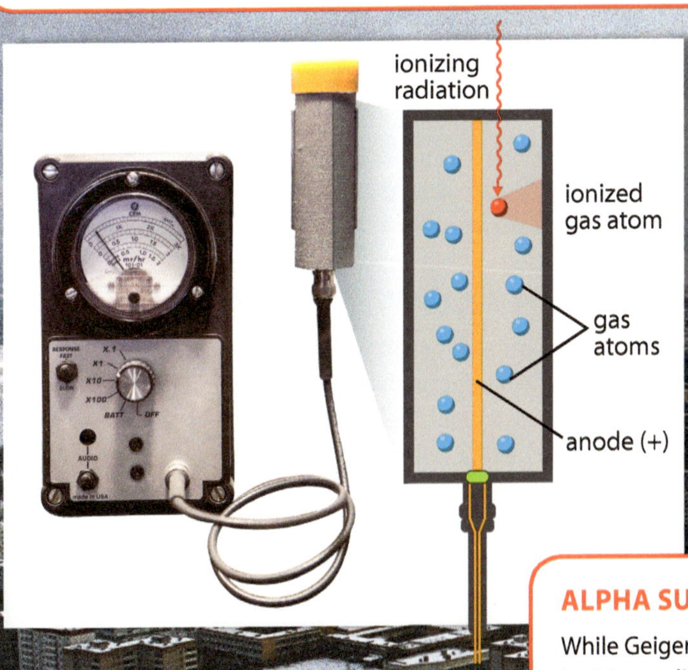

Radiation levels in the city of Chernobyl (background image) still remain high following the 1986 meltdown of its nuclear power plant. Workers in the area wear dosimeters to monitor their radiation exposure.

ionizing radiation

ionized gas atom

gas atoms

anode (+)

ALPHA SURVEY METERS

While Geiger counters measure the amount of all ionizing radiation, *alpha radiation survey meters* are designed to measure the presence and amount of alpha particles being emitted. Some can even identify the isotope that produced the radiation.

DOSIMETERS

Some people work in environments that routinely have higher levels of radiation than the normal environment does. Shown at right is a personal dosimeter, which a worker wears somewhere on his body. The devices collect data on exposure to radiation. Some dosimeters contain a readout that allows the user to monitor his exposure in real time. Other dosimeters must be analyzed by a lab and don't provide real-time data.

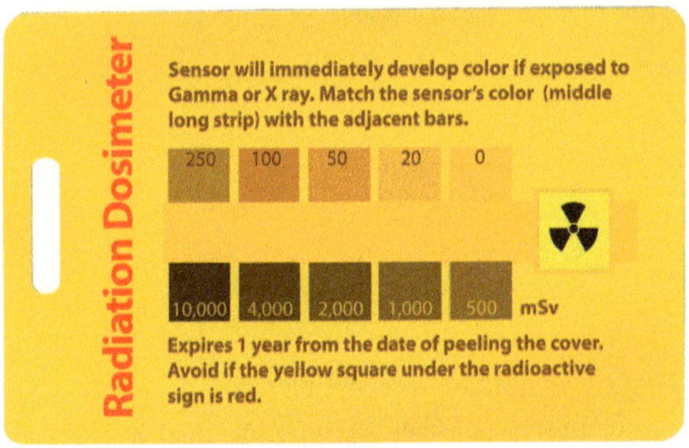

Radiation Dosimeter

Sensor will immediately develop color if exposed to Gamma or X ray. Match the sensor's color (middle long strip) with the adjacent bars.

| 250 | 100 | 50 | 20 | 0 |

| 10,000 | 4,000 | 2,000 | 1,000 | 500 | mSv |

Expires 1 year from the date of peeling the cover. Avoid if the yellow square under the radioactive sign is red.

WORLDVIEW SLEUTHING: NUCLEAR WASTE

There is a give-and-take related to many things in our lives. The use of radioactive materials is one of those things. While there are many beneficial uses of radiation, we must limit the harm to people and the environment. Most applications of radioactive materials generate waste that remains radioactive. What do we do with this material?

TASK
You are a member of the city planning board. A company has proposed building a new facility in the city to process, store, and dispose of nuclear waste. This facility will create many jobs and boost the local economy. But the company will be transporting, processing, storing, and disposing of large amounts of radioactive waste. You have volunteered to report on the issues related to radioactive wastes and have agreed to give a five-minute presentation at the next board meeting before the vote regarding this proposal.

PROCEDURE
1. Research the issue by doing a keyword search for "radioactive waste management," "radioactive waste storage," "radioactive waste hazards," and "nuclear waste disposal."
2. Plan your presentation and collect any required materials. Remember to cite your sources.
3. Show your presentation to a classmate or friend for feedback.
4. Present your findings to your classmates.

8.11 RADIATION AND WORLDVIEW

The study of nuclear chemistry began as a pursuit of pure science, seeking scientific knowledge for its own sake. Often people think that science glorifies God only when it helps others, but even pure science glorifies God. Many of the technologies that help people were made possible because of discoveries in pure science. In most cases, we protect ourselves from radiation by avoiding exposure to it. In Chapter 3 you had the opportunity to investigate radon-222 exposure in homes. Homeowners that identify issues with radon in their home can install a system to remove the radon. By doing this, they avoid exposure. Think about getting x-rays at the dentist. As a patient, you sit there because it's the only way to get images of your teeth, but your dental hygienist leaves the room. For a patient, the once- or twice-a-year exposure to x-rays during dental exams is acceptable, but a hygienist would be exposed to too much radiation if she stayed in the room for every x-ray. Both our study of and use of radioactivity fulfill the Creation Mandate by glorifying God and helping others.

8C | REVIEW QUESTIONS

1. Name three natural sources of radiation.
2. Briefly describe two applications of radiation in medical technology.
3. List three ways outside of the medical field that radiation is used to benefit people.
4. What form of radiation is especially dangerous to living cells?
5. What type of radiation damage has the potential to be passed on to an organism's offspring?
6. Why is it beneficial to detect radiation?
7. Why are dental patients covered with a heavy, lead-lined blanket when getting x-rays?

CHAPTER 8 REVIEW

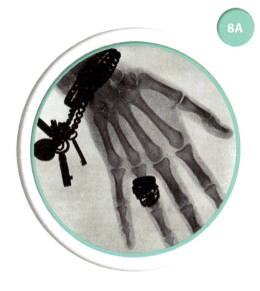

8A RADIOACTIVE DECAY

- Radioactivity was accidentally discovered when uranium salts were observed to emit radiation spontaneously.
- The ratio of neutrons to protons in the nucleus determines whether isotopes are stable or unstable.
- Unstable isotopes move toward greater stability by emitting particles, energy, or both, a change known as radioactive decay.
- Three forms of radioactive decay are alpha, beta, and gamma.
- Each isotope decays at a specific rate. The time needed for half of a sample of radioactive material to decay is called its half-life.
- Scientists estimate the age of objects by using the half-life of radioactive elements.

8A Terms

strong force	178
radioactive decay	178
alpha decay	180
beta decay	180
gamma decay	181
half-life	182

8B FISSION AND FUSION

- Scientists can trigger changes within the nucleus through artificial transmutations.
- Fission is a nuclear reaction in which a large nucleus splits into two or more smaller nuclei.
- Nuclear power plants rely on products of one reaction to start additional reactions, a process known as a chain reaction.
- A chain reaction needs a minimum amount of radioactive fuel—the critical mass—to sustain the reaction.
- Fusion is a nuclear reaction in which two small nuclei combine to form a larger nucleus.
- The energy in a nuclear reaction comes from converting some of the mass into energy.
- Fission produces over a million times the energy of fossil fuels. Fusion produces three to four times the energy of fission reactions.

8B Terms

fission	186
chain reaction	187
critical mass	187
fusion	188

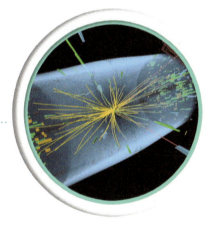

8C NUCLEAR CHANGES: BENEFITS AND RISKS

- Scientists have discovered a number of beneficial applications for radiation, including medical, food safety, security, and power production uses.
- Radiation can cause damage to cells, including both genetic and somatic damage.
- We must balance the benefits of using radiation with the risks of overexposure.
- Studying radiation has led to many technologies that have benefited people.

8C Terms

radiotracer	193
somatic damage	194
genetic damage	194

NUCLEAR CHANGES 197

CHAPTER 8

Recalling Facts

1. Summarize the discoveries that led to the discovery of radioactivity.
2. What happens to an atom during radioactive decay?
3. Why are some isotopes stable while other isotopes are unstable and experience radioactive decay?
4. Which type of radioactive decay does not result in a new element?
5. What happens to a sample of an isotope during one half-life?
6. Why were scientists surprised that they produced barium after crashing a neutron into a uranium atom?
7. What is a chain reaction?
8. In what application of fission would supercritical mass be desired? Explain.
9. What does it mean for a fissionable material to have subcritical mass?
10. How does fusion impact our lives on Earth?
11. Name a manmade source of radiation.
12. How does CT technology differ from x-rays?
13. How are radiotracers used in medicine?

Understanding Concepts

14. Compare physical, chemical, and nuclear changes.
15. Why is it surprising that protons remain close together in the nucleus? What force is believed to hold these protons close to each other?
16. Complete the following equation of a decay event.

    $$^{210}_{84}\text{Po} \rightarrow\ ^{A}_{Z}X + ^{4}_{2}\text{He}$$

17. Using a graphic organizer, summarize what we know about alpha, beta, and gamma decay. Include the symbol, charge, mass, and penetrating power.
18. Can we predict when a single atom of a radioactive isotope will decay? Explain.
19. A 179.2 g sample of silicon-31, which decays into phosphorus-31 (half-life: 2.6 h), has decayed, leaving 5.6 g of silicon-31. How much time has passed?
20. How much of 11.2 g of iodine-135 (half-life: 6.6 h) would remain after 19.8 hours?
21. How do alpha and beta decay differ from artificial transmutations?
22. Complete the equation for the fission of plutonium-239.

    $$^{239}_{94}\text{Pu} + ^{1}_{0}n \rightarrow\ ^{134}_{54}\text{Xe} + ^{A}_{Z}X + 3^{1}_{0}n$$

23. Compare subcritical, critical, and supercritical mass.
24. Compare nuclear fission and fusion.
25. Create a concept map using the terms *radioactive decay*, *stable nucleus*, *unstable nucleus*, *fusion*, *fission*, *critical mass*, and *chain reaction*.
26. Explain why nuclear reactions don't violate the law of conservation of matter.
27. Compare somatic and genetic damage.
28. Respond to the following statement: "The purely scientific study of radiation is not a valuable use of time because it doesn't help others."

REVIEW

Critical Thinking

29. Would you expect an unstable isotope that was above atomic number 83 to emit an alpha particle (^4_2He) or a beta particle ($^{\ 0}_{-1}e$) to move toward the belt of stability?

30. Would you expect an isotope that undergoes beta decay to be above or below the belt of stability?

31. A 10.0 g sample of radioactive material has decayed to 2.0 g after 6 h. Estimate the half-life of the material.

32. If a neutron collided with a plutonium-239 nucleus and produced cerium-140 and krypton-97, how many neutrons would also be produced? Write the nuclear equation that supports your answer.

33. Why does a dental hygienist ask her female patient whether she is pregnant before taking x-rays?

ETHiCS: NUCLEAR POWER GENERATION

THE ISSUE: CLEAN ENERGY?

Nuclear reactions can produce large amounts of energy without generating any air pollution. Some people feel that nuclear power can meet our growing energy needs and reduce air pollution, including greenhouse gases. Others fear that the possibility of a mishap, like at Fukushima in Japan or Chernobyl in Ukraine, makes the use of nuclear power too risky a proposal. A third group likes the reduction in air pollution but wonders what we will do with the leftover radioactive material.

34. What information can you find about this issue?
35. What does the Bible say about this issue?
36. What are some of the acceptable and unacceptable options?
37. What might be the consequences of adopting the acceptable and unacceptable options that you have listed?
38. What are the motivations for the acceptable options?
39. Write a one-page position paper indicating whether you believe that a Christian should support an initiative to develop more energy from nuclear sources.

Use the rubric below to help you determine whether you have completed all the necessary tasks for writing your response.

Self-Assessment Rubric

	Task Completed	Task Partially Completed	Task Not Completed
Additional Information	two or more sources	just one source	none
Biblical Principles	two or more passages	just one passage	none
Acceptable and Unacceptable Options	both	only one	neither
Consequences	yes	—	no
Biblical Outcomes	all three	only one or two	none
Biblical Motivations	all three	only one or two	none
Opinion Statement	yes	—	no

NUCLEAR CHANGES

CHAPTER 9
Solutions

FROTHY FOUNTAIN OF FOAM

You reach into the refrigerator and grab an ice-cold bottle of soda. As you unscrew the top, you hear the hiss and see bubbles forming inside the bottle. But every so often something goes wrong: the hiss is louder and the bubbles form rapidly, spraying soda everywhere. Why did that happen? Most of us have seen someone add Mentos™ to soda, creating spraying soda to the extreme.

Soda is a mixture of water, sweetener, and carbon dioxide. Normally when we open a bottle, the carbon dioxide slowly comes out of solution. In the occasional mishap or in the Mentos and soda demonstration, the gas comes out of solution rapidly. We will learn through this chapter the what and how of solutions. You'll see how mixtures and solutions are a vital part of understanding and using the world that God has given us.

9A	Mixtures and Solutions	202
9B	Solution Concentration	212

9A Questions

- What are suspensions and colloids?
- What are solutions?
- What materials can we use to make solutions?
- How do solutions form?
- How can I separate a mixture?

9A Terms

suspension, colloid, solution, solute, solvent, alloy, dissociation, solubility

9A | MIXTURES AND SOLUTIONS

How do things dissolve?

In Chapter 2 we learned about matter. We learned about pure substances, which can be elements or compounds, and mixtures, which are either heterogeneous or homogeneous. In this chapter we will look more closely at mixtures, especially homogeneous mixtures. Recall that a mixture is two or more substances that are physically combined in a changeable ratio. Mixtures can form from substances that are in the same state of matter (e.g., liquid in liquid or gas in gas) or in different states (e.g., gas in liquid or solid in liquid). Some common examples of mixtures include trail mix and tossed salad (heterogeneous) as well as salt water and tea (homogeneous).

9.1 HETEROGENEOUS MIXTURES

Recall from Chapter 2 that heterogeneous mixtures do not have a uniform appearance since the substances are unevenly distributed—the materials are not evenly spread out. In a trail mix, you can point to areas in the mixture that are only raisins or only peanuts. Italian salad dressing is another great example since the seasonings can be seen floating around in the mixture. Scientists classify heterogeneous mixtures as suspensions or colloids according to particle size.

TYPES OF HETEROGENEOUS MIXTURES

SUSPENSIONS

A **suspension** is a heterogeneous mixture of a fluid and large particles that will *settle out*—separate naturally—over time. The particles are typically larger than 1 μm in diameter. Human hair has a diameter of 10 to 200 μm.

An aerosol spray is a liquid suspended in a gas.

The foam on your root beer is a gas suspended in a liquid.

Mud consists of a solid suspended in a liquid.

Airborne dust is a solid suspended in a gas.

Oil and vinegar salad dressing is a liquid suspended in a liquid.

COLLOIDS

A **colloid** is a heterogeneous mixture in which the particles are between 1 nm and 1 μm in diameter, which means that they are too small to settle and thus remain dispersed. (Note that a helium atom has a diameter of about 0.1 nm.) The particles in colloids are so small that they are difficult to see. Scientists use the Tyndall effect to identify colloids. The *Tyndall effect* is the scattering of light by the particles in the colloid. In the image below, the light from a laser passes through a homogeneous mixture on the left without the beam being seen but is scattered by the dispersed particles in the colloid on the right.

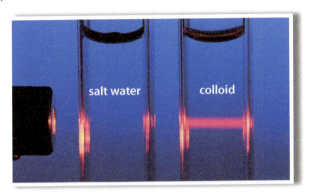

Gases are dispersed in a liquid to make whipped cream.

Mayonnaise is liquid oil dispersed in vinegar.

Paints are liquids with dispersed solid particles.

A marshmallow's soft consistency is due to air dispersed in a solid.

Opals are gemstones in which water is dispersed in an amorphous silicate solid.

Cranberry glass consists of dispersed gold nanoparticles in glass (solid).

Liquid water can be dispersed in air (gas) as fog or clouds.

Smoke and exhaust contain solid particulate matter dispersed in air.

SOLUTIONS 203

9.2 HOMOGENEOUS MIXTURES

Recall from Chapter 2 that homogeneous mixtures, also known as **solutions**, are mixtures with a uniform appearance throughout. This uniform appearance occurs because the dissolved particles are smaller than the particles in colloids and are uniformly distributed throughout the mixture. The particles are so small that we can't see them and they don't scatter light. The particles are said to have dissolved into the mixture. The material that dissolves in a solution is called the **solute**. The substance into which the solute dissolves is the **solvent**. In some cases these definitions don't work well, such as mixtures of two gases, two liquids, or two solids. In these cases the solvent is the material in greater amount, while the solute is the material in lesser amount.

Salt water is a familiar example of a solution. We make it by dissolving salt (the solute) in water (the solvent) to form the saltwater solution. They are physically combined but not chemically. The sodium and chloride ions are uniformly and completely distributed among the water molecules.

When people think about solutions, they often think of things like salt water and sugar water. These are considered liquid solutions because the solvent is liquid. Solutions can also have either gases or solids as the solvent. Earth's atmosphere, for example, is a solution. Nitrogen gas is the solvent, and other gases such as oxygen and argon are dissolved in it.

WORLDVIEW SLEUTHING: SPORTS DRINKS

Watch a sporting event and you will see sports drinks. They are on the sidelines. They are in advertisements. You can find shelves and shelves of these drinks in the supermarket. Manufacturers tell you that they will help you before, during, and after the event.

TASK

You are the newly elected student body president at your school. A group of students has asked you to convince the administration that having sports drinks available throughout the school day will help students. They tell you that they will help improve the performance of student athletes. They also tell you that the drinks will help students pay better attention in class and therefore do better in their studies. Your principal tells you to do some research and come back with a one-page justification for offering sports drinks in school. She also said that you have to be able to defend your findings.

PROCEDURE

1. Research the issue by doing keyword searches for "sports drinks," "do you really need sports drinks," and "what do sports drinks do."

2. Plan, organize, and write your paper. Support each claim in your paper with specific evidence. Remember to cite your sources.

3. Show your paper to a classmate or friend for feedback.

4. Turn in your paper.

CONCLUSION

Do sports drinks do what they claim? Do they benefit all people or only certain groups? When in a leadership position, we want to meet the needs of those we represent. The challenge can be in recognizing true needs, not mere wants.

Alloys are another very common type of solution. An **alloy** is a solid solution made of two or more elements, at least one of which is a metal. Bronze (tin dissolved in copper) and steel (carbon dissolved in iron) are common alloys. Let's look at some other examples of solutions.

SOLUTIONS

Solutions are homogeneous mixtures with particles that are tiny and evenly distributed so that they have a uniform appearance throughout. Solutions can be made with solutes and solvents in any state of matter.

Pewter is a solution of tin with other metals (e.g., copper, lead, antimony) dissolved in it.

Palladium hydride is a solid solution of solid palladium with metallic hydrogen dissolved in it.

Amalgam is a liquid solute (mercury) dissolved in a solid solvent (silver) and has been used in dentistry applications.

Air is a solution in which the solute (oxygen) and the solvent (nitrogen) are both gases.

Sports drinks are made by dissolving solids (salts) in liquid (water).

Soda is carbonated because it is a mixture of liquid (soda) with a gas (carbon dioxide).

Rubbing alcohol is a homogeneous mixture of two liquids: isopropyl alcohol and water.

SOLUTIONS 205

9.3 FORMING SOLUTIONS

Each of the different types of solution (solid, liquid, and gaseous) forms by different methods. We will look at the most familiar solution form, a solid dissolving in a liquid. Think about making salt water. We know that salt dissolves in water, but what happens to allow that?

SOLVATION

The process of dissolving a solute into a liquid solvent is called *solvation*. The solvation process involves attractive forces between the solvent particles and the solute particles, similar to but weaker than those that cause bonding. The result is that the solute particles become surrounded by particles of the solvent. Consider the dissolving of table salt in water.

Na^+ Cl^- H_2O

① The positive ends of water molecules are attracted to and exert a force or a pull on the negative chloride ions within the crystalline salt solid. At the same time, the negative ends of water molecules are attracted to and pull on the positive sodium ions.

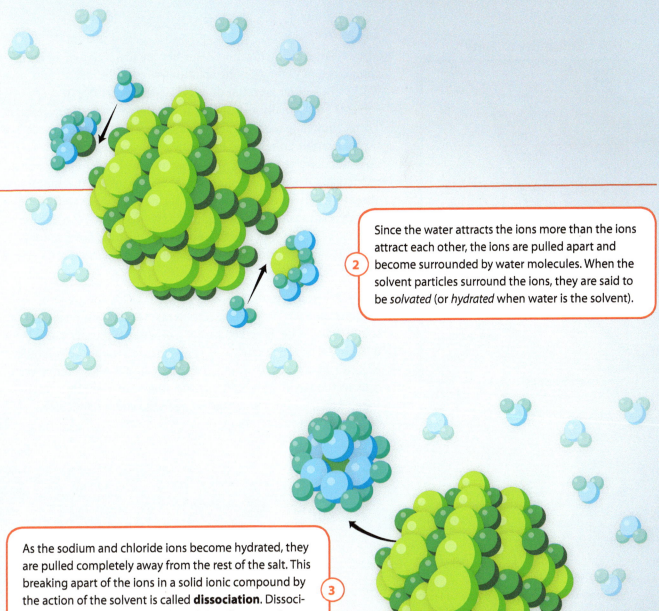

2. Since the water attracts the ions more than the ions attract each other, the ions are pulled apart and become surrounded by water molecules. When the solvent particles surround the ions, they are said to be *solvated* (or *hydrated* when water is the solvent).

3. As the sodium and chloride ions become hydrated, they are pulled completely away from the rest of the salt. This breaking apart of the ions in a solid ionic compound by the action of the solvent is called **dissociation**. Dissociation is a physical process—the salt is still present but as separated sodium and chloride ions.

This whole process involves energy. Energy is needed to separate the solvent particles from each other ($E_{solvent}$). More energy is needed to move the solute particles away from each other (E_{solute}). The process also releases some energy as the solvent and solute particles attract each other ($E_{solution}$).

$$E_{solvation} = E_{solute} + E_{solvent} - E_{solution}$$

If $E_{solvation} > 0$, then the process is endothermic.

If $E_{solvation} < 0$, then the process is exothermic.

SOLUTIONS

Solubility of Salts

[Graph showing solubility curves for KI, NaNO₃, KNO₃, NH₄Cl, NH₃, KCl, NaCl, and KClO₃ plotted as grams of solute/100 g H₂O versus Temperature (°C) from 0 to 100 °C. A highlighted line traces from 60 °C up to the KNO₃ curve and across to approximately 113 g.]

Solubility

The explanation of solvation describes the dissolving of an ionic compound in water. Making solutions with molecular solutes works in a similar way. But does every solute dissolve in every solvent? You actually already know the answer to that question. Think about any ocean beach around the world. While the sea is a saltwater solution made of different salts dissolved in water, sand does not dissolve in water. Why do some substances dissolve in a particular solvent while others don't? It's because of the amount of attraction between the solute and solvent particles as well as the energy relationship outlined on the previous page. We measure the degree of dissolving with **solubility**—the maximum amount of solute that can dissolve in a given amount of solvent at a certain temperature.

The graph at left shows the solubility curves for a number of substances. Each curved line represents the maximum amount of solute that can be dissolved in 100 g of water at temperatures from 0 °C to 100 °C. For example, if you wanted to know the solubility of KNO_3 at 60 °C, you would enter the graph at 60 °C, move vertically until reaching the KNO_3 curve, then move left until reaching the vertical axis. Now you can read the solubility of KNO_3 at 60 °C, approximately 113 g.

HOW IT WORKS

Hot and Cold Packs

If you enjoy outdoor activities, you may have needed either an ice pack or hot pack. Many people have ice packs in their freezer and can soak in a hot bath at home. But these items are not available to you on a hike, at the ball field, or at the skating rink. Chemists have provided these in a form that gets cold or hot only when needed, and both operate according to the properties of solutions.

Instant Cold Pack. Many injuries cause pain and swelling. A common first-aid practice is to use an ice pack to ease the pain and reduce the swelling. Instant cold packs remain at room temperature until they are needed. The pack contains a solid—such as ammonium nitrate, calcium ammonium nitrate, or urea—and water. The two compounds are in separate chambers until needed. When cold is needed, you break the divider between the two compounds, usually by twisting the package, allowing the solid to dissolve in the water. This process is endothermic, so it absorbs heat from the environment, making it feel cold.

Chemical Heat Pack. Whether you participate in or watch winter sports, you probably don't want an ice pack, but a hand warmer would be nice. Chemical hand warmers can work either by chemical reactions or by the action of chemical solutions. The hand warmer above contains a *supersaturated* solution of sodium acetate. Supersaturated means that there are excess solute particles dissolved in the water (see Subsection 9.6). When the metal disk is bent, the solute begins to come out of solution. This is an exothermic process, so heat is released. These warmers can reach about 55 °C and maintain that temperature for about an hour.

208   CHAPTER 9

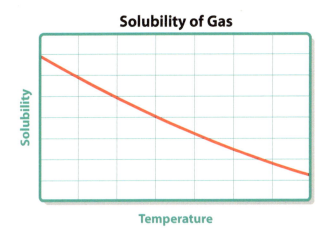
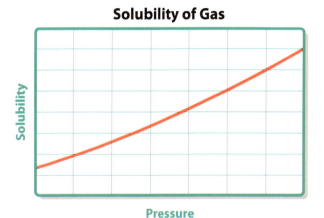

Solubility depends on the chemical and physical properties of the substances involved. The attraction between the solvent and solute has to be great enough to overcome the attractive forces between solvent particles and between solute particles. In general terms, polar and ionic solutes will dissolve in polar solvents while nonpolar solvents are needed to dissolve nonpolar solutes. Chemists remember this by saying "like dissolves like." The solubility of a solute in a certain solvent changes on the basis of solvent conditions, such as temperature and pressure (gaseous solutes only). For example, you can dissolve more sugar in hot tea than in iced tea.

The graphs above show the effect of temperature and pressure on the solubility of a gas in a liquid.

9.4 RATE OF DISSOLVING

On a hot summer day, you may like a nice tall glass of iced tea, and perhaps you like your tea sweetened. So you add sugar cubes and wait, and wait, and wait. Are there ways to speed up the process of dissolving? What you learned in Chapter 7 can help because many of the things that affect reaction rates also affect dissolving rates.

As temperature increases, not only does solubility increase for most solid solutes, but the rate of dissolving increases also. This is because the solvent particles are moving faster, which in turn allows them to come in contact with the solute particles more frequently. Dissolving the sugar in the tea while it is warm and then adding the ice afterward makes the process faster.

The process of stirring also speeds up the dissolving process. By stirring, you are physically moving more solvent particles into contact with the solute. So stir that tea but not too vigorously because you don't want to work up a sweat on an already hot day!

Finally, you could crush those sugar cubes. By crushing them, you expose more of the sugar to the water in the tea. You are providing more surface area for the dissolving process to occur.

SOLUTIONS 209

9.5 SEPARATING MIXTURES

So far we have looked at how to make a mixture, but what if we want to separate a mixture into its components? Since mixtures are physical combinations that are formed by physical processes, we use physical processes to separate them. We choose a method of separation on the basis of the physical properties of the substances in the mixture. While there are many methods of separation, we will look at just a few.

SEPARATING MIXTURES

EVAPORATION

Evaporation is a common process for removing solid solute from a liquid solvent. As the liquid evaporates, the solid is left behind. Salt production along oceanic coasts commonly uses evaporative separation, but it can also be done at inland locations.

MAGNETISM

The fact that some materials are magnetic allows us to separate them from nonmagnetic substances. This process is useful in mining magnetic ores or recovering magnetic wastes. Magnetic separation can also be used to separate ions.

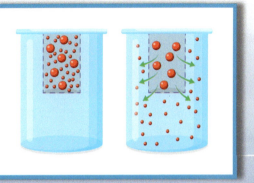

FILTERING/SIFTING

Filtration separates mixtures according to particle size. The filter is designed to allow small particles to pass through while blocking larger particles. Filtration is often used to remove solids from fluids. Our kidneys remove wastes from our blood by filtration.

DISTILLATION

Distillation is used to separate liquids that have different boiling points. As the mixture is heated, the substance with the lower boiling point boils first and its vapor can be collected and condensed. This process is used to separate the various hydrocarbons in crude oil. It's also used to make alcohols more concentrated.

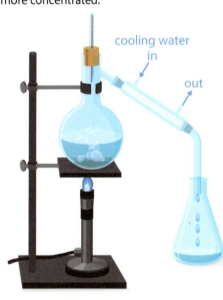

9A | REVIEW QUESTIONS

1. What is the Tyndall effect?
2. Give examples of two suspensions and two colloids.
3. Define *dissociation*.
4. Describe the process by which ionic solutes are dissolved to make a solution.
5. Define *solubility*.
6. According to the graph on page 208, what is the solubility of $NaNO_3$ in 100 g of water at 20 °C?
7. Why does increasing the surface area of a solute speed up the dissolving process?
8. Explain the process of distillation.

9B Questions

- What does "juice from concentrate" mean?
- How can we describe the amount of solute in a solution?
- How can solutions be helpful?

9B Terms

concentration, saturated solution, unsaturated solution, supersaturated solution, molarity, colligative property

9B | SOLUTION CONCENTRATION

How can we describe the amount of solute in a solution?

9.6 SATURATION

Have you ever tried to see how much salt or sugar you could dissolve in water? Just as there is a limit to how much water you can pour into a glass, there is a limit to how much solute will dissolve in a certain amount of solvent. Recall that solubility is the *maximum* amount of solute that can dissolve in a given amount of solvent at a certain temperature. We use the term **concentration** to describe the actual amount of solute dissolved in a given amount of solution.

We can use descriptive terms for how concentrated a solution is. When we have a solution with a relatively large amount of solute, we call it a *concentrated* solution. When the amount of solute is relatively low, we say that it is *dilute*. For example, frozen orange juice is a concentrated solution. To prepare the juice for drinking, we dilute it back into orange juice by adding water. We can also use the terms *saturated*, *unsaturated*, or *supersaturated* as descriptive ways to compare the amount of dissolved solute with the solubility of that solute.

A **saturated solution** contains the maximum amount of solute that the solution can hold at that temperature. If we were to dilute that solution, it would then be an **unsaturated solution**, meaning that the solution contains less than the maximum amount of solute that it could hold at that temperature.

In the case of a **supersaturated solution**, the solution contains more than the maximum amount of solute that it can normally hold at that temperature. How can that be? Under normal conditions, adding more solute to a saturated solution will result in that added solute just settling to the bottom since no more solute can dissolve. The solution is in equilibrium, as some solute continues to dissolve while some dissolved particles return to their undissolved form. But remember that solubility usually increases as temperature increases. So if we make a saturated solution of some solutes at a high temperature and then carefully cool the solution, we can make a supersaturated solution.

Concentration can vary from dilute (left) to concentrated (right), even to the point of being supersaturated.

Chemists are typically more interested in quantitative descriptions of concentration, and there are many ways to express concentration numerically. We will look at percent by mass, percent by volume, and molarity.

9.7 CALCULATING CONCENTRATION

Percent by Mass

Percent by mass is a comparison of the mass of the solute with the mass of the entire solution. For example, chlorine bleach is typically a 5% by mass solution of NaClO dissolved in water. That means that for every 100 g of bleach solution, there are 5 g of sodium hypochlorite. The formula for percent by mass is

$$\%_m = \left(\frac{m_{solute}}{m_{solution}}\right)100\%,$$

where $\%_m$ is the percent by mass, m_{solute} is the mass of the solute, and $m_{solution}$ is the mass of the entire solution (the mass of the solute plus the mass of the solvent).

EXAMPLE 9-1: Using Percent by Mass

How many grams of salt are needed to make 175 g of a 13.1% salt solution?

Write what you know.

$$\%_m = 13.1\%$$

$$m_{solution} = 175 \text{ g}$$

$$m_{solute} = ?$$

Write the formula and solve for the unknown.

$$\%_m = \left(\frac{m_{solute}}{m_{solution}}\right)100\%$$

$$\frac{\%_m}{100\%} = \left(\frac{m_{solute}}{m_{solution}}\right)\frac{\cancel{100\%}}{\cancel{100\%}}$$

$$\left(\frac{\%_m}{100\%}\right)m_{solution} = \left(\frac{m_{solute}}{\cancel{m_{solution}}}\right)\cancel{m_{solution}}$$

$$m_{solute} = \left(\frac{\%_m}{100\%}\right)m_{solution}$$

Plug in known values and evaluate.

$$m_{solute} = \left(\frac{13.1\cancel{\%}}{100\cancel{\%}}\right)175 \text{ g}$$

$$= 22.9 \text{ g}$$

SOLUTIONS

Percent by Volume

Percent by volume is used when both the solute and solvent are liquids. It is a comparison of the volume of the solute with the volume of the entire solution. You may have cleaned a cut with isopropyl alcohol—typically a 70% by volume solution. This mixture contains 70 mL of isopropyl alcohol in every 100 mL of solution. The formula for percent by volume is

$$\%_V = \left(\frac{V_{solute}}{V_{solution}}\right)100\%,$$

where $\%_V$ is the percent by volume, V_{solute} is the volume of the solute, and $V_{solution}$ is the volume of the entire solution (the volume of the solute plus the volume of the solvent).

EXAMPLE 9-2: Using Percent by Volume

What is the volume of solution if 24.3 mL of ethanol is used to make an 11.4% solution?

Write what you know.

$$\%_V = 11.4\%$$

$$V_{solute} = 24.3 \text{ mL}$$

$$V_{solution} = ?$$

Write the formula and solve for the unknown.

$$\%_V = \left(\frac{V_{solute}}{V_{solution}}\right)100\%$$

$$\%_V V_{solution} = \left(\frac{V_{solute}}{\cancel{V_{solution}}}\right)\cancel{V_{solution}}100\%$$

$$\frac{\cancel{\%_V} V_{solution}}{\cancel{\%_V}} = \frac{V_{solute}100\%}{\%_V}$$

$$V_{solution} = \frac{V_{solute}100\%}{\%_V}$$

Plug in known values and evaluate.

$$V_{solution} = \frac{(24.3 \text{ mL})(100\%)}{11.4\%}$$

$$= 213 \text{ mL}$$

Molarity

Probably the most common measure of concentration used by chemists is molarity. **Molarity** (symbol M) is the number of moles of solute per liter of solution. For example, in a lab activity, you may use 2 M HCl, which we would read as "two molar hydrochloric acid." This solution would contain 2 mol of hydrogen chloride in 1 L of HCl solution. The formula for molarity is

$$M = \frac{mol_{solute}}{V_{solution}},$$

where M is the molarity in mol/L, mol_{solute} is the number of moles of solute, and $V_{solution}$ is the total volume of the solution in liters.

EXAMPLE 9-3: Using Molarity

You need 2.75 mol of NaOH to make a batch of soap. How many liters of 8.4 M NaOH solution would you need to have the required moles of NaOH?

Write what you know.

$$M = 8.4 \text{ mol NaOH/L}$$

$$mol_{solute} = 2.75 \text{ mol NaOH}$$

$$V_{solution} = ?$$

Write the formula and solve for the unknown.

$$M = \frac{mol_{solute}}{V_{solution}}$$

$$MV_{solution} = \left(\frac{mol_{solute}}{V_{solution}}\right)V_{solution}$$

$$\frac{\cancel{M}V_{solution}}{\cancel{M}} = \frac{mol_{solute}}{M}$$

$$V_{solution} = \frac{mol_{solute}}{M}$$

Plug in known values and evaluate.

$$V_{solution} = \frac{2.75 \cancel{\text{mol NaOH}}}{8.4 \frac{\cancel{\text{mol NaOH}}}{L}}$$

$$= 0.33 \text{ L}$$

MINI LAB

MASS AND VOLUME IN SOLUTIONS

Essential Question:

How do mass and volume change as a solution is made?

Equipment

laboratory balance

weighing boat or paper

graduated cylinder, 10 mL

graduated cylinder, 25 mL

stirring rod

sugar, 5.0 g

We have seen throughout this textbook that chemical and physical properties often change when we make changes to a substance. This activity will allow you to investigate how making a solution affects the mass and volume of the substances.

1. What is the difference between mass and volume?

PROCEDURE

Ⓐ In the weighing boat, measure out exactly 5.0 g of sugar.

Ⓑ Using the dry 10 mL graduated cylinder, measure the volume of the sugar. Record your data on a piece of paper.

Ⓒ Put exactly 15 mL of water in the 25 mL graduated cylinder.

Ⓓ Measure the mass of the 25 mL graduated cylinder and water. Record your data on a piece of paper.

2. Predict the change in mass when you add the sugar to the 25 mL graduated cylinder.

3. Predict the change in the volume of the water when you add the sugar.

Ⓔ Add the sugar to the 25 mL graduated cylinder and stir until the sugar completely dissolves.

Ⓕ Measure the mass and volume of the 25 mL graduated cylinder and sugar solution. Record your data on a piece of paper.

CONCLUSION

4. Did the mass and volume change as expected?

5. Why did the mass change as it did?

6. Why did the volume change as it did?

GOING FURTHER

7. How do you think the density of the solution changed compared with the density of water alone?

216 CHAPTER 9

9.8 COLLIGATIVE PROPERTIES

As you saw in the in-text lab activity, a solution has different properties compared with its parent solvent. Some of the properties that change are called **colligative properties**, meaning that they are caused by a collection of particles. Colligative properties depend only on the concentration of dissolved particles and not on the identity or properties of the solute. We will look specifically at two colligative properties: *boiling point elevation* and *freezing point depression*.

Boiling Point Elevation

As the relative amount of solute increases, the boiling point of the solution becomes higher than that of the pure solvent. To understand why this happens we have to recall how liquids boil. As we warm a liquid, its vapor pressure increases. Boiling occurs at the temperature where the liquid's vapor pressure becomes equal to or greater than the atmospheric pressure. As solute dissolves in a solvent, the vapor pressure of the solvent becomes lower. Therefore we have to warm the solution to a higher temperature to get the liquid's vapor pressure to equal or exceed the atmospheric pressure.

In the past, chemists used boiling point elevation to measure both the molar mass of the solute and the degree to which ionic solutes dissociate in a solution. Today, other techniques have replaced these uses. But a common use of boiling point elevation is in candy making. As a cook boils a candy solution, it becomes more concentrated, and the boiling point increases. On the candy thermometer (right), you will notice the markings "thread," "soft ball," and "hard ball." The candy maker has to heat each type of candy to a particular temperature and uses the target temperatures on the thermometer for the type of candy that he is making. If you were making fudge, you would heat the mixture to 250°F ("softball").

Freezing Point Depression

A common misconception is that since the boiling point increases when a solute is added to a solvent, the freezing point must increase too. The opposite is true—the freezing point of a solution is actually lower than that of the solvent alone. This phenomenon, called *freezing point depression*, also happens because of the lowering of the vapor pressure in the solution.

As they do with boiling point, chemists can use freezing point depression to measure dissociation in a solution and the molar mass of the solute. Road crews in northern climates use freezing point depression to help keep roadways clear of ice as they apply solid salt or a salt solution to roads. The salt lowers the freezing point of the water on the roads, keeping the water from freezing, even though the water temperature is below 0 °C. Freezing point depression is also key to getting the very cold temperatures for making homemade ice cream.

As the boiling point rises and the freezing point decreases, the temperature range of the liquid phase increases. This again is very useful in our everyday lives. We use a solution of propylene glycol and water in the radiators of our cars. The solute raises the boiling point, which makes the solution useful as a coolant to prevent the engine from overheating in hot weather. The same solute also lowers the freezing point, enabling the same solution to act as an antifreeze in winter. Chemistry is both amazing and useful!

CASE STUDY: ROAD SALT

In many climates, winter means snow, and along with it fleets of snowplows. But snowplows alone can't keep the roads clear, so many locations have those same trucks apply salt or brine (salt water solution) to the roads. The application of salt to the roads melts the snow and ice and also prevents new snow from freezing on road surfaces. The most common salt used is sodium chloride, which can lower the freezing point of water to about −10 °C. Other salts are also used, such as potassium chloride, calcium chloride, and potassium acetate.

9B | REVIEW QUESTIONS

1. Define *concentration*.
2. While making a salt solution, you notice that after stirring the mixture for a long period, there is still salt on the bottom of the glass and it won't dissolve. Explain why this occurs.
3. How much sugar is in 275 g of a sugar solution that is 10.5% sugar by mass?
4. What is the molarity of a solution made by dissolving 0.35 mol of $MgCl_2$ in enough water to make 250 mL of solution?
5. How are colligative properties related to measures of concentration? Give an example.

Use the Case Study above to answer Questions 6–8.

6. Calcium chloride ($CaCl_2$) lowers the freezing point of a solution more than sodium chloride (NaCl) does. Why do you think this is true?
7. Why does salt or brine melt the snow and keep it from freezing even when the temperature is below the freezing point of water?
8. What are some of the hidden costs of using road salt?

CHAPTER 9 REVIEW

9A MIXTURES AND SOLUTIONS

- Heterogeneous mixtures are classified on the basis of solute particle size. Two examples are suspensions and colloids.
- Solutions are homogeneous mixtures made by dissolving a solute in a solvent.
- Solutes and solvents can be in any of the states of matter.
- Solutions form due to the electrostatic attractions between the solute and solvent particles. This process can be either exothermic or endothermic.
- Solubility is the maximum amount of solute that can be dissolved in a given amount of solvent at a particular temperature. Solubility of solids in liquids generally increases with increased temperature. Solubility of gases in liquids generally decreases as temperature increases.
- The rate of dissolving can be altered by changing the temperature of the solution, increasing the surface area of the solute, or by stirring the mixture.
- Because mixtures are physical combinations, they are separated by physical means. The technique needed to separate a particular mixture depends on the properties of the substances in the mixture.

9A Terms

suspension	202
colloid	203
solution	204
solute	204
solvent	204
alloy	205
dissociation	207
solubility	208

9B SOLUTION CONCENTRATION

- We can describe solutions as unsaturated, saturated, or supersaturated, depending on the amount of dissolved solute compared with its solubility.
- Concentration can be measured in many ways, including percent by mass, percent by volume, and molarity.
- Colligative properties depend on the number of dissolved particles in a solution, not their identity or properties.
- Boiling point elevation is a colligative property in which the boiling point of a solution is increased above the normal boiling point of the pure solvent.
- Freezing point depression, another colligative property, is the lowering of the freezing point of a solution below the normal freezing point of the pure solvent.

9B Terms

concentration	212
saturated solution	212
unsaturated solution	212
supersaturated solution	212
molarity	215
colligative property	217

SOLUTIONS 219

CHAPTER 9

CHAPTER REVIEW QUESTIONS

Recalling Facts

1. Relate the terms *solute*, *solvent*, and *solution* in a sentence.
2. Classify the following as heterogeneous or homogeneous.

    a. salt water   **b.** smog   **c.** chalk in water   **d.** brass alloy

3. During the dissolving process, the solute becomes solvated. Explain.
4. Why does the definition of solubility include the temperature?
5. Why does a warm soda fizz more than a cold soda?
6. What does "like dissolves like" mean?
7. Why do people add sugar to the tea while it is hot when making sweetened iced tea?
8. Which method would work to separate
    a. iron filings from sand?
    b. pollen from the air?
    c. table salt from water?
9. What is a dilute solution?
10. Define *molarity*.
11. Why does the boiling point of a solution rise as solute is added?

Understanding Concepts

12. Would a mixture of a liquid with a solid with particle size 2 μm be considered a suspension, colloid, or solution? Explain.
13. Create a hierarchy chart relating the terms *heterogeneous mixture*, *homogeneous mixture*, *colloid*, *suspension*, and *solution*.
14. Dissolving sodium hydroxide (NaOH) is a highly exothermic process. Where does this energy come from?
15. According to the graph (left), what is the solubility of $NaNO_3$ in 100 g of water at 50°C?
16. According to the graph (left), what happens to the solubility of NH_3 as the temperature increases?
17. Refer to the graphs on pages 208 and 209 to create a table summarizing the solubility trends with changes to temperature and pressure. Include both solid and gaseous solutes.
18. Evaluate the following statement. "I am going to separate this mixture with a chemical reaction."
19. Decanting is a technique used to separate mixtures in which a scientist pours off a less dense material from atop a more dense material. Give an example of a mixture that could be separated by decanting.
20. Explain how to calculate percent by mass.
21. Calculate the mass of solution prepared when 51.8 g of solute is used to make a 14.7% solution.

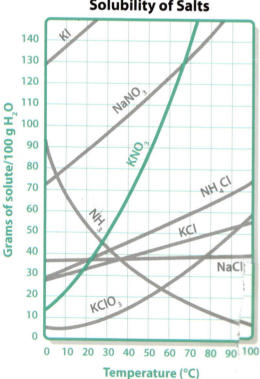

Solubility of Salts

REVIEW

22. What is the percent by volume of a solution made by mixing 345 mL of ethanol with enough water to form 1325 mL of solution?

23. Create a concept map using the terms *solute, solvent, solution, solubility, concentration, concentrated, dilute, unsaturated solution, saturated solution, supersaturated solution, percent by mass, percent by volume,* and *molarity*.

Critical Thinking

24. Where would saturated, unsaturated, and supersaturated solutions be on the graph on page 220?

25. Calculate the percent by mass of 155 g of a solution that contains 142 g of solvent.

26. How many grams of salt should be added to 100 g of water to make a 14.5% salt solution?

27. Show how you could convert a percent by mass to a percent by volume. (*Hint*: Remember that mass and volume are related by density [$d = m/V$].)

28. Some classmates tell you that if they mix 100 mL of a 3.5 M sugar solution with 100 mL of a 5.0 M sugar solution, they will have 200 mL of an 8.5 M solution. Are they correct? Explain.

Use the Case Study at right to answer Questions 29–32.

29. Which material is the solvent and which is the solute in the sap?

30. What process is being used to separate the water from the sucrose in the sap?

31. What is the percent by mass of sucrose when done?

32. Why does the boiling point rise while the water is boiling off?

Use the Ethics Box below to answer Question 33.

33. Using the strategy presented in Chapter 3, write a one-page essay about how Christians should approach this issue.

CASE STUDY: MAPLE SYRUP

Many people love the unique taste of maple syrup, which is a mixture made of primarily sucrose and water. The syrup is made from the sap of the sugar maple tree. The sap is collected in late winter and early spring. All of the collected sap is then boiled, removing much of the water. The process is complete when the boiling point has risen to 104.1 °C. The final syrup contains about 60 g of sucrose for every 100 g of syrup.

ETHICS: POLLUTION

The Issue: Air Pollution

Air pollution is an issue in many areas around the world. The compounds that cause air pollution come from sources that are both natural (volcanoes, plants and animals, and forest fires) and manmade (fossil fuel power plants, motor vehicles, and aerosols). While most people think of air pollution being outdoors in major urban areas, air pollution can be indoors also. The impact of this pollution is widespread. Air pollution can damage both the natural environment and manmade structures. The health impact is enormous, increasing the number of cases of asthma and allergies as well as lung and heart disease. The World Health Organization estimates that air pollution causes about 7 million deaths each year.

CHAPTER 10
Acids, Bases, and Salts

NOT YOUR IDEAL SWIMMING HOLE

Were it not for a hint of blue sky and clouds, you might think that the image on the left is from another planet—but it isn't. Lake Natron is located in Tanzania in East Africa. It's an example of a *soda lake*, a lake rich in sodium carbonate. The lake has no outflow. Rivers carry dissolved minerals into the lake, and evaporation increases their concentration. Between the high concentration of minerals and water temperatures that can reach 40° C, you might think that nothing could live here. But Lake Natron's red color is caused by a type of blue-green algae, whose color, despite the name, is due to a red pigment that helps with photosynthesis. The algae forms the base of a food chain that supports a large population of flamingoes. Even in such harsh conditions, some of God's creatures can find what they need to thrive.

10A Acids and Bases	224
10B Acidity and Alkalinity	228
10C Salts	232

10A | ACIDS AND BASES

10A Questions

- What are the characteristics of acids and bases?

10A Terms

acid, aqueous solution, hydronium ion, indicator, base, hydroxide ion

What's the difference between an acid and a base?

10.1 ACIDS

There's a good chance that you might start salivating when you see the lemon on the next page. You might even know that citric acid is the substance that gives lemons, oranges, and grapefruit their tart flavors. But what exactly is citric acid or any acid for that matter? Let's find out!

ACIDS

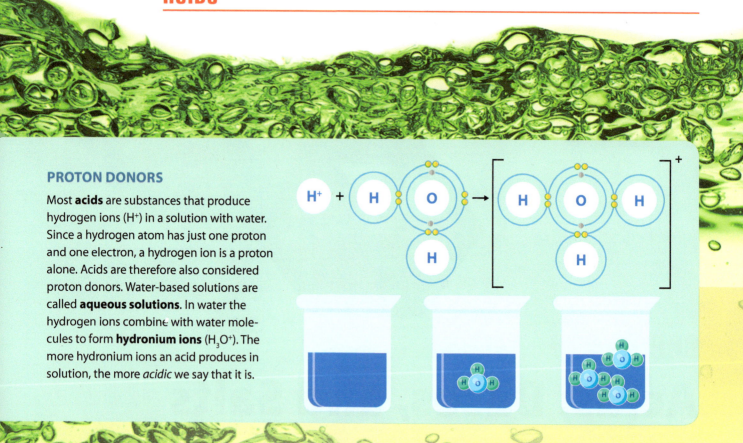

PROTON DONORS

Most **acids** are substances that produce hydrogen ions (H^+) in a solution with water. Since a hydrogen atom has just one proton and one electron, a hydrogen ion is a proton alone. Acids are therefore also considered proton donors. Water-based solutions are called **aqueous solutions**. In water the hydrogen ions combine with water molecules to form **hydronium ions** (H_3O^+). The more hydronium ions an acid produces in solution, the more *acidic* we say that it is.

REACTIVITY WITH METALS

Acids can chemically react with some metals in a single-replacement reaction that also releases hydrogen gas. A common use of this property is an industrial process called *etching*. Etching can be used to create delicate metal structures, such as the patterns found on electronic circuit boards.

CONDUCTIVITY

Have you ever wondered why you should never plug in your hair dryer near the bathtub? Pure water doesn't conduct electricity, but any ions dissolved in the water will. Since acids dissolved in water produce ions, an acid solution *will* conduct an electric current. Virtually all household water supplies have some dissolved ions in them, though not necessarily hydronium ions, so you should always remain alert for shock hazards.

COLOR CHANGE WITH INDICATORS

An **indicator** is a substance that will change color in the presence of an acid or base. A very common indicator used in chemistry classes is litmus, which is a kind of dye. Litmus-saturated strips of paper are easily dipped into solutions. Acidic solutions will turn a strip of blue litmus paper red.

SOUR TASTE

Many foods owe their sour taste to the presence of acids. In addition to the citric acid mentioned earlier, common acids found in foods include acetic acid, which is found in vinegar, and malic acid, the acid that produces the tart taste of apples and berries. Strong acids can cause chemical burns, so chemicals in a chemistry lab should *never* be tasted to see whether they are acids!

10.2 BASES

As you'll see, some characteristics of bases are similar to those of acids, but others are quite different. Take a look at some identifying features of bases.

BASES

PROTON ACCEPTORS

Many **bases** are substances that produce **hydroxide ions** (OH^-) in an aqueous solution. Such solutions are described as being *basic*, or *alkaline*. The more hydroxide ions that a base can produce in solution, the more basic it is. Hydroxide ions act as proton acceptors. Other molecules or ions that can accept protons are also considered bases. Hydroxide ions can react with hydrogen ions to form water.

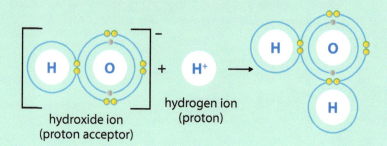

hydroxide ion (proton acceptor) + hydrogen ion (proton) →

SLIPPERY FEEL

Some strong bases in solution can react with the oil on your skin in much the same way that soap is made. Such bases are thus slippery to the touch, just as soap is slippery. Like acids, strong bases can cause chemical burns, so there is no safe "touch and taste" test for unknown chemicals!

CONDUCTIVITY

The ions produced by bases in solution conduct electrical current just as those in acidic solutions do.

226   CHAPTER 10

COLOR CHANGE WITH INDICATORS

Red litmus paper is often used to test whether a solution is basic. If it is, the color of red litmus paper will change to blue.

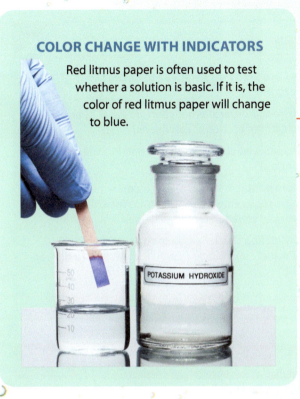

CASE STUDY: THE KING OF CHEMICALS

Chemical companies in the United States produce more sulfuric acid (H_2SO_4) than any other single chemical. This acid is used in almost every industrial process. In fact, economists consider sulfuric acid so essential to industry that they sometimes measure the economic condition of a country by how much sulfuric acid it uses. Generally, when a nation's usage of sulfuric acid drops, its whole economy is headed for a downturn. You can easily understand why sulfuric acid is often referred to as the king of chemicals.

A dense, oily substance with a high boiling point, concentrated sulfuric acid is highly corrosive and can dissolve many metals in a matter of minutes! Most importantly, sulfuric acid can react with numerous other chemicals to produce thousands of useful products. More than 60% of the sulfuric acid used worldwide goes into the manufacture of fertilizers and detergents. In addition, chemists use sulfuric acid to produce paints, dyes, plastics, fibers, and a vast array of other products.

BITTER TASTE

Bases tend to taste bitter. The caffeine in coffee beans and a chemical called *theobromine* in unsweetened chocolate are both proton acceptors and thus bases.

The chemical formulas for acids and bases often hint at their acid or base nature. Look at the formulas for some common acids and bases in the table below and see whether you can spot the sources of hydrogen or hydroxide ions.

Common Acids and Bases

Source	Common Name	Formula
vinegar	acetic acid	$HC_2H_3O_2$
fruits	ascorbic acid	$H_2C_6H_6O_6$
soft drinks	carbonic acid	H_2CO_3
citrus fruit	citric acid	$H_3C_6H_5O_7$
insects	formic acid	$HCHO_2$
sour milk	lactic acid	$H_2C_3H_4O_3$
rhubarb, spinach	oxalic acid	$H_2C_2O_4$
household ammonia	ammonium hydroxide	NH_4OH
milk of magnesia	magnesium hydroxide	$Mg(OH)_2$
lye	sodium hydroxide	$NaOH$

ACIDS, BASES, AND SALTS

10A | REVIEW QUESTIONS

1. What is an aqueous solution?
2. What ions do many acids produce in aqueous solutions?
3. List three characteristics of acids.
4. Suppose you find a beaker in the laboratory that contains an unidentified aqueous solution. Describe one *safe* way to test whether the solution might be an acid.
5. What ions do many bases produce in aqueous solutions?
6. List three characteristics of bases.
7. Why would testing the solution to see whether it will conduct electrical current not be an acceptable answer for Question 4?

Use the Case Study on page 227 to answer Questions 8–11.

8. Why is sulfuric acid production looked at as an indicator of economic conditions?
9. What ions would you expect to find in a sulfuric acid solution?
10. Dilute sulfuric acid reacts with aluminum in a single-replacement reaction. Write the balanced chemical equation for this reaction.
11. Considering the definition of an acid, why do you think that sulfuric acid (H_2SO_4) is considered a *diprotic* acid?

10B | ACIDITY AND ALKALINITY

10B Questions

- What makes an acid or base strong or weak?
- What is pH?
- How does pH relate to acid and base strength?
- Is acid and base strength related to concentration?
- How do buffers work?

10B Terms

strong acid, weak acid, strong base, weak base, pH, buffer

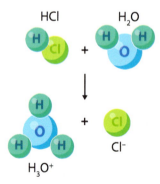

Why are strong acids and bases dangerous?

10.3 ACID AND BASE STRENGTHS

It might be confusing to hear your science teacher talk about the dangers of acids in the laboratory but then learn that there are acids in our foods. Why is sulfuric acid dangerous but not vinegar? The answer is found in an understanding of two concepts: *strength* and *concentration*. Let's look first at what determines the strength of an acid or base.

Ionization of Acids

You learned in Subsection 10.1 that acids produce hydrogen ions in aqueous solutions. But not all acids give up their hydrogen ions easily. As in other covalently bonded compounds, the bonds between the atoms in some acids are stronger than in others. This means that some acids ionize in water more easily than others. If all or most of the molecules in an acid will ionize in water, then that acid will produce many hydrogen ions in solution. Such acids are called **strong acids**. In contrast, most of the molecules in **weak acids** do not ionize in water, so fewer hydrogen ions are produced.

The more ions an acid produces in solution, the more reactive it is. Vinegar is not particularly dangerous because acetic acid, the acid in vinegar, is a weak acid. Sulfuric acid is far more hazardous because it is a strong acid. Below, see a list of strong and weak acids.

Strong Acids	Weak Acids
perchloric acid ($HClO_4$)	hydrosulfuric acid (H_2S)
sulfuric acid (H_2SO_4)	carbonic acid (H_2CO_3)
hydrochloric acid (HCl)	acetic acid ($HC_2H_3O_2$)
nitric acid (HNO_3)	nitrous acid (HNO_2)

Dissociation of Bases

Bases vary in their strength, just as acids do. Some, like sodium hydroxide, are very hazardous. Others, like baking soda (sodium bicarbonate), pose far less danger. Many common bases are ionic compounds. As you learned in Chapter 9, ionic compounds dissociate in water. **Strong bases** include those that readily dissociate, producing large numbers of hydroxide ions, which in turn act as proton acceptors. The metal hydroxides shown in the table below are easily identifiable as bases because of the presence of the OH in their chemical formulas.

Ionization of Bases

As you can see in the table below, not all bases contain a readily identifiable OH^- ion. Covalent compounds such as ammonia (NH_3) do not dissociate in water. Instead, these **weak bases** produce hydroxide ions by ionizing in a reaction with water. In this reaction, a hydroxide anion is formed when a water molecule donates one of its protons (H^+) to a weak base. The weak base that accepts the proton becomes a cation. Weak bases do not ionize completely and thus produce fewer hydroxide ions in aqueous solutions than strong bases. The ionization of ammonia in water, which is a reversible reaction, is shown in the following equation.

$$NH_3 + H_2O \rightleftharpoons NH_4^+ + OH^-$$

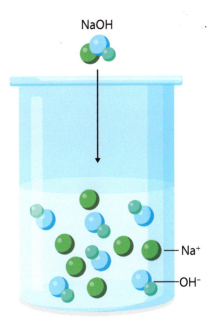

Strong Bases	Weak Bases
lithium hydroxide (LiOH)	ammonia (NH_3)
sodium hydroxide (NaOH)	methanamine (CH_3NH_2)
potassium hydroxide (KOH)	aniline ($C_6H_5NH_2$)
rubidium hydroxide (RbOH)	pyridine (C_5H_5N)

Concentration

Acetic acid is a weak acid, so chugging large amounts of your favorite balsamic vinegar should be okay, right? Not exactly. Vinegar is relatively safe only because it's not pure acetic acid. Vinegar is actually a solution of acetic acid in water. It's usually only about 5% acetic acid by mass. That's good, because pure acetic acid is rather hazardous and can damage your lungs, eyes, and skin.

Like all solutions, acid and base solutions can vary in their concentration—the amount of solute per unit of solution. So if we're talking about a concentrated solution of acetic acid, we mean that it has a lot of acetic acid dissolved in it. Similarly, sodium hydroxide is a very hazardous strong base, but if it is made into a very dilute solution—one with a small amount of solute in a large amount of solvent—then it will pose less risk than a more concentrated solution. Your favorite balsamic vinegar is thus doubly safe because it is both a weak acid *and* a dilute solution.

10.4 THE pH SCALE

Since the strength of an acid or base compound and its solution concentration can vary widely, how can we compare one such solution to another? Chemists do this using the *pH scale*, a gauge developed to make measuring and reporting acidity easier. The **pH** of a substance is a unitless number between 0 and 14 that tells how acidic or basic a substance is. The pH value is determined by the abundance of hydrogen ions that are produced in a solution of the substance. The greater the abundance of hydrogen ions a solution has, the lower its pH will be (this is an inverse mathematical relationship). Values on the pH scale that are less than 7 indicate acidic solutions. Values greater than 7 are basic. A pH equal to 7 is neutral—neither acidic nor basic. Pure water has a pH of 7. When comparing the acidity of two solutions, the one with the lower pH is more acidic; if comparing two bases, the one with the higher pH is more basic. Thus, lemon juice (pH 2) is more acidic than brewed coffee (pH 5), and ammonia (pH 11.6) is more basic than milk of magnesia (pH 10).

A slight difference between two values on the pH scale means a big difference in the number of hydrogen ions. A change of one on the pH scale indicates a tenfold change in the amount of hydrogen ions. So an acid with a pH of 1 has ten times as many hydrogen ions as an acid with a pH of 2. A basic solution with a pH of 10 has one one-hundredth as many hydrogen ions as a solution of pH 8.

Since the pH of a solution is related to its concentration of hydrogen ions, it is affected by the strength and concentration of the acid or base. Strong acids produce more hydrogen ions in solution than weak acids do. So a strong acid like sulfuric

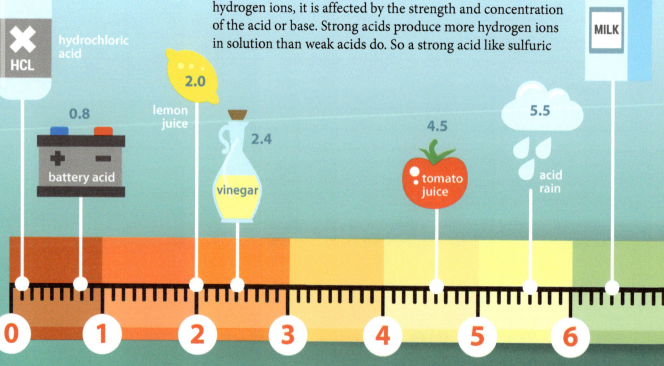

acid will have a lower pH (more acidic) in solution than a weak acid in a solution of the same concentration. Concentrated solutions (i.e., those with greater amounts of solute) contain more ion-producing particles than dilute solutions do. This means that a highly concentrated solution of sulfuric acid will be more acidic than a dilute solution and will thus have a lower pH.

10.5 BUFFERS

Human blood is a chemical solution with a pH of about 7.4. Our health depends on our blood remaining at or very near this particular pH value (7.35–7.45). Blood pH levels that are too low (acidemia) or too high (alkalemia) can cause some very serious medical issues. Happily for us, our blood is an example of a *buffered system*. The **buffers** in a buffered system are weak acids or bases in solution. Buffers respond to changes in pH in order to maintain the pH level at a certain value. If excess acid is added to the system, it is neutralized by the weak base. If there is too much base in the system, the weak acid neutralizes it. In human blood, the primary weak acid is carbonic acid (H_2CO_3), and the primary weak base is bicarbonate ion (HCO_3^-). Similar buffering systems work to maintain a stable pH inside your cells.

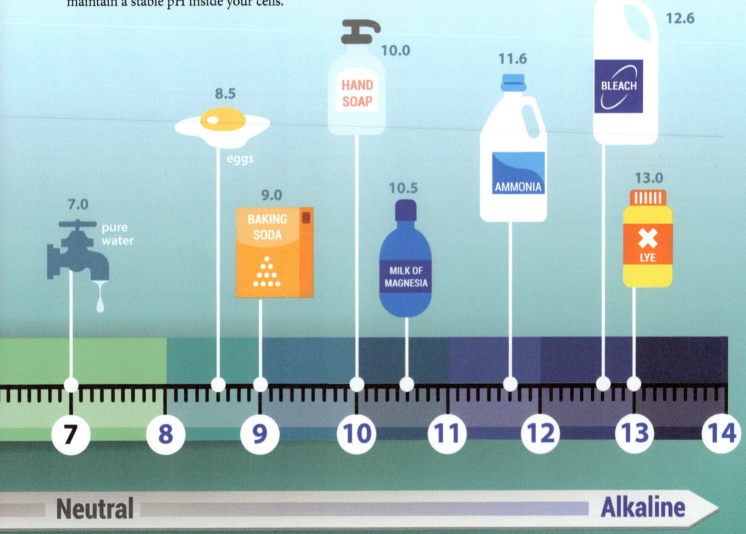

10B | REVIEW QUESTIONS

1. What determines the strength of an acid or base?
2. What is the difference between a *strong acid* and a *concentrated solution* of that acid? Is it possible to have a concentrated weak acid? Explain.
3. Are the chemical properties of an acid or base solution based on the strength of the acid or base, the concentration of the acid or base, or both? Explain.
4. Your friend says that a solution with a pH of 6 is more acidic than one with a pH of 4. Is your friend correct? Explain.
5. How does the amount of hydrogen ions in a nitric acid solution with a pH of 1 compare with the amount of hydrogen ions in a nitric acid solution with a pH of 3?
6. If a strong base and a weak base are each used to make a solution of equal concentration, which is more likely to have a higher pH value? Explain.
7. Why do buffered systems require the presence of both acids and bases?

10C Questions

- How is an acid or base neutralized?
- How can I identify the ions in an acid or base?
- What are the products of a neutralization reaction?

10C Terms

salt, neutralization

10C | SALTS

What happens when acids and bases mix?

10.6 WHAT ARE SALTS?

The white stuff that you sprinkle on your food, table salt, is only one of a large class of compounds that chemists call *salts*. **Salts** are electrically neutral, ionic compounds formed by a combination of cations and anions. In many salts, known as *binary salts*, the cation is a metal and the anion is a nonmetal. Table salt (NaCl), composed of sodium cations and chloride anions, is one such salt. There are many other kinds of salts as well. Let's take a look at some examples.

SALT SURVEY

MAGNESIUM SULFATE: $MgSO_4$

Magnesium sulfate, also known as *Epsom salt*, is the main component in mineral bath salts. Soaking in a tub containing a solution of magnesium sulfate feels great, but claims for its medicinal value are not well supported by scientific studies.

AMMONIUM DICHROMATE: $(NH_4)_2Cr_2O_7$

Early photographers used ammonium dichromate in a process called *gum bichromate photography*. Using this method, they were able to produce photographs with painting-like qualities.

CALCIUM CARBONATE: $CaCO_3$

Plain, white calcium carbonate crystals aren't much to look at, but calcium carbonate is the main component of pearls and seashells. Animals aren't the only ones who use calcium carbonate to make stuff either. In the form of limestone, it's used by humans to build stone structures and to make concrete.

MALACHITE GREEN: $C_{23}H_{25}ClN_2$

Malachite green is a member of a large group of salts called *triarylmethane dyes*. While malachite green obviously produces a green color, other members of the group yield shades of blue, red, or violet. Malachite green solutions are also used to treat fungal diseases in aquarium fish.

ACIDS, BASES, AND SALTS

MINI LAB
BASIC PROBLEM

Turmeric is an orange-yellow spice that is used in Asian cuisine. Spicy yellow curries get their color from it. Turmeric is also an indicator. In this activity, you'll use turmeric indicator to test some common household products.

Essential Question:

Is there a simple way to determine which of two solutions is more basic?

Equipment

goggles

pipette

small bowls (4)

stirring rod or spoon

turmeric indicator

lemon juice

basic solutions to test

PROCEDURE

A Use the small bowls to obtain some turmeric indicator, lemon juice, and samples of two basic solutions from your teacher. You'll need to either use a separate stirring rod for each solution or rinse the stirring rod in between testing each solution.

1. Which two basic solutions will you be testing?
2. What color is the turmeric indicator?
3. Which of the two basic solutions do you think is more basic? Write your answer in the form of a hypothesis.

B Add one drop of turmeric indicator to the lemon juice. Gently stir the indicator and juice together.

4. Does the turmeric indicator change color in lemon juice?

C Add one drop of turmeric indicator to one of the bases. Gently stir the indicator and base together.

5. Does the turmeric indicator change color in the base? If so, what color is produced?
6. How does turmeric indicator show whether a substance is an acid or base?

D Now add drops of lemon juice to the base, stirring gently after each drop. Keep count of how many drops are added. Stop when the color changes back to a shade that matches that of the indicator in lemon juice alone.

7. What is happening to the basic solution as you add drops of lemon juice?

8. How many drops did you add to obtain the desired color change?

E Repeat Steps **C** and **D** for the second base.

9. How many drops did you add before obtaining the desired color change in the second basic solution?

ANALYSIS

10. Which of the two basic solutions is more basic? How do you know?

11. Did this result support your hypothesis?

CONCLUSION

The technique that you used to compare the basicity of two solutions is called *titration*. It is a common laboratory procedure. You'll see it again in a standard chemistry course.

GOING FURTHER

12. The two images below show hydrangea flowers, which act as natural indicators. What would explain their two colors?

13. What would a gardener need to do to if she desired the same plant to have blue flowers one year and pink the next?

10.7 NEUTRALIZATION

Salts can be produced by the chemical reaction of an acid with a base, called a *neutralization reaction*. **Neutralization** is a double-replacement reaction that generally produces a salt and water. The salt consists of the cation from a base and the anion from an acid. For example, sodium chloride contains a sodium cation (Na^+) and a chloride anion (Cl^-). The sodium cations for a sodium chloride–forming neutralization reaction come from a sodium base, such as sodium hydroxide. The chloride anions come from a chloride acid, such as hydrochloric acid. Below is the neutralization reaction that results when sodium hydroxide and hydrochloric acid are combined.

$$HCl + NaOH \rightarrow NaCl + H_2O$$

Since these compounds completely ionize or dissociate in water, this equation could be written the following way.

$$H^+ + Cl^- + Na^+ + OH^- \rightarrow Na^+ + Cl^- + H_2O$$

The double-replacement reaction produces water, which mingles with the water in the original solution, and the soluble table salt. The table salt remains dissociated as long as it is in solution, but it can be crystallized if the water is removed by evaporation.

We can use the previous example of neutralization to create a general formula for such reactions.

$$acid + base \rightarrow salt + water$$

This general formula can be used to predict the products of a neutralization reaction.

10.8 PREDICTING SALT COMPOUNDS

We've already seen that sodium hydroxide, a strong base, reacts with a strong acid like hydrochloric acid. What happens when sodium hydroxide (NaOH) reacts with a weak acid such as carbonic acid (H_2CO_3)? Let's use what we know about the general formula for neutralization reactions, along with the ionization of acids and dissociation of bases, to predict the products formed by a reaction of those two compounds. The following process will work well for neutralization reactions involving strong bases.

EXAMPLE 10-1: Predicting a Salt

Begin writing out a chemical equation by writing down the reactants on the left side of the equation. Identify which compound is the acid and which is the base.

$$H_2CO_3 + NaOH \rightarrow salt + water$$

The OH^- group in sodium hydroxide alerts us to its being the base in this reaction. It will serve as the proton acceptor.

Remember, water will be formed by the anion from the base (OH^-) and a hydrogen ion (a single H^+ ion, that is, a proton) from the acid.

$$H_2CO_3 + NaOH \rightarrow salt + H_2O$$

The salt will be formed from the leftover cation from the base (Na^+ in this example) combined with the anion from the acid (H_2CO_3 less one H^+ ion leaves HCO_3^-). The finished equation thus looks like this:

$$H_2CO_3 + NaOH \rightarrow NaHCO_3 + H_2O$$

Remember that list of common polyatomic ions back in Chapter 5? HCO_3^- is bicarbonate, so the salt formed by this reaction is called *sodium bicarbonate*.

10.9 PUTTING NEUTRALIZATION TO WORK

Have you ever eaten too much spicy food at one sitting? Perhaps afterward you felt a burning sensation in your chest and esophagus—acid indigestion. It's caused by the excess digestive juices that your stomach produces in response to a large meal. These juices are highly acidic and have a pH value around 1. How can the symptoms associated with too much stomach acid be treated? One way is to take an antacid, a base compound used to relieve the symptoms of indigestion. Common bases found in antacids are calcium carbonate, sodium bicarbonate, magnesium hydroxide, and aluminum hydroxide. These bases work to neutralize stomach acid. Many people find great relief through these kinds of products. But as with any medicine, antacids must be taken in limited doses.

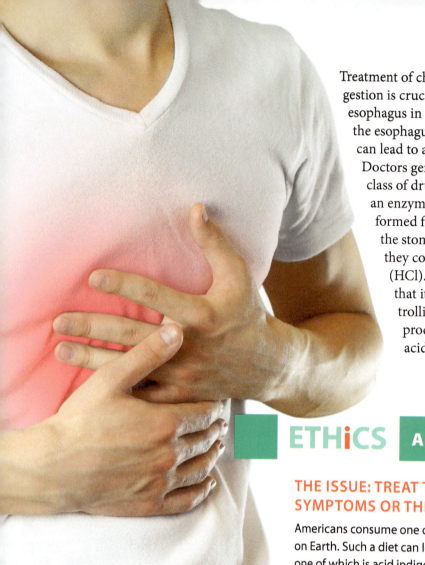

Treatment of chronic (i.e., continuous or repeated) acid indigestion is crucial since excess stomach acid can damage the esophagus in cases where the muscular valve that shuts off the esophagus at the top of the stomach leaks. This damage can lead to a very deadly kind of cancer in the esophagus. Doctors generally treat chronic acid indigestion with a class of drugs known as *proton pump inhibitors*. Normally, an enzyme known as a proton pump transports H^+ ions formed from water molecules within the cells that line the stomach into tiny ducts in the stomach wall. There they combine with Cl^- ions to form hydrochloric acid (HCl). Proton pump inhibitors block the enzyme so that it can't carry the H^+ ions into the ducts. Controlling the H^+ ions here stops or greatly reduces the production of hydrochloric acid, preventing chronic acid indigestion.

ETHiCS — ANTACIDS

THE ISSUE: TREAT THE SYMPTOMS OR THE CAUSE?

Americans consume one of the richest diets of any nation on Earth. Such a diet can lead to a variety of problems, one of which is acid indigestion, also called heartburn. In 2017 alone, Americans spent over $2.6 billion on over-the-counter (OTC) treatments for acid indigestion. But the overuse or long-term use of antacids can cause its own set of problems. Some issues, such as kidney stones, are serious.

10C | REVIEW QUESTIONS

1. Of what two parts is a salt composed?
2. What two kinds of substances react in a neutralization reaction?
3. What are the products of a neutralization reaction?

Use the following neutralization reaction to answer Questions 4–7.

$$Mg(OH)_2 + 2HF \rightarrow ? + ?$$

4. Which of the reactants is an acid?
5. What is the name of the cation that is contained in the base? What is the name of the anion that is part of the acid?
6. Write out the complete balanced equation.
7. What is the name of the salt produced by this reaction?

Use the Ethics box above to answer Question 8.

8. Using the strategy presented in Chapter 3, write a one-page essay about how Christians should approach the usage of OTC antacids.

CHAPTER 10 REVIEW

10A ACIDS AND BASES

- Most acids produce hydrogen ions in aqueous solutions. Most bases produce hydroxide ions.
- Acids are proton donors; bases are proton acceptors.
- Acids taste sour, react with metals, and turn blue litmus paper red.
- Bases taste bitter, feel slippery, and turn red litmus paper blue.
- Both acids and bases produce ions that will conduct electricity in water.

10A Terms

acid	224
aqueous solution	224
hydronium ion	224
indicator	225
base	226
hydroxide ion	226

10B ACIDITY AND ALKALINITY

- Strong acids or bases are those that ionize or dissociate completely in water. Weak acids and bases do not ionize or dissociate completely.
- The pH scale assigns values between 0 and 14 that indicate how acidic or basic a solution of a substance is. A value of 7 is neutral, while values less than 7 are acidic and more than 7 are basic.
- The pH of a solution depends both on the strength of its acid or base solute and on its concentration.
- A buffered system contains both a weak acid and a weak base that work to maintain a particular pH value.

10B Terms

strong acid	228
weak acid	228
strong base	229
weak base	229
pH	230
buffer	231

10C SALTS

- Acids and bases neutralize each other in a double-replacement reaction that produces a salt and water.
- Salts are electrically neutral compounds formed by a combination of cations and anions.
- Antacids contain bases that neutralize excess stomach acid.

10C Terms

salt	232
neutralization	236

ACIDS, BASES, AND SALTS

CHAPTER 10

CHAPTER REVIEW

Recalling Facts

1. How are hydronium ions formed when an acid is dissolved in water?
2. What is an indicator?
3. Create concept definition maps for the terms *acid* and *base*.
4. A classmate tells you that acids are dangerous but bases are not. Is he correct? Explain.
5. What does the word *strong* mean when describing an acid or base?
6. If the pH of a solution increases, does its acidity increase or decrease?
7. A change from pH 5 to pH 2 represents what change in the amount of hydrogen ions?
8. What are buffers?
9. What is the difference between salts in general and binary salts?
10. What type of chemical reaction is a neutralization reaction?

Understanding Concepts

11. One of your classmates says that acids produce hydronium ions in solution. Another says that acids produce hydrogen ions. Which is correct? Explain.
12. A chemist accidentally drops a small piece of metal into an unmarked test tube of clear liquid (left). Small bubbles begin to form on the metal. Is the liquid in the test tube most likely an acid or a base? Explain.

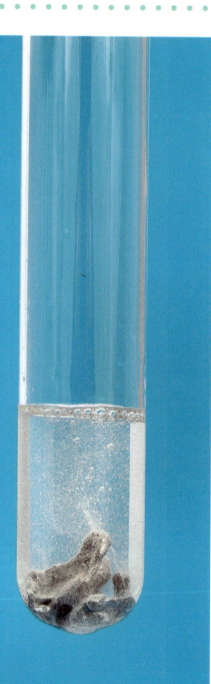

13. Classify each of the following substances as either an acid or a base.
    a. KOH
    b. H_3BO_3
    c. NH_4OH
    d. $Zn(OH)_2$
    e. $HClO_2$
14. Is the substance being tested at right an acid or a base? Explain.
15. Why is it incorrect to say that sodium hydroxide (NaOH) ionizes in an aqueous solution?
16. A chemist attempts to make a basic solution by adding an unknown base to a liter of distilled water. A relatively large amount of solute shifts the pH of the solution only from 7.0 to 7.4. What can the chemist conclude about the strength of the unknown base?
17. Is pH affected by the strength of an acid or base, the concentration of the acid or base, or both? Explain.
18. Two acid solutions of equal concentration have different pH values. How is this possible?
19. What would happen if human blood contained bicarbonate as a buffer but not carbonic acid?

REVIEW

20. Nitric acid (HNO_3) forms hydrogen cations and nitrate anions in solution. Why is it not considered a salt?

21. Predict the salt that will form during a neutralization reaction between cesium hydroxide (CsOH) and hydrobromic acid (HBr). Write the balanced chemical equation for the reaction.

Critical Thinking

22. Water is an example of an *amphiprotic* substance, one that can either donate or accept a proton. Use information and examples from the textbook to explain how this is possible.

23. Your study buddy tells you that 1 M hydrochloric acid is weaker than 2 M hydrochloric acid. What is the flaw in your partner's statement? What would be a better way to compare the two solutions?

24. Do you agree or disagree with the statement, "Vinegar tastes strong, so it must be a strong acid"? Defend your choice.

25. Scientists use conductivity meters to measure how well a solution can conduct an electrical current. The higher the conductivity measured by the meter, the better the solution is at conducting electricity. If one basic solution conducts electrical current better than another basic solution of equal concentration, what can be inferred about the strengths of the two bases? Explain.

26. Carbonic acid is a weak acid that forms when carbon dioxide dissolves in water. Why does this make it nearly impossible to maintain samples of pure water?

27. Zinc reacts with hydrochloric acid (HCl) as shown in the following equation. State at least two reasons why the reaction is not considered a neutralization reaction.

$$Zn + 2HCl \rightarrow ZnCl_2 + H_2$$

28. How is the presence of blood buffers an indication of God's creative design?

We do not know what the rules of the game are; all we are allowed to do is to watch the playing. Of course, if we watch long enough, we may eventually catch on to a few of the rules. The rules of the game are what we mean by fundamental physics.

Richard P. Feynman (1918–88), American physicist

UNIT 3
MATTER IN MOTION

CHAPTER 11: **KINEMATICS**

CHAPTER 12: **DYNAMICS**

CHAPTER 13: **WORK AND MACHINES**

CHAPTER 14: **ENERGY**

CHAPTER 15: **THERMODYNAMICS**

CHAPTER 16: **FLUIDS**

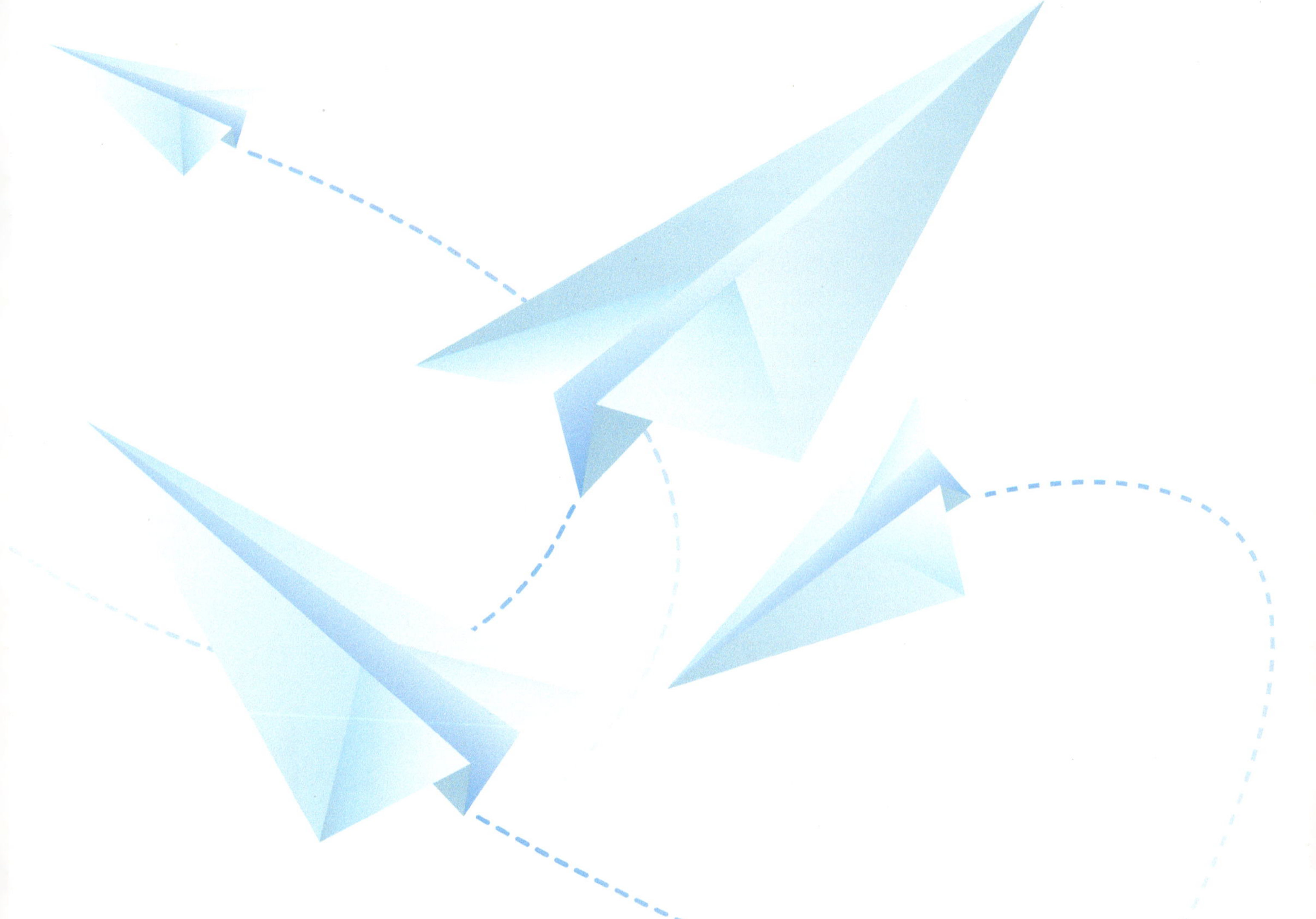

CHAPTER 11
Kinematics

MOVING MOMENTS

Click, click, click, click, click! Anticipation grows as you slowly climb that huge first hill of the rollercoaster. An amazing panoramic view greets you as you slowly crest the top of the hill, and then the view changes to a blur as you speed down the hill. What follows is two and a half minutes of high speeds, abrupt accelerations, and rapid twists, turns, and loops. Most people don't think about mechanics—the study of motion—while riding a rollercoaster, but the designers of the ride thought a lot about it. For the rider, rollercoaster rides provide a crash course in mechanics—well, not literally!

11A Describing Position — 246
11B Describing Motion — 253
11C Changing Motion — 261

11A Questions

- Why use Newtonian mechanics if newer models are better?
- How do we study motion?
- How can tools help us study motion?
- Aren't distance and displacement the same thing?

11A Terms

mechanics, kinematics, frame of reference, system, distance, displacement, scalar, vector

11A | DESCRIBING POSITION

How can we describe where an object is?

11.1 WANDERING PLANETS

God created the universe and set it in motion. On Day 4, God purposefully created the sun, moon, and stars so that their motion would mark out the days, seasons, and years (Gen. 1:14). As man began to study the motion of these heavenly bodies, the motion looked orderly. But some stars didn't move as expected; they seemed to wander. Today we know that these are planets, moving in a regular, orderly, but different-than-the-stars pattern. The desire to understand these wandering planets led scientists to begin their study of motion—**mechanics**. In this chapter, we will study how things move, a study known as **kinematics**. In Chapter 12 we will continue our study of motion with *dynamics*—why things move.

HISTORY OF MECHANICS

GREEK PHILOSOPHICAL THOUGHT

Studying motion, like matter, began with the ancient Greeks. Greek thought about motion focused on the natural state of the particular body. Aristotle believed that things moved as they did because it was their nature. Earth and water moved downward because it was in their nature to be closer to the ground. Air and fire moved upward because it was natural for them to do so. Heavenly bodies moved around the earth because, by nature, they moved in circular motion. But some appeared to wander, so scientists called them *planets*, from the Greek word meaning "wanderers."

EARLY SCIENTIFIC THOUGHT

In the early 1600s, Tycho Brahe and Johannes Kepler worked together, studying the motion of planets. Brahe worked on making highly accurate observations of the planets' positions over time. Kepler used Brahe's observations and applied mathematics to the study of motion. He used Brahe's work to explain *retrograde motion*, the apparent reversal of the planets in their orbits. Kepler recognized that he could predict the motion of planets as they moved in elliptical, not circular, orbits. His work resulted in his identifying three laws of planetary motion.

Kepler

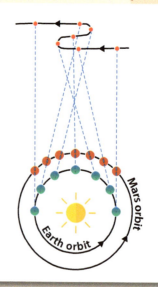

Galileo conducted many experiments related to the motion of objects, especially falling objects. His experiments led him to the conclusion that all objects in a vacuum, regardless of mass, would fall at the same rate. Brahe, Kepler, and Galileo made great progress in our understanding of motion, but the imprecise instruments that they had limited their study.

CLASSICAL MECHANICS

Isaac Newton developed the first truly mathematical model of motion in the late 1600s. His three laws of motion, along with his law of universal gravitation, form the basis for *classical mechanics*—the study of how and why things move. You will learn about his three laws in Chapter 12.

MODERN MOTION MODELS

As we learned more about the world around us, other models were developed. The discovery of subatomic particles led to the development of *quantum mechanics* to describe the behavior of tiny objects. Albert Einstein developed *relativistic mechanics* to explain the motion of high-speed (near the speed of light) objects. While these two models are more workable than Newtonian mechanics, they involve mathematics that is much more complex. We still use Newtonian mechanics to study the motion of the most common, everyday objects because Newtonian mechanics explains observations and makes predictions accurately enough for most objects without being overly complex.

tracks of subatomic particle movement in a bubble chamber

KINEMATICS 247

The woman's book is not moving in her frame of reference, but the man sees the book as moving because of the motion of the earth and moon.

11.2 FRAMES OF REFERENCE

As you read this textbook, are you moving? Regardless of whether you are sitting, standing, or even walking as you read, the answer is yes. No matter how still you remain, you are moving because the earth is spinning on its axis as it revolves around the sun. At the same time, the sun is moving within the Milky Way, which in turn is moving through space. If we instead ask whether you are moving around the room, the answer may be different. By referring to the room, the question establishes a **frame of reference**, which is a coordinate system that we can use to describe the motion of an object.

To better understand this concept, imagine that you are taking a train trip with your family. You and your brother sit playing a game while your sister walks toward the back of the train. You would claim that you and your brother are *not* moving and your sister *is* moving because your frame of reference is the train. But an observer outside the train would see all of you moving along with the train because their reference frame is the ground.

You have been using frames of reference for quite a while in math class. Number lines and coordinate planes are both frames of reference. A coordinate plane is a two-dimensional frame of reference consisting of two number lines connected at their origins. We can use either of them to describe the position of an object. In the coordinate plane below, where is the fly? The fly is five units to the right of zero, the origin. Where is the spider? The spider is seven units to the left of the origin, or at −7. The fly is probably glad that the spider is twelve units to his left. What is the position of the beetle on the coordinate plane?

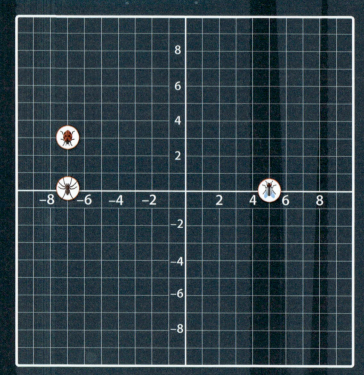

To fully describe position we need to know the point of reference, as well as our distance and direction from that point.

Often when we study motion, we find that the motion is very complex. We can simplify things by just studying part of the more complex motion. Consider a basketball game. What would be moving during the game? There are ten players, two referees, and the ball. The players could be running, jumping, spinning, or standing still and waving their arms. The ball might be dribbled, passed, or shot. Trying to study all that motion at the same time would be challenging. So instead we can identify a **system**—a portion of the larger motion that we are interested in studying. Maybe we are just interested in the ball. We would ignore the other motion and focus just on the ball system. Anything that is not part of the selected system is considered part of the environment.

11.3 DISTANCE AND DISPLACEMENT

Distance

By using a frame of reference, we can identify the position of different objects. We can also describe how far apart the positions are or how far a moving object has traveled during a time interval. Each of those quantities represents a **distance** (d). For example, the library may be 4 km from your house.

If you travel from your home to the library, you will travel 4 km. When you return home, you again will travel 4 km. This should make sense because if the library is 4 km from your house, then your house is 4 km from the library.

Displacement

But sometimes it is not enough to know the distance between two points. If a friend invites you over to his house and tells you only that his house is 5 km from your house, will it be easy for you to find? No. You need to know the direction to the friend's house too. When a distance is combined with a direction, the result is called a **displacement**—a change in position. The formula for calculating displacement is

$$\Delta x = x_f - x_i,$$

where Δx is displacement (or change in position), x_f is the final position, and x_i is the initial position.

Let's assume that his house is east of your house. To get to his house, you have to travel 5 km east as indicated by the red arrow below. The direction could be described in other ways, such as to the right or in the positive direction. To return home, you reverse direction and travel 5 km to the west (see the blue arrow). Again the direction could be to the left or in the negative direction. The two arrows are equal in magnitude but in opposite directions.

> **Symbols**
>
> The Greek letter delta (Δ) is used to indicate "change in" a particular quantity. So Δx is a change in x or change in position.
>
> The f (for *final*) and i (for *initial*) subscripts are there to remind you which position goes first. Other textbooks may use different subscripts.

It is possible for distance and displacement to have equal values, but distance is typically larger than displacement. When traveling to your friend's house, unless there is a road that goes straight between the two houses, your distance traveled will probably be more than 5 km. You will follow roads and sidewalks as you travel. But remember that displacement is the direct change in position between our starting and ending points. It's the actual distance between those points, or as we might say, "as the crow flies."

UNDERSTANDING DISTANCE AND DISPLACEMENT

Let's look at two more examples to better understand distance and displacement.

MOVING IN ONE DIMENSION

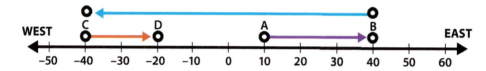

Notice that the number line has an origin, has positive and negative numbers, and is labeled with directions, east (+) to the right and west (−) to the left. When we describe direction, we can use right and left, east and west, or "+" and "−." Point A is 10 m to the east (right) of the origin.

A→B

Starting from A (10 m east) and moving to B (40 m east), we travel 30 m, and our displacement is

$$\Delta x = 40 \text{ m} - 10 \text{ m} = 30 \text{ m},$$

which means 30 m east, or 30 m to the right.

A→B→C

Continuing from B (40 m east) to C (40 m west), our distance traveled from A to C is 110 m and our displacement is

$$\Delta x = -40 \text{ m} - 10 \text{ m} = -50 \text{ m},$$

which means 50 m west, or 50 m to the left.

A→B→C→D

Turning around again and traveling from C (40 m west) to D (20 m west), our total distance from A to D is 130 m, and our displacement is

$$\Delta x = -20 \text{ m} - 10 \text{ m} = -30 \text{ m},$$

which means 30 m west, or 30 m to the left.

Notice that distance is always positive; the sum of several distances is always a positive number, regardless of the direction we move. But displacement can be positive or negative. Note too that displacement is always less than or equal to the distance.

MOVING IN TWO DIMENSIONS

The first example was on a number line. Let's look at a similar example in two directions.

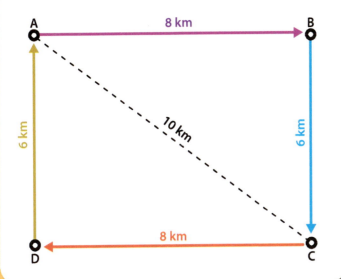

A→B

Starting from A and moving to B (8 km east), we have traveled 8 km and our displacement is 8 km east.

A→B→C

Continuing from B (8 km east) to C (8 km east, 6 km south), our distance traveled from A to C is 14 km. Our two displacement vectors are the legs of a right triangle. According to the Pythagorean theorem, the square of the hypotenuse is equal to the sum of the squares of the two legs.

$$c^2 = a^2 + b^2$$

So we can calculate the length of the hypotenuse.

$$c^2 = a^2 + b^2$$
$$\begin{aligned}c &= \sqrt{a^2 + b^2} \\ &= \sqrt{(8 \text{ km})^2 + (6 \text{ km})^2} \\ &= \sqrt{100 \text{ km}^2} \\ &= 10 \text{ km}\end{aligned}$$

Our displacement is 10 km to the southeast.

A→B→C→D

Turning west and continuing to D (6 km south), our total distance traveled from A to D is 22 km, but our displacement is 6 km to the south.

A→B→C→D→A

If we finish the trip by returning to our starting point, our total distance is 28 km, but our displacement is zero, because we are back at our starting point. Our position has not changed.

KINEMATICS 251

EXAMPLE 11-1: Calculating Distance and Displacement

While hiking in the national forest, you start at the ranger station and hike 4.5 km east. On the second leg of the hike, you travel 2.3 km south. Finally, you travel 5.7 km to the west. What is your total distance traveled? What is your displacement from the ranger's station?

Write what you know.

Δx_1 = 4.5 km east

Δx_2 = 2.3 km south

Δx_3 = 5.7 km west

Draw three displacement vectors end to end, using arrows to indicate direction.

Your total distance traveled is 12.5 km. Our movement leaves us 2.3 km south and 1.2 km west of the ranger station. Use the Pythagorean theorem.

$$c^2 = a^2 + b^2$$

$$c = \sqrt{a^2 + b^2}$$

$$= \sqrt{(2.3 \text{ km})^2 + (1.2 \text{ km})^2}$$

$$= \sqrt{6.73 \text{ km}^2}$$

$$= 2.6 \text{ km}$$

Your total displacement (red arrow) from the ranger station is 2.6 km to the southwest.

11.4 SCALARS AND VECTORS

As you can see, the concepts of distance and displacement are closely related. Distance and displacement differ in two ways. One is that distance depends on the path traveled, while displacement depends only on the starting and ending position. Also, distance is a *scalar* while displacement is a *vector*.

A **scalar** is any quantity that consists of magnitude, or size, only. Examples of scalar quantities include distance (35 m), temperature (22 °C), pressure (101 325 Pa), and speed (115 kph). **Vectors**, on the other hand, have both magnitude and direction. Vectors are drawn as an arrow with the length representing the size of the vector and the arrow pointing in the vector's direction. Throughout this textbook we will indicate vectors by using boldface symbols, such as **F** (for a force). Some common vector quantities are *displacement* (35 m north), *velocity* (115 kph northwest), *acceleration* (9.8 m/s² downward), and *force* (45 N upward).

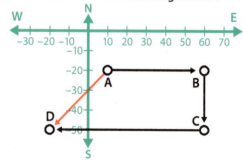

Notice all the vectors associated with this airplane. The lift, drag, and weight forces are vectors, as are velocity and acceleration.

11A | REVIEW QUESTIONS

1. Summarize the progression of thought about motion from the time of the ancient Greeks to modern times.
2. Why is a frame of reference important for describing motion?
3. Compare distance and displacement.
4. What is the distance and displacement for the motion, from A to B to C to D, shown in the image below?

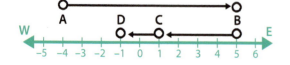

5. What is the distance and displacement (red arrow) for the motion shown in the image below?

6. Compare vectors and scalars.

11B | DESCRIBING MOTION

How can we study motion?

11.5 SPEED AND VELOCITY

Speed

When you are late for class, you may wonder why everyone else is walking so slowly. Other times, as people seemingly fly by you on the highway, you may ask why everyone is in such a hurry. In each case, we are thinking about the speed of the other people. **Speed** (*s*) is the distance traveled in an amount of time. For example, a highway speed of 95 kph means that during each hour you will travel 95 km. Speed is probably the most common way we talk about motion. The formula to calculate speed is

$$s = \frac{d}{t},$$

where *s* is speed in meters per second, *d* is distance in meters, and *t* is the time in seconds to travel that distance. Our cars are all equipped with a speedometer to let us know how fast we are going.

11B Questions

- How can we describe how something is moving?
- Are speed and velocity the same thing?
- What is momentum?
- Is momentum conserved?

11B Terms

speed, velocity, momentum

EXAMPLE 11-2: Calculating Speed

While on vacation, you travel by car from your home to Florida. You travel 1232 km in 14 h. What is your average speed?

Write what you know.

$d = 1232$ km

$t = 14$ h

$s = ?$

Write the formula and solve for the unknown.

$$s = \frac{d}{t}$$

Plug in known values and evaluate.

$$s = \frac{1232 \text{ km}}{14 \text{ h}}$$

$$= 88 \frac{\text{km}}{\text{h}}$$

Your average speed on your trip to Florida was 88 kph.

Notice in the above example that you calculated the *average* speed for the trip. Of course you didn't drive exactly 88 kph the entire time since you would have stopped for gas, food, or red lights. You may have also slowed for traffic congestion or highway construction. Average speed is based on the total distance and time. But don't look to your speedometer for this measure! It tells you only *instantaneous* speed—how fast you're going at a particular point in time.

Velocity

As we saw in the case of distance and displacement, information about what direction an object is traveling is important. In cases that involve both speed and direction, we use **velocity** (**v**), which is the rate at which an object's position changes. The formula to calculate velocity is

$$\mathbf{v} = \frac{\Delta \mathbf{x}}{\Delta t},$$

where **v** is velocity in meters per second, $\Delta \mathbf{x}$ is the displacement in meters, and Δt is the time interval in seconds. The formula can also be written with the displacement and time interval expanded to include positions and times.

$$\mathbf{v} = \frac{\mathbf{x}_f - \mathbf{x}_i}{t_f - t_i}$$

In this formula, $\mathbf{x}_f$ is the final position, $\mathbf{x}_i$ is the initial position, t_f is the final time, and t_i is the initial time.

EXAMPLE 11-3: Calculating Velocity from Positions and Time

A horse moves from 10 m east of the gate to 45 m east of the gate in 15 s. What is the average velocity of the horse?

Write what you know, including a sketch.

x_i = 10 m east

x_f = 45 m east

Δt = 15 s

v = ?

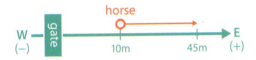

Write the formula and solve for the unknown.

$$v = \frac{\Delta x}{\Delta t} = \frac{x_f - x_i}{\Delta t}$$

Plug in known values and evaluate.

$$v = \frac{45 \text{ m east} - 10 \text{ m east}}{15 \text{ s}}$$

$$= \frac{35 \text{ m east}}{15 \text{ s}}$$

$$= 2.3 \frac{\text{m}}{\text{s}} \text{ east}$$

The horse's average velocity is 2.3 m/s to the east.

We can also use the velocity formula to calculate a change in position, as the following example shows.

EXAMPLE 11-4: Calculating Position from Velocity and Time

A ship was 185 km north of the port at noon and moves with an average velocity of 17 kph south. Where will the ship be at 3:30 p.m. (3.5 h later)?

Write what you know, including a sketch.

x_i = 185 km north

v = 17 kph south

Δt = 3.5 h

x_f = ?

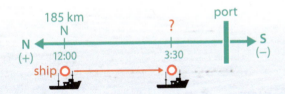

The sketch shows north as positive.

Write the formula and solve for the unknown.

$$v = \frac{\Delta x}{\Delta t} = \frac{x_f - x_i}{\Delta t}$$

$$v \Delta t = \left(\frac{x_f - x_i}{\Delta t}\right) \Delta t$$

$$v \Delta t + x_i = x_f - \cancel{x_i} + \cancel{x_i}$$

$$x_f = v \Delta t + x_i$$

Plug in known values and evaluate.

$$x_f = \left(-17 \frac{\text{km}}{\cancel{\text{h}}}\right)(3.5 \cancel{\text{h}}) + 185 \text{ km}$$

$$= -59.5 \text{ km} + 185 \text{ km}$$

$$= 126 \text{ km}$$

The ship is at +126 km, or 126 km north of the port at 3:30.

ADDING VELOCITIES

Velocities, like displacements, can add together. Because velocities are vectors, we have to account for direction. This is called *vector addition*. Let's look at how this works.

MOVING IN THE SAME DIRECTION

Notice how wind can affect an airplane.

$v_{airplane}$ = 720 kph west

v_{wind} = 83 kph west

$v_{airplane} + v_{wind} = v_{total}$ = 720 kph west + 83 kph west = 803 kph west

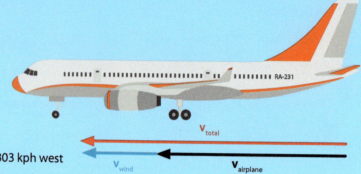

$v_{airplane}$ = 720 kph east

v_{wind} = 83 kph west

$v_{airplane} + v_{wind} = v_{total}$ = 720 kph east + 83 kph west

West is the negative direction.

720 kph + (–83 kph) = 637 kph = 637 kph east

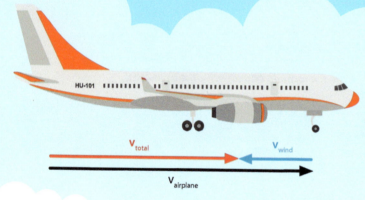

MOVING IN TWO DIMENSIONS

$v_{airplane}$ = 720 kph east

v_{wind} = 83 kph north

$v_{airplane} + v_{wind} = v_{total}$

Use the Pythagorean theorem to solve.

$c^2 = a^2 + b^2$

$c = \sqrt{a^2 + b^2}$

$= \sqrt{(720 \frac{km}{h})^2 + (83 \frac{km}{h})^2}$

$= \sqrt{525\,289 \left(\frac{km}{h}\right)^2}$

$= 725 \frac{km}{h}$

720 kph east + 83 kph north = 725 kph northeast

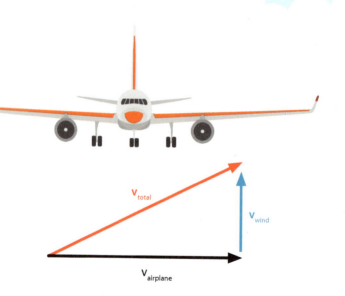

256 CHAPTER 11

11.6 GRAPHING MOTION

While descriptions and formulas can give us a good understanding of motion, graphs are also a great tool for understanding motion. One common graph for depicting motion is a *position versus time graph* (sometimes called a *p-t graph*). Consider a truck that starts at position 0 m and drives at a constant velocity for 5 seconds and ends up at position 40 m. We can create a p-t graph for that motion (see right).

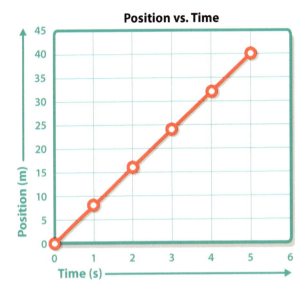

What can we see on this graph? The starting position is (0, 0), and the final position is (5, 40). What is the slope of this graph? Recall from math class that the slope between any two points can be found by dividing Δy by Δx. For this graph that would be change in position—displacement—divided by change in time. So the slope of the p-t graph is the velocity. In the example above, the truck moved at a constant 8.0 m/s in the positive direction.

We can also create a *velocity versus time graph* (v-t graph) of the truck (see right). Again, the graph shows that the truck is traveling at a constant 8.0 m/s because the horizontal line represents the velocity. But we can also determine the displacement of the truck using this graph. Think about the formula for velocity.

$$\mathbf{v} = \frac{\Delta \mathbf{x}}{\Delta t}$$

Now solve for displacement.

$$\Delta \mathbf{x} = \mathbf{v}\Delta t$$

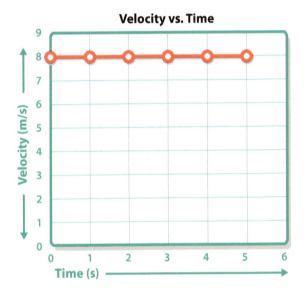

Displacement is just velocity multiplied by the time interval. Look again at the graph below: the velocity and time intervals are the sides of a rectangle with the velocity as the height and the time interval as the base. Therefore the displacement is the area between the curve of the v-t graph and the *x*-axis. Mathematicians call this the area under the curve. In the graph below, we can see that the area under the curve is 40 m, which matches the displacement described above as the truck moved from 0 m to 40 m.

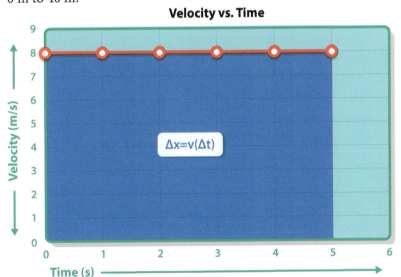

KINEMATICS 257

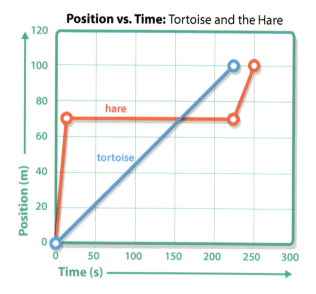

Position vs. Time: Tortoise and the Hare

Let's look at an example in which an object changes velocities. Do you remember the story "The Tortoise and the Hare"? In the story, the hare bragged about how fast he was, which led the tortoise to challenge him to a race. On race day, as expected, the hare took off at a great speed. He had such a lead that he stopped to rest. Having fallen asleep, he didn't realize that the tortoise had passed him until the cheers of the crowd awoke him. The hare continued, but the tortoise with his slow and steady pace had already won the race.

The slope of the graph indicates that the tortoise arrived at the finish line first by walking at 0.44 m/s. Additionally, in the two segments in which the hare was running, he averaged 18 m/s and 20 m/s. So why did he lose the race? Let's calculate his average velocity for the entire race.

EXAMPLE 11-5: Calculating the Hare's Average Velocity

What was the hare's average velocity for the entire race?

Write what you know.

$x_i = 0$ m

$x_f = 100$ m

$t_i = 0$ s

$t_f = 250$ s

$v = ?$

Write the formula and solve for the unknown.

$$v = \frac{\Delta x}{\Delta t} = \frac{x_f - x_i}{\Delta t}$$

Plug in known values and evaluate.

$$v = \frac{100 \text{ m} - 0 \text{ m}}{250 \text{ s} - 0 \text{ s}}$$

$$= \frac{100 \text{ m}}{250 \text{ s}}$$

$$= 0.4 \frac{\text{m}}{\text{s}}$$

The hare lost because his average velocity for the race was slower than the tortoise's.

The momentum of the RMS *Titanic* prevented its crew from stopping or turning before hitting an iceberg on its maiden voyage.

11.7 MOMENTUM

Another term used to describe motion is one that you may be familiar with in a different context. Sports announcers will often mention the momentum of a game. They may talk of a shift in momentum toward one team or the other. So a team with the momentum is moving toward the win, and it will be hard to stop them.

In physics, **momentum (p)** is a property of a moving system that is equal to its velocity multiplied by its mass. The formula for momentum is

$$\mathbf{p} = m\mathbf{v},$$

where **p** is momentum in kilogram-meters per second, *m* is mass in kilograms, and **v** is velocity in meters per second.

EXAMPLE 11-6: Calculating Momentum

Compare the momentums of the three vehicles.

29 m/s east
1272 kg

29 m/s east
12 980 kg

175 m/s east
12 980 kg

Write what you know.

v_{car} = 29 m/s east

m_{car} = 1272 kg

v_{truck} = 29 m/s east

m_{truck} = 12 980 kg

$v_{airplane}$ = 175 m/s east

$m_{airplane}$ = 12 980 kg

Write the formula and solve for the unknown.

$$p = mv$$

Plug in known values and evaluate.

Car

p_{car} = (1272 kg) (29 $\frac{m}{s}$ east)

= 37 000 kg $\frac{m}{s}$ east

Truck

p_{truck} = (12 980 kg) (29 $\frac{m}{s}$ east)

= 370 000 kg $\frac{m}{s}$ east

Airplane

$p_{airplane}$ = (12 980 kg) (175 $\frac{m}{s}$ east)

= 2 270 000 kg $\frac{m}{s}$ east

Even though the car and truck have the same velocity, the truck has more momentum because of its greater mass. And though the truck and the airplane have the same mass, the airplane has more momentum because of its higher velocity. The airplane has more momentum than the car because its mass and velocity are both higher than the car's.

Conservation of Momentum

Momentum is another physical quantity that is conserved. The *law of conservation of momentum* states that within a closed system—a system that is isolated from outside pushes or pulls—the total momentum remains constant. This law is helpful in analyzing collisions and is also the basis for rocket science.

11B | REVIEW QUESTIONS

1. Define *speed* and *velocity*.
2. How long will it take you to travel 647 m if you travel at 47.2 m/s?
3. A boat is cruising north at 15 kph. Add the boat vector to the given tide vector for each of the following.
   a. 5 kph north
   b. 5 kph south
   c. 5 kph east
4. Are speed and velocity the same thing? Explain.
5. What two factors affect momentum?
6. Under what conditions is momentum conserved?

Use the table below to answer Questions 7–8.

Time (hr)	Position (km)
0	40
2	70
3	70
6	90
12	−30

7. Draw the p-t graph.
8. What is the average velocity between 2 h and 6 h?

MINI LAB

GRAPHING MOTION

Essential Question:

How can we graphically represent motion?

Equipment

stopwatch

masking tape

centimeter ruler

computer paper

graph paper (2)

The ability to map our movements is critical for any type of navigation. Pilots, navigators, and hikers all have to be able plot their progress on maps. They then have to be able to calculate distances and displacements from the mapped data. In this activity, you will mark your progress on a number line and then model the motion on a graph.

1. What do we need to know to be able to describe position?
2. How do velocity and speed compare?

PROCEDURE

A Using a ruler, draw a number line on the computer paper with the origin about 1/3 of the way across the page.

B Start by numbering the origin "0." When the timekeeper tells you to start, move your pencil along the number line from the origin and at a constant velocity. When the timekeeper tells you to change, number the spot and then change your motion. You can change direction, change speed, or you can stop.

C Timekeeper: Say start and start the stopwatch. Say change at 5 s, 12 s, and 20 s. At 25 s, say stop.

D Using a piece of graph paper, plot these positions on a p-t graph (see graph on page 258).

CONCLUSION

3. How do you calculate average velocity between any two points on a p-t graph?
4. What is the total distance traveled and displacement?

GOING FURTHER

E Calculate the average velocity between each pair of positions (0 to 1, 1 to 2, etc.). Using the second sheet of graph paper, create a v-t graph for your motion. See the v-t graph on page 257.

5. The v-t graph shows motion that is physically impossible. Explain why that is.

11C | CHANGING MOTION

How do objects move in the real world?

11.8 ACCELERATION

Up to this point, we have considered only things that move at a constant velocity, but that is usually not what happens in real life. We say that an object is accelerating when its velocity changes. **Acceleration (a)** is the change in velocity in a given amount of time. Acceleration occurs when an object's speed changes, when an object changes direction, or when both speed and direction change. The formula for acceleration is

$$\mathbf{a} = \frac{\Delta \mathbf{v}}{\Delta t},$$

where **a** is the acceleration in meters per second squared, $\Delta \mathbf{v}$ is the change in velocity in meters per second, and Δt is the time interval in seconds. Like velocity, acceleration is also a vector quantity, which means that it includes direction. Acceleration can be either positive or negative. Positive acceleration means that a velocity is becoming more positive. If a velocity becomes more negative, then the object has a negative acceleration. Both positive and negative accelerations can make an object speed up or slow down. Speeding up occurs when the velocity and acceleration vectors are in the same direction, while slowing occurs when the velocity and acceleration vectors are in opposite directions.

The units for acceleration are m/s². Remember that acceleration is a change in velocity per unit of time. So if a hummingbird has an acceleration of 2.0 m/s², then its velocity will change 2.0 m/s *each second*. If initially moving at 10.0 m/s, a second later it will be moving at 12.0 m/s and at 14.0 m/s another second later. Given the formula for acceleration above, the units work out to be (m/s) ÷ s. In acceleration calculations, it's important to remember that when dividing by a fraction, *we multiply by the reciprocal*, so our units become $\left(\frac{m}{s}\right)\left(\frac{1}{s}\right) = \frac{m}{s^2}$.

11C Questions

- How do we describe motion if velocity changes?
- How does circular motion differ from linear motion?
- What causes projectiles to follow a curved path?

11C Terms

acceleration, free fall, circular motion, centripetal acceleration, projectile motion, trajectory

EXAMPLE 11-7: Calculating Acceleration from Velocity and Time

At the end of a race, a top fuel dragster accelerates from 147 m/s west to 119 m/s west in 3.1 s. What is the acceleration of the car?

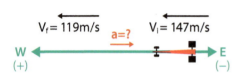

Write what you know.

v_i = 147 m/s west

v_f = 119 m/s west

Δt = 3.1 s

a = ?

Write the formula and solve for the unknown.

$$a = \frac{\Delta v}{\Delta t}$$

Plug in known values and evaluate.

$$a = \frac{\Delta v}{\Delta t} = \frac{v_f - v_i}{\Delta t}$$

$$= \frac{119 \text{ m/s} - 147 \text{ m/s}}{3.1 \text{ s}}$$

$$= \frac{-28 \text{ m/s}}{3.1 \text{ s}}$$

$$= -9.0 \frac{m}{s^2}$$

The car accelerates at -9.0 m/s², or 9.0 m/s² to the east.

EXAMPLE 11-8: Calculating Velocity from Acceleration and Time

A bus is traveling down the road at 12.4 m/s. If it accelerates at 5.1 m/s² for 2.7 s, what is its new velocity?

Write what you know.

v_i = 12.4 m/s

a = 5.1 m/s²

Δt = 2.7 s

v_f = ?

Write the formula and solve for the unknown.

$$a = \frac{\Delta v}{\Delta t} = \frac{v_f - v_i}{\Delta t}$$

$$a \Delta t = \left(\frac{v_f - v_i}{\Delta t}\right) \Delta t$$

$$a \Delta t + v_i = v_f - v_i + v_i$$

$$v_f = a \Delta t + v_i$$

Plug in known values and evaluate.

$$v_f = \left(2.7 \frac{m}{s^2}\right)(5.1 \text{ s}) + 12.4 \frac{m}{s}$$

$$= 13.8 \frac{m}{s} + 12.4 \frac{m}{s}$$

$$= 26 \frac{m}{s}$$

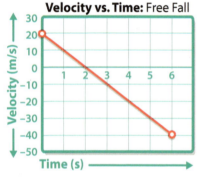

Free Fall

Recall from Subsection 11.1 that Galileo discovered that all objects fall at the same rate—if there is no air resistance. This motion is called **free fall**—the motion of objects that fall due to gravity alone with no other forces acting on them. An object can experience this condition only in a vacuum. Objects in free fall change speed with a constant acceleration, regardless of their masses. The acceleration of a body, near the surface of the earth, in free fall is approximately 9.81 m/s² downward. This means that for every second the object falls, it is moving 9.81 m/s faster. So after one second, it is moving at 9.81 m/s, after two seconds, at 19.62 m/s, and after three seconds, at 29.43 m/s. The symbol for free fall due to gravity is **g**.

Graphing Motion

Let's look again at graphing motion. What do the v-t and p-t graphs look like for constant accelerated motion? Let's think about standing on a 58.8 m tall building and throwing a ball upward at 19.6 m/s. The ball is going to move upward, even as it is slowing down. It will momentarily come to a stop, and then it will begin to fall downward, gaining speed. Observe the two graphs at left to see how the graphs look for this ball in free fall.

In the first graph, notice the *x*-intercept at about 2 seconds. Since this is the v-t graph, the *x*-intercept indicates that the velocity is zero. This point is where the ball reaches its highest point, the apex, and is about to start falling. What is the acceleration of the ball at this point? The correct answer is that the acceleration is still 9.81 m/s² downward. Since it stopped momentarily, you may be tempted to think that its acceleration must be zero. But if that were true, then the ball would hover there. We know that doesn't happen. Remember, in free fall on Earth, the acceleration is a constant 9.81 m/s² downward. The slope on the v-t graph is the acceleration of the object.

On the p-t graph, we can see the apex. What would be the slope at that point? The slope would be zero, which matches the *x*-intercept from the v-t graph. Notice that the graph stops at 0 m. That is because the ball has landed on the ground next to the building. We will *drop in* on the topic of free fall again in Chapter 12 when we discuss force due to gravity.

What was the change in the position of the ball? We know that it started at 58.8 m and landed at 0 m, so Δx = 0 m – 58.8 m = – 58.8 m. Remember that the area between the v-t graph and the *x*-axis should be the displacement. Look at the v-t graph at right to see that it does work.

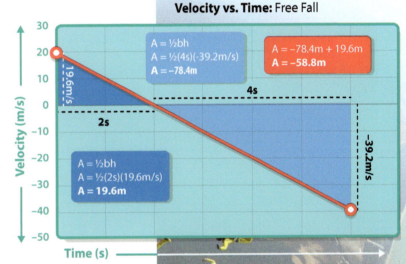

Velocity vs. Time: Free Fall

$A = \frac{1}{2}bh$
$A = \frac{1}{2}(4s)(-39.2m/s)$
$A = -78.4m$

$A = -78.4m + 19.6m$
$A = -58.8m$

$A = \frac{1}{2}bh$
$A = \frac{1}{2}(2s)(19.6m/s)$
$A = 19.6m$

The motion that we have discussed so far is called *linear motion*. Linear is a term that you know from math class. Linear motion is a description of motion when an object moves along straight lines or in straight paths. But very little motion in our daily lives moves in this way. Most objects move along curved paths. We will consider two types of curved motion: circular motion and projectile motion.

11.9 TWO-DIMENSIONAL MOTION

Circular Motion

Imagine that you are riding the Round Up at the amusement park. You are moving along a circular path—**circular motion**. If the ride were spinning at a constant speed, would you be accelerating? The answer is yes! Recall from Subsection 11.7 that acceleration is a change in velocity that could include a change in speed, a change in direction, or both. So even if the ride spins at a constant speed, the riders would be accelerating because they are constantly changing direction. When acceleration causes an object to move around a circular path, we see **centripetal acceleration** in action. This kind of "center seeking" always accelerates an object toward the center of a circle.

SERVING AS AN IMAGINEER:
MAKING MAGIC

Derek Howard grew up in southern California, home of Disneyland® Resort in California. After attending a college of design and working as both a graphic and industrial designer, he decided that he could mesh art and industrial design as an imagineer. Today Derek is the Industrial Designer Principal for Disney.

The job of a creative designer is to mesh the thrills of motion with the wonder of art. They incorporate displacement, velocity, and acceleration with sensory inputs. Many of us think of the visual and sound effects of a roller coaster ride, but ride designers also think about the senses of touch and smell. The creative team can even give the sensation of motion without riders even moving an inch.

God gives each of us different gifts and talents. Some have great scientific and mathematical abilities while others have amazing creative talents. And some have the opportunity to combine their mathematical, engineering, and creative abilities as imagineers.

You experience circular motion quite often in your daily life. For example, as you ride in the car, you experience circular motion each time you take a turn. As the driver turns the wheel of the car, the car moves in a circular path along the road. The car causes you to accelerate toward the inside of the curve so that you keep moving in the same direction as the car. You may feel as if something is pulling toward the outside of the curve, but that is just your body wanting to continue in the direction it was already going.

Projectile Motion

Have you ever watched expert applied physicists on TV? Without realizing it, you very likely have. During the fall season, you may watch athletes throw, catch, and kick a football. In the winter, you may love to watch the majestic arc of a three-point shot in basketball. Spring and summer provide you the opportunity to watch your favorite baseball players catch fly balls and hit towering home runs. Each time a player kicks, hits, or throws a ball, he is demonstrating **projectile motion**. This motion is the two-dimensional motion of any flying object whose path is determined by the influence of external forces only, such as gravity. Projectile motion looks complex, but in reality, it is two simple forms of linear motion occurring at the same time.

The projectile—the flying object—follows a curved path called its **trajectory**. This path is the combination of constant velocity motion in the horizontal direction and accelerated motion in the vertical direction. Without air resistance, these two motions combine to follow a path that is defined as a *parabola*. But in most cases there is air resistance, so the path is changed as shown in the graph at left. The consistency of this motion is what allows athletes to predict where the ball will end up so that they can catch it.

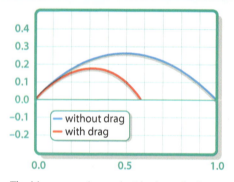

The blue curve shows the ideal parabolic trajectory of projectile motion. The red curve shows a more realistic path that is due to air resistance slowing down the object.

11C | REVIEW QUESTIONS

1. Define *acceleration*.
2. The SpaceX Falcon Heavy on page 261 accelerates from launch to 78 000 m/s in 561 s. What is its acceleration?
3. A boat is moving at 27.5 m/s south and accelerates at 2.51 m/s² north for 12.1 s. What is its velocity after it accelerates?
4. Give an example of a positive acceleration and a negative acceleration.
5. What is free fall?
6. Define *circular motion*.
7. How do we describe motion when velocity changes?
8. What do we call the curved path of a projectile?
9. What are some reasons that projectiles in the real world don't perfectly follow the shape of a parabola?

CHAPTER 11 REVIEW

11A DESCRIBING POSITION

- Motion can be modeled using Newtonian, quantum, or relativistic mechanics. While less workable, Newtonian mechanics works well for the motion in our daily lives.
- A frame of reference provides a coordinate system by which we can describe motion.
- A system is the portion of the universe that we are studying. Anything that is not part of a system is its environment.
- Distance is the number of units traveled by an object. Displacement is a change in position and includes direction.
- All quantities are either scalars or vectors. Scalars show magnitude only, while vectors include both magnitude and direction.

11A Terms

mechanics	246
kinematics	246
frame of reference	248
system	248
distance	249
displacement	249
scalar	252
vector	252

11B DESCRIBING MOTION

- The rate at which an object moves is its speed, or its velocity. Speed, a scalar, is the distance traveled per unit of time. Velocity, a vector, is the displacement per unit of time.
- In vector addition, quantities are added together while taking into account their directions.
- Motion can be modeled well with position versus time and velocity versus time graphs.
- Momentum, a vector quantity, is an object's mass multiplied by its velocity. Momentum is conserved if the system is isolated.

11B Terms

speed	253
velocity	254
momentum	258

11C CHANGING MOTION

- Acceleration, a vector quantity, is the change of velocity in a unit of time. The units for acceleration are m/s^2.
- Free fall is the constant accelerated motion of an object falling with only gravity working on it. The free-fall acceleration of all objects near the earth's surface is 9.81 m/s^2 downward.
- Motion can also occur in two dimensions, such as in circular motion and projectile motion.
- Circular motion is accelerated motion in which the object moves in a circular path. The acceleration that causes the motion is always toward the center—a centripetal acceleration.
- Projectile motion is two-dimensional motion that combines constant velocity motion in the horizontal direction with free fall. The path that a projectile follows is called its trajectory. Without air resistance, the shape is that of a parabola.

11C Terms

acceleration	261
free fall	262
circular motion	263
centripetal acceleration	263
projectile motion	264
trajectory	264

KINEMATICS 265

CHAPTER 11

CHAPTER REVIEW QUESTIONS

Recalling Facts

1. What is a frame of reference?
2. What are the positions of the car, truck, and airplane in the graph below?

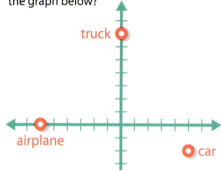

3. Identify the quantity being measured by each of the following and identify it as either a scalar or a vector.
   a. 45 m/s to the south
   b. 45 m
   c. 23.4 °C
   d. 3.2 m/s² downward
4. Give an example of a speed and a velocity.
5. How can two objects with the same mass have different momentums?
6. What is the acceleration rate for objects in free fall near the surface of the earth?
7. In what direction does centripetal acceleration act on the driver in the image below?

8. Sketch the trajectory of projectile motion with and without air resistance.
9. What causes projectiles to follow a curved path?

Understanding Concepts

10. Explain why we still use Newtonian mechanics for studying motion even though more-workable models are available.
11. Explain why people developed latitude and longitude as a common frame of reference.
12. As you are working on a lab, your lab partner tells you that distance and displacement are the same thing. Do you agree? Explain.
13. You walk three blocks west, then three blocks north, and finally seven blocks south. What is your distance traveled? your displacement?
14. You are traveling west on the highway. At 1:00 you pass kilometer marker 485. At 4:30, you pass kilometer marker 154. What has your average velocity been?
15. If a pitcher throws a baseball with a mass of 0.145 kg at 43.3 m/s west, what is the ball's momentum?
16. At what velocity would a softball with a mass of 0.188 kg have to be thrown to have the same momentum as the baseball in Question 15?

Use the p-t and v-t graphs below to answer Questions 17–23.

17. What is the position of the object after 8.5 s?

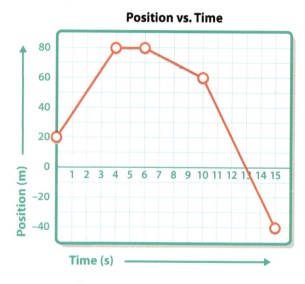

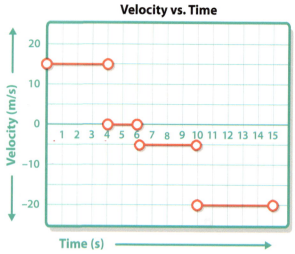

REVIEW

18. During what time interval is the object stopped?
19. At what time is the object at position zero?
20. What is the object's average velocity between 4 s and 10 s?
21. What is the object's displacement during the motion graphed?
22. What is the velocity of the object at 7 s?
23. What is the average acceleration of the object from 5 s to 15 s?
24. Explain why m/s² are the units for acceleration.
25. A major league baseball player hits a pitch for a home run. It was moving at 40.0 m/s toward the batter but moves at 51.0 m/s away from the batter after being hit. If the change in velocity happened in 0.34 s, what was the acceleration of the ball? (*Note*: Away from the batter is positive.)
26. Create a concept map for motion. Include the terms ~~distance~~, ~~displacement~~, ~~scalar~~, ~~vector~~, ~~speed~~, ~~velocity~~, ~~acceleration~~, ~~linear motion~~, ~~circular motion~~, ~~centripetal acceleration~~, projectile motion, and ~~trajectory~~.
27. Compare linear, circular, and projectile motion.

Critical Thinking

28. In the image at right, what frame of reference would allow the fly to be moving in a straight line?

29. Give an example for each kind of motion listed.
    a. The distance and displacement are the same.
    b. The distance and displacement are exactly opposite.
    c. An object has traveled but has no displacement.
30. On a two-day trip, you average 80 kph for 11 h on the first day and 120 kph for 7 h on the second day. What is your average speed for the two days? (*Hint*: Remember the definition for speed.)

31. Create a story to fit the graphs below.

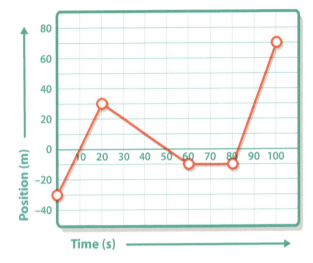

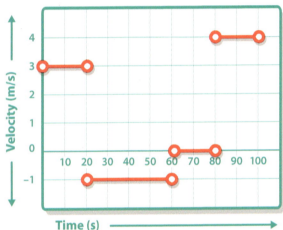

Use the Ethics box below to answer Question 32.

32. Write a three-paragraph response on the ethical use of radar detectors. In the first paragraph, discuss principles from Scripture that may guide Christians in this area. In the second paragraph, discuss the possible outcomes of heeding or ignoring the guidelines of Scripture. In the third paragraph, give your opinion on whether we should use radar detectors.

ETHiCS — RADAR DETECTORS

THE ISSUE: OBEYING THE LAW

Along our highways there are speed limits, both maximums and minimums. A speed limit is determined by what is considered a safe speed for a section of road on the basis of road design, traffic volume, and other conditions.

According to the National Highway Traffic Safety Administration, speeding caused 27% of roadway fatalities nationwide (over 10,000 people) in 2016. Regardless of how high the speed limit is, people exceed it. Radar detectors are designed to alert drivers to police speed detection devices so that the driver can slow down before being ticketed.

CHAPTER 12

Dynamics

SIMPLY SMASHING!

Crashing headlong into a carload of your friends is fun, as long as it's bumper cars at the fair that we're talking about. Crashing in a car traveling at 30 m/s down the freeway is another story. Every year, motor vehicle accidents are one of the leading causes of death in the United States and elsewhere. Much research is being done to find ways to make collisions more survivable. Crash tests, like the one shown at left, collect data that is used to analyze the forces at work during a collision. Engineers use this data to help them design better safety features for cars. By the end of this chapter, you'll have a better understanding of the basic principles of physics that are at work in car crashes and other changes in motion. Buckle up!

12A	Classifying Forces	270
12B	Newton's Laws of Motion	275
12C	Types of Forces	282

12A Questions

- What is a force?
- Can a force be exerted at a distance?
- Do all forces change motion?
- How can I illustrate the forces acting on an object?

12A Terms

force, dynamics, contact force, field force, free-body diagram

12A | CLASSIFYING FORCES

What causes a change in motion?

12.1 FORCE

What makes bumper soccer fun? Crashing into another player! Such a crash is simply a very hard and abrupt kind of push. And that's what a **force** is—it's a push or pull on an object. A force is needed to change the motion of an object, so all the accelerations that you read about in Chapter 11 are caused by forces. Throwing a baseball, stopping a runaway shopping cart, and sliding a casserole dish toward you at the dinner table are all examples of forces at work. **Dynamics** is the branch of physics that studies these forces and how they can change an object's motion. As you'll see throughout this chapter, there are many different kinds of forces, but remember that they are all some form of pulling or pushing.

In the SI system, the unit of force is the newton (N). You first saw this unit back in Chapter 2 in reference to weight. Later in this chapter you'll see that weight is itself a type of force. One newton of force is the amount of push or pull needed to accelerate a 1 kg object at 1 m/s². Thus, one newton is equal to one kilogram meter per second squared.

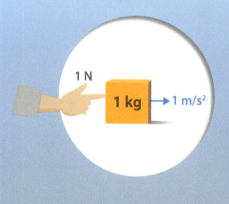

You may have noticed that all the examples in the first paragraph involved a person actually touching an object whose motion is changed. But do all forces require such contact? We can classify forces into two broad categories on the basis of whether the thing exerting the force is in contact with the object being acted on. Let's take a closer look at these two kinds of forces.

Contact Force Versus Field Force

As their name implies, **contact forces** act only when one object touches another. Contact forces originate at the atomic level of matter. Atoms resist the intrusion of other particles into their space, so when the atoms of another object push on them, they push back. You exert a contact force when you open a drawer, step on the pavement, twist a jar lid, or apply a strip of adhesive tape.

Field forces, also known as *forces at a distance*, act between objects that are not touching. Though such forces originate in matter, they do not need to act through matter. Physicists explain that such forces act through a force field. So what is a field? A field is the area around an object in which other things may feel a force exerted by the object. In other words, everywhere in the field objects will have a force acting on them. The principal field forces are the gravitational, magnetic, and electric forces. You'll read more about these in Section 12C.

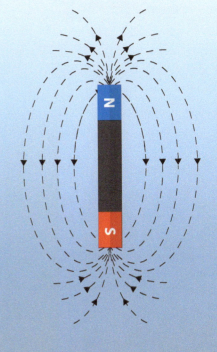

Balanced Force Versus Unbalanced Force

Most objects in the universe have more than one force acting on them. For example, the forces acting on an airplane flying at 10,000 m include the thrust of its engines, the lift on its wings, air resistance (drag), and, of course, Earth's gravity (weight). If forces like these are acting on every object all the time, why isn't everything in continual motion? It's because most objects experience *balanced forces*—multiple, simultaneous forces whose pushes and pulls cancel each other out.

A common and easily understood example of this is a tug of war. Let's see how this works. Assume that two teams are pulling on a tug-of-war rope. Each team exerts a force by pulling on the rope, but they are pulling in opposite directions. If the teams pull equally hard, then the two forces are balanced and the rope doesn't move. As far as the rope is concerned, there is no *net force* acting on it, because the sum of all the forces on the rope is zero. It behaves as though there were no forces applied to it at all. In the same manner that a positive number and a negative number of the same absolute value add up to zero, the two forces in this example are equal in magnitude but opposite in direction, producing a *zero net force*. On the other hand, if one team pulls harder than the other does, then an *unbalanced force* acts on the rope. The net force vector equals the sum of all the forces acting on the rope. (Refer back to Chapter 11 to review adding vectors.) The stronger team will eventually drag the other team over the line in the direction of their pull. In the same fashion, any acceleration—a change in motion—must be initiated by an unbalanced force.

12.2 FREE-BODY DIAGRAMS

In the real world, there are usually more than two forces acting on a system, often in multiple directions. If these forces cancel each other out, they are balanced and no net force exists (a zero net force). If the forces don't cancel, there is a single net force exerted in a specific direction. The result of an unbalanced force on a system is an acceleration. A convenient way to analyze the forces acting on an object is to do a quick sketch. Such a sketch, called a **free-body diagram**, or *force diagram*, shows the object and the forces acting on it as vectors. Vectors are represented by arrows: the vector's magnitude is represented by the arrow's length and its direction by the arrow. Don't worry if you aren't an artist; many physicists represent the object with a dot and then add the vectors. By comparing the magnitude and direction of the vectors, one can quickly determine the magnitude and direction of any resulting net force.

Take the tug of war example, for instance. We can use a free-body diagram to analyze the forces acting on the rope. If the two teams are pulling equally hard on the rope, then our diagram might look something like the one below.

If one team pulls with more force than the other, then that team's vector is shown larger than the other team's vector.

If we know the amount and direction of each force acting on the object represented in a free-body diagram, then we can also calculate the magnitude and direction of any resulting net force.

- rope
- F_1 force applied by Team 1
- F_2 force applied by Team 2

EXAMPLE 12-1: Calculating Net Force—Forces Working Together

Problem: John pushes a lab cart with a force of 135 N. Sally comes along to help him and pushes in the same direction with a force of 115 N. What is the net force acting on the lab cart?

Draw a diagram. Your diagram doesn't need to show a great deal of artistic ability or even any artistic ability at all. In this example, all that is needed is a rectangle to represent the lab cart. We'll also add a vector to represent John's pushing force. We'll include a label for the vector to show how much force John is exerting on the cart.

Next we'll add Sally's vector. Remember that Sally is pushing in the same direction as John.

Analyze. Recall that the net force is the sum of the forces on the object. Thus,

$$\mathbf{F}_{net} = \mathbf{F}_{John} + \mathbf{F}_{Sally}$$

$$= 135 \text{ N to the left} + 115 \text{ N to the left}$$

$$= 250 \text{ N to the left}$$

John and Sally push with a net force of 250 N in the same direction as their original forces.

EXAMPLE 12-2: Calculating Net Force—Opposing Forces

Problem: John again pushes the lab cart with a force of 135 N. Sally comes along but pushes in the opposite direction with a force of 115 N. What is the net force acting on the lab cart?

Draw a diagram. In this example, a dot is used to represent the lab cart. We add a vector to represent John's pushing force. Next we'll add Sally's vector. Remember that Sally is pushing in the opposite direction as John.

Analyze. Again we add the vectors. Thus,

$$\mathbf{F}_{net} = \mathbf{F}_{John} + \mathbf{F}_{Sally}$$

$$= 135 \text{ N} + (-115 \text{ N})$$

$$= 20 \text{ N to the left}$$

John and Sally push with a net force of 20 N in the same direction as John's original force.

As you'll soon see, there are other forces acting on the cart besides just John and Sally. Physicists can choose the forces to model in a free-body diagram, whether all of them or just some of them. Modeling *all* of them can get rather complex! Since physicists are usually interested in those forces that can change an object's motion, they often omit balanced forces from their analyses.

12A | REVIEW QUESTIONS

1. What is a force?
2. How does dynamics differ from kinematics (see Chapter 11)?
3. What is the unit of force in the SI system?
4. Do objects need to be touching in order for one to exert a force on the other? Explain.
5. What are the three principle field forces?
6. Explain the difference between balanced and unbalanced forces.
7. Identify whether each of the scenarios shown in the free-body diagrams below will result in acceleration. If acceleration results, include the direction that the object will accelerate.

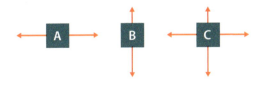

12B | NEWTON'S LAWS OF MOTION

How can we predict changes in motion?

12.3 FROM GALILEO TO NEWTON

In ancient times, people believed that all things would slow to a stop if left to themselves because, for most objects, being at rest was their natural state. This idea seems logical enough. After all, if a car engine stops, the car will roll to a standstill. If you shove a book across a table, it slides to a stop. But it wasn't until the 1500s that anyone tested the idea. Galileo is the first person known to have rigorously tested ideas about motion. He spent a great deal of time experimenting by rolling balls down inclined planes, measuring projectile motion, and timing pendulums. From his experiments, Galileo concluded that objects continue their motion indefinitely unless some external force changes their motion. He was able to infer this even though he couldn't directly observe such a situation on Earth. Johannes Kepler first gave this property of matter a name. He called the tendency of matter to resist change in its motion **inertia**.

12B Questions

- Why is it harder to change the motion of heavier objects?
- How are force, mass, and acceleration related?
- How do rockets fly?
- Can a stationary object exert a force?

12B Terms

inertia, law of inertia, law of acceleration, law of action-reaction, tension, normal force

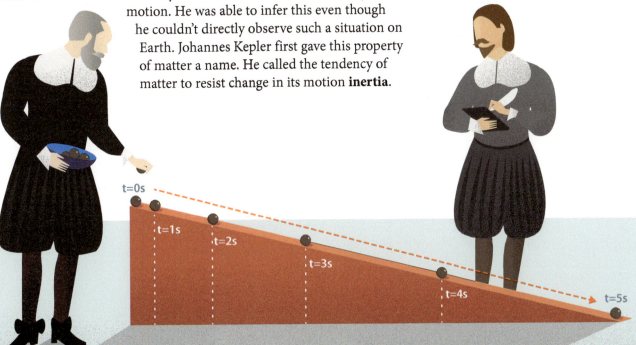

Early scientists, including Galileo, never settled on an explanation for how forces caused changes in motion. One reason for this was the lack of the right technology. To study motion, one must be able to measure time intervals accurately. Galileo developed techniques to subdivide time for his work but without the necessary accuracy. It wasn't until a few years after his death that the invention of accurate and precise pendulum clocks solved the problem. The new technology allowed Isaac Newton to build on Galileo's concept of inertia. The results of this important work were his three laws of motion.

12.4 NEWTON'S FIRST LAW

Newton merged Galileo's principle of inertia into a more general statement of physics that included his own discoveries about the actions of forces. Newton's first law or the **law of inertia** states:

> *Objects at rest remain at rest, and objects in motion continue in a straight line at a constant velocity (speed and direction) unless acted on by a net external force.*

We can't measure inertia directly and so it has no units associated with it. Physicists quantify inertia by mass. In fact, the concept of mass was "invented" by physicists as a measure of inertia.

object at rest

We can infer from Newton's first law of motion that any change in an object's motion requires a net force acting on it. Recall from Subsection 12.1 that a zero net force means that all of the forces acting on a system are balanced. As long as this condition exists, no change to the system's motion will occur. For example, a car driver maintains a constant speed by balancing the slowing forces of friction, air resistance, and gravity (on hills) with the driving force of traction between the tires and the road. The driver regulates this force by adjusting engine speed—pushing down or letting up on the gas pedal.

Newton's first law is not easy to prove here on Earth. Familiar motions such as kicking a soccer ball are subject to two ever-present forces—friction and gravity. A kicked soccer ball doesn't sail away in a straight line. It curves gracefully toward the ground because of gravity. Only deep-space probes seem to be nearly free of external forces. But even in the vastness of deep space, the sun's weak gravity and even the gravity of the galaxy itself can affect the motion of distant objects. Thus, any combination of unbalanced contact or field forces acting on a system will result in an acceleration. You will learn more about friction and gravity later in this chapter.

Balanced forces leave an object's motion (moving or at rest) unchanged.

An object acted on by unbalanced forces changes speed, direction, or both.

12.5 NEWTON'S SECOND LAW

Have you ever tried to hit a bowling ball with a golf club? Such an effort would not only be pointless, it might actually be painful! A small, lightweight golf ball requires relatively little force to accelerate it. A good golf swing can send the ball hundreds of meters. Larger, heavier bowling balls need much more force applied to them to get them to move. They have more mass, so they have more inertia. The same swing that sends a golf ball flying may only move the bowling ball a few meters. Newton's most significant accomplishment in developing his three laws of motion was relating force, mass, and acceleration in an equation that works for describing our normal, everyday experiences of motion. His **law of acceleration** declares:

> *The acceleration of an object is directly proportional to the net force acting on the object and is inversely proportional to its mass.*

Thus, a net force on an object of a given mass results in a certain acceleration of the object. If you double the net force on the object, you double its acceleration. On the other hand, if the original net force acts on another object with twice the mass of the original object, it results in only half the acceleration. Newton originally thought of his law in the form of this equation:

$$\text{acceleration of object} = \frac{\text{net force on object}}{\text{mass of object}}$$

$$\mathbf{a} = \frac{\mathbf{F}}{m}$$

Physicists usually write Newton's second law in symbols as

$$\mathbf{F} = m\mathbf{a},$$

which is simpler to remember. In this equation $\mathbf{F}$ is the net external force on an object, in newtons, m is the mass of the object, in kilograms, and $\mathbf{a}$ is the acceleration of the object, in meters per second squared.

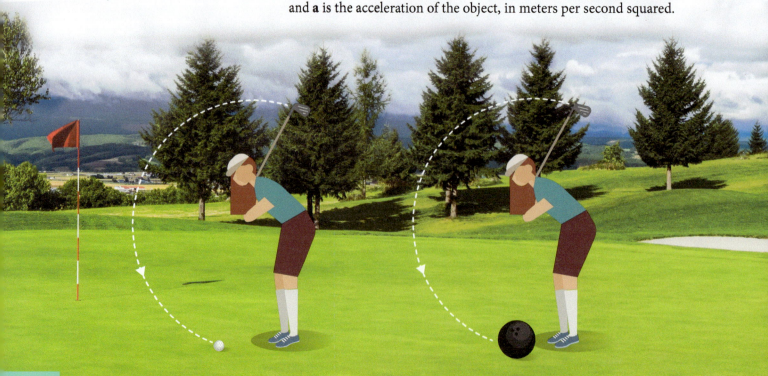

Newton's first law explains what happens when there is no net force acting on a system—no acceleration occurs. Newton's second law shows how the motion of a system changes when the net force acting on it is not zero. Let's take a look at an example.

EXAMPLE 12-3: Acceleration and Newton's Second Law

Imagine every car enthusiast's dream—a flashy sports car and a long stretch of highway. Stomping on the gas pedal causes the engine to accelerate the car at 12 m/s² to the west. If the car's mass is 1500 kg, what net force must be generated to achieve this acceleration?

Write what you know.

$$a = 12 \text{ m/s}^2 \text{ to the west}$$

$$m = 1500 \text{ kg}$$

$$F = ?$$

Write the formula and solve for the unknown.

$$F = ma$$

Plug in known values and evaluate.

$$F = (1500 \text{ kg})(12 \text{ m/s}^2 \text{ to the west})$$

$$= 18\,000 \text{ N to the west}$$

To achieve an acceleration of 12 m/s² to the west, a net force of 18 000 N must be applied.

We can also use the law of acceleration to calculate the acceleration that will result if a certain amount of force is applied to an object. Suppose you don't stomp on the gas pedal of your sports car but instead apply a more modest force to it—4500 N to the west. What acceleration would you expect?

EXAMPLE 12-4: Using Newton's Second Law to Calculate Acceleration

Write what you know.

$$F = 4500 \text{ N to the west}$$

$$m = 1500 \text{ kg}$$

$$a = ?$$

Write the formula and solve for the unknown.

$$F = a$$

$$\frac{F}{m} = \frac{\cancel{m}a}{\cancel{m}}$$

$$a = \frac{F}{m}$$

Plug in known values and evaluate.

$$a = \frac{4500 \text{ N to the west}}{1500 \text{ kg}}$$

Remember that a newton is a kilogram meter per second squared.

$$a = 3.0 \, \frac{\frac{\text{kg} \cdot \text{m to the west}}{\text{s}^2}}{\text{kg}} = 3.0 \text{ m/s}^2 \text{ to the west}$$

Not surprisingly, as this example shows, if you apply one-fourth of the force in the previous example to the car, it accelerates one-fourth as fast.

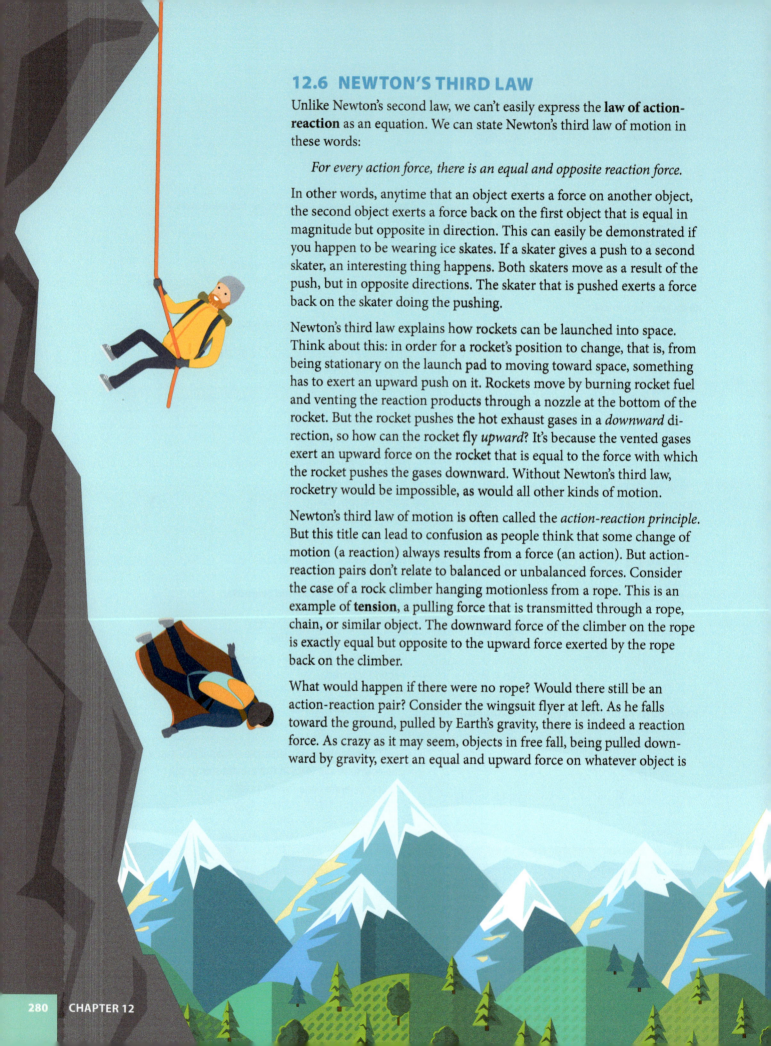

12.6 NEWTON'S THIRD LAW

Unlike Newton's second law, we can't easily express the **law of action-reaction** as an equation. We can state Newton's third law of motion in these words:

For every action force, there is an equal and opposite reaction force.

In other words, anytime that an object exerts a force on another object, the second object exerts a force back on the first object that is equal in magnitude but opposite in direction. This can easily be demonstrated if you happen to be wearing ice skates. If a skater gives a push to a second skater, an interesting thing happens. Both skaters move as a result of the push, but in opposite directions. The skater that is pushed exerts a force back on the skater doing the pushing.

Newton's third law explains how rockets can be launched into space. Think about this: in order for a rocket's position to change, that is, from being stationary on the launch pad to moving toward space, something has to exert an upward push on it. Rockets move by burning rocket fuel and venting the reaction products through a nozzle at the bottom of the rocket. But the rocket pushes the hot exhaust gases in a *downward* direction, so how can the rocket fly *upward*? It's because the vented gases exert an upward force on the rocket that is equal to the force with which the rocket pushes the gases downward. Without Newton's third law, rocketry would be impossible, as would all other kinds of motion.

Newton's third law of motion is often called the *action-reaction principle*. But this title can lead to confusion as people think that some change of motion (a reaction) always results from a force (an action). But action-reaction pairs don't relate to balanced or unbalanced forces. Consider the case of a rock climber hanging motionless from a rope. This is an example of **tension**, a pulling force that is transmitted through a rope, chain, or similar object. The downward force of the climber on the rope is exactly equal but opposite to the upward force exerted by the rope back on the climber.

What would happen if there were no rope? Would there still be an action-reaction pair? Consider the wingsuit flyer at left. As he falls toward the ground, pulled by Earth's gravity, there is indeed a reaction force. As crazy as it may seem, objects in free fall, being pulled downward by gravity, exert an equal and upward force on whatever object is pulling them down. The earth barely moves in such cases because of the huge difference in mass between itself and almost any falling object. The effect is much more noticeable when the objects' masses are more similar, such as for binary stars.

Now suppose that our climber has descended and is standing safely on the ground. Is Newton's third law in effect even there? Absolutely! As the climber stands on the ground, he exerts a downward force on the ground. At the same time, the ground is exerting an equal but opposite force back up on the climber! This upward force acts in a direction that is perpendicular to the surface where two objects make contact and is called the **normal force**. Consider the image at right. The man is pushing on the wall, but the wall isn't moving. Strange as it may sound, Newton's third law says that the wall is pushing back! This is an example of a normal force because the place where the man's hands meet the wall is vertical, and the force exerted by the wall is perpendicular to the vertical surface (i.e., to the right, in the opposite direction of the push). The man's push force on the wall and the wall's normal force on the man are not the only action-reaction pair in the image. As in the previous example, the man is also experiencing a downward force—his weight. The force of his weight pushing on the ground is balanced by an upward-acting normal force exerted by the ground. As you can see, the pairs of vectors are equal in magnitude and opposite in direction.

12B | REVIEW QUESTIONS

1. Why was Galileo unable to formulate equations to describe motion?

2. According to Newton's law of inertia, what must happen in order for an object's motion to change?

3. Explain how an object can be in motion if no net force is acting on it.

4. Which of the following can a physicist do to double the acceleration on an object? (Choose all that apply.)
   a. Double the mass of the object.
   b. Double the force acting on the object.
   c. Halve the mass of the object.
   d. Halve the force acting on the object.

5. Identify the action-reaction force pair in each of the following scenarios.
   a. A skier uses her ski poles to start moving downhill.
   b. A boat propeller spins rapidly in the water.
   c. A baseball player hits a pitched ball with a bat.
   d. A party balloon contains rapidly moving helium atoms.

6. Do any of the force pairs suggested in Question 5 *not* produce an acceleration? If so, which one(s)?

7. Which of the following diagrams shows the correct direction of the normal force acting on a box sitting on a ramp?

8. Draw a free-body diagram for the unbalanced forces acting on the car in the image below.

9. Refer back to the normal force diagram at the top of this page. Describe the acceleration that might occur if the vector representing the man's push were greater in magnitude than the vector representing the normal force of the wall.

DYNAMICS 281

12C Questions

- Why do things fall down instead of up?
- How does friction slow things down?
- What kind of force causes centripetal acceleration?
- Are some forces more fundamental than others?

12C Terms

gravity, friction, centripetal force

12C | TYPES OF FORCES

How do different forces affect our daily experiences?

So far we've seen what a force is and how forces work. In this section, we'll look at some different kinds of forces and study some examples of how they affect our everyday experiences.

12.7 GRAVITY

Gravitational Fields

Isaac Newton was supposedly inspired to formulate a law of gravity after observing an apple fall from a tree. Whether there is any truth to the story remains shrouded in mystery. What is *not* a mystery is that all objects exert a kind of field force on every other object. This was suspected even before Newton developed his *law of universal gravitation*. This law describes **gravity** as a field force that acts between any two objects. The strength of this force varies in direct proportion to the masses of the objects involved and inversely to the square of the distance between them.

Put simply, this means that massive objects exert more gravitational force than less massive objects, and the force gets rapidly weaker as the distance between the objects increases. Stated mathematically, this formula is

$$\mathbf{F}_g = G\frac{m_1 m_2}{r^2},$$

where $\mathbf{F}_g$ is the gravitational force, m_1 and m_2 are the masses of the two objects between which the force is acting, r is the distance between the centers of mass of the two objects, and G is the gravitational constant.

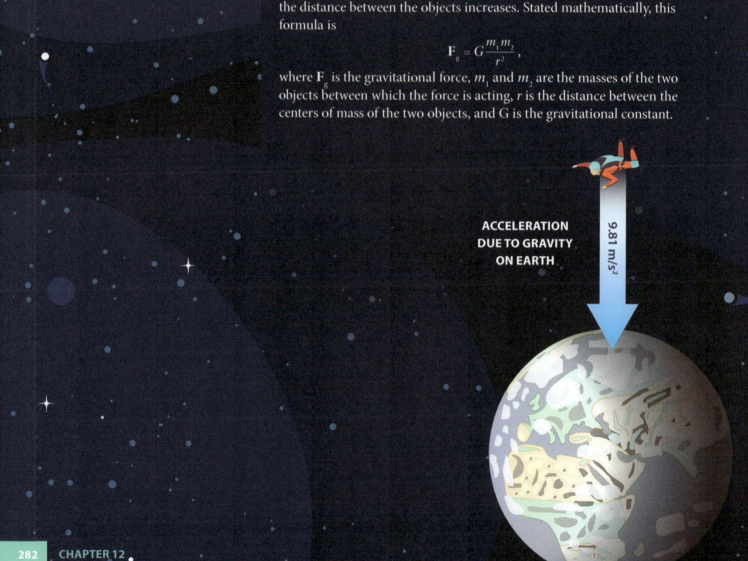

ACCELERATION DUE TO GRAVITY ON EARTH

9.81 m/s²

We still don't know exactly what causes gravity, but we can model how it operates. And like Newton's laws of motion, gravity is better modeled by Einstein's general theory of relativity. But Newton's model works well to describe most gravitational interactions that we can observe.

As you learned in Chapter 11, when only gravity is acting on an object, the object is in free fall. Small objects in the neighborhood of massive objects, such as skydivers near Earth, are accelerated toward the center of mass of the larger object. This acceleration averages 9.81 m/s² on Earth. Planets that have smaller or larger radii or are more or less massive than Earth will accelerate objects faster or slower. The gravity of Jupiter, the most massive planet in the solar system, accelerates objects at 25.89 m/s². That's nearly three times as fast as on Earth. Dwarf planet Pluto, being far less massive than Earth, much less Jupiter, can accelerate an object at only 0.58 m/s². There is actually a gravitational force between *any* two objects in the universe—even between you and this book! You can't sense the gravitational tug of your book, though, because the force is incredibly tiny. A book simply does not have sufficient mass to generate a significant gravitational field.

ACCELERATION DUE TO GRAVITY ON JUPITER

25.89 m/s²

ACCELERATION DUE TO GRAVITY ON PLUTO

0.58 m/s²

DYNAMICS 283

Gravitational Acceleration and Weight

In everyday language, we often use the terms *mass* and *weight* interchangeably, but they are really two different things. Mass, as you know, is a measure of how much matter is in an object. Weight is something different. We know that the earth's gravity accelerates objects near its surface. As a skydiver free falls, he experiences a sensation of weightlessness, even though his weight is still pulling him toward the ground.

We don't experience that weightlessness while standing on Earth's surface. The force that we do feel, weight (**W**), is a product of both our mass and acceleration due to gravity. The acceleration due to gravity, **g**, is also known as the gravitational field strength. If we substitute weight for force in the formula for Newton's second law, **F** = *m***a**, along with **g** for acceleration due to gravity, we get

$$\mathbf{W} = m\mathbf{g},$$

where **W** is the weight of an object in newtons, *m* is the mass of the object in kilograms, and **g** is gravitational acceleration in meters per second squared. As you can see, multiplying *m* times **g** produces units of kg·m/s², the units of force defined as newtons.

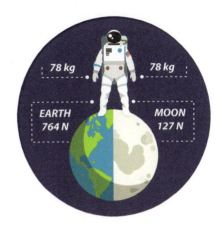

The weight formula derived from Newton's second law is a convenient way to quickly calculate the weight of any object if we know its mass. You've probably heard that astronauts weigh less on the moon. Let's use the weight formula to see how the weight of an object on Earth compares with its weight on other bodies in our solar system.

EXAMPLE 12-5: Finding Weight on Different Planets

The Mars rover *Curiosity*, which landed on the Red Planet in 2012, has a mass of 899 kg. What is *Curiosity*'s weight on Earth?

Write what you know.

$$W = ?$$
$$m = 899 \text{ kg}$$
$$g_{Earth} = 9.81 \text{ m/s}^2$$

Write the formula and solve for the unknown.

$$W = mg$$

Plug in known values and evaluate.

$$W = (899 \text{ kg})(9.81 \text{ m/s}^2)$$
$$= 8820 \text{ N}$$

Mars has much less mass than Earth, so its gravity is weaker. The acceleration due to gravity there is only 3.71 m/s². How much does *Curiosity* weigh on Mars?

$$g_{Mars} = 3.71 \text{ m/s}^2$$
$$W = (899 \text{ kg})(3.71 \text{ m/s}^2)$$
$$= 3340 \text{ N}$$

Not surprisingly, *Curiosity* weighs less than half as much on Mars.

As you might expect, the weight formula derived from Newton's second law is related to Newton's law of universal gravitation. Let's look at that relationship.

When calculating weight on Earth, we use the acceleration due to gravity, **g**. Where does this come from? It actually is part of Newton's law of universal gravitation.

$$\mathbf{g}_{Earth} = G \frac{m_{Earth}}{r^2}$$

By using the mass and radius of any object, you can calculate the gravitational field strength caused by that object.

DYNAMICS   285

12.8 FRICTION

Did you do any walking today? Friction makes motions like walking, running, skating, and driving a car possible. Friction is everywhere and affects all motion here on Earth. It slows things down and wears parts out. But what is friction? **Friction** is a contact force that works against the motion of objects moving past each other. You can think of the term "objects" in a broader sense here. Friction affects not only objects touching along surfaces (such as a pencil on paper), but also fluids flowing around or against a rigid object (such as a stream flowing around a boulder in a river channel). The direction of the friction force acting on an object is opposite to its direction of motion or attempted motion, as shown in the free-body diagram at left. If you slide a box down a ramp off the back of a truck, the ramp exerts a friction force acting up the ramp.

The magnitude of the friction force between solid objects is proportional to the force holding the two objects together. In the box and ramp scenario, this force depends on the weight of the box and the angle of the ramp. For instance, the friction on the box decreases if the angle of the ramp is increased. The box slides more easily. Similarly, a lighter box experiences less friction than a heavier box because the weight of the heavier box exerts a greater downward force at the point of contact. The friction force magnitude is also affected by the types of surfaces in contact with each other. You know this from your daily experiences! Your hand slides much more easily across a sanded and polished piece of wood than it does across a rough-hewn piece of lumber—and with fewer splinters too!

Physicists recognize several different kinds of friction. Let's examine some of these.

TYPES OF FRICTION

STATIC FRICTION

Can you stand on the slope of a hill? Why doesn't gravity just pull you all the way to the bottom? Static friction is the friction that exists between two objects that are attempting to move but are not yet in motion relative to each other. It's the "sticky" force that holds things in place.

SLIDING FRICTION

Have you ever noticed that pushing or dragging a heavy object is easier once you get the thing moving? Sliding friction, the friction that exists when two objects are in motion relative to each other, is always less than the static friction that exists when the objects aren't moving.

ROLLING FRICTION

Refrigerators and other large furnishings are often placed on casters, small wheels that make moving the item easier. Wheels roll, of course, greatly reducing the friction force between two objects. Wheels make riding a bicycle much more fun than riding a bicycle without wheels!

Friction can be both a blessing and a curse. There are many situations for which scientists, engineers, and even people just trying to move around safely work to increase the amount of friction between surfaces. Extra friction, or *traction*, between tires and road surfaces is considered beneficial. On the other hand, a turning wheel on an axle creates a lot of wear and tear on machinery. Lubricating oils and ball bearings are used to reduce friction in such instances.

MINI LAB

A WEIGHTY PROBLEM

Essential Question:

Can I change my weight instantly?

Equipment

paper and pencil

Did you know that your bathroom scale doesn't actually measure your weight? It's true! Bathroom scales measure the strength of the normal force acting upward on a person, which is typically equal to the person's weight acting downward on the scale. You know that this must be true because you don't accelerate up or down when you stand on the scale—there's no net force acting on you. Since the two forces are equal in magnitude, the scale can report its measurement of the normal force as your weight. But would the scale display a different value if it were moving while weighing you? Let's use some free-body diagrams to think about this question.

PROCEDURE

1. Imagine a person whose mass is 70 kg standing on a bathroom scale inside a motionless elevator. Draw a free-body diagram that shows the weight and normal forces acting on this person. Label each vector with its force name and magnitude. (*Reminder*: You'll need to use **F** = m**a** to calculate the net force and **W** = m**g** to calculate the person's weight force.)

2. Now imagine that the elevator begins to accelerate upward at 2 m/s². Calculate the net force acting on the person (remember to include both magnitude and direction).

3. Draw a free-body diagram that shows the forces acting on the person as the elevator moves upward. Label the vectors and include their magnitudes. (*Note*: The force that causes the elevator to move upward is caused by the tension of the elevator's cable. But the upward force of the scale pushing upward on the person causes him to accelerate.)

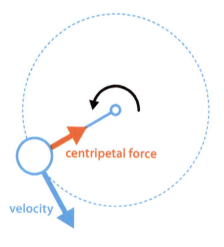

12.9 CENTRIPETAL FORCE

Remember centripetal acceleration from Chapter 11? It's the center-seeking acceleration of objects that move in circular motion. Since any acceleration requires a force, centripetal acceleration requires a force too. Not surprisingly, such a force is called a **centripetal force**—a force that accelerates an object toward the center of a circular motion.

Your yearly experience of the changing seasons is due in part to a centripetal force. Gravity acts as a centripetal force as it keeps Earth moving in its orbit around the sun. All orbiting objects experience a similar centripetal force tug, whether it's Earth's moon or the satellite that broadcasts your favorite TV programming. Centripetal force even makes the playground game of tetherball possible. The tension force exerted by the tether keeps the ball moving in a circular path around the pole.

4. Will the normal force (apparent weight) displayed by the scale be higher, lower, or the same? Explain.

5. Now imagine that instead of going up, the elevator begins to descend with an acceleration of 2 m/s². What is the net force acting on the person?

6. Draw a free-body diagram that shows the forces acting on the person as the elevator descends (*Hint*: Think carefully about which vector is actually changing). Label the vectors and include their magnitudes.

7. Will the normal force (apparent weight) displayed by the scale be higher, lower, or the same? Explain.

CONCLUSION

8. Look at the free-body diagrams. Explain why we feel heavier as the elevator accelerates upward and lighter as the elevator accelerates downward.

GOING FURTHER

9. Imagine a bottomless elevator in which a person could descend indefinitely. What would happen to the weight of the person if the elevator's acceleration continued to increase as it descended?

12.10 FUNDAMENTAL FORCES

So far we've looked at examples of forces operating at scales that we can easily observe. But physicists have also identified forces that operate on atomic scales. In total, there are four forces that physicists consider to be *fundamental forces*—the forces that appear to underlie all the other known forces in nature. These forces hold together the structure of all matter in the universe, from subatomic particles to galaxies. One of these fundamental forces is gravity, which we've already discussed. Gravity is the weakest of the four fundamental forces. Let's take a brief look at the other three.

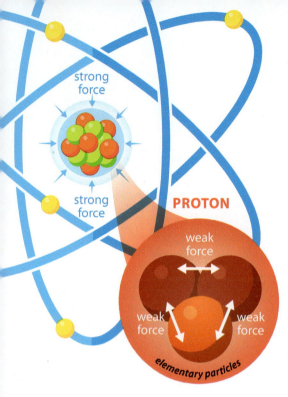

Nuclear Forces

Strong Nuclear Force

The strong nuclear force strongly attracts at distances equal to the diameter of a proton. It holds the protons and neutrons together inside atomic nuclei. But at much shorter distances it repels strongly. The strong nuclear force is the most powerful of all fundamental forces.

Weak Nuclear Force

The weak nuclear force is an extremely short-range force, attracting at distances smaller than the radius of a proton. It holds together the elementary particles of matter that make up protons and neutrons. It is also involved in some forms of nuclear decay. The weak nuclear force is about one millionth the strength of the strong nuclear force.

Electromagnetic Force

The electromagnetic force is a force that exists between electrically charged particles. It can be an attractive or repulsive force depending on the charges of the involved particles. This force can act over distances ranging from the subatomic to the very large. The electromagnetic force originates in static (not moving) and moving electrical charges. This force is 1/137 of the strength of the strong nuclear force. You'll learn more about electromagnetism in Chapter 20.

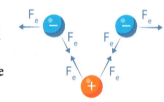

12C | REVIEW QUESTIONS

1. How does the gravitational force between two objects relate to their masses and the distance between them?
2. Explain the relationship between gravity and weight.
3. In what direction does friction always act?
4. List the factors that can affect the magnitude of a friction force.

Use the information in the photo at right to answer Questions 5–7.

5. Describe at least two types of friction that can be seen at work.
6. Using a free-body diagram, draw and label the vectors for gravity, normal force, and friction acting on the longboard rider.
7. If the longboard rider is accelerating downhill, what must be true about the forces acting on him? How is this shown by the magnitudes of the three vectors?

8. How do sliders (placed on the bottoms of the feet of heavy furniture) help make moving furniture easier?
9. Is a car traveling along Earth's surface experiencing any centripetal force? Explain.
10. Compare the strong and weak nuclear forces.

CHAPTER 12 REVIEW

12A CLASSIFYING FORCES

- A force is a push or pull on an object.
- Dynamics is the branch of physics that studies how forces change motion.
- The SI unit of force, the newton, is the amount of force needed to accelerate a 1 kg mass at 1 m/s^2.
- Contact forces act only when objects are touching.
- A field force can be exerted over a distance.
- Only unbalanced forces can change the motion of an object.
- Free-body diagrams are used to analyze the forces acting on an object.

12A Terms

force	270
dynamics	270
contact force	271
field force	271
free-body diagram	273

12B NEWTON'S LAWS OF MOTION

- Newton's first law states that objects at rest remain at rest and objects in motion remain in motion unless acted on by a net external force.
- Newton's second law states that the acceleration of an object is directly proportional to the net force acting on the object and is inversely proportional to its mass.
- Newton's third law states that for every action force, there is an equal and opposite reaction force.
- Tension is a pull force that acts through a rope, chain, or similar object.
- A normal force is one that acts in a direction perpendicular to the surface where two objects make contact.

12B Terms

inertia	275
law of inertia	276
law of acceleration	278
law of action-reaction	280
tension	280
normal force	281

12C TYPES OF FORCES

- Gravity is an attractive field force that acts between the masses of any two objects.
- A gravitational force is proportional to the masses of the objects involved and inversely proportional to the square of the distance between them.
- An object's weight is the product of its mass and the acceleration due to gravity.
- Friction is a contact force that acts in the opposite direction of an object's motion.
- The strength of a friction force is affected by the magnitude of the force holding two objects together, the types of surfaces in contact, and the kind of motion taking place.
- Centripetal force accelerates an object toward the center of a circular motion.
- The four fundamental forces in decreasing order of strength are the strong nuclear force, the electromagnetic force, the weak nuclear force, and gravity.

12C Terms

gravity	282
friction	286
centripetal force	288

DYNAMICS 291

CHAPTER 12

CHAPTER REVIEW QUESTIONS

Recalling Facts

1. How much force is one newton?
2. Create a T-chart that compares field forces and contact forces.
3. What conditions must be met in order to produce a zero net force?
4. (True or False) A force acting on an object will always accelerate that object.
5. Why don't physicists usually include all of the forces acting on an object in a free-body diagram?
6. How do physicists depict the magnitude of a force in a free-body diagram?
7. Define *inertia*.
8. According to Newton's first law, if an object is in motion but has no net external force acting on it, what will the object do?
9. What kind of force do we normally mean when we say "force is equal to mass times acceleration"?
10. What force prevents you from falling to the center of the earth when standing on a sidewalk?
11. Give an example of a tension force that doesn't involve a rope.
12. Define *gravity*.
13. State the universal law of gravitation in your own words.
14. (True or False) Newton's law of universal gravitation explains what causes gravity.
15. Define *friction*.
16. Rank *rolling*, *sliding*, and *static friction* in order from greatest to least force.
17. How can you determine the direction of a friction force?
18. Why is it harder to ride a bike up a hill than on a flat surface? What changes when you are riding down a hill? Draw a free-body diagram for each situation.
19. In what direction does centripetal force always act?

Understanding Concepts

20. You need to accelerate a 15 kg box of books across the floor at 0.50 m/s². Ignoring the effect of friction, how much force must you push with?
21. If the floor in Question 20 exerts a sliding friction force of 5 N, how much force would be needed to achieve the same acceleration on the box of books?
22. Draw a free-body diagram of the box from Question 20. Include your push force, friction, weight, and normal forces.
23. In the Space Shuttle Program, a fully loaded space shuttle had a mass of 2 000 000 kg and was designed to never exceed an acceleration of 29.4 m/s². How much force would be needed to achieve that acceleration?
24. A typical BMX bike has a mass of about 12 kg. A 60 kg rider exerts a force of 150 N. How quickly will the bike and its rider accelerate?

REVIEW

25. A farmer pulls on the rope tied to his obstinate mule with 250 N of force to the right, but the mule won't budge. Your study partner believes that the mule isn't moving because Newton's third law says that the mule exerts an equal and opposite force back on the farmer. Explain the flaw(s) in your partner's reasoning.

26. Ceres is a dwarf planet located in the asteroid belt of our solar system. Explain why the gravitational force on Ceres is only 1/338th that of Earth's even though Ceres is a very massive object at 9.39×10^{20} kg.

27. Two satellites of equal mass orbit the earth. One orbits at an altitude of 500 km and the other orbits at 2500 km. Compared with the satellite in the lower orbit, does the satellite at 2500 km experience the same amount of gravitational force, less gravitational force, or far less gravitational force? Explain.

28. A friend asserts that an object on the moon has only one-sixth the mass of the same object on Earth's surface. Do you agree? Explain.

29. Explain why football players change to shoes with longer studs if field conditions are wet or muddy.

30. Why is the existence of the strong nuclear force particularly critical for the existence of matter?

Critical Thinking

31. Draw a free-body diagram of the car on the bridge shown at right showing any forces acting on the car and bridge system that you can identify, along with their magnitudes. Assume that the car is traveling at constant speed.

32. What would happen to Earth if the sun's gravity were to cease acting on it?

Use the Ethics box below to answer Question 33.

33. Write a three-paragraph response on the ethics of mandatory helmet laws. In the first paragraph, discuss principles from Scripture that may guide Christians in this area. In the second paragraph, discuss the possible outcomes of heeding or ignoring the guidelines of Scripture. In the third paragraph, give your opinion on whether the wearing of helmets is a personal choice or should be mandated by law.

ETHiCS — MANDATORY HELMET LAWS

Many people enjoy the thrill of riding a motorcycle. But motorcycle accidents are particularly dangerous. The forces experienced in a car crash can be deflected away from a car's occupants by crumple zones and other features designed into the car's body. But motorcycles, of course, lack this layer of protection. Head injuries are the greatest risk that a motorcycle rider faces, yet most states in the United States don't require all riders to wear helmets. Is there a proper balance between allowing people to make their own choices about personal safety and society making the decision for them? Does society even have the right to make that decision?

Norwegian rider Ole Hem is seen here competing in the fourteenth running of the Hero MTB Himalaya mountain bike race in 2018.

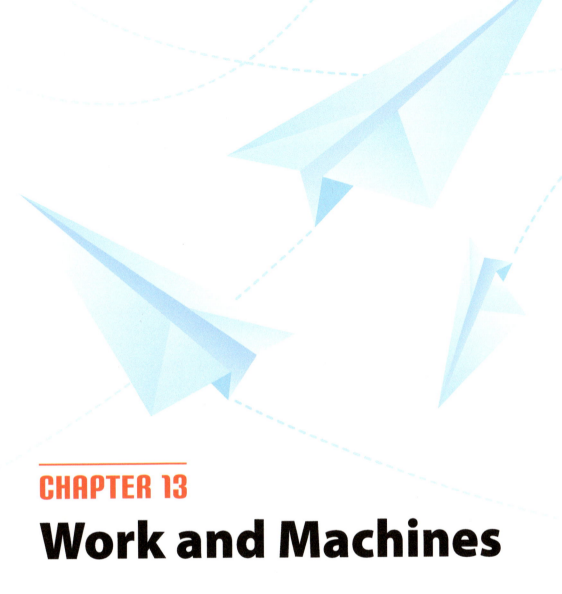

CHAPTER 13
Work and Machines

MOUNTAIN MACHINE

Bicycles can be used for transportation, for racing on roads, tracks, or mountains, for doing jumps, or for taking quiet rides around your neighborhood. Different types of bikes are designed for different tasks. The bike at left is designed to be ridden in mountainous terrain. If you are really into mountain biking, what better place to do it than the Himalayas!

Each year competitive mountain bikers take part in the Hero MTB Himalaya race. The 650 km race from Shimla to Dharamshala in India challenges racers in this eight-day event. The bikes that these riders use are not your ordinary mountain bikes. These bikes are specially designed to give the riders an advantage to help them climb, descend, and maneuver on this grueling course. You could say that these bikes have reached great heights in design.

Bicycles are compound machines that help us to achieve a goal. God has given people the ability to design and build machines that help us to work more easily. When we create this way, we're living out God's image in us. Let's learn more about work, power, and the machines that we make to help us with daily tasks.

13A Work and Mechanical Advantage	296
13B Lever	303
13C Wheel and Axle	309
13D Inclined Plane	313

13A Questions

- What are work and power?
- How can I determine the work and power done by a machine?
- If a machine makes work easier, then what is the cost?

13A Terms

work, power, simple machine, mechanical advantage, ideal mechanical advantage, actual mechanical advantage, efficiency

13A | WORK AND MECHANICAL ADVANTAGE

How do simple machines make work easier?

Have you ever worked hard doing something? Maybe it was practicing or playing a sport. Maybe it was mowing lawns as a summer job. Perhaps it was doing your science homework last night. While there are many things that we think of as work, physicists have a very specific definition of work. Some of the activities mentioned above wouldn't be considered "work" by physicists.

13.1 WORK AND POWER

Work

Physicists say that work is done any time a force acts to move an object. **Work** (W) is the energy transferred to an object when an external force acts on the object to move it. The quantity of work done on an object depends on two factors: the distance the object moves and the force acting in the direction of the movement. Work is a scalar, and its formula is

$$W = Fd,$$

where W is work, F is the force parallel to the distance, and d is the distance traveled.

By looking at the formula for work, we can determine its units. Force is measured in newtons and distance is measured in meters, so work must have units of newton-meters (N·m), which physicists call the joule (J) in honor of James Prescott Joule, an English physicist who did much of the early research on work and energy.

WORK OR NO WORK?

Recall that the work done on an object depends on the force applied to the object *in the direction the object travels*. So in which of the following images do you think the people are doing work?

The man is doing work on the suitcase. But only part of the force he is applying is doing work because he is pulling at an angle to the direction of travel. Only the force that is parallel to the direction of travel ($F_{\parallel}$) produces work.

The skateboarder is not doing any work because all of the force he applies to the board is perpendicular to the distance traveled.

The man is doing work on the van, and all his force is doing work because he is pushing parallel to the direction that the vehicle is moving.

Let's look at an example of calculating the work done.

EXAMPLE 13-1: Calculating Work

Ships traveling through a canal must use a ship lift because there is a significant elevation difference between sections of the canal. How much work does the lift do if it applies a force of 13 230 000 N to lift a ship the 73.15 m elevation difference from one section of the canal to the next?

Write what you know.

$F = 13\ 230\ 000$ N

$d = 73.15$ m

$W = ?$

Write the formula and solve for the unknown.

$W = Fd$

Plug in known values and evaluate.

$W = (13\ 230\ 000\ \text{N})(73.15\ \text{m})$

$W = 967\ 800\ 000$ N·m

Recall that the unit for work, N·m, is a joule (J). The lift does 967 800 000 J of work to lift the ship.

Power

Look at the images on this page and on the following page. In both cases, a farmer is plowing a field. If the fields are the same size, then the horse and the tractor will accomplish the same amount of work. But what do you think the difference is between using a horse to plow that field and using a tractor for the same task? If you are thinking about time, then you are correct. Why? The tractor is a 160 horsepower machine, while the horse is, well, just a one horsepower machine. Horsepower is an English unit of **power** (P), which is the rate of doing work. The SI unit of power is the watt (W), named for Scottish inventor James Watt.

Power is also a scalar and has the formula

$$P = \frac{W}{t},$$

where P is power in watts, W is work in joules, and t is time in seconds. We can see from the formula that a watt must be a J/s. In other words, power in watts represents the quantity of work that can be done each second.

EXAMPLE 13-2: Calculating Power and Time

If the engine from a Ford Model T can do 13 428 000 J of work in 15 min, how much power does the engine produce?

Write what you know.

$$W = 13\,428\,000 \text{ J}$$

$$t = 15 \text{ min}\left(\frac{60 \text{ s}}{1 \text{ min}}\right) = 900 \text{ s}$$

$$P = ?$$

Write the formula and solve for the unknown.

$$P = \frac{W}{t}$$

Plug in known values and evaluate.

$$P = \frac{13\,428\,000 \text{ J}}{900 \text{ s}} = 15\,000 \frac{\text{J}}{\text{s}}$$

$$= 15\,000 \text{ W}\left(\frac{1 \text{ kW}}{1000 \text{ W}}\right) = 15 \text{ kW}$$

The Model T engine produces 15 kW of power.

How long would it take the Boss 429 engine in a 1970 Ford Mustang, which can produce 370 kW of power, to do the same amount of work?

Write what you know.

$$W = 13\,428\,000 \text{ J}$$

$$P = 370 \text{ kW} = 370\,000 \text{ W}$$

$$t = ?$$

Write the formula and solve for the unknown.

$$P = \frac{W}{t}$$

$$Pt = \frac{W}{t}t$$

$$\frac{Pt}{P} = \frac{W}{P}$$

$$t = \frac{W}{P}$$

Plug in known values and evaluate.

$$t = \frac{13\,428\,000 \text{ J}}{370\,000 \text{ W}} = 36.3 \frac{\text{J}}{\text{W}}$$

Since a watt is a J/s,

$$t = 36.3 \frac{\text{J}}{\frac{\text{J}}{\text{s}}} = 36.3 \text{ s.}$$

The Boss 429 could do that same work in just 36.3 seconds!

13.2 MACHINES

After the Fall, God told Adam that he would have to work to get food from the ground. As man has worked, he has often sought ways to make that work easier to do. If a farmer needed to remove a large stone from his field, he might have used a branch to help pry that rock out. Noah, as he was building the Ark, would have quickly consumed much timber. To move timber to his construction site, he may have fashioned a wheel to help transport it. We can imagine Noah using a ramp to help build the Ark and probably even load it. Over time, man has used wood, then rock and metal, and finally plastics to develop tools that make his work easier to do. Many of these tools are what we call **simple machines**, basic mechanical devices that change the magnitude, direction, or distance traveled of the force used when doing work. The examples above highlight the general types of simple machines. The farmer used a *lever* to pry the rock from the ground. Noah used a *wheel and axle* to move the lumber and an *inclined plane* to reach greater heights as construction on the Ark progressed. We will look more closely at these simple machines throughout this chapter.

We use machines to make it easier to do the work that we have to do. These machines give us an advantage over doing the work without them. We call this **mechanical advantage**—the measure of the change in input needed to do a certain amount of work (output) when using a simple machine. Typically, the advantage gained by using a machine is that the machine multiplies the effect of our input force. In cases like this, the mechanical advantage is greater than one. The higher the mechanical advantage, the more the machine multiplies our force. Because of the conservation of energy, the cost of gaining this force multiplication is that we have to apply our force over a greater distance. But in other situations we want the machine to multiply the input distance. In this case, the mechanical advantage will be greater than zero but less than one. The cost of utilizing this machine is that we will have to apply more force to gain the added distance. Occasionally we will want the machine to change only the direction of the force, in which case the mechanical advantage is one.

In a perfect world, machines would do the same amount of work as a person puts in. The mechanical advantage of this machine is called the *ideal mechanical advantage*. But in reality, all machines have internal friction that wastes some of the work put in, producing less output work than the work input. The *actual mechanical advantage* represents the fact of this wasted work.

MECHANICAL ADVANTAGE AND EFFICIENCY

The man at right is utilizing a simple machine to help him lift a crate. He is applying a 240 N input force (F_{in}) by pulling the rope. He pulls the rope an input distance (d_{in}) of 6.0 m. The machine applies an output force (F_{out}) of 400 N to lift the crate, and it lifts the crate a 3.0 m output distance (d_{out}). Let's look at mechanical advantage and efficiency for this machine.

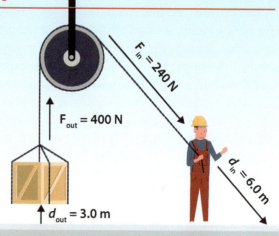

IDEAL MECHANICAL ADVANTAGE

Ideal mechanical advantage (IMA) is the measure of mechanical advantage for an ideal machine—100% efficient with no friction. It is the ratio of the input distance to the output distance.

$$IMA = \frac{d_{in}}{d_{out}}$$

The ideal mechanical advantage of the machine is

$$IMA = \frac{6.0 \text{ m}}{3.0 \text{ m}} = 2.0.$$

The machine would ideally make the work twice as easy. Notice that mechanical advantage has no units—we call this a *dimensionless quantity*.

ACTUAL MECHANICAL ADVANTAGE

Actual mechanical advantage (*AMA*) measures the mechanical advantage taking into account all mechanical losses sustained throughout the machine. It is the ratio of the output force to the input force.

$$AMA = \frac{F_{out}}{F_{in}}$$

The actual mechanical advantage of the machine is

$$AMA = \frac{400 \text{ N}}{240 \text{ N}} = 1.67.$$

The machine actually makes the work less than twice as easy.

EFFICIENCY

Because some work is wasted by using any machine, no machine is ideal. We calculate how well a machine does at not losing the input work as its **efficiency**, which is the effectiveness of a machine or process at converting energy from one form to another.

$$\text{efficiency} = \left(\frac{W_{out}}{W_{in}}\right) 100\%, \text{ or efficiency} = \left(\frac{AMA}{IMA}\right) 100\%$$

The efficiency of the machine depicted here is shown below.

$$\text{efficiency} = \left(\frac{1.67}{2.0}\right) 100\% = 84\%$$

The machine is 84% efficient, meaning that 16% of the man's work was wasted by the machine.

As you can see from the illustration above, all machines waste work, which means that we will actually do *more* work when using a machine. This fact startles many people, and so they may wonder why we use the machine. The machine is used to make the task easier even though we end up doing more work. Consider the example on the next page.

EXAMPLE 13-3: Calculating Mechanical Advantage

When using a simple machine, you exert 125 N of force through a distance of 15 m. The machine, in turn, applies a force of 1155 N and lifts a load 1.5 m. What is the *IMA*, *AMA*, and *efficiency* of this system?

Write what you know.

$d_{in} = 15$ m

$d_{out} = 1.5$ m

$IMA = ?$

Write the formula and solve for the unknown.

$$IMA = \frac{d_{in}}{d_{out}}$$

Plug in known values and evaluate.

$$IMA = \frac{15 \text{ m}}{1.5 \text{ m}} = 10$$

Write what you know.

$F_{in} = 125$ N

$F_{out} = 1155$ N

$AMA = ?$

Write the formula and solve for the unknown.

$$AMA = \frac{F_{out}}{F_{in}}$$

Plug in known values and evaluate.

$$AMA = \frac{1155 \text{ N}}{125 \text{ N}} = 9.24$$

Write what you know.

$IMA = 10$

$AMA = 9.24$

$efficiency = ?$

Write the formula and solve for the unknown.

$$efficiency = \left(\frac{AMA}{IMA}\right)100\%$$

Plug in known values and evaluate.

$$efficiency = \left(\frac{9.24}{10}\right)100\% = 92\%$$

13A | REVIEW QUESTIONS

1. Define *work*.
2. Determine whether work is being done in each of the following images.
   a. a tow truck pulling a car up a ramp
   b. a man sitting on an elephant as the elephant moves some logs to the left
   c. a woman pushing against a wall
3. A 755 W motor is used to lift a crate. If it takes 45 s for the motor to lift the crate, how much work does the motor do?
4. What is the difference between a 12 kW machine and a 24 kW machine?
5. What is a simple machine?
6. Compare *IMA* and *AMA*.
7. A girl uses a simple machine to lift a cooler into her treehouse. She applies a 15 N force over a distance of 28 m, and the machine applies an 85 N force to lift the cooler 4.7 m. Calculate the *IMA* and *AMA*.
8. If I have to do more work to do a task with a machine than I would have to do without it, then what is the benefit of using a machine?

a.
b.
c.

13B | LEVER

How can a man move a train?

13.3 SIMPLE MACHINE: LEVER

Did you ever play on a seesaw in school? Have you ever had a splinter removed with tweezers? Have you ever pushed anything in a wheelbarrow? If you have done these things, then you are familiar with levers. A **lever** is a simple machine that consists of a rigid bar that turns about a pivot point. The point about which a lever turns, or pivots, is called the **fulcrum**. Don't limit your thinking to strictly bar shapes because a lever can be almost any shape—it just has to be able to pivot about a fulcrum.

So how does a lever make work easier? Look at the kids on the seesaw below. Do you think that either of those kids has the ability to lift the other kid over her head? Probably not, but using the seesaw, either of these kids could lift the other off the ground, perhaps over her head. The seesaw gives her a mechanical advantage, which allows her to do work that she couldn't do without the seesaw "machine."

13B Questions
- How is a lever a simple machine?
- How do levers work?
- Why are there different types of levers?

13B Terms
lever, fulcrum, torque

HOW LEVERS WORK

Levers work on the basis of rotational motion, which is the motion of an object around some axis. The application of any force not in line with the axis has the tendency to cause the object to rotate. Physicists call this tendency of a force to cause rotation about a pivot point *torque*. Let's look at how torques work with levers by examining a child and an adult balanced on a seesaw.

A **torque** is any force that tries to cause an object to rotate around its axis. The distance from the rotational axis to the location where the force is applied is called the *moment arm* (ℓ). The torque on an object is equal to the force times the moment arm. Only the portion of the force that is perpendicular to the moment arm will cause a rotation. The unit for torque is newton-meters (N·m). (Note that the joule is not used for torque.) According to the *law of torques*, for a system to be in rotational equilibrium, the sum of the torques on the system must be zero.

The torque due to the adult is $F_A \ell_A$. If the adult applies her full weight of 760 N a distance of 0.700 m from the pivot point, her torque is (760 N)(0.700 m) = 532 N·m. Her torque is trying to rotate the seesaw counterclockwise.

The torque due to the child is $F_c \ell_c$. Since the child can apply only his full weight of 380 N, he will have to be farther from the pivot point to achieve equilibrium. To equal the torque of the adult, the child will have to be 1.40 m from the axis. His torque is (380 N)(1.40 m) = 532 N·m and is trying to rotate the seesaw clockwise.

The system is in equilibrium because the child's force is trying to rotate the seesaw clockwise with a torque of 532 N·m while the adult's force is trying to rotate the system counterclockwise with the same torque.

304   CHAPTER 13

MINI LAB

LAW OF TORQUES

Kids on the playground quickly gain an understanding of the law of torques. They know that if they want to balance the seesaw with someone who weighs more or less than they do, they must adjust their positions on the seesaw. This activity gives you the opportunity to investigate the law of torques with a simple seesaw.

Essential Question:

How do I get a picture to hang balanced?

Equipment

meter stick

pencil

masses, 100 g (6)

1. If you want to balance the meter stick by itself on the side of the pencil, under what number should you place the pencil? Explain.

PROCEDURE

A Lay the pencil on the table and balance the meter stick on the pencil.

2. If you place a 100 g mass on the very end (100 cm) of the meter stick, how much mass would you have to place at 25 cm (half the distance from the other end to the fulcrum) to balance the meter stick?

B Place a 100 g mass at 100 cm and test your hypothesis (your answer to Question 2 above).

3. If you place a 200 g mass on the very end (100 cm) of the meter stick, where would you have to place 300 g to balance the meter stick?

ANALYSIS

4. Using grams and centimeters, calculate the torques for Step **B**.

CONCLUSION

5. Explain the balancing of the meter stick according to the law of torques.

GOING FURTHER

C Move the meter stick so that the pencil is at 60 cm. Place 100 g at 100 cm and 100 g at 20 cm.

6. Why didn't the meter stick balance this time?

13.4 CLASSIFYING LEVERS

Levers are all around us. Whether you realize it or not, you have used levers today. When you lift your arm up, bring your hand close to your face, or stand on your toes, you are using levers. There are three types of levers. On the basis of your conclusions from the mini lab, do you know what might change from one type of lever to another? Let's take a look.

CLASSES OF LEVERS

When a lever is used as a simple machine, someone moves a distance (d_{in}) as he applies an input force (F_{in}) perpendicular to the moment arm (ℓ_{in}). The lever produces an output force (F_{out}) perpendicular to its moment arm (ℓ_{out}) and moves the object some distance (d_{out}).

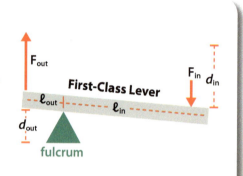

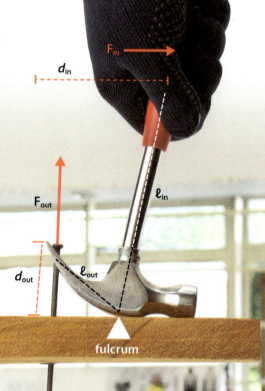

FIRST-CLASS LEVERS

When most people think of levers, they think of what we call a *first-class lever*. You can recognize a first-class lever because the fulcrum is between the input force and the output force. Because the fulcrum can be anywhere between the two forces, the mechanical advantage can be any number greater than zero.

Examples: crowbar, pliers

SECOND-CLASS LEVERS

When the output force is between the fulcrum and the input force, you are dealing with a *second-class lever*. A wheelbarrow is a perfect example. The fulcrum is located within the wheel, and the load is between you and the fulcrum. The mechanical advantage of a second-class lever is always greater than one.

Examples: brake pedal on a car, bottle opener

THIRD-CLASS LEVERS

A *third-class lever* is interesting because it multiplies your distance while decreasing your force. A third-class lever has the input force between the fulcrum and output. The mechanical advantage will always be between zero and one.

Examples: your arms, salad tongs

Can you identify the first-class lever on the nail clippers?

EXAMPLE 13-4: Calculating Work In and Work Out

When using a first-class lever to lift a rock, you exert 87.3 N of force through a distance of 1.75 m. The lever applies a force of 389 N and lifts the rock 0.38 m. How much work did you do? How much work did the lever do?

Write what you know, including a diagram.

$F_{in} = 87.3$ N

$d_{in} = 1.75$ m

$F_{out} = 389$ N

$d_{out} = 0.38$ m

$W_{in} = ?$

$W_{out} = ?$

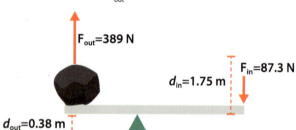

Write the formula and solve for the unknown.

$$W = Fd$$

Plug in known values and evaluate.

Calculating W_{in}:

$$W_{in} = F_{in} d_{in}$$
$$= (87.3 \text{ N})(1.75 \text{ m})$$
$$= 153 \text{ N·m} = 153 \text{ J}$$

Calculating W_{out}:

$$W_{out} = F_{out} d_{out}$$
$$= (389 \text{ N})(0.38 \text{ m})$$
$$= 148 \text{ N·m} = 148 \text{ J}$$

As expected, you did more work than the machine, but the work was easier to do.

13B | REVIEW QUESTIONS

1. What is a lever?
2. Explain the law of torques using the image below.

3. Identify the class of lever in each image.

a.
b.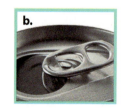
c.
d.

4. Give an example of each type of lever other than any shown in Question 3.

For Questions 5–7, review Example 13-3 if needed.

5. If you apply a 37.5 N input force on a lever that has an *AMA* of 2.50, how much force does the lever apply to the load?

6. You are using a lever with an *IMA* of 2.70. If you apply the input force over a 0.75 m distance, how far is the load lifted?

7. If your lever has an *AMA* of 2.50 and an *IMA* of 2.70, what is its efficiency?

13C | WHEEL AND AXLE

Is a wheel all by itself a machine?

13.5 SIMPLE MACHINE: WHEEL AND AXLE

The drawback to levers is that they often have a limited degree of movement. Recognizing this limitation led to the development of the **wheel and axle**, which is a simple machine consisting of a wheel with a rod running through its axis that acts as the pivot point. You can think of the wheel and axle as an extension of the lever concept. Thinking about the spokes on a bike can help you see the connection between a wheel-and-axle system and a lever. While the car tire attached to the axle of the car is an obvious example of a wheel and axle, anything with a crank handle is also a wheel and axle.

The mechanical advantage of a wheel and axle can be any number greater than zero. In the case of the wheels on the car, the output force is reduced, meaning that the mechanical advantage is less than one. But this also means that the output distance increases, so you can travel farther for each rotation of the axle. In the case of the steering wheel, we get a multiplying of the force, but we have to move our hands farther than the axle in the steering column moves. In this case, we have a mechanical advantage greater than one.

13C Questions

- How is a wheel a machine?
- How do pulley systems help us do work?

13C Terms

wheel and axle, gear, pulley

13.6 GEARS

The gears on a bicycle allow us to easily climb hills and then fly down the other side. A **gear**, a common variation on the wheel and axle, is a simple machine that consists of a wheel with teeth around its perimeter that mesh with teeth on other gears to do work while producing rotational motion. Gears can be used as force multipliers (mechanical advantage > 1), which make us input the force over a greater distance; as force reducers (mechanical advantage between zero and one), which allow the output to travel a greater distance; or as a means to change the direction of the force (mechanical advantage of one), which allows us to apply the force in a better direction.

For two gears to mesh, their teeth must be the same size. That means that as the size of the gears changes, the number of teeth changes too. If one gear is half the size of another gear, it will have half as many teeth, which provides us a modified formula for ideal mechanical advantage.

$$IMA = \frac{teeth_{in}}{teeth_{out}}$$

EXAMPLE 13-5: Calculating Teeth on an Output Gear

You are using gears as a simple machine. If your input gear has 72 teeth and you want an ideal mechanical advantage of 3.0, how many teeth must be on the output gear?

Write what you know.

$teeth_{in}$ = 72 teeth

IMA = 3.0

$teeth_{out}$ = ?

Write the formula and solve for the unknown.

$$IMA = \frac{teeth_{in}}{teeth_{out}}$$

$$IMA\,(teeth_{out}) = \frac{teeth_{in}}{teeth_{out}}(teeth_{out})$$

$$\frac{IMA\,(teeth_{out})}{IMA} = \frac{teeth_{in}}{IMA}$$

$$teeth_{out} = \frac{teeth_{in}}{IMA}$$

Plug in known values and evaluate.

$$teeth_{out} = \frac{teeth_{in}}{IMA}$$

$$= \frac{72 \text{ teeth}}{3.0}$$

$$= 24 \text{ teeth}$$

13.7 PULLEY SYSTEMS

Pulleys are another variation on the wheel and axle. A **pulley** is a simple machine that consists of a wheel-and-axle system with a groove around the perimeter of the wheel in which a rope, cable, or belt moves with the wheel as it rotates. We can use pulleys alone or in groups. The mechanical advantage when using pulleys is greater than or equal to one. In general, the number of rope segments holding up the load is the ideal mechanical advantage of a pulley system.

PULLEY SYSTEMS

SINGLE PULLEYS

A single fixed pulley is the simplest pulley system. It has a pulley attached to a fixed structure. The rope goes over the pulley and attaches to the load. Work is done by applying the input force to the free end of the rope. This pulley has an ideal mechanical advantage of one, and it changes only the direction of the force. Why is that helpful? Imagine that you are trying to lift a crate that weighs a little less than you do. Most likely you can't lift it. But by using a single fixed pulley, you can pull downward with all your weight to lift the crate.

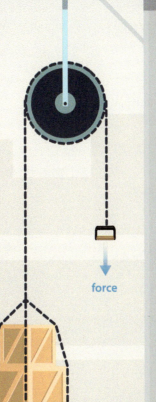

We can also utilize a single pulley as a movable pulley. The rope is attached to a fixed structure, and the pulley is attached to the load. The rope feeds through the pulley, and the free end is pulled upward. Note that there are two sections of the rope holding up the load. Therefore the ideal mechanical advantage is 2.0.

BLOCK AND TACKLE

While you can gain advantage using a single pulley, you can multiply the mechanical advantage by combining pulleys as in the image on the left. In this case, we have combined two fixed pulleys and two movable pulleys. An arrangement like this is called a *block and tackle*. Notice that there are four segments of rope holding up the load, which means that the ideal mechanical advantage is 4.0. The additional pulleys add more friction, which in turn reduces efficiency. Typically the more pulleys that a block and tackle has, the less efficient it becomes.

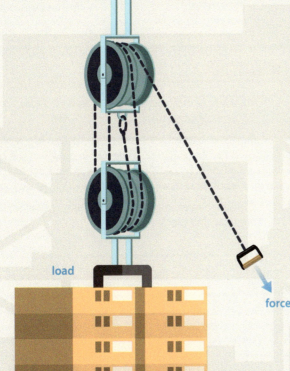

WORK AND MACHINES

EXAMPLE 13-6:
Calculating Mechanical Advantage and Force Needed for a Block and Tackle

What is the *AMA* if the efficiency of the block and tackle on the left is 93.5%? (*Hint*: Notice how many rope segments are supporting the load.)

Write what you know.

$$IMA = 6.00 \text{ (6 rope segments)}$$
$$\text{efficiency} = 93.5\%$$
$$AMA = ?$$

Write the formula and solve for the unknown.

$$\text{efficiency} = \left(\frac{AMA}{IMA}\right)100\%$$

$$\frac{(\text{efficiency})\,IMA}{100\%} = \left(\frac{AMA}{\cancel{IMA}}\right)\left(\frac{\cancel{100\%}}{\cancel{100\%}}\right)\cancel{IMA}$$

$$AMA = \frac{(\text{efficiency})\,IMA}{100\%}$$

Plug in known values and evaluate.

$$AMA = \frac{(\text{efficiency})\,IMA}{100\%}$$
$$= \frac{(93.5\%)(6.00)}{100\%}$$
$$= 5.61$$

What is the input force needed to lift the load if the output force is 475 N?

Write what you know.

$$F_{out} = 475 \text{ N}$$
$$AMA = 5.61$$
$$F_{in} = ?$$

Write the formula and solve for the unknown.

$$AMA = \frac{F_{out}}{F_{in}}$$

$$AMA(F_{in}) = \frac{F_{out}}{\cancel{F_{in}}}\cancel{F_{in}}$$

$$\frac{\cancel{AMA}(F_{in})}{\cancel{AMA}} = \frac{F_{out}}{AMA}$$

$$F_{in} = \frac{F_{out}}{AMA}$$

Plug in known values and evaluate.

$$F_{in} = \frac{475 \text{ N}}{5.61}$$
$$= 84.7 \text{ N}$$

13C | REVIEW QUESTIONS

1. What is a wheel and axle?
2. The wheels on your car travel 3.5 m every time the axle turns 0.37 m. What is the *IMA* of the wheel and axle?
3. How do you gain a mechanical advantage with gears?
4. How do pulley systems help us do work?
5. A single fixed pulley has an *IMA* of one. How does it make work easier?
6. A single movable pulley has an *IMA* of two. What can you conclude about the pulley's *AMA*?
7. You are using a pulley system to lift a 415 N object to a height of 2.70 m. If the *IMA* and *AMA* of the system are 5.00 and 4.61 respectively, what force must you apply and how far will you have to pull the rope?
8. Give one example each of a pulley, wheel and axle, gear, and block and tackle.

13D | INCLINED PLANE

How does a screw do work?

13.8 SIMPLE MACHINE: INCLINED PLANE

Imagine trying to move a piano into a truck. Even a large number of people probably wouldn't be able to lift it that high. This situation is a great application for a simple machine. A block and tackle would work if we could attach it to the ceiling. Another option would be to use an inclined plane, what most of us call a ramp. An **inclined plane** is a simple machine consisting of a flat surface whose opposite ends are at different heights.

An inclined plane makes work easier by allowing us to push with a decreased force over the extended distance of the ramp. The mechanical advantage of an inclined plane is always greater than one. The steeper the inclined plane is, the closer the mechanical advantage is to one, while shallower inclined planes have higher mechanical advantages.

13D Questions

- How does a ramp make work easier?
- Can I combine simple machines?
- How do I calculate the mechanical advantage of a compound machine?

13D Terms

inclined plane, wedge, screw, compound machine

WORK AND MACHINES 313

EXAMPLE 13-7: Calculating IMA and AMA of a Ramp

To move a piano that weighs 3125 N onto a truck, you have to push with 865 N of force along a 6.3 m long ramp. The truck is 1.5 m above the ground. What is the IMA and AMA of the ramp?

Write what you know, including a diagram.

F_{out} = 3125 N (weight of piano)

F_{in} = 865 N

d_{in} = 6.3 m

d_{out} = 1.5 m

IMA = ?

AMA = ?

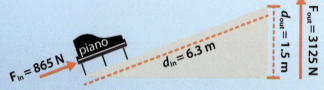

Write the formula and solve for the unknown.

$$IMA = \frac{d_{in}}{d_{out}}$$

Plug in known values and evaluate.

$$IMA = \frac{d_{in}}{d_{out}}$$

$$= \frac{6.3 \text{ m}}{1.5 \text{ m}}$$

$$= 4.2$$

Write the formula and solve for the unknown.

$$AMA = \frac{F_{out}}{F_{in}}$$

Plug in known values and evaluate.

$$AMA = \frac{F_{out}}{F_{in}}$$

$$= \frac{3125 \text{ N}}{865 \text{ N}}$$

$$= 3.61$$

As expected, the AMA is less than the IMA due to the friction acting against the piano as it moves up the ramp.

13.9 OTHER USES OF INCLINED PLANES

Inclined planes, in the form of ramps, are extremely common, but there are other applications of inclined planes that may be even more common but not as easily recognized. Those applications are the *wedge* and the *screw*. Let's see how inclined planes are part of these two simple machines.

WEDGES AND SCREWS

WEDGES

A **wedge** consists of two planes connected with an acute angle that is thick at one end and thinner at the other. It is used for splitting materials. The wedge spreads material apart as it is forced into the material. The mechanical advantage of a wedge depends on the angle between the two planes. If the angle is small, such as on a knife, the mechanical advantage is high. If the angle is larger, like on an ax, the mechanical advantage is lower. The aerodynamic shape of sports cars and the V-shaped bow of a ship are two examples of wedges.

SCREWS

An inclined plane that is wrapped around a cone or cylinder in a spiral pattern is a simple machine called a **screw**. We typically use screws as fasteners, such as bolts and wood screws. The mechanical advantage comes from the steepness of the inclined plane, called the *thread*, around the cone or cylinder. The steeper the thread, the less mechanical advantage you gain. The screw will go in quickly, but will require much more force. A screw with a shallow slope will be much easier to put in, but you will have to turn the screw many times.

inclined plane

13.10 COMPOUND MACHINES

Simple machines show the ability of man to use his intellect to make work easier to do. While the simple machines we've considered can do a lot, we have also figured out how to combine two or more simple machines into a **compound machine**. The mechanical advantage of a compound machine is the product of the mechanical advantages of the simple machines that make it. Most of the items that we consider machines are compound machines. Mechanical clocks are compound machines made of gears, wheel-and-axle systems, and levers. Another compound machine that many of us are familiar with is the bicycle. Let's take a closer look.

BICYCLE—A COMPOUND MACHINE

A bicycle is a compound machine that is made up of levers, wheel-and-axle systems, gears, and pulleys.

LEVERS

Levers can give us a force advantage or a distance advantage. On the bicycle, there are numerous levers. On many bikes, we actuate the brakes by using the levers on the handlebars. The brake levers apply a force on the brakes, which are also levers. Many gear changers incorporate levers. Some bikes have quick release levers for removing wheels and adjusting seat height. Even the handlebars are levers. You turn the handlebars to turn the front wheel.

WHEEL-AND-AXLE SYSTEMS

Wheel-and-axle systems work with two or more circular systems with different radii attached to the same axis. We all know that bikes move when we push the pedals. But not everyone recognizes that the pedals are a wheel-and-axle system. The pedals with a large radius turn the axle, which in turn rotates the chain sprockets, giving the rider a mechanical advantage. This work is transmitted through a chain to the rear derailleur gears, turning the rear axle, which turns the rear wheel. This arrangement gives us a mechanical advantage much less than one, which means that each rotation of the pedals produces a huge output distance for the wheel and the bicycle itself.

GEARS

The chain sprockets, connected to the pedals, and rear derailleur gears, in the back, work as a sequence of gears that are indirectly connected through the chain. By selecting various combinations of chain sprockets and rear gears, road bikes have a great variety of mechanical advantage. The low gear gives a high mechanical advantage, meaning that you will pedal a lot but will climb those hills with ease. The highest gear will give you a low mechanical advantage; you will have to push harder on the pedals, but you will fly along the flats and downhills.

HOW IT WORKS

Clocks

On Day 4 of Creation, God created the sun, moon, and stars to mark the passage of time. The rising and falling of the sun marks days. The movement of the moon marks the months. The changes in the location of sunrise and sunset mark the seasons and years. But man found that he needed a method of dividing the day into smaller time increments to better use time and coordinate work. So he used his God-given talent to invent different types of clocks.

One type of clock is a mechanical clock, which is a compound machine. All mechanical clocks have an energy source. Throughout history, people have used water, suspended masses, and springs to provide the energy to run their mechanical clocks. All mechanical clocks also utilize some rhythmically moving part to control the time indicators accurately. While many early clocks had only hour hands, people eventually created more precise clocks with a minute and even a second hand. Today we use the term *clock* to denote most timekeeping devices, but technically all clocks have a bell or ringers. The word *clock* comes from the Celtic word for bell. Let's look inside and see how it works.

You can see the large spiral balance spring on the right side. The spring is the energy source for this clock. You can also see the numerous gears that cause each hand to turn at a certain rate. Also, you can see some levers that control the rate at which the gears can turn and activate the striker for the bells. All clocks are wonderful examples of man's creativity making our lives better. Well, maybe not the alarm clock!

13D | REVIEW QUESTIONS

1. What is an inclined plane?
2. How does an inclined plane give you a mechanical advantage?

Use the following information to answer Questions 3–5.

You are rolling a large cart up a ramp. You have to apply a force of 78.5 N up the 4.32 m ramp. The ramp lifts the cart 1.13 m by applying a force of 285 N.

3. What is the *IMA* of the ramp?
4. What is the *AMA* of the ramp?
5. What is the *efficiency* of the ramp?
6. You want a screw that will be easy to put in, even if you have to turn it many times. Do you want a screw with a steep or shallow thread?
7. What is a compound machine?

CHAPTER 13 REVIEW

13A WORK AND MECHANICAL ADVANTAGE

- Work is the energy transferred to a system by an external force when it acts on the system to move it. Work is measured in units called newton-meters (N·m) or joules (J).
- Work is done only by forces parallel to the direction of motion.
- Power is the rate at which work is done. Power is measured in joules per second (J/s) or watts (W).
- A simple machine is a basic mechanical device that makes work easier by changing the magnitude, distance, or direction of the input force. Simple machines give a user a mechanical advantage.
- Mechanical advantage is the degree to which the input force is changed by the machine.

13A Terms

work	296
power	298
simple machine	299
mechanical advantage	299
ideal mechanical advantage	300
actual mechanical advantage	301
efficiency	301

13B LEVER

- Levers are rigid bodies that can rotate about a pivot point called the fulcrum.
- Levers work by torque—the tendency of a force to cause a body to rotate about an axis. The degree of rotation depends on the size of the force and the moment arm—the perpendicular distance to the pivot point.
- The law of torques states that if a system is in rotational equilibrium, then the sum of the torques must be zero.
- There are three classes of levers, depending on the relative positions of the input force, the output force, and the fulcrum.
- In a first-class lever, the fulcrum is between the input and output forces.
- In a second-class lever, the output force is between the fulcrum and the input force.
- In a third-class lever, the input force is between the fulcrum and the output force.

13B Terms

lever	303
fulcrum	303
torque	304

13C WHEEL AND AXLE

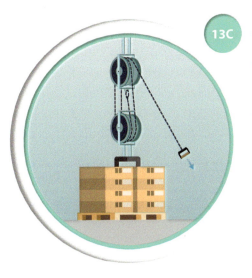

- A wheel and axle is a simple machine consisting of a wheel with a rod through its rotational axis.
- Gears (wheels with teeth along their circumferences) and pulleys (grooved wheels with a rope or belt) are types of wheel-and-axle systems.
- The mechanical advantage of gears is a result of their different numbers of teeth.
- Pulleys can be used singly or combined.
- The mechanical advantage of pulleys depends on the number of rope segments supporting the load.

13C Terms

wheel and axle	309
gear	310
pulley	310

WORK AND MACHINES 319

CHAPTER 13

13D INCLINED PLANE

- A ramp, or inclined plane, is a simple machine consisting of a plane with its opposite ends at different heights.
- The mechanical advantage of an inclined plane depends on the steepness of the plane. The shallower the plane is, the higher the mechanical advantage.
- A wedge is two planes attached at an angle.
- A screw is an inclined plane wrapped around a cone or cylinder.
- Simple machines can be combined to form a compound machine.
- The mechanical advantage of a compound machine is the product of the mechanical advantages of the simple machines of which it is made.

13D Terms

inclined plane	313
wedge	315
screw	315
compound machine	316

CHAPTER REVIEW QUESTIONS

Recalling Facts

1. Define *power*.
2. What is mechanical advantage?
3. What is torque?
4. Describe the relative positions of the input force, output force, and fulcrum in each class of lever.
5. What is a gear?
6. Describe the two ways that a single pulley can be used.
7. What is a wedge?
8. How is a screw an inclined plane?

Understanding Concepts

9. How long will it take for a 1238 W motor to do 64 896 J of work?
10. If the motor in Question 9 lifted a crate 3.8 m, how much force did the motor apply to the crate?
11. Explain how force, work, and power are related.

Use the following information to answer Questions 12–14.

While using a simple machine, you apply 132 N of force over 1.77 m. The machine applies an 878 N force to the load and moves the load 0.25 m.

12. What is the *IMA* of the machine?
13. What is the *AMA* of the machine?
14. What is the *efficiency* of the machine?

15. If a machine allows us to apply less force to do work, what is the cost?
16. How do levers work?
17. Daniel applies a 471 N force 1.81 m to the right of the fulcrum of a seesaw, while Isaiah applies a 355 N force 2.25 m to the left of the fulcrum. Will the seesaw be in rotational equilibrium? (*Hint*: Draw a picture and label it before you do any calculations.)

REVIEW

18. Explain how your lower arm can be considered a lever when you are lifting a glass of water. What class of lever is it?
19. Give an example of when the mechanical advantage of a lever is less than one. Explain.
20. Why can a third-class lever have a mechanical advantage only between zero and one?
21. How is a wheel and axle a machine?
22. Compare wheel-and-axle systems, pulleys, gears, and blocks and tackles.
23. The pulley system shown on the right applies an output force of 917 N while lifting an object 1.15 m. The efficiency of the pulley system is 92.5%. What is the *AMA* of the system? Notice the five rope segments holding up the load.

Use the following information to answer Questions 24–25.

A person applies 64.9 N of force over a distance of 12.77 m while using a block and tackle. The machine applies a force of 763 N of force over a distance of 1.04 m.

24. How much work did the person do?
25. How much work did the machine do?
26. Why does a person using a pulley system, or any machine for that matter, have to do more work than the machine does?

Use the following information to answer Questions 27–29.

You are using a wedge to split firewood. You apply a force of 265 N, and the log splits after the wedge separates the two halves 0.0652 m (d_{out}). The *IMA* of the wedge is 3.35, and the efficiency is 93.5%.

27. What is the *AMA* of the wedge?
28. What force did the wedge apply to the log?
29. How far did you have to move the wedge?

Critical Thinking

30. Your friend tells you about an engineer who has built a machine that is 100% efficient. What is your response?
31. How does torque differ from work?
32. Evaluate the statement, "All forces produce torque."
33. A wheelbarrow is a classic example of a second-class lever. Describe how a wheelbarrow might be used (very inefficiently) as either a first- or third-class lever.
34. The wheels on your car roll 3.5 m every time the axle turns 0.37 m. What would you need to do if you wanted to make the car travel farther for each turn of the axle?
35. Why can't you have a ramp with a mechanical advantage of one?
36. A ramp has no moving parts, so why can't an inclined plane have an efficiency of 100%?
37. Justify the use and development of simple machines using the Creation Mandate.

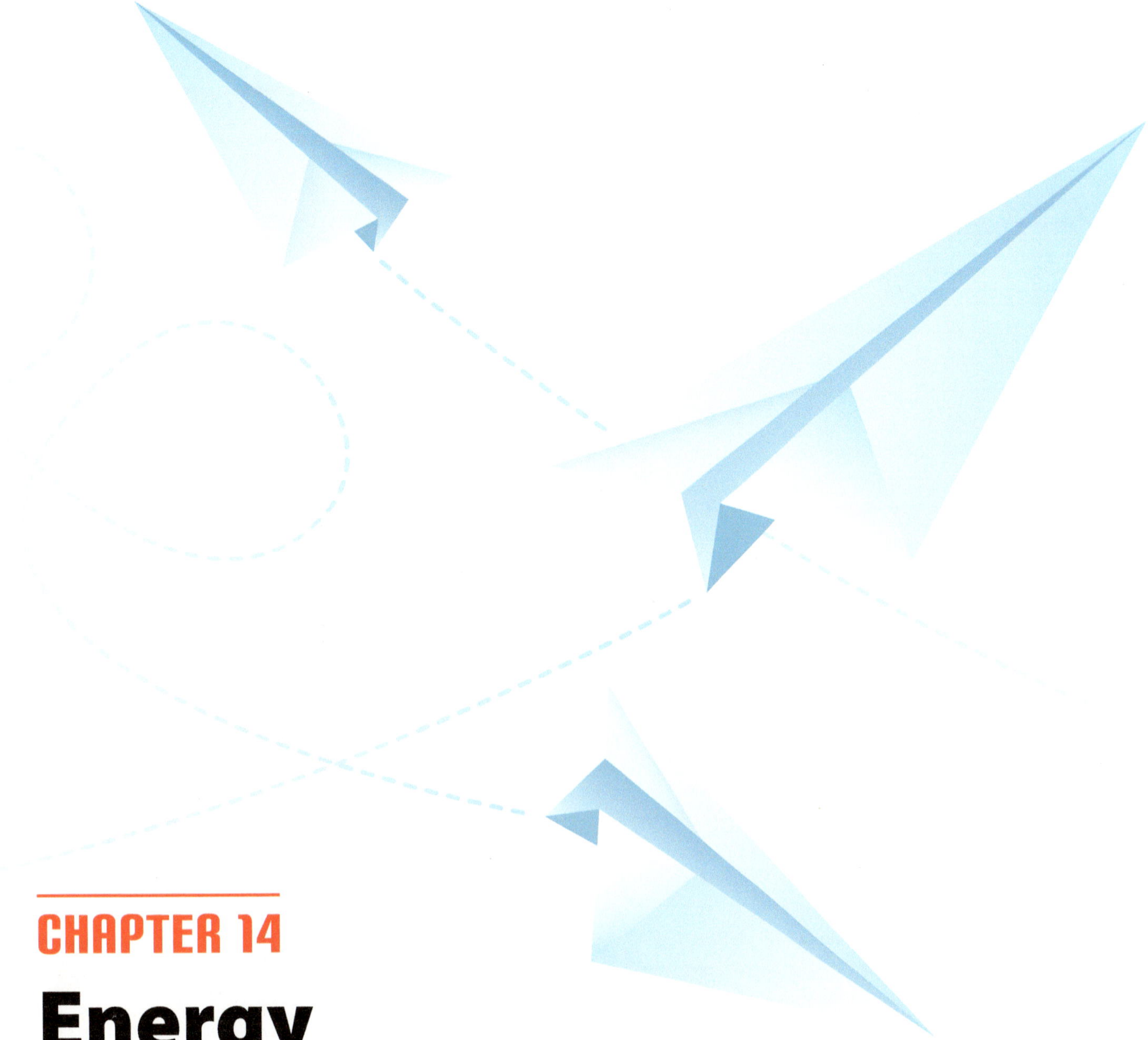

CHAPTER 14

Energy

A NEW KIND OF FAKE PLANT

Cheap, abundant energy—it's been the dream of humankind for centuries. But we still haven't found the ideal answer. Cheap energy often pollutes, and clean energy is often expensive. One potential solution is artificial photosynthesis. You may recall that plants have special organelles called chloroplasts that use energy from the sun to form sugars from carbon dioxide and water. Humans have developed ways to do this too, but the technology is still in its infancy. It can't yet produce energy cheaply or sufficiently to meet the needs of a growing human population. What can we do in the meantime until technology like this is perfected?

God created leaves with the amazing ability to convert light energy into chemical energy. Man can imitate God's creativity, as when scientists at Eindhoven University in the Netherlands designed an artificial leaf (inset) in 2016.

14A Classifying Energy	324
14B Energy Changes	332
14C Energy Resources	337

14A Questions

- What is energy?
- Is energy conserved?
- Are there different kinds of energy?
- How can I calculate the amount of energy in an object?

14A Terms

energy, kinetic energy, potential energy

14A | CLASSIFYING ENERGY

Where does energy come from?

14.1 ENERGY

We talk about energy all the time. We might say, "I don't have any energy today," in which case we might indulge in an "energy drink." But what exactly is this mysterious thing we call energy? Like other fundamental properties of creation, energy is not easy to define, though we can describe what it does. In fact, we define **energy** as the ability to do work. We may not know what energy is, but we do know what it does or has the potential to do. It is a property of matter that can sometimes be measured, though measuring some forms of energy is very difficult or impossible with our current level of technology.

The universe would cease to exist without energy. Just as space, matter, and time could not exist without each other, so matter could not exist without energy. Energy holds matter together. In fact, as Albert Einstein deduced, matter and energy are *interconvertible*—they

are different forms of that which makes up the universe! But not only are energy and matter interconvertible, energy can also be changed from one form to another as well as used to do work. To understand some aspects of the nature of energy, you can think of it as being like wealth. Material wealth can exist in different forms: savings, stocks, bonds, real estate, precious metals, or possessions. Just as you may have many forms of wealth, a system may have many forms of energy that contribute to its total energy. When you buy something, you exchange one form of your wealth for another form of wealth. Similarly, energy can be exchanged for work, for other forms of energy, or even for matter.

Because work is the transferring of energy between objects, energy is measured in joules (J) just as work is. As you saw in Chapter 13, a joule is the amount of work that is done when a force of 1 N is applied for a distance of 1 m. One joule of energy is transferred when one joule of work is done.

Conservation of Energy

Energy is another conserved property of matter. As you learned in Chapter 7, the law of conservation of energy states that the total energy within a closed system remains constant over time. "Closed system" simply means that no energy is either leaving or entering the system, that is, the objects within the system are considered in isolation. You may have heard this law stated in another fashion. The idea of conservation is what is meant when someone says that energy can neither be created nor destroyed, only changed from one form into another. During such energy transformations, the total amount of energy within a system remains the same. Before we consider some examples of how the conservation of energy works, we'll look at two general forms of energy, *kinetic energy* and *potential energy*.

14.2 KINETIC ENERGY

You actually know a little bit about different kinds of energy already, though you may not have realized it. Take the two hammers on the next page, for instance. Which one do you think would be more useful for doing demolition work prior to remodeling a kitchen? A sledge hammer, obviously! But why is a sledge hammer more useful than a tack hammer for demolition work? The answer is **kinetic energy**, the energy that objects possess due to their motion. When a hammer is swung, its kinetic energy does work on whatever the hammer strikes—hopefully not your thumb!

KINETIC ENERGY

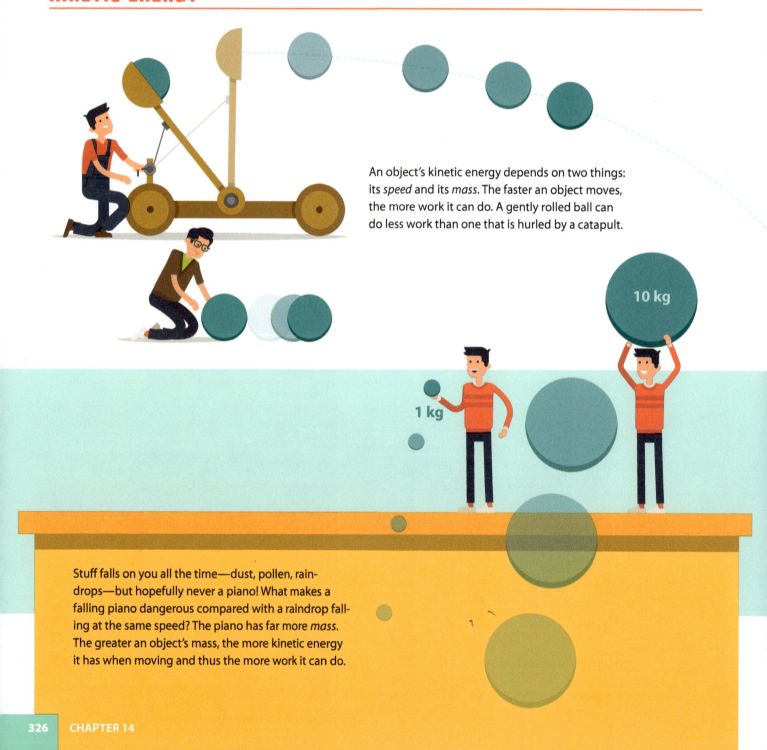

An object's kinetic energy depends on two things: its *speed* and its *mass*. The faster an object moves, the more work it can do. A gently rolled ball can do less work than one that is hurled by a catapult.

Stuff falls on you all the time—dust, pollen, raindrops—but hopefully never a piano! What makes a falling piano dangerous compared with a raindrop falling at the same speed? The piano has far more *mass*. The greater an object's mass, the more kinetic energy it has when moving and thus the more work it can do.

So thinking back on our hammer example, how can you get a hammer to do more work on a nail? To do more work, you need to increase the hammer's kinetic energy. You could swing the hammer faster, use a heavier hammer, or both. This implies that there is a mathematical relationship between kinetic energy, mass, and velocity (speed), and indeed there is. The formula for finding the kinetic energy (KE) of an object (in joules) with a mass of m (in kilograms) and a speed of v (in meters per second) is shown below.

$$KE = \tfrac{1}{2}mv^2$$

If you had to choose one, which would you rather be hit by, a train or a feather? Sounds silly, right? In truth, it's a tricky question because the safe answer depends on how fast each object is moving. Let's use the kinetic energy formula to help us decide which option to choose.

EXAMPLE 14-1: Slow Train Coming

A slow-moving locomotive has a mass of 107 000 kg and is moving at 0.250 m/s. How much kinetic energy does it have?

Write what you know.

$m = 107\,000$ kg

$v = 0.250$ m/s

$KE = ?$

Write the formula and solve for the unknown.

$$KE = \tfrac{1}{2}mv^2$$

Plug in known values and evaluate.

$$KE_{locomotive} = \tfrac{1}{2}(107\,000 \text{ kg})(0.250 \tfrac{m}{s})^2$$

$$= 3343.75 \, \tfrac{kg \cdot m^2}{s^2} = 3343.75 \, \tfrac{kg \cdot m}{s^2}(m)$$

$$= 3340 \text{ N} \cdot \text{m} = 3340 \text{ J} = 3.34 \text{ kJ}$$

EXAMPLE 14-2: Tickle Me with a Feather?

On average, the mass of a chicken feather is 0.000 008 20 kg. The speed of the fastest ever man-made object is 98 900 m/s. How much kinetic energy would an average chicken feather have if it were moving at that speed?

Write what you know.

$m = 0.000\,008\,20$ kg

$v = 98\,900$ m/s

$KE = ?$

Write the formula and solve for the unknown.

$$KE = \tfrac{1}{2}mv^2$$

Plug in known values and evaluate.

$$KE_{feather} = \tfrac{1}{2}(0.000\,008\,20 \text{ kg})(98\,900 \tfrac{m}{s})^2$$

$$= 40\,100 \text{ N} \cdot \text{m} = 40\,100 \text{ J} = 40.1 \text{ kJ}$$

So in our admittedly far-fetched example, the fast-moving feather has twelve times as much kinetic energy as the slow-moving locomotive. Did you choose wisely?

Transferring Energy

The fact that energy is conserved as it is transferred or transformed allows us to make some interesting calculations. In the formula for kinetic energy, it should be apparent that if we know the amount of energy present in a moving object, we can predict how the motion of another object will be changed if that kinetic energy is transferred. For example, baseball fans these days hear a lot about the "exit velocity" of batted balls. Can we predict how fast a batted ball will travel?

EXAMPLE 14-3: Slugfest

A baseball bat with a mass of 1.00 kg is swung at 33.5 m/s and hits a baseball that has a mass of 0.150 kg. If all the bat's kinetic energy is transferred to the ball, how fast will the batted ball travel?

Write what you know.

$m_{bat} = 1.00$ kg

$v_{bat} = 33.5$ m/s

$m_{ball} = 0.150$ kg

$v_{ball} = ?$

Write the formula and solve for the unknown.

$$KE = \tfrac{1}{2}mv^2$$

Plug in known values and evaluate to find the kinetic energy of the bat.

$$KE_{bat} = \tfrac{1}{2}(1.00 \text{ kg})(33.5 \text{ m/s})^2$$

$$KE_{ball} = \tfrac{1}{2}(0.150 \text{ kg})(v_{ball})^2$$

Since the KE for both the bat and ball is assumed to be constant, we can set the two mathematical expressions equal to each other.

$$\tfrac{1}{2}(1.00 \text{ kg})(33.5 \text{ m/s})^2 = \tfrac{1}{2}(0.150 \text{ kg})(v_{ball})^2$$

$$(2)\tfrac{1}{2}(1.00 \text{ kg})(33.5 \tfrac{m}{s})^2 = (2)\tfrac{1}{2}(0.150 \text{ kg})(v_{ball})^2$$

$$\frac{(1.00 \text{ kg})(33.5 \tfrac{m}{s})^2}{(0.150 \text{ kg})} = \frac{(0.150 \text{ kg})(v_{ball})^2}{(0.150 \text{ kg})}$$

$$\sqrt{7481.666\,66 \tfrac{m^2}{s^2}} = \sqrt{v_{ball}^2}$$

$$v_{ball} = 86.5 \text{ m/s}$$

14.3 POTENTIAL ENERGY

Potential energy (PE) is sometimes called the *energy of condition* or the *energy of position*. **Potential energy** is stored energy that can be used later. A drawn bow string, for example, stores *elastic potential energy* in the limbs of the bow. That stored energy is used to do work on an arrow, sending it toward a target. Storing energy in a drawn bow is an example of energy of condition. Perhaps you play a stringed instrument such as a guitar. As a guitar string is plucked, it is stretched slightly from its resting position. As it returns to this position after being played, some of its elastic potential energy is converted to sound energy.

Living things rely on *chemical potential energy*. Where do the animals in an ecosystem, for instance, get the energy that they need to survive? From their food, of course! As you learned in Chapter 7, chemical energy can be released during chemical reactions, such as those that happen during cellular respiration. The energy stored in food molecules is an example of chemical potential energy.

Gravitational Potential Energy

Let's revisit our hammer example and think about what happens when a hammer is dropped instead of swung. You know that a dropped hammer can do work on your toe if you're not quick enough to get out of the way. To do work on your toe, the dropped hammer must have energy. The hammer's potential energy is due to its *position* relative to your toe—it's above your toe, and Earth's gravity will pull the hammer downward if you let go of it. We call this form of potential energy *gravitational potential energy* (GPE).

Now let's consider what might make the dropped hammer potentially more dangerous. What factors would increase the hammer's potential energy? One obvious factor is the *height* from which the hammer is dropped. A hammer falling from a rooftop has more energy than one that is dropped only a few inches above ground. The *weight* of the hammer is also a big factor. If the sledge and tack hammers we saw earlier were dropped from the same height, the sledge hammer would definitely pose a greater danger.

So if we know an object's weight (the product of its mass and its acceleration due to gravity) and its height above a reference position (usually the ground), we should be able to calculate its *GPE* (in joules). We use the formula

$$GPE = mgh,$$

where *m* is the object's mass in kilograms, **g** is its acceleration due to gravity in meters per second squared, and *h* is the object's height above the reference position, often the ground, in meters. Let's use this formula to calculate how much gravitational potential energy Newton's famous (but likely mythical) apple might have had.

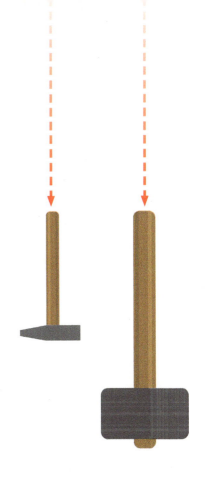

EXAMPLE 14-4: Calculating GPE

How much gravitational potential energy does a 0.25 kg apple on a tree have if it is 3.0 m above the ground (reference position)?

Write what you know.

$$m = 0.25 \text{ kg}$$

$$g = 9.81 \text{ m/s}^2$$

$$h = 3.0 \text{ m}$$

Write the formula and solve for the unknown.

$$GPE = mgh$$

Plug in known values and evaluate.

$$GPE = (0.25 \text{ kg})(9.81 \tfrac{m}{s^2})(3.0 \text{ m})$$

$$= 7.4 \tfrac{kg \cdot m^2}{s^2} = 7.4 \text{ N} \cdot \text{m} = 7.4 \text{ J}$$

Now let's suppose that the apple in Example 14-4 does in fact fall to the ground. Could we calculate how fast it would be moving when it hit the ground? We can, in fact. We'll assume that all the apple's gravitational potential energy is conserved as it is converted to kinetic energy. That means it will have 7.4 J of kinetic energy just at the moment of contact. By plugging the apple's kinetic energy and mass into the kinetic energy formula, we can solve for the apple's velocity.

EXAMPLE 14-5: Calculating Velocity

How fast is a 0.25 kg apple moving when it hits the ground after falling 3.0 m from a tree?

Write what you know.

$$KE = 7.4 \text{ J}$$
$$m = 0.25 \text{ kg}$$
$$v = ?$$

Write the formula and solve for the unknown.

$$KE = \tfrac{1}{2}mv^2$$

$$2(KE) = \left(\tfrac{1}{\cancel{2}}mv^2\right)\cancel{2}$$

$$\sqrt{\frac{2(KE)}{m}} = \sqrt{\frac{\cancel{m}v^2}{\cancel{m}}}$$

$$v = \sqrt{\frac{2(KE)}{m}}$$

Plug in known values and evaluate.

$$v = \sqrt{\frac{2(7.4 \text{ J})}{0.25 \text{ kg}}}$$

$$= \sqrt{\frac{2\left(7.4 \, \frac{\cancel{\text{kg}} \cdot \text{m}^2}{\text{s}^2}\right)}{0.25 \, \cancel{\text{kg}}}}$$

$$= 7.7 \text{ m/s}$$

In this instance, the apple is traveling at 7.7 m/s when it strikes the ground.

There are many other forms of energy other than the ones that we have discussed so far. You'll read about some of these in the chapters to come.

14A | REVIEW QUESTIONS

1. Why are work and energy both measured in joules?
2. What does the law of conservation of energy say about the nature of energy?
3. Suppose a runaway shopping cart in a parking lot strikes another cart with 15 J of kinetic energy. What is the maximum amount of kinetic energy the second cart could have as a result of the collision?
4. What two things determine the amount of kinetic energy possessed by a moving object?
5. How much kinetic energy does a 50.0 kg rider and 65.0 kg motocross bike have if traveling at 55.0 kph?
6. A 250.0 kg bumper car carrying a 55.0 kg passenger and traveling at 1.50 m/s strikes an empty bumper car of the same mass. What is the fastest possible speed the second car could achieve as a result of the collision?
7. How much gravitational potential energy does a 0.75 kg flowerpot have while sitting on a 1.5 m high shelf (assume that the floor is the reference height)? How much kinetic energy would the pot have if it struck the floor after being knocked off the shelf?

MINI LAB

VISUALIZING POTENTIAL ENERGY

We humans instinctively know that mass affects the gravitational potential energy of objects (stored energy). That's why we scurry to get out of the way of large falling objects but worry less about smaller ones. Here's a quick and easy way to see how mass and height affect an object's gravitational potential energy—its potential to do work due to its height.

Essential Question:
Can we predict changes in potential energy?

1. Assuming that two objects are at the same height, which one do you think has more potential energy, one with more mass or one with less mass? Write your answer in the form of a hypothesis.

Equipment
large shallow pan
fine sand
masses, various
meter stick
centimeter ruler

PROCEDURE

A Fill a shallow pan with fine sand to a level depth of 3–5 cm.

B Copy Table 1 onto a sheet of paper.

C Select a mass to drop and record its mass in Table 1.

D Drop the mass straight down into the sand from a height of 0.5 m. Then measure the depth of the impact crater produced by the falling mass and enter it into Table 1.

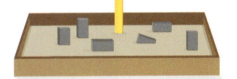

Table 1
Mass Drop Data

Mass (g)	Depth of Crater (cm)

E Repeat Steps **C** and **D** using different-sized masses.

ANALYSIS

2. Which mass produced the deepest crater?

CONCLUSION

3. What can you conclude about the relationship between mass and gravitational potential energy? Does this support your hypothesis?

GOING FURTHER

4. Why did you need to drop the masses from the same height?

5. Does this procedure directly measure the gravitational potential energy of the masses? Explain.

6. Describe a method for using this setup to test the relationship between height and gravitational potential energy.

ENERGY 331

14B Questions

- What is mechanical energy?
- Is energy transformation an efficient process?
- What happens to "lost" energy?

14B Term

mechanical energy

14B | ENERGY CHANGES

How is energy transformed?

14.4 MECHANICAL ENERGY

The two forms of energy we've considered so far—kinetic energy and potential energy—often work closely together. **Mechanical energy** is the sum of any kinetic energy and all the forms of potential energy in a system. Remember that energy within a closed system is conserved. It can change forms or be transferred, but the total amount remains constant. Let's look at an example of how this works.

HOP TO IT!

Extreme pogo, or Xpogo®, uses a souped-up version of the classic pogo stick to do a variety of aerial stunts. Xpogo sticks don't use a spring like their pogo stick ancestors; instead, they rely on compressed air. Here's what happens.

Gravitational Potential Energy. At the top of his jump, the rider has his maximum amount of gravitational potential energy. For the tiniest fraction of a second, he also has no kinetic energy. Gravity will now accelerate him downward.

HOP TO IT!

50% | 50%
GPE | KE

Going Up! When the gas's elastic potential energy has all been converted back to kinetic energy, the stick leaves the ground and heads upward. As it rises, its kinetic energy is converted back to gravitational potential energy.

100% | 0%
GPE | KE

One More Time. When all of the stick's kinetic energy has been converted back to gravitational potential energy, the rider reaches the top of his jump and begins the cycle again.

The example above assumes that none of the Xpogo stick's energy is transformed into other forms besides either potential or kinetic energy. That makes it sound like an Xpogo stick is a kind of perpetual motion machine, a device that can keep on moving forever. Of course, such a thing doesn't actually exist. Most systems lose some energy in various forms to their environment. Left on its own, the stick and its rider would eventually bounce to a standstill. In reality, the rider is constantly adding some energy to the system with his legs at the bottom of each jump.

14.5 ENERGY TRANSFORMATIONS

As you just saw, energy often changes from one type to another. We use energy transformations every day to light our homes, cook our food, and travel to school. For example, a light bulb transforms electrical energy into light and heat energy. A gas oven transforms chemical energy into heat energy. A car transforms the chemical energy in gasoline into mechanical energy. Our bodies transform chemical energy from food into heat energy in our cells, electrical energy in our nerves, and mechanical energy in our muscles. Every form of useful energy can change into at least one other form of useful energy. But the conversion is never 100%—some energy that can't be used always escapes the process. By comparing the amount of usable energy remaining after a process with the original amount of energy that went into it, we can measure the *efficiency* of an energy transformation. Most manmade energy transformations are not more than 20%–40% efficient. The following illustration will help you think about how much energy is "lost" during the transformation process.

POWER PLANT

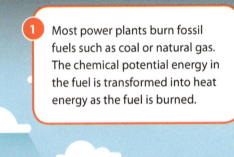

1. Most power plants burn fossil fuels such as coal or natural gas. The chemical potential energy in the fuel is transformed into heat energy as the fuel is burned.

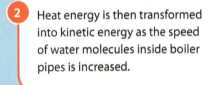

2. Heat energy is then transformed into kinetic energy as the speed of water molecules inside boiler pipes is increased.

3. The kinetic energy of the fast-moving molecules in steam is transferred to the kinetic energy of the turning blades and shaft within a turbine.

4. The kinetic energy of the turning turbine shaft is transformed into electrical energy by a generator. By this point in the process, well over half of the original energy in the fuel has been "lost"—it has been converted into forms that cannot be used to do useful work.

ENERGY 335

POWER PLANT, CONTINUED

5 As electrical energy travels through transmission lines, some is transformed into heat and sound. You've probably heard the hum of high-voltage power lines.

6 In your home, electrical energy is transformed into other forms of energy—for example, sound from your TV or light from a lamp. Modern LED lightbulbs are very efficient, but even they waste some electrical energy as heat.

As the prior example shows, the process of transforming energy from one form to another "wastes" a lot of energy. This missing energy hasn't somehow disappeared—that would be a violation of the law of conservation of energy. The lost energy has simply been converted into a form that is less available for doing work. While the chemical potential energy in fossil fuels is fairly easy to collect and store, the heat energy that has radiated into the power plant's environment can't be gathered again. This idea that energy becomes increasingly less available for work will be explored further in Chapter 15.

14B | REVIEW QUESTIONS

1. What is mechanical energy?
2. Compare the kinetic and gravitational potential energies of a model rocket as it reaches the apex (highest point) of its flight.
3. What happens to the rocket's kinetic and gravitational potential energies as it falls back to Earth?
4. Draw a diagram of the motion of a playground swing. Label the following portions of the swing's motion: point of maximum GPE/minimum KE, point of minimum GPE/maximum KE, GPE being transformed to KE, KE being transformed to GPE.
5. Describe the energy transformations that take place when a person plays a high striker carnival game. Be sure to identify where during the process that energy is being transferred to work or back to energy.

6. In Section 14.1, you learned that energy is conserved within closed systems. Why are closed systems more of an ideal rather than something encountered in the real world?

14C | ENERGY RESOURCES

How can we best generate energy?

14.6 NONRENEWABLE ENERGY

In 2018, the world's population stood at 7.7 billion. That number could rise to 10 billion by 2055. That's a lot of people, and people need energy—energy to do work, grow food, heat and cool homes, and get from place to place. Some of this energy comes from renewable energy sources. A **renewable energy resource** is one that is easily replaced by natural methods. Energy resources that are not replaced naturally, at least not on a human timescale, are called **nonrenewable energy resources**. As you can see in the pie chart below, most of the world's energy demands—over 85%—are presently being met by nonrenewable energy resources.

Fossil Fuels

About 80% of the energy from nonrenewable resources is in the form of coal, oil, and natural gas. These three resources are known as fossil fuels. **Fossil fuels** are believed by most to have been formed from the remains of plants and animals that lived in the past. These energy sources may be used directly, such as when heating oil is burned to heat a home. Often, though, the energy in a fossil fuel is transformed into electrical energy, as you saw in Subsection 14.5.

14C Questions

- What is the difference between renewable and nonrenewable energy?
- What are fossil fuels?
- Should people conserve energy?
- Can renewable energy sources meet the world's needs for energy?

14C Terms

renewable energy resource, nonrenewable energy resource, fossil fuel, nuclear energy, biomass energy, hydroelectric energy, geothermal energy, solar energy, wind energy, tidal energy

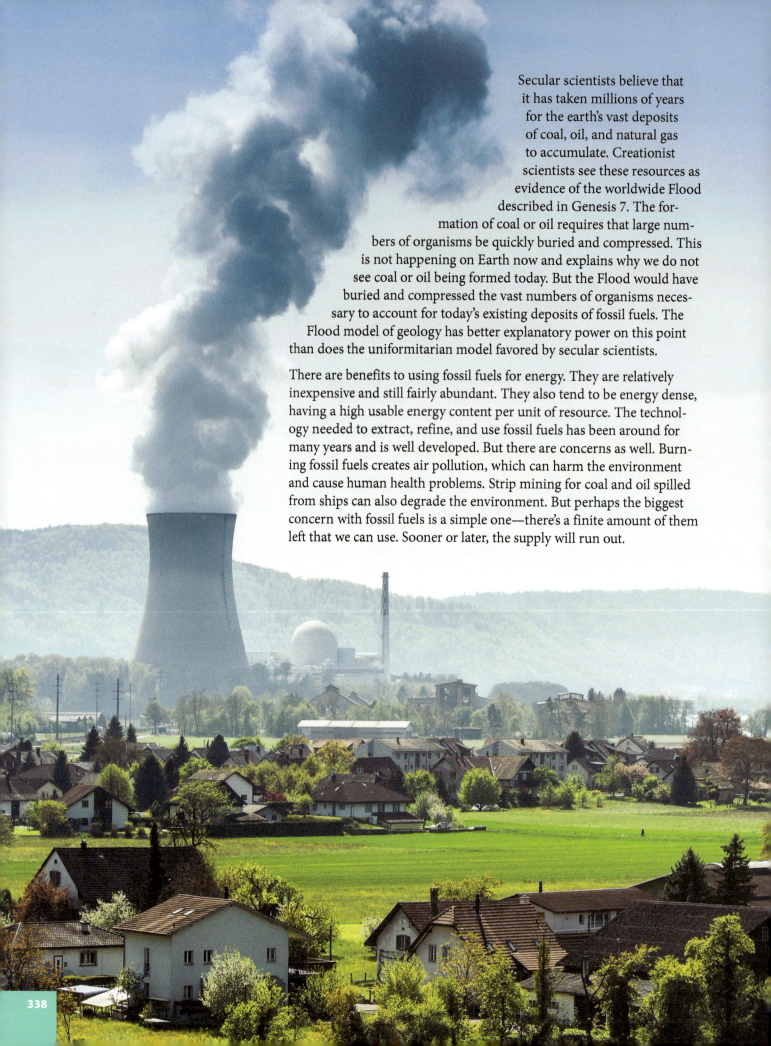

Secular scientists believe that it has taken millions of years for the earth's vast deposits of coal, oil, and natural gas to accumulate. Creationist scientists see these resources as evidence of the worldwide Flood described in Genesis 7. The formation of coal or oil requires that large numbers of organisms be quickly buried and compressed. This is not happening on Earth now and explains why we do not see coal or oil being formed today. But the Flood would have buried and compressed the vast numbers of organisms necessary to account for today's existing deposits of fossil fuels. The Flood model of geology has better explanatory power on this point than does the uniformitarian model favored by secular scientists.

There are benefits to using fossil fuels for energy. They are relatively inexpensive and still fairly abundant. They also tend to be energy dense, having a high usable energy content per unit of resource. The technology needed to extract, refine, and use fossil fuels has been around for many years and is well developed. But there are concerns as well. Burning fossil fuels creates air pollution, which can harm the environment and cause human health problems. Strip mining for coal and oil spilled from ships can also degrade the environment. But perhaps the biggest concern with fossil fuels is a simple one—there's a finite amount of them left that we can use. Sooner or later, the supply will run out.

Nuclear Energy

Roughly 10% of the world's energy need is met by nuclear energy. **Nuclear energy** is produced by the nuclear fission reactions that you learned about in Chapter 8. Uranium is used as the fuel source in an atomic power plant. As the uranium undergoes fission, the energy that it releases is used to heat water into steam. The steam turns the blades of a turbine just as in a coal- or oil-powered plant.

Nuclear energy has great potential. It produces a lot of energy from relatively small amounts of fuel. In France, over three-quarters of the country's energy comes from nuclear power. But radiation, of course, poses a danger to living things. Great care must be taken to ensure that it is kept sealed inside any nuclear facility. Even after the uranium in a plant is no longer useful as fuel, it remains radioactive for many years. The question of where to store radioactive waste long-term is the subject of much debate.

Over the years, several serious accidents have happened at nuclear power plants. A recent example is the accident that occurred at Fukushima, Japan, when three reactors melted down following a tsunami in 2011. As a result, large amounts of radioactive contamination were released into the environment. The need for safeguards against such accidents makes nuclear power plants expensive to build. Their potential for accidents tends to make them an unpopular option for generating electricity. Still, the world may need to use more nuclear energy as the population keeps growing and the supply of other nonrenewable energy resources declines.

WORLDVIEW SLEUTHING: CLEAN ENERGY

One often hears terms like "alternative energy" and "clean energy" these days. But are they the same thing? Are some fossil fuels cleaner than others? Are all renewable energy sources equally clean? An ideal energy source would be one that is both renewable *and* clean. Being inexpensive wouldn't hurt either! Some energy sources hit closer to this mark than others.

TASK

The town of Ideal Acres (population 11,000) needs to build a new electrical generating plant, and its citizens have demanded clean energy. You've been hired as an energy consultant to suggest what type of clean renewable energy source should be used to generate the town's electricity. The town's board of supervisors expects you to deliver a five-minute presentation of your findings and make a recommendation at the next board meeting.

PROCEDURE

You can assume that Ideal Acres has the means to implement any of the clean energy resources that you might investigate. Do the necessary research to inform your decision. The following questions will get you started.

1. How many people are in an average American household?
2. How much energy does an average American household use?
3. How much energy can electrical generation plants produce using different kinds of renewable energy (typically measured in megawatts)?
4. How cost-effective are the various kinds of plants?

Try starting with keyword searches for "per household energy consumption," "people per household," and "alternative energy costs."

1. Plan your presentation and collect any required materials. Remember to cite your sources.
2. Show your presentation to a classmate or friend for feedback.
3. Present your findings to your class.

ENERGY

14.7 RENEWABLE ENERGY

There are many different kinds of renewable energy resources, and scientists are working to develop more. Let's take a quick look at some that are already being put to widespread use.

RENEWABLE ENERGY RESOURCES

BIOMASS ENERGY

Biomass energy is chemical potential energy obtained from renewable organic materials. The wood in a campfire is a familiar example. Many home-heating stoves now burn wood pellets made from the sawdust waste produced by lumber mills. Animal dung can also be collected, dried, and used as biofuel. Many people in some underdeveloped nations still rely on firewood and animal dung for heating and cooking.

HYDROELECTRIC ENERGY

Hydroelectric energy is electrical energy generated by the movement of water. Hydroelectric power plants start with the gravitational potential energy of water stored in a reservoir behind a dam. The kinetic energy of the moving water turns a series of generators as it descends through pipes to the river below.

GEOTHERMAL ENERGY

Geothermal energy is heat energy that originates deep within the earth's interior. The heat energy can be used directly for indoor heating. It can also be used to produce steam for turning the turbines in an electrical generating plant.

SOLAR ENERGY

Every day, the sun bombards Earth with millions of joules of **solar energy** per square meter. Solar panels contain devices that can convert some of that energy into electricity. Solar energy can also be used to heat water; the thermal energy of the heated water can be transferred to the inside of a home.

WIND ENERGY

As a child, you may have had a pinwheel. A wind turbine works on the same principle. **Wind energy** turns the blades attached to the shaft of a generator, transforming the kinetic energy of the blowing wind into electricity.

TIDAL ENERGY

Every day, at many locations around the world, many cubic kilometers of water surge inland from the sea and then back again. The kinetic **tidal energy** in these rising and falling tides can be transformed by turbines into electrical energy. Moving water turns the turbine's blades in the same way that wind turns the blades on a wind turbine.

ENERGY

Benefits and Drawbacks

There are some advantages to using renewable energy sources to meet mankind's need for energy. One big plus is that renewable energy sources tend to be "clean" energy, meaning that they don't create the pollution associated with the burning of fossil fuels. And of course the main consideration with renewable energy is that it is indeed renewable—there's little danger that we could ever run out of it as we eventually will with fossil fuels. One particular form of renewable energy, hydroelectric power, is among the least expensive ways of generating electricity. Cheap electricity from hydropower made the large-scale smelting of aluminum possible. Prior to that, pure aluminum metal had been uncommon and expensive. Reservoirs built for hydroelectric dams also store water for irrigation, provide opportunities for recreation, and can help control flooding.

But there are disadvantages with renewable energy as well. The main one is that renewable energy simply can't produce enough energy to meet the world's needs at this time. Renewable energy technologies tend to be more expensive than those used for extracting energy from fossil fuels. And in many instances, a facility for generating power from renewable resources is much larger than a plant that uses fossil fuels. Wind farms, solar farms, tidal power plants, and the reservoirs behind hydroelectric dams take up far more space than a traditional fossil-fuel-powered facility. Lastly, some forms of renewable energy are not completely reliable. Wind farms need wind, solar farms need sunny days, and hydroelectric dams need enough rain to keep reservoirs full. Those conditions aren't always met.

Even if a renewable energy source doesn't create air pollution by being burned, as biomass fuels typically are, the resource may still not be free of environmental concerns. The turning blades of wind farms kill many birds. Solar farms and tidal power facilities alter the ecology of the lands and estuaries upon which they are built. The reservoirs behind hydroelectric dams eventually fill up with silt, rendering the dam useless. Removing a dam (lower left) in such an instance is expensive, and the environment may not fully recover for decades. On top of the environmental concerns, large renewable energy projects are often unsightly. No one wants a huge wind or solar farm spoiling the view!

Conserving Energy

Nonrenewable energy resources that pollute, or renewable energy resources that can't meet the world's energy needs—where is the balance? Two things are certain: nonrenewable energy sources will eventually run out, and the human population is growing. What can people in general and Christians in particular do to address this problem?

One thing we can do is conserve energy. There are many ways to do so. You can keep a household thermostat set a few degrees warmer in summer and cooler in winter. This lowers the demand for electricity and natural gas. On-demand hot water heaters save energy by heating water only when it is actually used. You can switch from incandescent light bulbs to longer-lasting LED lightbulbs that use less electricity. Tips like these not only conserve energy—they save money too.

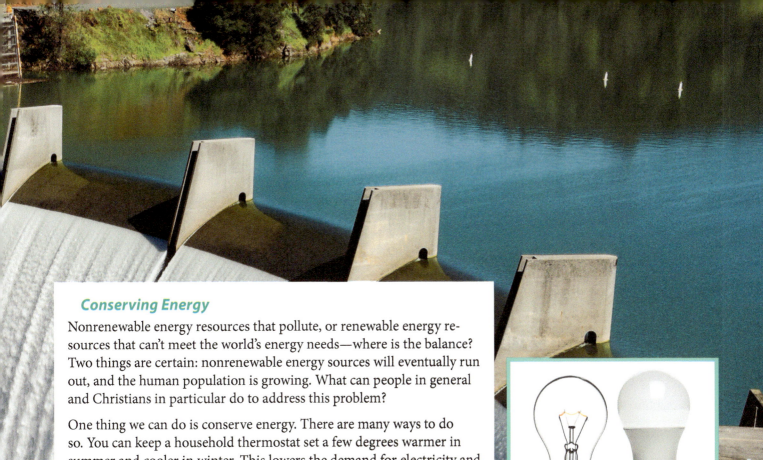
incandescent   LED

We also need engineers who can create new ways to use and generate energy more efficiently. Many household appliances, for example, use far less energy today than they used to. Engineers have also greatly reduced the size and increased the efficiency of solar collection panels. And new ways of obtaining energy are being explored, such as making synthetic oil from humble marine algae. God gave us the earth and its resources to use and to care for wisely. By using your talents to help conserve energy and develop new sources, you can help fulfill the Creation Mandate and demonstrate God's concern for both His creation and His image bearers.

14C | REVIEW QUESTIONS

1. Define *fossil fuels* and identify three kinds.
2. Why do you suppose that nuclear energy is classified as nonrenewable?
3. Create a T-chart to compare nonrenewable and renewable energy resources.
4. State two advantages of using renewable energy sources.
5. Why are wood pellets, which are made from the remains of plants, as is coal, not classified as a fossil fuel?
6. Choose two forms of renewable energy and describe a disadvantage of each.
7. Why should Christians be concerned about conserving energy?
8. Describe two things that you *personally* could do to help conserve energy.

CHAPTER 14

14A CLASSIFYING ENERGY

- Energy is defined as the ability to do work and is measured in joules. One joule of energy can do one joule of work.
- Energy can be transferred from one object to another, or transformed into another kind of energy, but the total amount of energy in a closed system remains constant over time.
- Kinetic energy is energy of motion. Potential energy is stored energy that can be used later.
- The kinetic energy possessed by an object is directly related to its mass and the square of its velocity.
- Gravitational potential energy is the potential energy possessed by an object due to its position above a defined reference height.
- Gravitational potential energy is directly related to the mass of an object and its acceleration due to gravity.

14A Terms
energy 324
kinetic energy 326
potential energy 328

14B ENERGY CHANGES

- A system's mechanical energy is the sum of its kinetic and potential energies.
- Within a system, mechanical energy can be transformed from potential energy to kinetic energy and from kinetic energy to potential energy.
- In reality, all systems lose some energy whenever it is transformed or used to do work. The "lost" energy, usually in the form of heat, becomes less available to do work.

14B Term
mechanical energy 332

14C ENERGY RESOURCES

- Energy resources can be either renewable—naturally replenished—or nonrenewable.
- Most of the world's energy needs are met by fossil fuels (coal, oil, and natural gas), which are believed to have formed from the remains of plants and animals that lived in the past.
- Nuclear energy is provided by the fission of radioactive elements.
- Renewable energy resources include biomass, hydroelectric, solar, wind, geothermal, and tidal energies.
- Each type of energy resource has particular advantages and disadvantages.
- Christians can help fulfill the Creation Mandate by conserving energy and helping to develop new sources

14C Terms
renewable energy resource 337
nonrenewable energy resource 337
fossil fuel 337
nuclear energy 339
biomass energy 340
hydroelectric energy 340
geothermal energy 340
solar energy 341
wind energy 341
tidal energy 341

REVIEW

CHAPTER REVIEW QUESTIONS

Recalling Facts

1. Define *energy*.
2. What two things can energy be converted or transformed into?
3. In theory, what specific condition(s) must exist in order for the energy in a system to be conserved?
4. Which kind of energy is energy of motion? Which kind is energy of condition or position?
5. What is usually considered to be the reference position when calculating the gravitational potential energy of an object?
6. What determines an object's mechanical energy?
7. How is the efficiency of an energy transformation determined?
8. Which of the following is *not* a fossil fuel?
    a. coal
    b. wood
    c. oil
    d. natural gas
9. From which specific energy resource do humans obtain the most energy?

Understanding Concepts

10. Which has more kinetic energy, a 205 kg pony running at 7.00 m/s or a 23.0 kg gazelle running at 19.0 m/s? Defend your answer.
11. Indicate whether each of the following examples is an example of kinetic energy (*KE*), gravitational potential energy (*GPE*), or both.
    a. an airplane drops fire retardant
    b. an axe is used to fell a tree
    c. a car travels along a flat road
    d. a boulder sits atop a 100 m high cliff
12. Why is it incorrect to say that a model rocket loses energy as it gains altitude?
13. How can a person add gravitational potential energy to an object?
14. Find the gravitational potential energy of a 0.443 kg soccer ball that has been kicked to a height of 3.5 m.
15. Is the answer that you calculated for Question 14 the total amount of energy in the soccer ball? Explain.

Refer to the image at top right and the information below to answer Questions 16–19.

Fred Freshman is enjoying an ice cream cone. The scoop of ice cream has a mass of 100.0 g. As Fred tries to lick the ice cream, the scoop falls out of the cone a distance of 1.75 m onto the ground. Assume that the ground is the reference height.

16. What was the ice cream's *GPE* before it fell?
17. What was the ice cream's *GPE* after it hit the ground?
18. If all of the ice cream's *GPE* was converted to kinetic energy just before it hit the ground, how much kinetic energy did it have?
19. What was the speed of the ice cream as it hit the ground?
20. Describe an example of a system in which energy goes through several transformations.
21. If a hammer strikes a nail with 15 J of kinetic energy, is the hammer likely to do 15 J of work on the nail? Explain.
22. Though it is nonrenewable, why is nuclear energy not considered a fossil fuel?

Critical Thinking

23. Imagine that you are the equipment manager of a baseball team. What advice would you give to a player who wants to increase his home-run power by trying a heavier bat?
24. How would the calculation of *GPE* on the moon be different from that on Earth? How would this affect the *GPE* of an object on the moon compared with one on Earth?
25. If the first bumper car and passenger in Review Question 6 in Section 14A (p. 330) were to continue moving at 0.25 m/s after the collision with the empty bumper car, what would be the maximum possible speed of the second car after the collision?
26. The baseball exit velocity of 86.5 m/s in Example 14-3 (p. 328) is actually absurdly fast—it's the equivalent of 193 mph! When professional baseball players hit home runs, the exit velocities are typically around 47 m/s. This means that not all of the bat's kinetic energy is transformed into the kinetic energy of the batted ball. Where might some of this "missing" energy have gone?
27. Write a paragraph defending the continued use of fossil fuels.
28. The textbook mentions using marine algae to produce synthetic oil. Would the resulting oil be considered an example of a fossil fuel? Explain.
29. Write a paragraph suggesting a Christian view of the use of renewable energy resources.
30. Which type of renewable energy resource do you think has the most potential for further use and development? Defend your answer.

CHAPTER 15
Thermodynamics

CITY-SAVING SUPERTREES

Have you ever been in a room with many people? When you entered the room, the temperature seemed reasonable. But after just a few minutes the temperature started to rise. The air conditioning system didn't seem to be able to keep pace.

If this can happen in a room, what must the effect of millions of people living in a city be like? All those people and the infrastructure to support them create an urban heat island. Cities spend millions of dollars each year to keep their citizens cool. In Singapore, developers have created huge gardens with manmade supertrees (left) to address their urban heat island problem.

Throughout this chapter, we will study the concepts of temperature, heat, and thermal energy. By the end of the chapter, you will have a better understanding of how these supertrees might help cities around the world.

15A Temperature	348
15B Heat	355
15C Thermodynamics	364

15A | TEMPERATURE

Is temperature the same thing as thermal energy?

15A Questions

- What is temperature?
- How do thermometers work?
- What are the different temperature scales?
- How can I convert from Celsius to Fahrenheit?

15A Terms

temperature, thermal energy, thermometric property

15.1 TEMPERATURE

What Is Temperature?

We often talk about objects being hot or cold. We may describe our soup as being cold or our soda as warm. Or we may say that it is hot outside on a summer day or cold on a rainy fall day. But these descriptions are comparative, not absolute. What we mean is that this or that is warmer or colder than we would like it to be or warmer or colder than something else. Scientists strive to be more precise in their descriptions of objects.

Scientists use temperature to describe the hotness or coldness of an object. By using temperature, the scientists have a standard with which to compare the object. Remember from Chapter 2 that the particles in an object are always moving. **Temperature** is the measure of the hotness or coldness of a substance and is proportional to the average kinetic energy of the particles within an object. As the particles move faster, they have more kinetic energy. A scientist would say that they have a higher temperature. Conversely, as the temperature of the object decreases, the particles move slower. They have less kinetic energy.

We measure temperature with a thermometer, which usually indicates temperature in units of degrees (°). Three temperature scales that we should be familiar with are the Fahrenheit, Celsius, and Kelvin scales. Scientists work in Celsius or Kelvin scales. We will look more closely at these three temperature scales later in this section.

Surprising as it may seem, something that has a high temperature doesn't necessarily have a lot of thermal energy in it. How can that be?

While temperature is related to the average kinetic energy of the particles in an object, the object's **thermal energy** is the sum of the kinetic energy of all of its particles. A teaspoon of water near boiling has a high temperature, but it has much less thermal energy than the Arctic Ocean with its water near freezing because there is much more water in the Arctic Ocean. Thermal energy is affected by both the temperature and the number of particles.

Temperature and Matter

Temperature and other properties of matter are often closely related. For instance, we know that the density of water changes as the temperature of the water changes. Any property that changes in a predictable way with changes in its temperature is called a **thermometric property**.

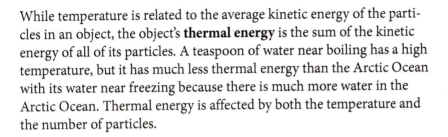

Most of us are familiar with liquid-filled thermometers, which consist of a glass bulb attached to a long, thin, hollow tube. These thermometers work on the basis of *thermal expansion*, a thermometric property. Most liquids expand as their temperature rises and contract as it decreases. As temperature increases, the liquid in the bulb expands, forcing some of the liquid into the thin tube that is next to the scale. The end of the column of liquid indicates the current temperature.

Thermal expansion can be a help but also a hindrance. Thermal expansion is helpful when used to make a bimetallic strip in a *thermostat* or when used by the pilot of a hot air balloon. But it can be a challenge on large construction projects. Engineers have to design bridges, buildings, and even roads and sidewalks with expansion joints that allow for the thermal expansion of a structure. You may recall the rhythmic thump-thump, thump-thump of driving across a bridge as you traveled over expansion joints. We will learn about other thermometric properties in later chapters, including pressure in gases and viscosity in Chapter 16 and electrical resistance in Chapter 19.

15.2 TEMPERATURE SCALES

The precursor to the thermometer was the thermoscope (left), which consisted of a gas-filled globe attached to a hollow tube. Thermoscopes lacked a scale for reading the *actual* temperature, but they could indicate *changes* in temperature. Scientists found it challenging to create accurate thermometers. For a thermometer to produce reproducible measurements, its scale had to be constructed on the basis of two known temperatures, known as *fiduciary points*. Early attempts used the freezing point of a solution, with equal parts salt and water, and the normal temperature of a human. These didn't work well because the freezing point of a solution depends on the concentration of the solute, and the temperature of the human body routinely changes.

The first accurate thermometer was developed in 1714 by Dutch scientist Daniel Gabriel Fahrenheit. In 1724, Fahrenheit developed a temperature scale, which, in a slightly modified form, is still used today. He calibrated his scale to the freezing point of pure water (32 °F) and the normal temperature for a person (which he labeled as 96 °F). He then discovered that water boiled at 212 °F on this scale. The Fahrenheit scale was later recalibrated to the freezing and boiling points of water.

In 1742, Swedish physicist Anders Celsius set out to develop a temperature scale with 100 degrees between the two known temperatures. He set the freezing point at 100 °C and the boiling point of pure water at sea-level pressure at 0 °C. After Celsius's death, Carl Linnaeus updated the scale, switching the numerical values for the two known points.

Scots-Irish physicist William Thomson developed the third temperature scale in 1848. Thomson, also known as Lord Kelvin, desired to establish a scale that was based on the absolute coldest temperature possible, known as *absolute zero*. Today, this scale uses absolute zero (0 K) and the *triple point* of water (273.16 K) as its two primary points. The triple point is the temperature and pressure at which the solid, liquid, and vapor states of a particular substance exist in equilibrium. Notice that we do not use the degree symbol for Kelvin temperatures.

Daniel Fahrenheit

Anders Celsius

William Thomson, Lord Kelvin

Scientists today use either the Celsius or Kelvin scale. In 1954, an international agreement redefined the Celsius scale using absolute zero and the triple point of water to align it with the Kelvin scale. Let's compare the three scales.

TEMPERATURE SCALES

FAHRENHEIT

There are 180 degrees between the freezing and boiling points of water on the Fahrenheit scale. Also notice that the freezing point of water is 32 °F. This temperate may seem arbitrary, but it resulted from Fahrenheit setting the coldest temperature in his hometown as his 0 °F temperature. The Fahrenheit scale is the official scale of only three nations—the United States, Liberia, and The Bahamas.

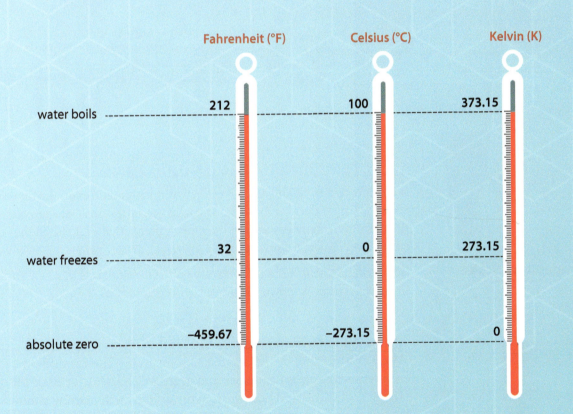

CELSIUS

On the Celsius scale, there are 100 degrees between the freezing and boiling points of water. Notice too that the freezing point of water is 0 °C. Since there are fewer degrees Celsius than degrees Fahrenheit between the two known points, a degree Celsius is larger than a degree Fahrenheit. The Celsius scale is used by most countries, and scientists use it for most scientific applications.

KELVIN

On the Kelvin scale, 100 units again separate the freezing point of water (273.15 K) and the boiling point of water (373.15 K). Therefore, a kelvin is the same size as a degree Celsius. Zero on the Kelvin scale is called absolute zero; accordingly, the Kelvin scale is called an *absolute temperature scale*. Scientists use it for some scientific applications.

THERMODYNAMICS 351

15.3 TEMPERATURE CONVERSIONS

It is important to be able to convert between temperatures on the three scales. We have derived formulas to convert temperatures between the scales from the temperatures of the known points that established the scales. Let's look at these formulas.

The formula to convert a temperature on the Celsius scale to a Fahrenheit temperature is shown below.

$$T_F = 1.8 T_C + 32$$

T_F is the Fahrenheit temperature and T_C is the Celsius temperature. Where do you think the 1.8 and the 32 come from? The size difference between degrees Fahrenheit and degrees Celsius is a factor of 1.8 (180 °F / 100 °C). The 32 comes from the difference in freezing point temperatures on the two scales. With a little algebra, we can solve the above equation for T_C to get the formula to convert a temperature on the Fahrenheit scale to one on the Celsius scale.

$$T_C = \frac{5}{9}(T_F - 32)$$

Again, the 5/9 is the size difference between degrees Celsius and degrees Fahrenheit, a ratio of 100 °C / 180 °F, and the 32 again comes from the difference in freezing point temperatures on the two scales.

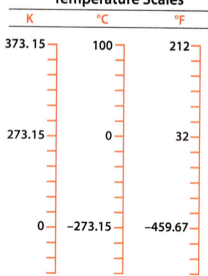

Temperature Scales

K	°C	°F
373.15	100	212
273.15	0	32
0	−273.15	−459.67

The formulas to convert between Celsius temperatures and Kelvin temperatures are simpler than those for converting between Celsius and Fahrenheit.

$$T_K = T_C + 273.15$$
$$T_C = T_K - 273.15$$

T_K is the Kelvin temperature. There are no coefficients because degrees Celsius and kelvins are the same size. The 273.15 comes from the difference in temperatures of the freezing point of water on each scale.

Let's practice a few of these conversions.

EXAMPLE 15-1: Converting from Celsius to Fahrenheit

While preparing to travel to Europe, you check the weather forecast to know what to pack. The expected daily high temperature is 28 °C. What is that in Fahrenheit?

Write what you know.

$T_C = 28 \,°C$

$T_F = ?$

Write the formula and solve for the unknown.

$T_F = 1.8 \, T_C + 32$

Plug in known values and evaluate.

$T_F = 1.8(28) + 32$

$= 50.4 + 32$

$= 82 \,°F$

EXAMPLE 15-2: Converting from Fahrenheit to Celsius

While working on a lab activity, you are trying to heat water to 125 °F, but all the thermometers are in Celsius. To what temperature should you heat the water?

Write what you know.

$T_F = 125 \,°F$

$T_C = ?$

Write the formula and solve for the unknown.

$T_C = \frac{5}{9}(T_F - 32)$

Plug in known values and evaluate.

$T_C = \frac{5}{9}(125 - 32)$

$= \frac{5}{9}(93)$

$= 52 \,°C$

EXAMPLE 15-3: Converting from Kelvin

The melting point of iron is 1811 K. What is that in Celsius and Fahrenheit?

Write what you know.

$T_K = 1811 \,K$

$T_C = ?$

$T_F = ?$

Write the formula and solve for the unknown.

$T_C = T_K - 273$

$T_F = 1.8 T_C + 32$

Plug in known values and evaluate.

$T_C = 1811 - 273$

$= 1538 \,°C$

Now use this answer to solve for T_F.

$T_F = 1.8(1538) + 32$

$= 2800 \,°F$

THERMODYNAMICS

HOW IT WORKS

Thermostats

Thermostats are devices that sense temperature and direct actions to maintain the temperature in the system at the setpoint. Thermostats are around us all the time! They control the heating and cooling systems in our houses. They enable our refrigerators and ovens to maintain their temperatures. Our bodies maintain temperature through a biofeedback system similar to a thermostat. Thermostats, just like thermometers, work on the basis of some of the thermometric properties mentioned in this section.

If you live in an older house, you may have a mechanical thermostat like the one at right. The setpoint is adjusted by the dial, which is connected to a coiled bi-metallic strip, which is two metal strips fused together. The two metals have different rates of thermal expansion. As the temperature in the room changes, the free end of the coil moves to either make or break the electrical connection with a fixed wire. When the connection is made, the heating or cooling system turns on; when the connection is broken, the system turns off.

Many newer or updated homes have electronic thermostats. A primary difference between mechanical and electronic thermostats is how the thermostat senses temperature. In the electronic thermostat, temperature is sensed by a thermistor—a device that senses temperature on the basis of electrical resistance.

Regardless of how a thermostat senses temperature, it maintains the temperature in a comfortable range while not overstressing the system.

15A | REVIEW QUESTIONS

1. Define *temperature*.
2. Give two examples of thermometric properties.
3. Why are there expansion joints in bridges?
4. What are the two primary temperatures that define the Kelvin scale?
5. Who uses each of the different temperature scales?
6. Explain how you know that one degree on the Celsius scale is larger than one degree on the Fahrenheit scale.
7. Convert 35 °C into a Fahrenheit temperature.
8. Convert 85 °C into a Kelvin temperature.

15B | HEAT

Why do metals warm faster than water?

15.4 WHAT IS HEAT?

We learned in Section 15A that temperature is the measure of how hot or cold an object is. But often when we touch objects and describe them as hot or cold, we don't mean their temperature. When you are running a fever, you feel your forehead but often can't detect whether it seems warm. If someone else feels your forehead, they can feel the warmth. If your temperature is elevated, why can't you feel it? It is because when you feel to see whether something is warm, you are not sensing temperature. Instead, you are feeling the movement of thermal energy between yourself and the object. Since your hand and forehead are approximately the same temperature, there is no noticeable movement of thermal energy. So to you, your forehead doesn't seem warm. But another person's hand is cooler than your forehead, so he feels the flow of energy from your forehead to his hand. This movement of thermal energy from an area of higher temperature to one of lower temperature is called **heat**.

15B Questions

- What is the difference between heat and thermal energy?
- How does thermal energy move?
- How can I calculate the energy required to warm an object?
- What happens to energy during changes of state?

15B Terms

heat, conduction, convection, radiation, thermal conductor, thermal insulator, specific heat

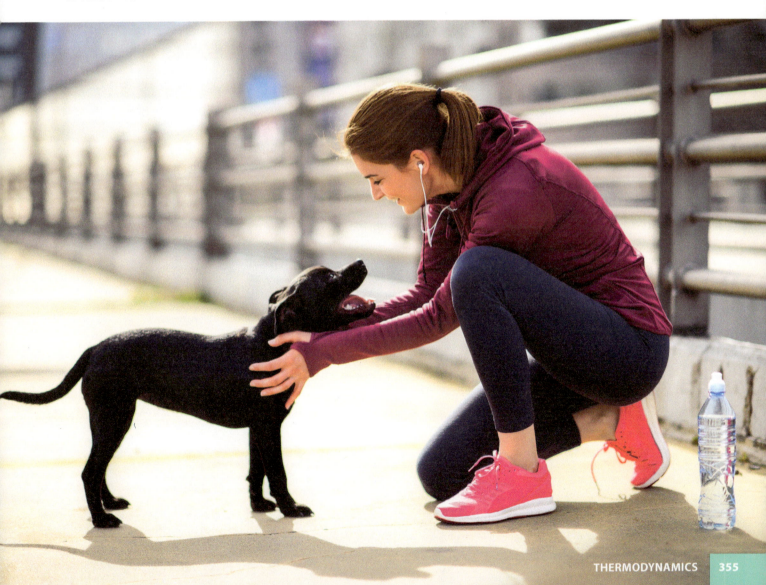

ENERGY TRANSFER

15.5 TRANSFERRING THERMAL ENERGY

We know that thermal energy will move from an area of higher temperature to one of lower temperature, but how does this happen? It can take place through *conduction*, *convection*, or *radiation*. Let's look at how these three occur.

CONDUCTION

When a warm and cold object are in contact with each other, thermal energy moves from the warmer object to the colder object by **conduction**. This flow of thermal energy will continue until the two objects are at the same temperature. The flow of energy occurs as the particles collide and warmer (faster-moving) particles transfer kinetic energy to the particles with lower kinetic energy. The flame heats the kettle, and the kettle heats the water inside by conduction.

CONVECTION

Scientists call the flow of thermal energy as fluids move **convection**. An example of this is seen in thunderstorms, which frequently occur in the summertime. They happen because warm air rises and cold air sinks, a movement known as *air convection*. The dense, cold air flows downward, and the warm air rises. As the warm air cools and the cold air warms, the cycle continues in a *convection current*. You can see the convection of the purple food coloring in the beaker.

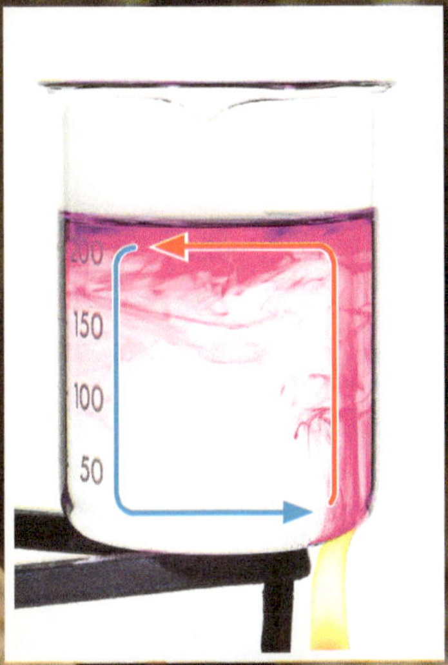

RADIATION

Most of the energy that we use here on the earth comes from the sun, but the sun and the earth are not in contact and there is no fluid in space. The nuclear reactions taking place within the sun produce tremendous thermal energy. As this thermal energy reaches the surface of the sun, it changes into *electromagnetic energy*. This energy is radiated out into space. The process of energy transfer in the form of electromagnetic waves is called **radiation**. The earth absorbs this energy and converts it back into thermal energy. The heated horseshoe glows orange as it emits radiant energy.

The rate at which thermal energy moves through a material is called *thermal conductivity*. A **thermal conductor** is a material through which thermal energy moves easily. A **thermal insulator** has the opposite effect.

THERMODYNAMICS 357

MINI LAB

UNDERSTANDING HEATING

Essential Question:
Why do room-temperature metals feel cold?

Equipment
tongs (2)
metal cube
wood cube
plastic cube

During the wintertime in cold climates, people learn that you shouldn't touch metal objects with your bare skin. Doing so can result in your skin freezing to the object. It could also cause frostbite. But what is it about metals that makes them particularly dangerous? Why isn't it a problem to touch plastic or wood?

1. Can we tell the temperature of an object by touching it? Explain.

PROCEDURE

A Using the tongs, simultaneously place the metal cube in one palm and the wood cube in the other palm of your partner's hands. Have your partner indicate which one feels colder. Record this in your notebook.

B Repeat Step **A** with the metal and plastic cubes paired and then with the plastic and wood cubes paired.

2. What is the order from coldest to warmest?
3. What do you think this tells us about the objects?

ANALYSIS

The tested objects are all at room temperature, and so their temperatures are the same.

C Find out from your teacher of what material each of your cubes is made.

D Use the internet to find the thermal conductivity for each of the materials you used. Record these in your notebook.

CONCLUSION

4. What is the relationship between thermal conductivity and how cold or warm an object feels?

GOING FURTHER

5. If you were making an insulated container, would you want a material with a high or low thermal conductivity? Explain.
6. The thermal conductivities of copper and steel are 401 W/m·K and 38 W/m·K respectively. Why do you think copper is used to coat the bottom of many steel cooking pans?

15.6 MEASURING HEAT

Reheating a plate of leftover food can be a frustrating task. After three minutes, the meat is completely warm, but the mashed potatoes are still cold. If you continue heating the plate, the meat will begin to get overcooked. Why does this happen? Each substance needs different amounts of thermal energy to raise its temperature. The thermal energy needed to warm the meat is less than the thermal energy required to warm the potatoes. It would be great to know how much energy is required to raise the temperature of each food! Scientists measure this energy using the **specific heat** of a substance—the energy required to raise the temperature of 1 g of the substance 1 °C. A material with a high specific heat needs a lot of energy to change its temperature. A material with a lower specific heat will warm up with a relatively small amount of energy added. We can calculate the energy required to raise the temperature of a sample with the formula

$$Q = c_{sp} m \Delta T,$$

where Q is the thermal energy transferred in joules, c_{sp} is the specific heat capacity in J/g °C, m is the mass in grams, and ΔT is the change in temperature in °C. A positive heat (Q) indicates that the system gained thermal energy, while a negative value represents a case in which the system lost thermal energy.

Let's practice a couple of energy problems.

Table 9-1

Specific Heats of Common Substances

Substance	c_{sp} (J/g °C)
acrylic	1.50
aluminum	0.90
brass	0.38
copper	0.39
gold	0.13
ice	2.11
nylon	1.67
oak	1.26
polyethylene	1.55
poplar	1.30
silver	0.24
steam	2.08
steel	0.49
water	4.18
willow	1.35

EXAMPLE 15-4: Calculating Energy

How much energy would be required to warm 74.8 g of brass from 23 °C to 255 °C?

Write what you know.

$m = 74.8$ g

$T_i = 23$ °C

$T_f = 255$ °C

$c_{sp} = 0.38$ J/g °C (from Table 9-1)

$Q = ?$

Write the formula and solve for the unknown.

$Q = c_{sp} m \Delta T$

$\quad = c_{sp} m (T_f - T_i)$

Plug in known values and evaluate.

$Q = (0.38 \, \frac{J}{g \, °C})(74.8 \, g)(255 \, °C - 23 \, °C)$

$\quad = (0.38 \, \frac{J}{°C})(74.8)(232 \, °C)$

$\quad = 6590 \, J \left(\frac{1 \, kJ}{1000 \, J}\right)$

$\quad = 6.59$ kJ

EXAMPLE 15-5: Calculating Temperature Change

How much will the temperature change if you heated the same mass, 74.8 g, of water with the energy that you calculated in Example 15-4?

Write what you know.

$$m = 74.8 \text{ g}$$

$$c_{sp} = 4.18 \text{ J/g °C (from Table 9-1)}$$

$$Q = 6590 \text{ J}$$

$$\Delta T = ?$$

Write the formula and solve for the unknown.

$$Q = c_{sp} \, m \, \Delta T$$

$$\frac{Q}{c_{sp} m} = \frac{\cancel{c_{sp}} \, m \, \Delta T}{\cancel{c_{sp}} \, \cancel{m}}$$

$$\Delta T = \frac{Q}{c_{sp} m}$$

Plug in known values and evaluate.

$$\Delta T = \frac{6590 \, \cancel{J}}{(4.18 \, \frac{\cancel{J}}{\cancel{g} \, °C})(74.8 \, \cancel{g})}$$

$$= 21.1 \, °C$$

Notice that the brass in Example 15-4 had a 232 °C temperature change, while the water in Example 15-5 had only a 21.1 °C change. You can see that water's higher specific heat allows it to absorb the same amount of energy without much change in its temperature. Because the masses were the same, this difference is completely due to the difference in the specific heats of water and brass.

Measuring Specific Heat

As heat flows from one object to another in an isolated system, the energy lost by the warmer object must equal the energy gained by the cooler object. Remember, the law of conservation of energy maintains that the energy in an isolated system must be conserved. On the basis of this law, scientists can determine the specific heat of a substance using a process called *calorimetry*. They use an instrument with an insulated cup and an inner cup to measure the temperature change when energy moves from the sample into the water in the inner cup. This device, called a *calorimeter*, enables scientists to calculate the specific heat using the following equation.

$$c_{sp} = \frac{Q}{m \Delta T}$$

This formula is the same as the heat formula above, solved for specific heat. Let's look at an example showing how a scientist in a lab would determine specific heat.

EXAMPLE 15-6: Determining Specific Heat from Calorimetry

A scientist is using calorimetry to determine the specific heat of a substance. The sample lost 2457 J of energy to water as it cooled from 95.8 °C down to 29.6 °C. If the sample has a mass of 95.2 g, what is the substance's specific heat?

Write what you know.

$Q = -2457$ J (negative because the sample lost energy)

$T_i = 95.8$ °C

$T_f = 29.6$ °C

$m = 95.2$ g

$c_{sp} = ?$

Write the formula and solve for the unknown.

$$c_{sp} = \frac{Q}{m\Delta T}$$

$$= \frac{Q}{m(T_f - T_i)}$$

Plug in known values and evaluate.

$$c_{sp} = \frac{-2457 \text{ J}}{(95.2 \text{ g})(29.6 \text{ °C} - 95.8 \text{ °C})}$$

$$= \frac{-2457 \text{ J}}{(95.2 \text{ g})(-66.2 \text{ °C})}$$

$$= 0.390 \frac{\text{J}}{\text{g °C}}$$

Compare the calculated value with those in Table 9-1. Can you identify the material that this sample may be made of?

15.7 HEAT AND CHANGES OF STATE

As you learned in Chapter 2, we can change the state of materials by adding or removing thermal energy. When we add energy, we can cause a solid to melt into a liquid, sublimate into a gas, or cause a liquid to vaporize into a gas. By removing energy we can cause a gas to deposit into a solid or condense into a liquid, or we can cause a liquid to freeze into a solid. Let's look at how energy is involved with changes of state by looking at a heating curve.

ENERGY TRANSFER BY HEATING

The graph at right shows the heating curve of water, with temperature on the *y*-axis and energy on the *x*-axis. The graph shows the energy needed to change 75.0 g of ice at −50 °C to steam at 150 °C. Notice the five sections of the curve. The three sloping sections are where a particular state is warming as thermal energy moves into the substance. The two horizontal sections are where phase changes (melting and vaporizing) occur.

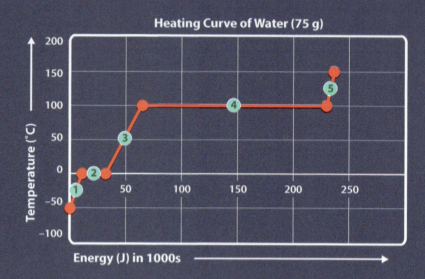

1 *Heating Ice.* Thermal energy flows into the ice, increasing the kinetic energy of the ice as its particles move faster until reaching the melting point. In Table 9-1, the specific heat for ice is 2.11 J/g °C.

$$Q = c_{sp} m(T_f - T_i)$$
$$= (2.11 \text{ J/g °C})(75.0 \text{ g})[0 \text{ °C} - (-50 \text{ °C})]$$
$$= 7900 \text{ J}$$

2 *Melting Ice.* Thermal energy still flows into the ice, but the temperature doesn't rise. The energy is going into melting the ice. The heat required to melt a quantity of a solid is called the *heat of fusion* (L_f). For water, the heat of fusion is 334 J/g.

$$Q = L_f m$$
$$= (334 \text{ J/g})(75.0 \text{ g})$$
$$= 25\,100 \text{ J}$$

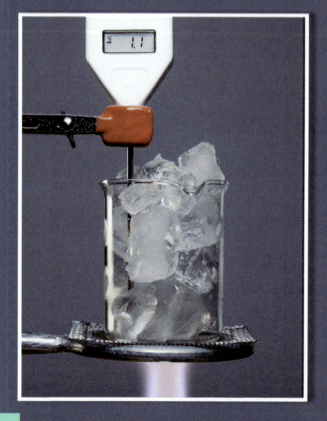
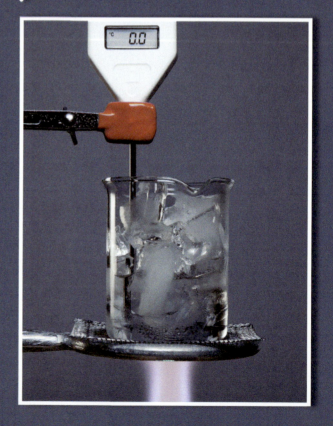

3 **Heating Water.** Again the thermal energy flowing into the water increases the kinetic energy of the water as its particles move faster. The water is warming from 0 °C to 100 °C.

$$Q = c_{sp}m(T_f - T_i)$$
$$= (4.18 \text{ J/g °C})(75.0 \text{ g})(100 \text{ °C} - 0 \text{ °C})$$
$$= 31\,400 \text{ J}$$

4 **Vaporizing Water.** Thermal energy continues to flow into the water, but the temperature doesn't rise. The energy is changing liquid to a gas. The heat required to vaporize a quantity of liquid is called the *heat of vaporization* (L_v). For water, the heat of vaporization is 2257 J/g.

$$Q = L_v m$$
$$= (2257 \text{ J/g})(75.0 \text{ g})$$
$$= 169\,000 \text{ J}$$

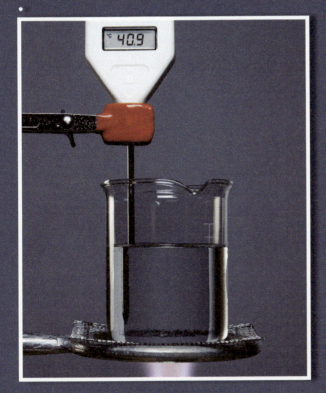
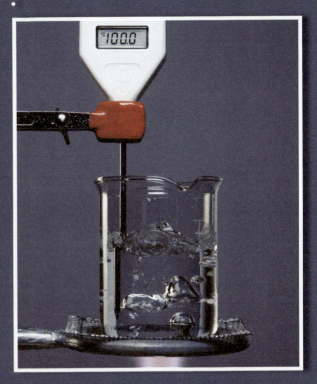

5 **Heating Steam.** Finally the steam warms as thermal energy flows into it, increasing the kinetic energy of the steam as its particles move faster. The specific heat for steam is 2.08 J/g °C.

$$Q = c_{sp}m(T_f - T_i)$$
$$= (2.08 \text{ J/g °C})(75.0 \text{ g})(150 \text{ °C} - 100 \text{ °C})$$
$$= 7800 \text{ J}$$

The curve shown on the facing page, when reading from right to left, is the *cooling curve*. Reading from right to left results in the heats being negative because energy is coming out of the substance.

15B | REVIEW QUESTIONS

1. What is heat?
2. What is a convection current?
3. Define *radiation*?
4. Identify whether conduction, convection, or radiation are occurring in each image below.

   a.   b.   c.

5. What is specific heat?
6. How much energy is needed to warm a 44.9 g sample of aluminum (c_{sp} = 0.90 J/g °C) from 35.2 °C to 74.6 °C?
7. What is the specific heat of a 75.4 g sample that is warmed 65.6 °C by adding 1187 J of energy?
8. In the graph on the facing page, as energy is removed from the system, what is happening in the substance represented by each section?

15C Questions

- How has our understanding of thermal energy changed?
- What is thermodynamics?
- What are the laws of thermodynamics?

15C Terms

thermodynamics, first law of thermodynamics, second law of thermodynamics, third law of thermodynamics

15C | THERMODYNAMICS

What is thermodynamics?

The industrial revolution sparked particular interest in the topics of temperature, thermal energy, and heat. The question most scientists were interested in was how to make engines more efficient. This led to the development of a field of science called **thermodynamics**, the study of thermal energy and heat and how they relate to work and other forms of energy. How did our models of heat and thermal energy change over time to bring us to our current understanding of thermodynamics?

15.8 EARLY THEORIES OF HEAT

In the sixth century BC, the ancient Greeks believed that there were four elements—earth, air, water, and fire. This concept was perpetuated in the writings of later philosophers, such as Socrates and Plato, and so people accepted the notion that heat and fire were the same thing for many centuries. In the eighteenth century, this thinking finally changed as Scottish chemist Joseph Black suggested that heat was an invisible, self-repelling fluid that adhered to other matter. This fluid could enter solids and liquids to warm them. This fluid was called *caloric* by the French chemist Antoine Lavoisier.

The *caloric theory* of heat was quite successful in explaining observations and making predictions. The self-repelling nature of caloric explained why warm objects cool off in a colder environment. Scientists easily explained the expansion of gases as they warm: as the caloric matter adhered to the substance, it occupied more space, requiring the sample gas to expand. Over time scientists explained all the gas laws using the caloric theory. Even the theory of heat engines, engines that turn thermal energy into mechanical energy, was developed using the caloric theory. In the caloric theory, caloric was thought to be an actual fluid, which is subject to the law of conservation of matter. The caloric model was workable in that it explained observations and made accurate predictions.

15.9 THE KINETIC-MOLECULAR THEORY OF HEAT

The first significant challenge to caloric theory came from a British soldier and scientist, Benjamin Thompson. Thompson, also known as Count Rumford, became interested in the study of heat after noticing that cannons fired without a cannonball heated up more than ones fired with a cannonball. This evidence seemed to contradict the caloric theory. After moving to Germany in 1785, Count Rumford was assigned to supervise the boring of new cannons. This assignment gave him the opportunity to observe the nature of heat.

The process of boring a cannon produces a tremendous amount of heat. According to the caloric theory, caloric attaches to other matter. So as metal is removed from the cannon, a proportional amount of caloric should have been released. Count Rumford noticed that when the drill became dull, it removed less metal, but the cannon became hotter than when the drill was sharp. Rumford decided that he needed to conduct a controlled experiment.

In his experiment, Rumford submerged a cannon in water during the boring process. He purposely used a dull bit and measured the amount of water that he was able to boil during the drilling process. He determined that drilling with the dull bit released more heat than would have been needed to melt the entire cannon! He repeated the experiment with the same cannon. Even though he started with less mass, he still generated the same amount of heat. It seemed that caloric was inexhaustible; it was not obeying the law of conservation of matter. Rumford concluded that the motion of the drill against the cannon produced the heat. The caloric theory needed to be modified or replaced.

Though not the first to think of it, British scientist James Prescott Joule is credited with connecting heat to mechanical energy in 1847. Through numerous experiments, he discovered that whenever a machine did one calorie of mechanical work, it produced 4.18 J of thermal energy. This is where we get the conversion factor 1 cal = 4.18 J. The unit *calorie*, which is a remnant of the caloric theory, is a unit of energy. Some of you may notice its similarity to the term we see on food labels. Understand that 1 Calorie (notice the capital C) on the nutrition label is 1000 calories, or 1 kilocalorie.

WORLDVIEW SLEUTHING: URBAN HEAT ISLANDS

Honeybees keep warm through the cold months of winter by gathering together in the hive. The combined warmth of all the individual bees protects the members of the hive. As more and more people move into urban areas, we see a similar phenomenon. In 1961, 34% of the world population lived in urban areas. By 2017, that figure had grown to almost 55%. The large population of a city along with its necessary infrastructure together generate huge amounts of thermal energy, creating an urban heat island effect. An overwhelming amount of energy is needed to keep everyone cool during the hot summer months. City planners are now looking into this phenomenon to see whether they can help alleviate the problem.

TASK
You are a city planner for Megalopolis. A 5% annual population growth is projected for the next ten years, so you expect the urban heat island effect to become a reality in Megalopolis. You are trying to be proactive to minimize the negative impacts of the extra thermal energy. Prepare a proposal for the other city planners to address this problem.

PROCEDURE
1. Do the necessary research to inform your decision. What is the impact of the urban heat island effect on financial, environmental, health-related, and other areas. How do some cities mitigate the effect? Can mitigation methods provide an unexpected gain for the community?
2. Plan your proposal and collect any required materials. Remember to cite your sources.
3. Show your proposal to a classmate or friend for feedback.
4. Submit your proposal.

CONCLUSION
City planners around the world seek to improve living conditions for people that live in their cities. In Singapore, city planners have built a series of gardens, covering 250 acres, around their city. A key feature of the largest of these gardens are supertrees (up to 50 m tall), which act as vertical gardens that absorb and dissipate huge amounts of thermal energy. The trees generate electricity by use of solar cells and collect rainwater for irrigating other plants in the garden. City planners seek to meet the needs of citizens in many ways.

According to the kinetic-molecular model, all particles are in constant motion and thus have kinetic energy. The sum of the kinetic energy of these particles is what we call thermal energy. While we can't measure thermal energy directly, we can measure changes in it. We can also change the thermal energy of a system by heating it or doing work on it.

15.10 STUDYING THERMODYNAMICS

Thermodynamics is still an important field of science for a variety of scientists. Chemists are interested in the changes of energy during chemical reactions. Physicists study the energy changes in physical systems. Engineers are interested in how energy can be used to do work most efficiently. The following three laws govern thermodynamics.

LAWS OF THERMODYNAMICS

FIRST LAW OF THERMODYNAMICS

The **first law of thermodynamics** states that energy cannot be created or destroyed but only transferred between objects or transformed into different forms of energy. In thermodynamics, we usually use the first law to state that the change in a system's energy is equal to the energy transferred to the system by heating plus the work done on the system. Throughout history, many people have tried to design perpetual motion machines, but all of these machines violate the first law.

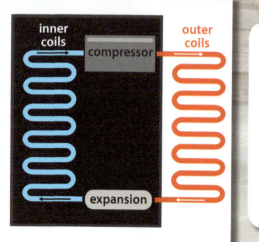

SECOND LAW OF THERMODYNAMICS

Your refrigerator is actually the second law of thermodynamics in action. The **second law of thermodynamics** states that energy can flow from a colder object to a warmer object only if something does work. We all know that if you remove food from the refrigerator and leave it on the counter, it is going to warm up; it never gets colder. But the inside of your refrigerator is colder than the room because energy is transferred to the room. This transfer can happen only because the mechanical system of the refrigerator is doing work.

Solar energy from the sun arrives on the earth, is absorbed by the earth and atmosphere, and is used by plants. That energy is then transformed into many different types of energy. The energy can then be used to do work, and through this work it is transferred between objects. While some of this energy can turn into less useful types of energy, none of it is destroyed. We can track the energy from the sun through its many changes within a system by using the principles of thermodynamics. Without this process of energy movement, life on Earth would cease to exist.

Our understanding of thermodynamics has changed significantly over time. And so our models must remain open to modification or replacement.

THIRD LAW OF THERMODYNAMICS

The third law of thermodynamics sets a limit on how cold things can get. As objects cool, their entropy—the measure of how spread out energy is—decreases. This decrease can happen only if the energy transferred to the environment increases the entropy of the environment. Entropy would be at its minimum value at absolute zero (0 K). But to cool something to this temperature, you would have to be able to move its energy to an even cooler environment, which is clearly impossible. Therefore absolute zero can never be achieved as stated by the **third law of thermodynamics**. As a material approaches absolute zero, a fifth state of matter, *Bose-Einstein condensate*, can be observed.

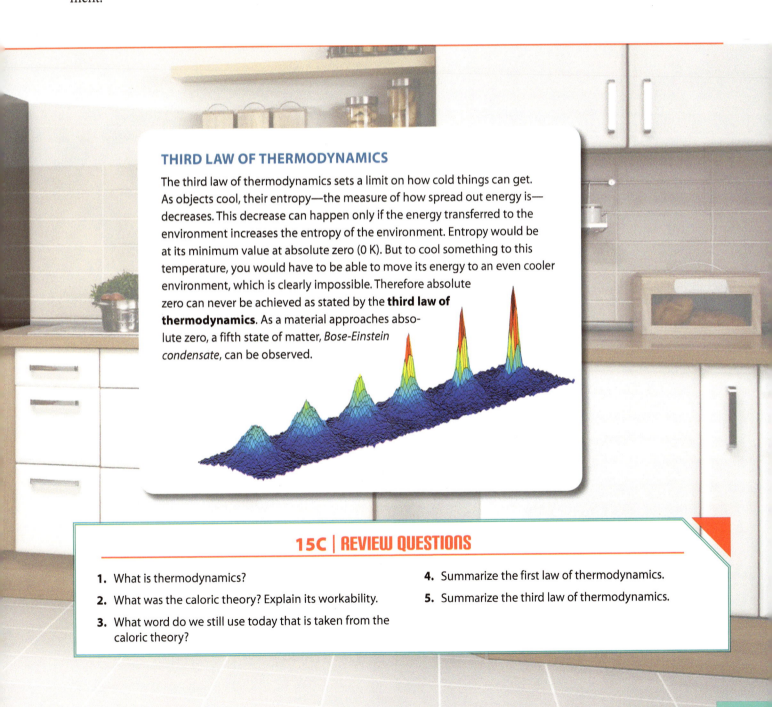

15C | REVIEW QUESTIONS

1. What is thermodynamics?
2. What was the caloric theory? Explain its workability.
3. What word do we still use today that is taken from the caloric theory?
4. Summarize the first law of thermodynamics.
5. Summarize the third law of thermodynamics.

CHAPTER 15

15A TEMPERATURE

- Temperature is the measure of hotness or coldness of an object. It is proportional to the average kinetic energy of the particles in an object.
- Thermal energy is the sum of all the kinetic energy of the particles in an object.
- Thermometric properties of matter are those that change predictably as temperature changes.
- Three common temperature scales are the Fahrenheit, Celsius, and Kelvin scales.
- Temperatures can be converted to other scales by using formulas that are based on the size of the degrees and the freezing point of water.

15A Terms
temperature	348
thermal energy	349
thermometric property	349

15B HEAT

- The flow of thermal energy from warm objects to cold objects is referred to as heat.
- Energy transfer occurs by conduction, convection, or radiation.
- Conduction is the transfer of thermal energy by direct contact between objects.
- Convection is energy transfer in a fluid by mass movement.
- Radiation is the emission of energy as electromagnetic waves.
- Specific heat is the amount of energy required to raise the temperature of 1 g of a specific material 1 °C.
- As we add thermal energy to a material, the material either warms as its particle's kinetic energy increases or the material changes state.

15B Terms
heat	355
conduction	356
convection	357
radiation	357
thermal conductor	357
thermal insulator	357
specific heat	359

15C THERMODYNAMICS

- Thermodynamics is the study of thermal energy and heating and how they relate to work and other forms of energy.
- Our understanding of thermodynamics has changed as models have been modified or replaced when they were no longer workable.
- According to the kinetic-molecular model, all particles are in constant motion and thus have kinetic energy.
- Energy cannot be created or destroyed; it can only be transferred or transformed.
- A cold object can transfer thermal energy to a warm object only if work is done.
- An object can never be cooled to absolute zero (0 K).

15C Terms
thermodynamics	364
first law of thermodynamics	366
second law of thermodynamics	366
third law of thermodynamics	367

REVIEW

CHAPTER REVIEW QUESTIONS

Recalling Facts

1. What is thermal energy?
2. What is a thermometric property?
3. What two temperatures did Fahrenheit and Celsius end up using to calibrate their thermometers?
4. What is a thermal insulator?
5. What did the ancient Greeks think about heat?
6. Which model of heat considered heat to be an invisible fluid?
7. What is the difference between a calorie and a Calorie?
8. Summarize the kinetic-molecular model of heat.

Understanding Concepts

9. Explain how a liquid-filled thermometer works.
10. Use a table to compare the Fahrenheit, Celsius, and Kelvin scales. Include the fiduciary points used, the units used, the relative size of the units, and the normal freezing and boiling points of water.
11. Write the formula to convert Celsius temperatures to Fahrenheit temperatures. Identify the variables and explain where the numbers come from.
12. Why aren't coefficients written in the formulas for converting between Celsius and Kelvin temperatures?
13. Convert 21 °F into a Celsius temperature.
14. How does heat relate to temperature and thermal energy?
15. Compare conduction, convection, and radiation.
16. What mass of copper (c_{sp} = 0.39 J/g °C) could be warmed from 46.5 °C to 135.3 °C by adding 955 J of energy?
17. What temperature change would be caused by removing 767 J of energy from a 115 g sample of water (c_{sp} = 4.18 J/g °C)?
18. How did Count Rumford's experiment discount the caloric theory?
19. Summarize the second law of thermodynamics.
20. Why has our understanding of thermal energy changed?

Critical Thinking

21. Convert 472 K into a Fahrenheit temperature.
22. At what temperature are the Fahrenheit and Celsius temperatures equal to each other?
23. What is the final temperature of a 135 g sample of steel (c_{sp} = 0.120 J/g °C) if it has an initial temperature of 23.2 °C and 4521 J of energy is added?

Refer to the information below to answer Questions 24–27.

In the laboratory, you have 100 g of water in a calorimeter at 20.0 °C and a 74.3 g mass of metal at 95 °C. You place the metal in the water. After a while, the setup ends at a final temperature of 24.7 °C.

24. Explain what is happening. Be specific about heat and temperatures.
25. Assuming that the insulated cup completely insulates the system, explain what will happen to the total thermal energy of the system.
26. What is the specific heat of the metal?
27. What metal do you think you are working with? Explain.
28. How many joules of energy will a 550-Calorie double cheeseburger supply you?

CASE STUDY: WATER AS A COOLANT

Many power plants, especially nuclear power plants, use an outside water source to cool the circulated water inside the plant. The cooling water is typically taken from a lake or river and then returned to that body of water. The water absorbs thermal energy from the power plant. The Indian Point Nuclear Power plant in New York gets its cooling water from the Hudson River.

29. What are some reasons that water is used for cooling this and other power plants?
30. If the water needs to absorb 12 500 000 000 000 J of energy, how much water is needed if the change in temperature must remain below 10 °C?
31. What are some possible negative impacts of using water from the Hudson River to cool the power plant?

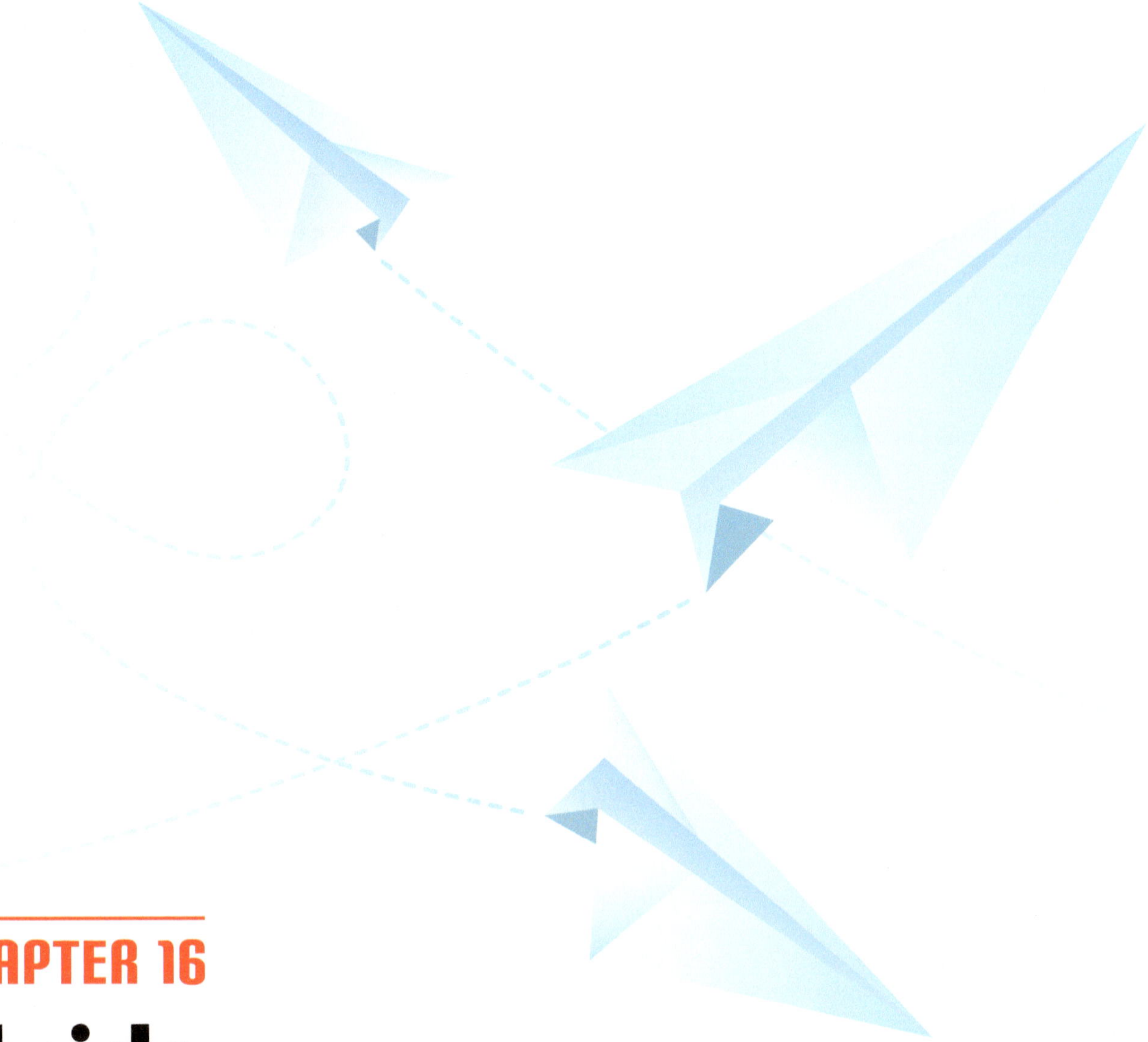

CHAPTER 16
Fluids

WIND SHEAR

Earth's restless atmosphere is always on the move. The gases in the atmosphere can move around easily. We feel the moving masses of air as wind. The moving air in a wind is a *fluid*—something that can flow. Sometimes a wind flows as a gentle breeze over the land's surface. At other times, though, a flowing air mass might take the form of a sudden and potentially very dangerous *microburst*. Microbursts are downdrafts from thunderstorms that occur over a small area. The rapidly sinking air in a microburst spreads out after it impacts the ground, creating strong winds and turbulence. These winds can flatten trees and damage homes. They are particularly hazardous for aircraft flying near the ground, such as when they are taking off or landing.

Microbursts can produce wind shear, instances when winds suddenly move in different directions over a short distance. They're just one of many types of weather events that are difficult to model. Improving our models of violent weather behavior can help us to protect lives by better predicting when and where such events might occur.

16A Properties of Fluids	372
16B Gas Laws	380
16C Fluid Mechanics	387

16A | PROPERTIES OF FLUIDS

Why does a hot air balloon rise?

16.1 FLUIDS AND PRESSURE

Most of us have experienced pouring water into a cup or seeing the wind blow away the seeds of a dandelion. Both of these experiences are possible because of a physical property shared by liquids, gases, and plasmas. These three states of matter are **fluids**—substances that can flow. They can move from one place to another. Fluids are not rigid, like solids, and that makes them a little different in the way that they exert, experience, and transmit forces and energy. Fluid mechanics is the branch of physics that studies these characteristics of fluids. Before we get into the details, though, we first need to look at the concept of pressure.

Pressure

Have you ever felt weighed down? You actually carry around a very large weight on your shoulders every day—the weight of Earth's atmosphere! Think about it—directly above your head is a very thick layer of gas molecules, and they are all being tugged down upon you by gravity. That means that the atmosphere pushes down upon the surface of your body, and as a result we experience pressure. **Pressure** (P) is a measurement of the amount of force (F) acting upon a given area (A), so the formula for calculating the pressure acting on a given area is

$$P = \frac{F}{A}.$$

16A Questions

- Is pressure transmitted the same way in all states of matter?
- Why do some things float and others sink?
- How do divers maintain a constant depth?
- Why does water flow more easily than syrup?

16A Terms

fluid, pressure, pascal, Pascal's principle, buoyant force, Archimedes's principle, viscosity

In the SI system, pressure is measured in newtons per square meter (N/m²). One N/m² of pressure is defined as one **pascal** (Pa), named after the French scientist and mathematician Blaise Pascal. In practical terms, a pascal is a very small amount of pressure. Everyday quantities of pressure, such as what you might measure in a bicycle tire, are measured in kilopascals (kPa).

Let's look at some examples of how the pressure formula can be used to solve problems.

EXAMPLE 16-1: Which Experiences Greater Pressure?

Picture a 10.0 N book lying on a table. Now picture the same book balanced atop a wooden dowel. If the book has an area of 0.058 m² and the dowel an area of 4.9×10^{-4} m², upon which item does the book exert more pressure, the table in the first instance or the dowel in the second? (Remember, weight is a force.)

Write what you know.

$F = 10.0$ N

$A_{book} = 0.058$ m²

$A_{dowel} = 4.9 \times 10^{-4}$ m²

$P = ?$

Write the formula and solve for the unknown.

$$P = \frac{F}{A}$$

Plug in known values and evaluate.

$$P_{table} = \frac{10.0 \text{ N}}{0.058 \text{ m}^2}$$

$$= 170 \text{ Pa}$$

$$P_{dowel} = \frac{10.0 \text{ N}}{4.9 \times 10^{-4} \text{ m}^2}$$

$$= 20\,000 \text{ Pa} = 20 \text{ kPa}$$

The book exerts almost 120 times more pressure on the dowel than it does on the table.

EXAMPLE 16-2: Calculating Area from Pressure

Now suppose the same book exerts 8.50 kPa of pressure on a third object. Over what area is that pressure exerted? Don't forget to convert kPa to Pa!

Write what you know.

$F = 10.0$ N

$P = 8.50$ kPa $= 8500$ Pa

$A_{object} = ?$

Write the formula and solve for the unknown.

$$P = \frac{F}{A}$$

$$AP = \frac{F}{\cancel{A}}\cancel{A}$$

$$\frac{\cancel{A}P}{\cancel{P}} = \frac{F}{P}$$

$$A = \frac{F}{P}$$

Plug in known values and evaluate.

$$A = \frac{10.0 \text{ N}}{8500 \text{ Pa}} = \frac{10.0 \text{ N}}{8500 \frac{\text{N}}{\text{m}^2}}$$

$$= 1.18 \times 10^{-3} \text{ m}^2 \text{ (an area 3.4 cm by 3.4 cm)}$$

Earth's atmosphere exerts 101.3 kPa of pressure at sea level. Seems like a lot, doesn't it? But the pressure exerted on you by the atmosphere is balanced by the fluid pressures acting on the inside of your body, so you

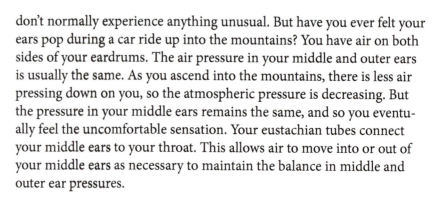

don't normally experience anything unusual. But have you ever felt your ears pop during a car ride up into the mountains? You have air on both sides of your eardrums. The air pressure in your middle and outer ears is usually the same. As you ascend into the mountains, there is less air pressing down on you, so the atmospheric pressure is decreasing. But the pressure in your middle ears remains the same, and so you eventually feel the uncomfortable sensation. Your eustachian tubes connect your middle ears to your throat. This allows air to move into or out of your middle ears as necessary to maintain the balance in middle and outer ear pressures.

Fluid Pressure

When you squirt hand lotion from a bottle, you're demonstrating the way that fluids experience pressure differently than solids do. A book lying on a desktop exerts a downward force on a certain amount of area. In other words, there's pressure on the desktop, but it exists only at the point of contact between the book and the desk. The rest of the desktop doesn't experience that pressure. But when you apply pressure with your hand to the lotion inside a bottle, the pressure is transmitted to *all parts* of the lotion throughout the bottle. This *fluid pressure* pushes some of the lotion out of the mouth of the bottle and onto your hand. Blaise Pascal first described this ability of fluids to transmit pressure throughout a sample, so it is called **Pascal's principle**. In Section 16C, you'll see how Pascal's principle has been put to good use in some familiar machines.

In Chapter 2, you learned about how particles move differently within solids, liquids, and gases. These differences in particle motion explain why solids and fluids experience pressure differently.

FLUID PRESSURE

When you squeeze a doorknob, you apply pressure to the outer surface of the knob. But the particles within the solid knob are rigidly locked into place. They have very little ability to move, so they can't effectively transmit the pressure to any particles that are deeper within the knob.

When you squeeze a balloon, you apply pressure to one part of the balloon. The pressure that you apply is transmitted to the freely moving fluid particles inside the balloon. This increases the number of collisions between the particles that transmit the pressure throughout the fluid.

16.2 BUOYANT FORCE

You may have noticed when squeezing a balloon that the balloon tends to bulge in areas that aren't being squeezed. As the pressure applied by your hand forces air molecules closer together, the number of collisions between them increases. The number of collisions between particles and the inside of the balloon increases as well. This, in turn, increases the pressure on the inside of the balloon. If you apply enough pressure, that is, you squeeze hard, the balloon may even burst.

A similar thing happens when objects are immersed into open bodies of water or other fluids. Take a boat being launched, for instance. As the boat is lowered into the water, it pushes, or *displaces*, particles out of its way, forcing the particles closer together. This causes the particles to push back on the boat, and the force increases with depth. Thus, the water along the bottom of the boat exerts more force than the water along the boat's sides near the surface. This upward force caused by the displacement of a fluid is called the **buoyant force**. *Buoyancy* is the effect caused by the buoyant force. It describes the tendency of an object to float when immersed in a fluid. The strength of the buoyant force depends in part on how many particles are displaced. As a launched boat settles deeper into the water, it displaces more particles, increasing the buoyant force acting on the boat.

Archimedes's Principle

In the third century BC, Greek mathematician and scientist Archimedes discovered a way to determine the buoyant force acting on an object. He learned that submerged objects displace an amount of fluid equal to the volume of the object itself. The weight of the displaced fluid is equal to the buoyant force acting on the object. Today we use the term **Archimedes's principle** to identify this relationship between displacement and buoyant force.

Archimedes's principle can be used to determine whether an object will float or sink. If the buoyant force acting on an object is greater than its weight, the object will float. If not, the object will sink. This explains why a wooden block will float in water, but an equal-sized lead block will not.

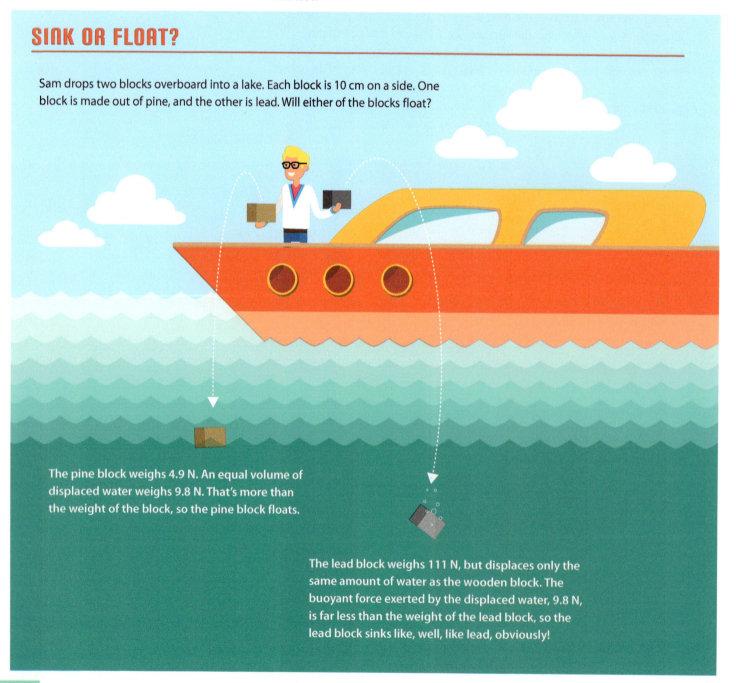

SINK OR FLOAT?

Sam drops two blocks overboard into a lake. Each block is 10 cm on a side. One block is made out of pine, and the other is lead. Will either of the blocks float?

The pine block weighs 4.9 N. An equal volume of displaced water weighs 9.8 N. That's more than the weight of the block, so the pine block floats.

The lead block weighs 111 N, but displaces only the same amount of water as the wooden block. The buoyant force exerted by the displaced water, 9.8 N, is far less than the weight of the lead block, so the lead block sinks like, well, like lead, obviously!

16.3 DENSITY AND BUOYANCY

There's another way to tell whether an object will sink or float. Water's density is about 1 g/cm³. Pine has a density around 0.5 g/cm³, and lead's density is 11.3 g/cm³. Did you notice something there? Pine is *less* dense than water (i.e., has less mass per unit of volume), and lead is *more* dense than water (it has more mass per unit of volume). This implies that an object whose density is less than that of the fluid into which it is immersed will float, and indeed that's true. If the object's density is greater than that of the fluid, the object will sink.

This relationship between density and buoyancy is true even if a fluid is placed into another fluid. Many salad dressings are mixtures of oil and vinegar. Left to sit, the oil and vinegar will usually separate, unless a chemical called an *emulsifier* is added to the dressing to keep it mixed. The oil is less dense than vinegar, so it floats to the top of the container.

Density also explains why hot air balloons rise. Air expands as it is heated, so it becomes less dense as it warms. The heated air inside the balloon is less dense than the cooler air outside the balloon, so the balloon floats. Adding more heat energy to the balloon warms the air further, decreasing its density and causing the balloon to gain elevation. Releasing some of the warm air from the balloon causes the overall density of the balloon to increase, so the balloon sinks.

Steel is almost eight times denser than water. So you might be wondering how ships made out of steel and weighing many tons can still float. The answer is in the shape of the ship! A ship's hull is hollow and has a form that pushes water aside. When a new ship is launched, it displaces water as it slides down the ways. Soon, the weight of the displaced water is equal to the weight of the ship, and the ship floats. The ship, being hollow, consists largely of air, making the ship's density less than that of the surrounding water.

Neutral Buoyancy

By carefully monitoring and adjusting the temperature and volume of warmed air inside his balloon, a balloonist can keep his craft at a constant altitude (i.e., within a column of Earth's atmosphere). At that point, the density of the balloon is equal to that of the air at that altitude. The craft neither ascends (climbs higher) nor descends (sinks lower). This condition is called *neutral buoyancy*.

Many fish must also be able to maintain their depth within a column of water. Most such fish do this by means of a swim bladder, a gas-filled sack within their bodies. A fish can adjust its buoyancy by adding or removing gas from its swim bladder. Scuba divers accomplish the same task by adding or removing air in a buoyancy compensator vest. The diver shown at left is releasing air from her vest, and thus increasing her density, so that she can descend into the water. To ascend later, she will add air to the vest from her compressed air tank.

16.4 VISCOSITY

Runny maple syrup is not very appealing, but trying to pour syrup that is too thick can be annoying. Maple syrup should flow easily—but not too easily! When we use words like *runny* or *thick* to describe maple syrup, we're really describing the syrup's viscosity. **Viscosity** is a measurement of a fluid's resistance to flowing. Highly viscous fluids, such as honey, do not flow easily, while less viscous fluids, like water, do. Viscosity is caused by friction between the particles in a fluid. The greater the friction force between particles, the more resistant to flowing the fluid will be. Some highly viscous fluids, such as pitch, flow so slowly that they can appear to be solids.

Viscosity is a thermometric property (see Chapter 15). It's an important quality of lubricants, such as motor oils. Most liquids become less viscous as their temperatures increase. In the high-temperature environment inside a car engine, a decrease in an oil's viscosity can lower its effectiveness at reducing friction between moving parts. The increased wear and tear on an engine will incur added repair costs and shorten the engine's useful life. Chemical engineers design motor oils that can retain their viscosities at high temperatures.

16A | REVIEW QUESTIONS

1. What is a fluid?
2. Define *pressure*.
3. How much pressure is exerted on the ground by a 535 N student if the soles of her shoes have an area of 0.0396 m²?
4. State Pascal's principle.
5. Summarize Archimedes's principle.
6. The density of beeswax is 0.96 g/cm³. Will beeswax float or sink in pure water? Explain.
7. Define *neutral buoyancy*.
8. Define *viscosity*.

MINI LAB

DEMONSTRATION: DENSITY STACK

The density stack at right was created by adding different liquids into a glass cylinder. The liquids then separated according to density, with the densest liquid at the bottom and each successive layer being less dense than the layer beneath.

PROCEDURE

A Your teacher will show and describe a group of liquids to you. Using what you know about these liquids, predict the order of the liquids within a density stack, from most dense (bottom) to least dense (top). Write your prediction in your notebook.

B Your teacher will then create the density stack by adding the liquids to the cylinder according to your predicted order of density.

ANALYSIS

1. Did the order within the stack agree with your prediction? Explain.

C Your teacher will provide you with the density value for each liquid in the stack.

2. Did the order of the densities, from most dense to least, match the order of the liquids within the density stack?

CONCLUSION

3. On the basis of the density values for two liquids, what can you predict about how those liquids will physically (not chemically) interact with each other?

GOING FURTHER

D How dense do you think rubber is? Make a prediction about where within the density stack a piece of rubber will come to rest. Write your prediction in your notebook. Your teacher will place a piece of rubber into the density stack to test your prediction.

4. Was your prediction correct? If not, then explain why the result is different than what you expected.

Essential Question:
Can the order of liquids in a density stack be predicted?

Equipment
graduated cylinder, 100 mL
liquids of varying densities

FLUIDS 379

16B Questions

- Are the volume, temperature, and pressure of a gas related?
- How do gases behave in response to changing conditions?
- How can we model the behavior of gases?

16B Terms

Boyle's law, Charles's law, Gay-Lussac's law, combined gas law

16B | GAS LAWS

How do changing conditions affect gases?

16.5 BOYLE'S LAW

In the previous section, we looked at properties that are common to all fluids. In this section, we'll look at some properties that are unique to gases. Remember from Chapter 2 that a gas will expand until it has completely and evenly filled its container. Any confined quantity of gas will have four measureable and interrelated properties:

1. the number of gas particles;
2. the volume of the gas;
3. the temperature of the gas; and
4. the pressure exerted by the gas on the container.

To properly understand how gases behave, we need to consider what's happening at the particle level. We'll start by looking at what creates gas pressure. Then we'll look at how different gas properties are related, starting with the relationship between pressure and volume.

PRESSURE AND VOLUME

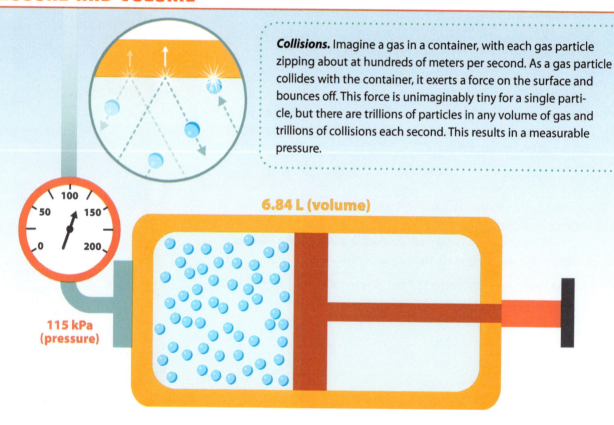

Collisions. Imagine a gas in a container, with each gas particle zipping about at hundreds of meters per second. As a gas particle collides with the container, it exerts a force on the surface and bounces off. This force is unimaginably tiny for a single particle, but there are trillions of particles in any volume of gas and trillions of collisions each second. This results in a measurable pressure.

6.84 L (volume)

115 kPa (pressure)

Gas Pressure. Gas pressure in a volume of gas depends on the number of particles present and their temperature. A gas enclosed in a container has a fixed number of particles in its volume. For each particular combination of volume and temperature, a unique pressure value is produced.

The Irish chemist Robert Boyle discovered the relationship between the volume and pressure of a gas on the basis of experiments done in 1659 and 1660. **Boyle's law** states that the pressure of a gas (P) is inversely proportional to its volume (V) if the amount of gas and its temperature are kept constant. Mathematically this can be expressed by the equation $PV = k$, where k is a constant. Usually we see Boyle's law expressed slightly differently. Since the products of any two volumes and pressures for a sample of a gas, shown as P_1V_1 and P_2V_2, are both equal to k, we can use the transitive property to set the two products equal to each other. The equation then looks like the one shown below.

$$P_1V_1 = P_2V_2$$

We can use this version of Boyle's law to calculate changes in pressure or volume in a sample of gas.

Volume Goes Down, Pressure Goes Up. Now imagine reducing the container's size without changing the number of gas particles or their temperature. This decreases the container's volume, the distance between particles, and the surface area inside the container. Since the temperature remains constant, the speed of the particles will still be the same. Each collision exerts the same amount of force, but there are more of them. More collisions means increased pressure.

Volume Goes Up, Pressure Goes Down. If the container's volume is enlarged, then the particles must travel farther between collisions, reducing the rate of collisions. Fewer collisions means decreased gas pressure.

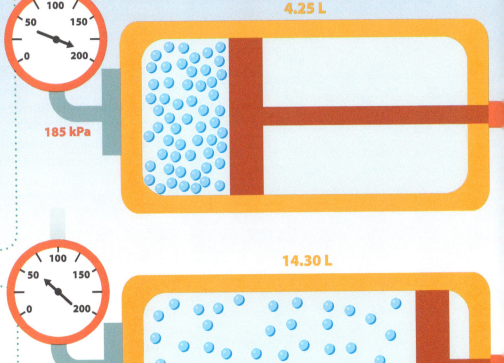

EXAMPLE 16-3: Pistons and Boyle's Law

Remember the piston in the Xpogo stick back in Chapter 14? Car shock absorbers are also built around a piston. When a car tire hits a bump, a piston compresses the gas inside the shock absorber. Once past the bump, the shock absorber returns to its resting volume and pressure. Suppose a shock absorber has a resting pressure of 101.3 kPa and volume of 1.30 L. What is the volume of the gas inside the shock absorber if its pressure when compressed is 658.6 kPa?

Write what you know.

$P_1 = 101.3$ kPa

$V_1 = 1.30$ L

$P_2 = 658.6$ kPa

$V_2 = ?$

Write the formula and solve for the unknown.

$$P_1 V_1 = P_2 V_2$$

$$\frac{P_1 V_1}{P_2} = \frac{\cancel{P_2} V_2}{\cancel{P_2}}$$

$$V_2 = \frac{P_1 V_1}{P_2}$$

Plug in known values and evaluate.

$$V_2 = \frac{P_1 V_1}{P_2} = \frac{(101.3 \cancel{\text{kPa}})(1.30\text{ L})}{(658.6 \cancel{\text{kPa}})} = 0.200 \text{ L}$$

The volume of gas in the compressed shock absorber is 0.200 L.

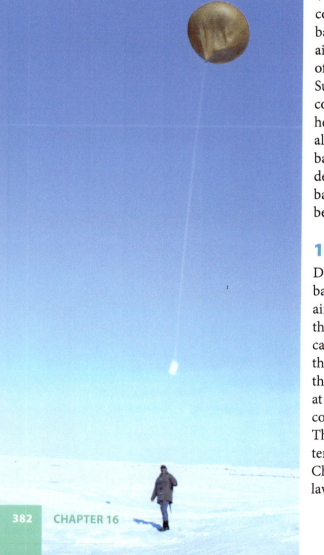

What are some everyday applications of Boyle's law? People use compressed air for all kinds of purposes, such as filling sports balls and operating powerful shop tools. Makers of compressed air tanks know that a small tank that contains the same amount of air as a larger tank has a greater pressure exerted on its walls. Such considerations are crucial. If the smaller tank were not constructed with thicker walls, it could explode! Boyle's law also helps explain why meteorologists only partially fill their high-altitude balloons with helium before they release them. As the balloon rises into the atmosphere, the surrounding air pressure decreases and the volume of the lifting gas increases to fill the balloon. If the balloon were full when released, it would burst before reaching the desired altitude.

16.6 CHARLES'S LAW

Decreasing atmospheric pressure may explain why a weather balloon expands as it ascends, but it doesn't explain why a hot air balloon expands while it's still on the ground. We already saw that the air in a hot air balloon expands as it is warmed. This can be explained at the particle level. The heat energy added to the balloon increases the kinetic energy of the air's particles, so the particles collide with more force. And since the air pressure at ground level remains constant, the increased force pushes the container outward. As a result, the volume of the gas increases. This example shows that there is a relationship between the temperature of a gas and its volume, a relationship described by Charles's law, named for the French scientist who formulated the law in the 1780s.

Charles's law states that the volume of a sample of gas at constant pressure is directly proportional to its temperature. Practically speaking, this means that if either the volume or temperature of a gas kept at constant pressure is changed, the other value will change in the same proportion. Mathematically this is expressed by the equation below.

$$\frac{V}{T} = k$$

Just as for pressure and volume, the quotient of any temperature and volume conditions for a gas are related by the constant *k*. This means that we can again set two sets of conditions equal to each other and use the formula

$$\frac{V_1}{T_1} = \frac{V_2}{T_2}$$

to solve for unknown conditions.

Because pressure is related to the motion of the particles, the Kelvin scale is used when doing gas law problems. This also avoids calculations with negative temperatures.

EXAMPLE 16-4: Weather Balloons and Charles's Law

A team of meteorologists fills a weather balloon with 525 L of helium prior to launch. It's 17 °C out when they fill the balloon. By the time everything is ready for launch, though, the temperature has risen to 25 °C. What is the new volume of the helium gas? (Remember, Celsius temperatures must first be converted to kelvins before using the formula.)

Write what you know.

$V_1 = 525$ L

$T_1 = 17$ °C = 290 K

$T_2 = 25$ °C = 298 K

$V_2 = ?$

Write the formula and solve for the unknown.

$$\frac{V_1}{T_1} = \frac{V_2}{T_2}$$

$$\frac{T_2 V_1}{T_1} = \frac{V_2 \cancel{T_2}}{\cancel{T_2}}$$

$$V_2 = \frac{T_2 V_1}{T_1}$$

Plug in known values and evaluate.

$$V_2 = \frac{(298 \text{ K})(525 \text{ L})}{290 \text{ K}} = 539 \text{ L}$$

The volume of gas in the warmed balloon is 539 L.

FLUIDS 383

16.7 GAY-LUSSAC'S LAW

By now you may have noticed a trend. So far we've seen a relationship between pressure and volume and between volume and temperature. Is there a relationship between pressure and temperature too? Yes, there is! Imagine adding thermal energy to a rigid container filled with a sample of gas. As heat is added, the gas particles gain kinetic energy. They move faster and collide harder with the walls of the container—harder collisions means greater pressure. This is exactly how modern steam turbine engines work, the kind that power ships all over the world. The superheated steam inside a turbine is contained within a rigid system, so the pressure inside such systems is high as well.

The French chemist Joseph Gay-Lussac was the first to demonstrate this relationship between pressure and temperature. **Gay-Lussac's law** states that the pressure of a sample of gas at a constant volume is directly proportional to its temperature. The equation for this is shown below.

$$\frac{P}{T} = k$$

Again, we can use a derived variation of this formula to solve pressure and temperature problems. The derived formula in this instance is

$$\frac{P_1}{T_1} = \frac{P_2}{T_2}.$$

EXAMPLE 16-5: Exploding Cans and Gay-Lussac's Law

Aerosol cans pose a risk of exploding if overheated. If such a can is exposed to a heat source, it may rupture and cause serious injury to anyone standing nearby. Suppose an aerosol can is being stored at sea level at 22 °C in a wooden shed. The can's internal pressure is 355 kPa. The shed catches fire, increasing the temperature inside to 400.0 °C. How does this change the pressure inside the can?

Write what you know.

$$T_1 = 22\,°C = 295\,K$$

$$P_1 = 355\,kPa$$

$$T_2 = 400.0\,°C = 673\,K$$

$$P_2 = ?$$

Write the formula and solve for the unknown.

$$\frac{P_1}{T_1} = \frac{P_2}{T_2}$$

$$\frac{T_2 P_1}{T_1} = \frac{P_2 \cancel{T_2}}{\cancel{T_2}}$$

$$P_2 = \frac{T_2 P_1}{T_1}$$

Plug in known values and evaluate.

$$P_2 = \frac{(673\,K)(355\,kPa)}{(295\,K)} = 810\,kPa$$

Not surprisingly, since the can experienced a roughly twofold increase in absolute temperature, it also experienced a twofold increase in pressure. Is that enough of an increase to burst the can? After all, some aerosol cans carry a warning that they shouldn't be stored above 120 °C (248 °F). But aerosol cans are also designed to withstand a minimum amount of burst pressure, often around 690 kPa. In this instance, the can would have exploded, since its pressure when heated far exceeded the burst pressure limit. In fact, at what temperature would our hypothetical spray can have reached burst pressure? We can use Gay-Lussac's law to find out.

Write what you know.

$$T_1 = 22\,°C = 295\,K$$

$$P_1 = 355\,kPa$$

$$P_2 = 690\,kPa$$

$$T_2 = ?$$

Write the formula and solve for the unknown.

$$\frac{P_1}{T_1} = \frac{P_2}{T_2}$$

$$\left(\frac{\cancel{P_1} T_2}{\cancel{P_1}}\right)\left(\frac{\cancel{P_1}}{\cancel{T_1}}\right) = \left(\frac{P_2}{\cancel{T_2}}\right)\left(\frac{T_1 \cancel{T_2}}{P_1}\right)$$

$$T_2 = \frac{T_1 P_2}{P_1}$$

Plug in known values and evaluate.

$$T_2 = \frac{(295\,K)(690\,\cancel{kPa})}{(355\,\cancel{kPa})} = 573\,K = 300\,°C$$

Our can would have exploded at 300 °C, long before the fire reached its 400.0 °C reported temperature.

16.8 THE COMBINED GAS LAW

Pressure, volume, and temperature—so far, each of the laws we have looked at relates only two of these quantities at a time. But they are obviously all related. What happens to the temperature of a gas sample if both its volume *and* pressure change? That's actually fairly easy to predict. We can combine Boyle's, Charles's, and Gay-Lussac's laws into a single formula. This **combined gas law** states that the ratio of the product of a gas sample's pressure and volume to its absolute temperature is equal to a constant. The equation for this is given below.

$$\frac{PV}{T} = k$$

FLUIDS

Remember those weather balloons we looked at earlier? It should be apparent that both the temperature and pressure of the balloon's gas are affected by atmospheric conditions as the balloon rises. As it ascends, the atmospheric pressure decreases and the air gets colder. How will these combined factors affect a balloon? We can once again set two quantities equal to each other and use the combined gas law to answer this question.

EXAMPLE 16-6: Up, Up, and Away with the Combined Gas Law

A balloon with a volume of 2.8×10^6 L of air is launched from sea level on a sunny 308 K summer day and rises to an altitude of 4600 m. At that altitude, atmospheric pressure is about 57 kPa, and the temperature is a chilly 259 K. What is the volume of the balloon at that altitude?

Write what you know.

P_1 (atmospheric pressure at sea level) = 101.3 kPa

$V_1 = 2.8 \times 10^6$ L

$T_1 = 308$ K

$P_2 = 57$ kPa

$T_2 = 259$ K

$V_2 = ?$

Write the formula and solve for the unknown.

$$\frac{P_1 V_1}{T_1} = \frac{P_2 V_2}{T_2}$$

$$\left(\frac{T_2}{P_2}\right)\left(\frac{P_1 V_1}{T_1}\right) = \left(\frac{\cancel{P_2} V_2}{\cancel{T_2}}\right)\left(\frac{\cancel{T_2}}{\cancel{P_2}}\right)$$

$$V_2 = \frac{T_2 P_1 V_1}{P_2 T_1}$$

Plug in known values and evaluate.

$$V_2 = \frac{(259 \text{ K})(101.3 \text{ kPa})(2.8 \times 10^6 \text{ L})}{(57 \text{ kPa})(308 \text{ K})} = 4.2 \times 10^6 \text{ L}$$

The balloon's volume increases by a factor of 1.5 by the time it reaches 4600 m.

16B | REVIEW QUESTIONS

1. State Boyle's law and write its mathematical form.
2. If the volume of a confined gas triples, what happens to its pressure (assuming that the temperature remains constant)?
3. State Charles's law and write its mathematical form.
4. A bouncy playground ball is left outside overnight. In the morning, the ball feels flat and doesn't bounce as well. Which gas law explains this? Defend your answer.
5. State Gay-Lussac's law and write its mathematical form.
6. According to Gay-Lussac's law, if the Kelvin temperature of a confined quantity of gas is doubled, what will happen to the gas's pressure?
7. State the combined gas law and write its mathematical form.
8. Which one of the gas laws best describes the behavior of gases? Explain.
9. A sample of helium gas has a volume of 5.0 L at 250 kPa of pressure. Assuming that its temperature remains constant, which gas law would you use to find the volume of the sample at 150 kPa? Find the new volume.
10. A sample of gas has an initial temperature of 375.0 K and pressure of 215 kPa. Assuming that its volume remains constant, which gas law would you use to find the sample's pressure if the temperature is raised 50.0 K? Find the new pressure.

16C | FLUID MECHANICS

How can a person lift a car?

16.9 HYDRAULICS

What do the brakes on the family car, a bucket on a backhoe, and the lift in a car repair shop have in common? Each produces a mechanical advantage through **hydraulics**, the magnification of a small input force on a confined liquid to produce a much larger output force. The output force can be used to do work, such as scooping and lifting a large amount of earth. To show how hydraulic systems work, let's examine the basic function of a hydraulic lift found in most any auto shop.

16C Questions
- How do hydraulics work?
- How is lift generated?
- What quantity is conserved in flowing fluids?

16C Terms
hydraulics, Bernoulli's principle

NEED A LIFT?

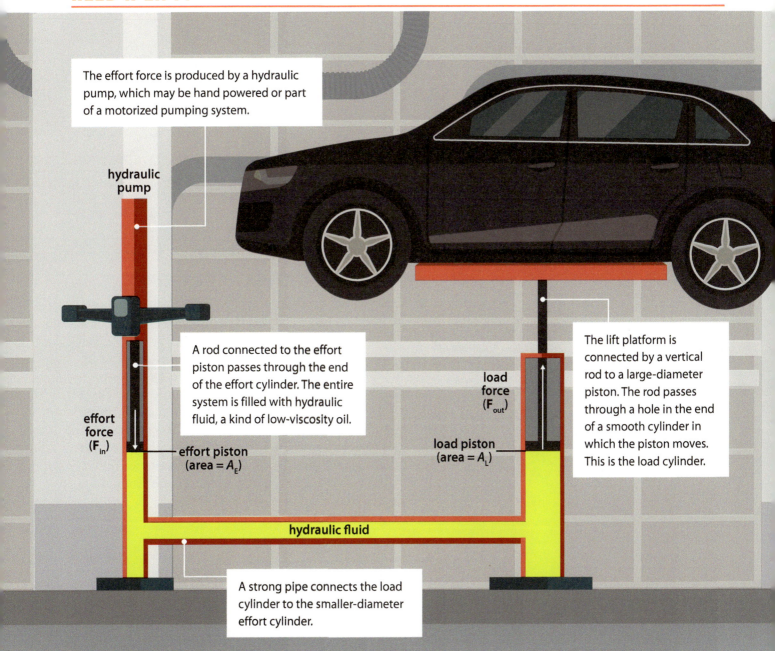

The effort force is produced by a hydraulic pump, which may be hand powered or part of a motorized pumping system.

hydraulic pump

A rod connected to the effort piston passes through the end of the effort cylinder. The entire system is filled with hydraulic fluid, a kind of low-viscosity oil.

effort force (F_{in})

effort piston (area = A_E)

load force (F_{out})

load piston (area = A_L)

The lift platform is connected by a vertical rod to a large-diameter piston. The rod passes through a hole in the end of a smooth cylinder in which the piston moves. This is the load cylinder.

hydraulic fluid

A strong pipe connects the load cylinder to the smaller-diameter effort cylinder.

FLUIDS

According to Pascal's principle, the pressure change due to the force exerted on a car lift's effort piston is transmitted throughout the entire system by the hydraulic fluid. Now recall that the formula for calculating pressure is $P = F/A$. This formula shows that if the pressure (P) inside the system is constant, then any difference in the sizes of the piston faces (A) must be matched by an equal difference in the amount of force (F) acting on them. For example, if the face of the load piston has an area that is five times larger than that of the effort piston, the output force will be five times larger than the input force. Further increasing the ratio of the two areas can greatly reduce the effort force needed to do a large amount of work—like lifting a car.

16.10 BERNOULLI'S PRINCIPLE

If you've ever held your thumb over the open end of a garden hose and turned on the water, then you know that the water squirts out much faster than when it flows out of the hose unhindered. There's a reason for this, of course. Putting one's thumb over the hose doesn't make any less water flow out of the hose—it's the same amount as that flowing inside the hose. Otherwise, water would be backing up in the hose! What changes is the *velocity* of the moving water. The water moving out of the narrowed opening moves much faster than the water flowing through the larger inside diameter of the hose. The result of this is that the *flow rate* of the water—the volume of water moved per unit of time—remains constant.

SERVING AS A PIPING ENGINEER:
KEEPING THE FOOD FLOWING

Natural gas, ketchup, ice cream—each is a fluid product that is transported through a complex system of pipes. Piping engineers are mechanical or materials engineers who specialize in designing pipe systems. Designing a facility with different-sized vats, storage tanks, and a maze of pipes connecting them is not as simple as it may sound. Many factors come into play, such as the temperature and viscosity of the product, pressure changes that occur when the diameter of a pipe changes, and choosing the most efficient route for pipes through a plant. Pipes full of liquid can be heavy, too, so the design must include sufficient support for all the equipment.

Prior to his retirement, Dave Lombard served as a piping engineer in California's agriculturally rich Central Valley. He helped design many piping systems used in the processing and packaging of foods and beverages. Dave says, "I really enjoy food and beverage work because everyone needs to eat. It's very fulfilling to be a part of that!" If you like challenging work, a career in piping engineering might be a good choice for you!

You might expect that faster-flowing fluids have higher fluid pressures, but in fact they don't. They actually have lower fluid pressures. This relationship between the increasing speed of a fluid and its decreasing pressure is known as **Bernoulli's principle**. A Swiss mathematician, Daniel Bernoulli, described the phenomenon in 1738.

de Laval nozzle

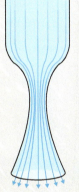

Bernoulli's principle can be put to many practical uses. It partly accounts for the lift generated by aircraft wings. A curved upper wing surface is shaped so that air flows faster over the top of the wing, creating lower pressure on that side. Hose-end sprayers, used for applying fertilizers to lawns and gardens, operate on Bernoulli's principle too. Water passing over a tube inside the sprayer creates low pressure that draws the fertilizer up the tube and into the stream of water. The hourglass shape of a de Laval nozzle creates a region of low pressure in the exhaust gas of a rocket motor. The low pressure increases the speed of the vented gas, providing more thrust for a rocket.

16C | REVIEW QUESTIONS

1. How does Pascal's principle explain the operation of a hydraulic lift?
2. What quantity remains constant within a fluid system regardless of the fluid's velocity or pressure?
3. State Bernoulli's principle.
4. The powerhead shown at right is a type of aquarium pump that circulates water. If a piece of plastic tubing is inserted into the discharge pipe on the powerhead, a stream of air can be drawn into the flowing water. Use Bernoulli's principle to explain how this is possible.

CHAPTER 16

16A PROPERTIES OF FLUIDS

- Fluids are substances that can flow, including liquids, gases, and plasmas.
- Pressure is the amount of force acting on a given area and is measured in pascals. One pascal is defined as 1 N/m².
- The pressure exerted anywhere on a fluid is transmitted throughout all parts of the fluid. This property of fluids is known as Pascal's principle.
- The fluid displaced by a submerged object exerts an upward force on the object. This is known as buoyant force.
- Archimedes's principle states that the buoyant force on an object is equal to the weight of the fluid displaced by the object.
- Viscosity is a measure of a fluid's resistance to flowing.

16A Terms
fluid	372
pressure	372
pascal	373
Pascal's principle	374
buoyant force	375
Archimedes's principle	376
viscosity	378

16B GAS LAWS

- The behavior of a gas is affected by the number of particles in the sample and its pressure, volume, and temperature.
- Boyle's law states that the pressure of a gas is inversely proportional to its volume at constant temperature.
- Charles's law states that the volume of a gas is directly proportional to its temperature at constant pressure.
- Gay-Lussac's law states that the pressure of a gas is directly proportional to its temperature at constant volume.
- The combined gas law integrates Boyle's, Charles's, and Gay-Lussac's laws into a single law and states that the product of the pressure and volume of a sample of gas is directly proportional to its temperature.

16B Terms
Boyle's law	381
Charles's law	383
Gay-Lussac's law	384
combined gas law	385

16C FLUID MECHANICS

- Hydraulic systems are a type of machine that produces mechanical advantage according to Pascal's principle.
- The ratio of output force to input force in a hydraulic system is equal to the ratio of the area of the load cylinder's piston face to that of the effort cylinder.
- Bernoulli's principle states that as the speed of a flowing fluid increases, the pressure within the fluid decreases. In a fluid system, the volume of fluid moved per unit of time, or flow rate, remains constant.

16C Terms
hydraulics	387
Bernoulli's principle	389

REVIEW

CHAPTER REVIEW QUESTIONS

Recalling Facts

1. Which states of matter are considered fluids?
2. What is the SI unit for pressure and how is it defined?
3. How is fluid pressure different from pressure exerted between solids?
4. What is buoyant force?
5. What relationship is described by Archimedes's principle?
6. What three factors contribute to the pressure of a gas?
7. What two quantities does Boyle's law relate?
8. What is assumed to remain constant when Gay-Lussac's law is applied to a gas?
9. What condition of a gas is assumed to remain constant when applying the combined gas law?
10. On what property of fluids are hydraulic systems based?
11. Give an example of one device that operates on Bernoulli's principle.

Understanding Concepts

12. Why are pascals not usually used for measuring everyday sorts of pressure?
13. A force of 15 N is applied to an area of 0.25 m². How much pressure does the area experience?
14. What property of fluids accounts for their ability to transmit pressure compared with solids?
15. As an object is lowered into a fluid, does the buoyant force acting on it increase, decrease, or remain the same?
16. Describe the relationship between density and buoyancy.
17. Polystyrene, a type of plastic, sinks in fresh water but floats in seawater. What does this tell you about the relative densities of polystyrene, fresh water, and seawater?
18. If the friction force between particles in a fluid is reduced, will the fluid's viscosity increase or decrease?
19. Relate viscosity to the temperature of a fluid.
20. A sample of gas occupies 0.500 L at a pressure of 30.5 kPa. Its volume is then reduced to 0.250 L at a constant temperature. What gas law would you use to find its new pressure? Find the new pressure.
21. A sample of gas occupies 1.75 L at a pressure of 50.20 kPa. If the gas is compressed at constant temperature until its pressure is 80.50 kPa, what will its new volume be?
22. Five liters of gas are heated at constant pressure to a temperature of 355 K, producing a final volume of 9.5 L. What gas law would you use to find the starting temperature of the gas? Find the starting temperature.
23. Four liters of gas are sealed in a container at a temperature of 25 °C and 101 kPa of pressure. The temperature is then raised to 212 °C. What gas law would you use to find the resulting pressure? Find the resulting pressure.
24. A balloon is filled with 35.0 L of gas at sea level. When its temperature is reduced to −2.0 °C, its volume decreases to 34.0 L, and the pressure falls to 98.0 kPa. What gas law is needed to find the original temperature of the gas? Find the original temperature.
25. Compare a hydraulic lift to the simple machines that you learned about in Chapter 13.
26. If the force exerted on the effort cylinder of a hydraulic lift produces 1500 kPa of pressure within the system, what will the pressure be at the load cylinder?
27. The piston within an effort cylinder has a face with an area of 0.2 m². If 500.0 N of force is applied to the piston, how much force will be produced on a load piston with an area of 0.5 m²?

Critical Thinking

28. The Amazon River has the largest discharge of any river on Earth. The Amazon's discharge remains fresh enough to drink even many miles out to sea. Explain this in terms of the densities of fresh water and seawater.
29. Write the following liquids in the correct order, bottom to top, according to how they would sort out after being poured into a graduated cylinder to create a density stack: coconut oil (0.92 g/cm³), corn syrup (1.33 g/cm³), glycerin (1.26 g/cm³), olive oil (0.80 g/cm³), rubbing alcohol (0.79 g/cm³).
30. Using the concept of density, explain why the RMS *Titanic* sank in 1912.
31. On a frosty fall morning, the equipment manager of a soccer team notices that the soccer balls, which were stored outside overnight, feel a little flat, so he adds air to them. Why might that not be such a good idea?
32. A friend tells you that according to the combined gas law, if the temperature of a gas increases, its pressure must also increase. What is the flaw in your friend's understanding of the combined gas law?
33. Why are Boyle's, Charles's, and Gay-Lussac's laws by themselves not completely satisfactory for modeling the behavior of gases?
34. Imagine a hydraulic lift working in reverse. If a large force were to be applied to the load cylinder, what would be true about the amount of force at the effort cylinder?
35. What would the formula for comparing the input and output forces of a hydraulic lift look like?
36. The structure shown at right is an aircraft venturi. It's used to create a vacuum that operates certain instruments inside the aircraft. Use Bernoulli's principle to explain how a venturi works.

UNIT 4
WAVES AND ENERGY

CHAPTER 17: **PERIODIC MOTION AND WAVES**

CHAPTER 18: **SOUND**

CHAPTER 19: **ELECTRICITY**

CHAPTER 20: **MAGNETISM**

CHAPTER 21: **ELECTROMAGNETIC ENERGY**

CHAPTER 22: **LIGHT AND OPTICS**

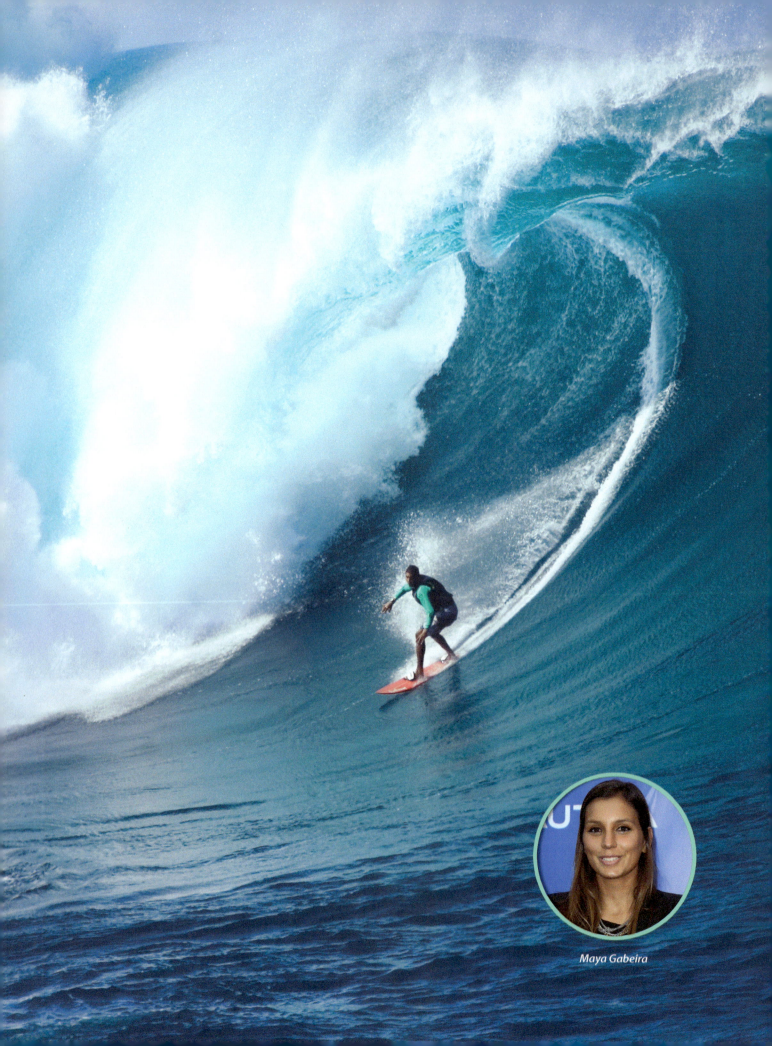

Maya Gabeira

CHAPTER 17

Periodic Motion and Waves

WILD WAVES

Surfers understand that larger waves have more energy, and to a surfer this means that really big waves give them a more exciting ride. They also know that bigger waves can do more damage if an accident occurs. This knowledge became reality for Maya Gabeira on October 28, 2013, when she nearly died riding a massive (> 18.3 m) wave in Nazaré, Portugal.

Maya worked her way back through rehabilitation to surf again. On January 18, 2018, Maya was again in Nazaré and had another opportunity to ride a massive wave. Things turned out better this time, and Maya was awarded a Guinness World Record for the largest wave surfed by a woman. The wave (20.72 m) was even larger than the 2013 wave that almost killed her.

Waves can travel through many different materials, and they carry energy from place to place. Throughout this chapter we will learn about periodic motion and waves.

17A Periodic Motion	396
17B Waves	403
17C Wave Behavior	410

17A | PERIODIC MOTION

What affects the swing of a pendulum?

17A Questions

- What is periodic motion?
- What are some examples of periodic motion?
- What causes a spring system to change directions?
- What is simple harmonic motion?

17A Terms

periodic motion, period, amplitude, simple harmonic motion, damping, resonance

17.1 DEFINING PERIODIC MOTION

Have you ever been eager for a particular time of day? You may have been waiting for dinnertime because your mom is making your favorite meal tonight. Or you may have anticipated the arrival of family from out of town. If you were watching an analog clock, you may have watched the second, minute, and hour hands making their paths around the clock repeatedly. Each hand repeats its motion in a set amount of time. Motion that repeats in equal time intervals is called **periodic motion**, and the time interval for the motion to repeat is called the **period** (T). For example, the period of the movement of the minute hand is sixty minutes.

There are many examples of periodic motion. On Day 4 of Creation, God made the sun, moon, and stars to mark signs, seasons, days, and years (Gen. 1:14–19). The motion of the earth moving around the sun repeats every year, while the relative movements of the sun and moon each month cause the moon's phases. The daily rising of the sun is caused by the continuous rotation of the earth on its axis. Kids of all ages love the rhythmic motion of a swing, while your grandparents may be soothed by quietly rocking in a rocking chair. The springs and shocks on your car's suspension absorb imperfections on the road by allowing the car to move up and down separately from the wheels. If your family owns a grandfather clock, you may have watched the slow, steady swing of the pendulum that moves the internal mechanism to keep time. Let's learn more about periodic motion by looking at an oscillating (i.e., moving back and forth) spring system.

The image below shows an analemma, which depicts the position of the sun at the same time of day throughout the year. You can see the periodic motion of the sun in this image. Notice that the angle of the analemma matches the angle of the gnomon of the sundial.

17.2 EXAMINING PERIODIC MOTION

A spring system is an excellent example of periodic motion. The simple spring system shown below consists of a mass attached to a spring positioned horizontally on a frictionless surface. When it is stationary, it

PERIODIC MOTION OF A SPRING SYSTEM

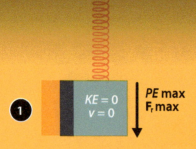

Compressing the Spring. By moving the mass and compressing the spring, we do work on the system. We are transferring energy to the system, storing potential energy in the spring. At the maximum displacement from the equilibrium position, the potential energy is at its maximum value and the kinetic energy is zero because the mass has zero velocity. When we release the mass, the potential energy begins a transformation into kinetic energy. The spring force acts to restore the mass to its equilibrium position, so this force is called the *restoring force* (F_r).

rest position (equilibrium)

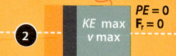

The Expanding Spring. As the mass reaches its equilibrium position, all of its potential energy has changed into kinetic energy, and the restoring force has decreased to zero. Some people would think that the mass would stop, but it has momentum, so it continues to move past the equilibrium position. In the equilibrium position, the potential energy is zero while the kinetic energy is at its maximum value, giving the mass its maximum velocity.

Momentum Stretches the Spring. As the mass continues moving to a position beyond its equilibrium position, the kinetic energy is being converted back into potential energy. Also, as the spring stretches, the restoring force increases, but now it is pointing in the opposite direction, back toward the equilibrium position. The mass will come to a stop at a position beyond *the rest position* that is equal to *the original displacement from the equilibrium position*. The potential energy is again at its maximum value, while the velocity and kinetic energy are zero.

is at its rest position, or equilibrium position. If we move the mass away from its equilibrium position and then release it, it will exhibit periodic motion. It repeatedly moves back and forth, passing through its equilibrium position. Let's investigate how this motion relates to conservation of energy.

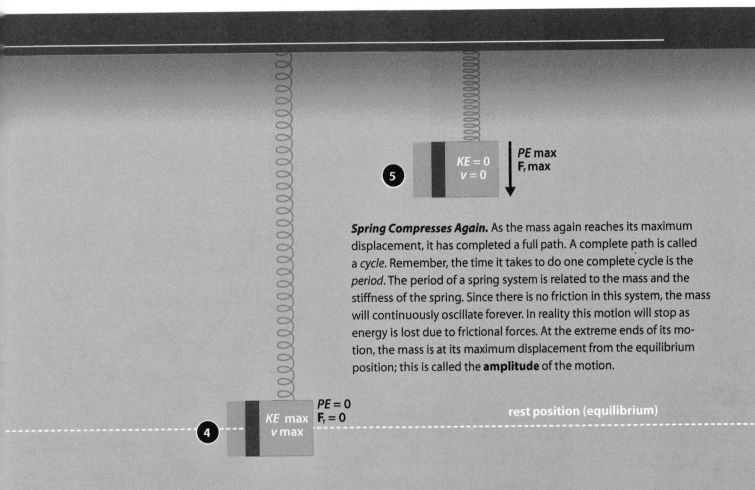

Spring Compresses Again. As the mass again reaches its maximum displacement, it has completed a full path. A complete path is called a *cycle*. Remember, the time it takes to do one complete cycle is the *period*. The period of a spring system is related to the mass and the stiffness of the spring. Since there is no friction in this system, the mass will continuously oscillate forever. In reality this motion will stop as energy is lost due to frictional forces. At the extreme ends of its motion, the mass is at its maximum displacement from the equilibrium position; this is called the **amplitude** of the motion.

The Recoiling Spring. The restoring force starts the mass back toward the equilibrium position. As the mass reaches its rest position once more, its momentum again makes it pass through the equilibrium position.

The restoring force causes the periodic motion of the system on the previous page spread. As you can see, the restoring force changes as the mass moves. The restoring force in spring systems is proportional to the displacement from the rest position. When the mass is farthest from the rest position, the force is greatest, and it decreases linearly to zero at the rest position. Periodic motion that is caused by a restoring force that is proportional to its displacement is called **simple harmonic motion**. The restoring force always points back to the equilibrium position.

17.3 USING PERIODIC MOTION

There are many ways that we use periodic motion in our daily lives. Most of our clocks and watches keep accurate time by the use of oscillators—devices that experience periodic motion. Many early clocks keep time by the working of a pendulum, which is a freely swinging body suspended from a support. Watches keep track of time by either a spring system or the vibration of a tiny crystal within them. All light and sound occur as a result of the repeated movements of waves (see Section 17C). In some applications, we want to have greater control over an object's oscillations.

Though we often want the periodic motion to continue, there are times when we want it to stop. When we do, we sometimes use a *damped oscillator*. The movement of such a system is said to experience **damping**, which is a force that intentionally works against the motion. Damping is often done with a frictional force because, as you learned in Chapter 8, friction always resists motion. The damping force does work on the system by transferring energy out of the system. Typically it turns this energy into thermal energy. As the damping force removes energy, the oscillations get smaller and smaller, until finally, the system comes to a stop.

As with many skyscrapers, the architect of the Taipei 101 building shown below included systems to reduce oscillations of the building. The large metal damping mass (inset) is both useful and beautiful.

HOW IT WORKS

Car Suspension

Driving along streets with potholes helps you recognize the importance of good suspension on your car. The suspension system consists of the tires, the air in the tires, springs, shock absorbers, and linkages. The purpose of this system is to isolate the passenger compartment from the drive portion of the car. If the system works correctly, the wheels may move rapidly up and down over bumps and potholes, but you experience a relatively smooth ride.

One key part of the system is the springs. The springs allow the wheels to move up and down without disrupting the smooth ride for the passengers. Automotive designers select springs to correspond to the vehicle's intended use. A race car or truck needs relatively stiff springs. On a race car, stiff springs permit better control of the car, but the ride will not be as smooth. A truck needs stiff springs to allow it to carry heavy loads. Stiff springs make for an uncomfortable ride when a truck is unloaded. Luxury cars have softer springs to deliver a smooth ride. But the cost of the softer springs is a loss of road feel and control for the driver. Automotive engineers spend a significant amount of time balancing vehicle control with ride comfort.

Shock absorbers are key for achieving a comfortable ride by damping oscillations. While there are many designs, hydraulic shocks are very common. Shocks, as they are called, consist of a piston that can slide in and out of a cylinder. The cylinder is filled with a viscous fluid that resists the up and down motion of the vehicle. As bumps in the road apply forces that cause the car to oscillate, the shock absorbers apply a frictional force that turns some of the energy into heat. Shock absorbers can be adjusted to balance vehicle performance and ride comfort. Shocks that are stiff give you better control but not a very smooth ride. Softer shocks give a nice smooth ride, but you give up some vehicle control. If the shocks fail, the car will continue oscillating as you try to drive down the road.

At other times we may want to keep the oscillations going and will have to provide an input force to overcome any natural losses of energy. In these cases, the input force is driving the oscillator to maintain the size of the oscillations. We call this a *driven oscillator*. Often with a driven oscillator, the result is **resonance**—when the driving force causes the object to vibrate at greater and greater amplitudes. Resonance will occur only at rates of vibration specific to that object. Musical instruments produce beautiful sounds as the musician applies an input force that causes the instrument to *resonate*. You will learn more about resonance as you study sound in Chapter 18.

17A | REVIEW QUESTIONS

1. Define *periodic motion*.
2. Where did the initial potential energy come from in the spring system example on pages 398–99?
3. What caused the spring system to slow to a stop at the extremes of its motion?
4. What is amplitude?
5. Using the spring system, explain conservation of energy.
6. How could you increase the amplitude of a spring system?
7. What is a damped oscillator?

MINI LAB

MAKING WAVES

Waves are all around us: we observe water waves, seismic waves, light waves, and sound waves. How are these waves similar to each other and how are they different? In this inquiry lab activity, you will try to form different types of waves.

Essential Question:

How can we form different types of waves?

1. Without looking up the definition, use your own words to describe a wave.

Equipment

Slinky* or long spring

PROCEDURE

A Stretch the Slinky between two lab partners. Be careful not to overstretch the Slinky or release either end of it.

B Have one of the two students disrupt the Slinky by moving it suddenly.

2. In what direction did the wave pulse move?

C Have one of the two students try to put another wave pulse into the Slinky by moving it differently.

D See how many different ways you can produce a wave pulse.

ANALYSIS

3. Sketch each of the waves that you were able to create. Label the direction of disruption and the direction that the wave traveled.

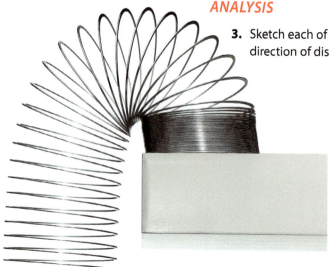

CONCLUSION

4. What can you conclude about the direction of the disruption and the direction that the wave traveled?

GOING FURTHER

5. See whether you can create two waves from opposite ends. Do they bounce off each other or pass through each other? (*Hint*: Consider whether sound waves pass through each other or bounce off each other when two people talk at the same time.)

17B | WAVES

What moves in a wave?

17.4 DEFINING WAVES

Waves surround us. Sound waves allow us to talk to each other, enjoy music, or listen to a ball game. We can see objects around us because they emit or reflect light waves. Radio waves, microwaves, and infrared waves carry information around our house, around the world, and even through space. The constant of almost any visit to the beach is the movement of ocean waves. Those waves can gently lap at the shore or cause great devastation when storm waves or tsunamis come ashore. Earthquake waves can cause destruction many miles from the epicenter of the quake. These are all waves, but they are not all the same. So what is a wave?

A **wave** is a disruption that carries energy from one location to another. A key to understanding waves is to realize that as the particles of the medium are disrupted, energy moves from place to place through the medium. After a wave has passed, the matter remains in its original position. Waves can occur as a series of disruptions, called a *wave train*, or as a single disruption, known as a *wave pulse*.

There are two general types of waves: mechanical and electromagnetic. A **mechanical wave** is a wave that moves through a physical medium and carries energy by disturbing the matter of the medium. These waves can move through solids, liquids, and gases, as long as there is matter that these waves can move through. Sound waves, earthquake waves, and water waves are all mechanical waves. *Electromagnetic waves* are disruptions in an electromagnetic field and can travel, unlike mechanical waves, through the vacuum of space, where there is no physical medium. You will learn more about electromagnetic waves in Chapter 21.

17B Questions

- What is a wave?
- What are the different types of waves?
- What is the difference between a transverse and a longitudinal wave?
- What affects wave speed?

17B Terms

wave, mechanical wave, crest, trough, wavelength, transverse wave, longitudinal wave, compression, rarefaction, frequency

17.5 DESCRIBING WAVES

Regardless of the type of wave that is occurring, many waves have a similar appearance. Most of us are familiar with water waves, so let's look at some to understand the parts of a wave better.

PARTS OF A WAVE

In this background image, you can see waves around this sailboat. The diagram below shows the parts of a wave. The dotted line in the diagram shows where the level of the water would be if there were no waves—the equilibrium position. The shape of the wave occurs as the water is disrupted up and down as the wave moves energy from left to right.

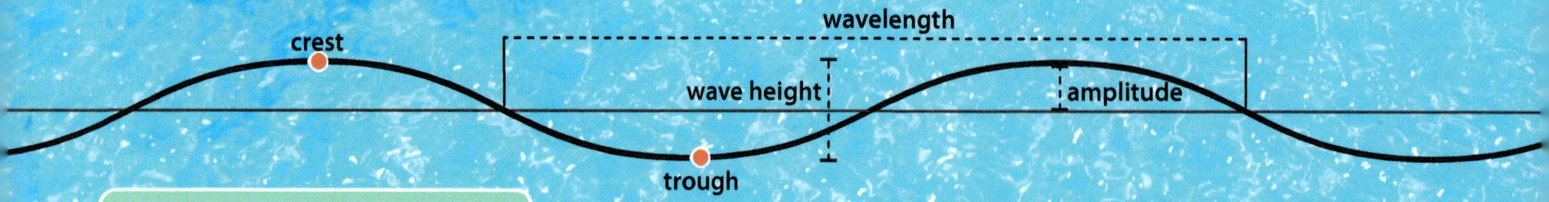

crest — *wavelength* — *wave height* — *amplitude* — *trough*

The highest point of each wave is called the **crest**.

Halfway between two crests is the lowest point of a wave, called the **trough**.

The length of a wave is the distance between two identical points on successive waves. We call this the **wavelength**.

Often we draw the wavelength between two crests or two troughs, but we can draw it between any two identical points. We measure wavelength in meters and use the Greek letter *lambda* (λ) to represent it.

Remember that waves are a type of periodic motion. Therefore the time interval for one complete cycle is still called the period (T). For a wave, this is the time for one complete wave to move past a particular position.

WORLDVIEW SLEUTHING: WAVE POWER GENERATION

All along the coast, you can see evidence of the work done by ocean waves on the shoreline. Ocean waves and tides relentlessly modify its shape. Through wave action that removes sediment and slowly erodes rock, the ocean eats into the landmass, causing the shoreline to retreat farther inland. There must be a lot of energy in that water to affect the land as much as it does.

TASK

You are a resident in Oceanside, and you live in a house that overlooks the ocean, of course. One morning an article in the newspaper grabs your attention. The article says that the state is interested in building a power generating plant in your town. The article goes on to say that your city is working to become the site of a wave-power generating plant. You are interested in alternative energy sources, but you're not sure that such a plant would be successful. You decide to research this form of energy generation and then produce a brochure to share what you find out.

PROCEDURE

1. Do the necessary research to inform your decision. You'll need to know several things: Is there a viable way to get energy from water waves? What impact could a wave-energy plant have on the environment? What conditions make a location a good site for wave-power generation? How does wave-power generation compare with other methods of power generation?
2. Plan your brochure and collect any required materials. Remember to cite your sources.
3. Show your brochure to a classmate or friend for feedback.
4. Submit your brochure.

CONCLUSION

The concept of using waves to generate energy is not a new one. The first patent for wave power generation was filed in 1799 in Paris. Since then numerous people have endeavored to make the technology workable, with the greatest interest occurring during the oil crisis of the 1970s. Interest in wave-power generation is on the rise again in response to concerns about climate change. The first commercial wave-power generating plant, Islay LIMPET, was opened in 2000. The first wave-power generator connected to a national power grid was the Pelamis Wave Energy Converter. Eight years later, the first wave farm opened at the Agucadoura Wave Park in Portugal. Today numerous companies continue to study waves as an energy source.

Scientists measure the vertical size of a wave two different ways. The vertical distance between the trough and the crest is called the *wave height*. Sailors use this term as they describe the conditions at sea. More often we talk about the amplitude of the wave. Similar to the spring system from Section 17A, the *amplitude* is the maximum displacement (vertical distance) from the undisturbed or equilibrium position. The amplitude of a wave is related to the amount of energy that the wave is transferring. Waves with larger amplitudes transfer more energy than waves with smaller amplitudes.

PERIODIC MOTION AND WAVES

17.6 CLASSIFYING WAVES

Scientists classify waves by how the disruption moves relative to how the energy moves in the wave. We classify waves as transverse, longitudinal, or surface.

TYPES OF WAVES

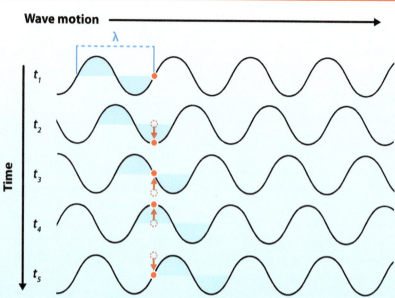

Transverse waves are waves in which the disruption moves perpendicular to the direction of wave travel. In the image at right, the disruption is up and down—notice the red particle—while the wave is moving from left to right. Examples of transverse waves include the waves on a guitar string, light waves, and secondary seismic waves (S waves).

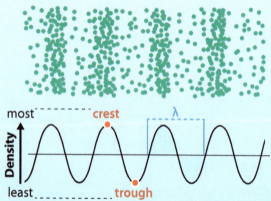

When the disruption occurs parallel to the direction of wave movement, the wave is called a **longitudinal wave**. Longitudinal waves, also known as *compression waves*, push the particles in the material closer together so that there are regions of high density (and pressure) called **compressions**. The restoring force then causes regions of lower density (and pressure) called **rarefactions**. The compressions are equivalent to the crest of a transverse wave, and the rarefactions correlate to the troughs. Examples of longitudinal waves are sound waves and primary seismic waves (P waves).

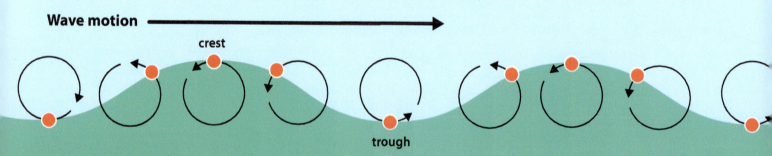

Surface waves occur along the surface between two mediums. The disruption in a surface wave typically moves both parallel and perpendicular to the direction of wave travel. Water waves are an example of surface waves. The water molecules are disturbed in a circular motion as depicted in the image above. This motion changes from circles to ellipses as the waves approach the shore. Other examples of surface waves include internal water waves between density layers and Rayleigh and Love seismic waves.

17.7 MEASURING WAVES

Maya Gabeira, the surfer in the Chapter opener, was very selective of the waves she chose. She knew that the wave with the biggest amplitude had the most energy and would provide her with the best ride. As you saw in the description of waves, there are some aspects of waves that we can measure. These include the wavelength, the amplitude, and the period. We can also measure the frequency of a wave and the speed at which it travels.

Frequency

If you were standing on a pier watching the waves, you could time how long it would take for each wave to pass by. As we noted before, this time is called the period (T) and is measured in seconds. You could also count how many waves would go by in a given amount of time. The number of waves, or cycles that occur each second is called the **frequency** (f) and is measured in Hertz (Hz)—cycles per second, or s^{-1}.

The frequency and period of waves, as well as other forms of periodic motion, are related to each other. Recall that period is the time for one wave, while frequency is the number of waves for one second. These two quantities are reciprocals of each other and are related according to the formula below.

$$f = \frac{1}{T} \text{ or } T = \frac{1}{f}$$

EXAMPLE 17-1: Finding Period from Frequency

Before an orchestra plays a concert, their instruments must be in tune with each other. All the instruments will tune to a note played by one instrument. The New York Philharmonic orchestra tunes to 442 Hz (musical note: A-4). How long does it take for one wave to pass the conductor?

Write what you know.

$f = 442$ Hz

$T = ?$

Write the formula and solve for the unknown.

$T = \frac{1}{f}$

Plug in known values and evaluate.

$T = \frac{1}{442 \text{ Hz}}$

$= \frac{1}{442 \frac{\text{cycles}}{\text{s}}}$

$= 2.26 \times 10^{-3}$ s

A wave will take just over 2 thousandths of a second to pass the conductor.

Speed of Waves

How fast do waves travel? That depends on the type of wave and the medium through which it is traveling. Sound waves in air travel at approximately 343 m/s, but that speed depends on conditions such as air temperature, density, and pressure. (Sound waves move faster in solids than in liquids, and faster in both than in air.) While the speed of sound seems fast, it does not compare with how fast electromagnetic waves move. Electromagnetic waves travel through space at 3.00×10^8 m/s, though different media can slow down electromagnetic waves. For instance, electromagnetic waves move slowest in diamond, through which they travel at *only* 1.24×10^8 m/s.

If wave speed depends only on the type of wave and the medium, how does wave speed relate to other measured quantities of waves? Recall from Chapter 11 that velocity is displacement divided by time.

$$v = \frac{\Delta x}{\Delta t}$$

Now think back to the description of waves earlier in this section. We learned that the time it takes for each wave to pass a particular point is called the period. The distance the wave will travel in that time is called the wavelength. Therefore we can write the velocity formula specifically for waves,

$$v = \frac{\lambda}{T},$$

where v is the speed in meters per second, λ is the wavelength in meters, and T is the period in seconds. We are using the term *speed* here because the direction of the wave is typically not significant; often waves radiate out from the source. Remember, period and frequency are reciprocals of each other.

$$f = \frac{1}{T}$$

So we can rewrite our formula as

$$v = \lambda f,$$

where v is the speed in meters per second, λ is the wavelength in meters, and f is the frequency in cycles per seconds or hertz.

EXAMPLE 17-2: Finding Wave Speed from Wavelength and Frequency

Waves that are 1.1 m long pass a pier. If 3.5 waves pass each second, how fast are the waves moving?

Write what you know.

$\lambda = 1.1$ m

$f = 3.5$ waves/s $= 3.5$ Hz

$v = ?$

Write the formula and solve for the unknown.

$$v = \lambda f$$

Plug in known values and evaluate.

$$v = (1.1 \text{ m})(3.5 \text{ s}^{-1})$$

$$= 3.9 \frac{\text{m}}{\text{s}}$$

The waves are moving at 3.9 m/s.

EXAMPLE 17-3: Finding Wavelength from Speed and Frequency

Your teacher is using a laser pointer in class. She tells you that the light is red (453 THz). Since you know that light travels at 3.00×10^8 m/s, how long is each wave?

Write what you know.

$f = 453 \text{ THz}\left(\dfrac{1 \times 10^{12} \text{ Hz}}{1 \text{ THz}}\right) = 4.53 \times 10^{14}$ Hz

$v = 3.00 \times 10^8$ m/s

$\lambda = ?$

Write the formula and solve for the unknown.

$v = \lambda f$

$\dfrac{v}{f} = \dfrac{\lambda \cancel{f}}{\cancel{f}}$

$\lambda = \dfrac{v}{f}$

Plug in known values and evaluate.

$\lambda = \dfrac{3.00 \times 10^8 \;\frac{m}{s}}{4.53 \times 10^{14} \;\frac{1}{s}}$

$= 6.62 \times 10^{-7} \text{ m}\left(\dfrac{1 \text{ nm}}{1 \times 10^{-9} \text{ m}}\right) = 662$ nm

The waves are 662 nm long.

EXAMPLE 17-4: Finding Period from Wave Speed and Wavelength

Tsunamis are huge ocean waves caused by earthquakes that happen underwater. If a tsunami moves at 890 kph and has a wavelength of 185 km, how much time will pass between the arrivals of two successive waves?

Write what you know.

$v = \left(\dfrac{890 \text{ km}}{\text{hr}}\right)\left(\dfrac{1000 \text{ m}}{1 \text{ km}}\right)\left(\dfrac{1 \text{ hr}}{3600 \text{ s}}\right) = 247 \;\frac{m}{s}$

$\lambda = 185 \text{ km}\left(\dfrac{1000 \text{ m}}{1 \text{ km}}\right) = 1.85 \times 10^5$ m

$T = ?$

Write the formula and solve for the unknown.

$v = \dfrac{\lambda}{T}$

$vT = \dfrac{\lambda \cancel{T}}{\cancel{T}}$

$\dfrac{\cancel{v}T}{\cancel{v}} = \dfrac{\lambda}{v}$

$T = \dfrac{\lambda}{v}$

Plug in known values and evaluate.

$T = \dfrac{1.85 \times 10^5 \text{ m}}{247 \;\frac{m}{s}}$

$= 749$ s

There would be 749 s (more than 12 min) between waves.

17B | REVIEW QUESTIONS

1. What is a wave?
2. Compare mechanical and electromagnetic waves.
3. Match the following terms with the labels from the image of the wave below: *amplitude, crest, trough, wave height,* and *wavelength*.

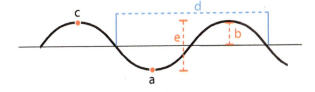

4. Relate medium motion with energy motion in a wave.
5. Explain the relationship between frequency and period.
6. What affects the speed of a wave?

Use the following information for Questions 7–8.

You are watching a series of waves pass by you. You notice that a wave passes by you every 0.806 s and that there are 34.6 m between each crest.

7. How many waves will pass by you each second?
8. How fast are these waves moving?

17C Questions

- What are wave behaviors?
- How do waves interfere with each other?
- How does the Doppler effect change a wave?

17C Terms

reflection, refraction, diffraction, interference, standing wave, Doppler effect

17C | WAVE BEHAVIOR

Why does the sound of an ambulance change as it speeds past?

17.8 HOW DO WAVES BEHAVE?

Have you ever been to a wave pool or tried to make huge waves in a swimming pool? These activities involve different wave behaviors. Waves can bounce off objects and even be bent by them. Waves can also interfere with each other. Let's look more closely at four wave behaviors.

BEHAVIOR OF WAVES

REFLECTION

You are familiar with the concept of reflection from bouncing a ball to a friend. The ball leaves the floor at the same angle that you threw it toward the floor. When a wave arrives at a surface, the new material absorbs some of the energy, while the rest bounces off the surface and continues to move in the original medium. This bouncing of waves off a surface is called **reflection**.

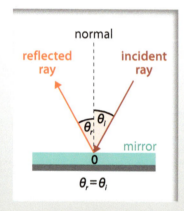

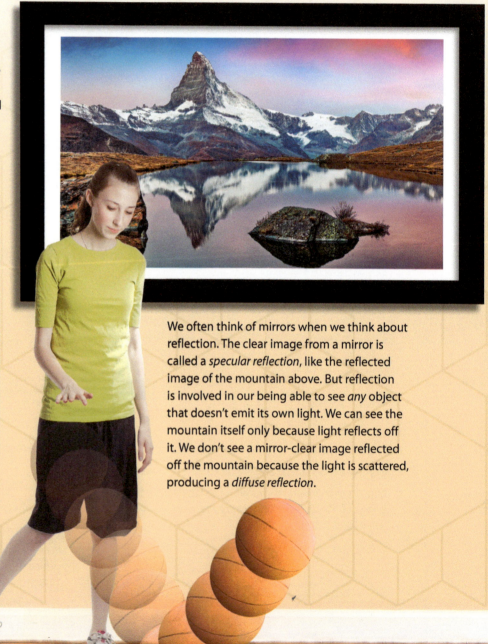

We often think of mirrors when we think about reflection. The clear image from a mirror is called a *specular reflection*, like the reflected image of the mountain above. But reflection is involved in our being able to see *any* object that doesn't emit its own light. We can see the mountain itself only because light reflects off it. We don't see a mirror-clear image reflected off the mountain because the light is scattered, producing a *diffuse reflection*.

Wave reflection is similar to the bouncing of a ball. According to the law of reflection, the *reflected angle* of the reflected wave is equal to the *incident angle*—the angle at which the wave arrived at the surface. The Greek letter *theta* (θ) in the image above represents a variable angle.

REFRACTION

In the image at right, notice how the straw appears to be broken. That is because light waves that are coming from the straw move from the water into the glass and then from the glass into the air. As the waves move from one medium to another, they are bent. The laser light (far right) bends as it enters the plastic block and it bends in the opposite direction as it moves back into the air.

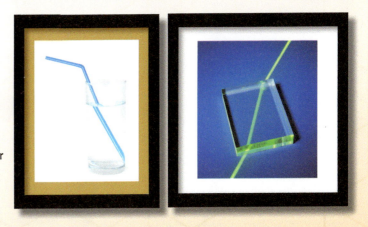

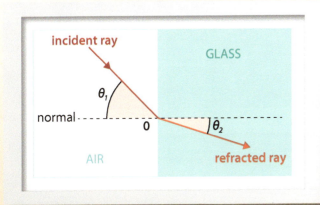

Recall that when a wave arrives at a surface, some of the energy goes into the new material. Often the wave changes direction due to the change in medium. This is called **refraction** and occurs because the wave moves at different speeds in different media. In the image at left, the wave is entering a material in which the wave moves slower. As the left edge of the wave reaches the surface first, the wave bends to the left. If the wave were moving into a material in which the wave moved faster, then the wave would bend toward the right. The law that governs the refraction of light is called *Snell's law*, a mathematical equation that describes how the wave changes direction as it moves into a new medium.

Our eye lens (left) refracts light to project a clear image on our retina. Our eye lens is amazing because it can change shape to focus on objects over a wide range of distances. When our eye lens is unable to focus a clear image, we can use eyeglasses or contact lenses to improve our vision.

Other waves refract too. If you have ever been to the beach, you may have noticed that the waves seem to come straight into the shore. The wind blowing across the surface of the water forms the ocean waves. Since the wind doesn't always blow toward the shore, the waves must change direction by refraction. As the waves approach the shore, the interaction of the waves with the ocean bottom turns the waves so that they are parallel to the shore.

BEHAVIOR OF WAVES

DIFFRACTION

Diffraction is the bending of waves around an obstacle or through an opening. Interaction between the wave and the obstacle or edges of the opening causes the wave to change direction. You can see the water waves at right being diffracted through the narrow openings of a log boom. Even light can be diffracted if the obstacles are small enough. Laser light (below) is diffracted by the thin lines etched in a *diffraction grating*.

A tombolo (above) is a landform that connects an island with the shore. Tombolos develop because of the refraction and diffraction of ocean waves. As the waves approach the shore, they turn (are refracted) to move toward the island from seaward. The water bends (diffracts) around the island. The waves that move behind the island are slower, allowing for more deposition of sediments, connecting the island to the mainland.

INTERFERENCE

Waves pass through each other without changing at all. But where parts of the waves overlap, they interfere with each other. **Interference** is the combining of waves where they overlap.

Mathematically, the combining of the waves is straightforward because the waves simply add together. The red and blue wave pulses below are both deflected upward. As they overlap, their amplitudes add to make a larger wave. Once they pass each other, the waves continue unaffected by the encounter. The adding of waves to form a larger amplitude wave is called *constructive interference*.

The wave pulses below have displacements that are in opposite directions. As they overlap, their amplitudes still add together, but since one of the amplitudes is negative, the resultant wave is smaller. The adding of waves to form a smaller amplitude wave is called *destructive interference*.

Interference can make waves more or less intense. We use the property of wave interference in a variety of applications. Noise-canceling headphones use the concept of destructive interference. A microphone in the headphones takes in the ambient sound. The headphones then produce an inverted matching wave to cancel the ambient sound. Interference causes the spectrum of colors, the rainbow, observed in a soap bubble or in oil floating on a puddle. As the thin film changes thickness, different wavelengths (colors) of light constructively interfere.

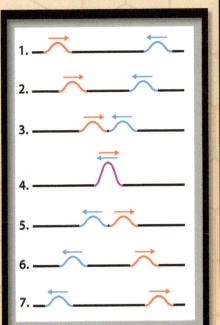

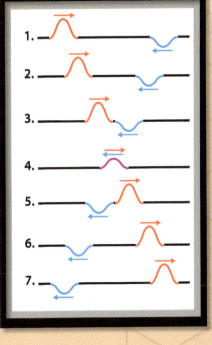

17.9 STANDING WAVE

An interesting application of constructive interference is a **standing wave**, a wave that is moving even though the locations of the crests and troughs appear to be stationary. Two identical waves *of a specific wavelength* moving through a medium in opposite directions can cause a standing wave. The length of the medium must be a multiple of one-half the wavelength.

The length of this string is 3/2 λ.

Look at the standing wave depicted at left. Notice the three crests and three troughs. You can also see four locations—called *nodes*—that don't seem to move at all. The location of the crests and troughs are the points of maximum motion, and these are called *antinodes*. To help you remember these two terms, remember that there is NO movement at the NOde.

CASE STUDY: GALLOPING GERTIE

The first Tacoma Narrows Bridge spanned the Tacoma Narrows, a strait of Puget Sound in Washington State. During construction, the bridge oscillated vertically whenever any significant wind blew, resulting in the construction workers nicknaming the bridge Galloping Gertie. Despite the unsettling motion of the bridge, construction continued and the bridge opened on July 1, 1940.

After its opening, visitors enjoyed traveling across the bridge as it made gentle (0.3 to 0.7 meter) oscillations. On November 7, 1940, windy conditions in the strait began to produce stronger oscillations. The winds were blowing at 64 kph, causing the bridge to move between 0.7 m and 1.0 m. Throughout the morning the oscillations grew as the wind continued blowing across the span of the bridge.

By mid-morning, the oscillations had grown to 3.7 m. It was a spectacle to see. From the side, you could see one-half of the bridge moving one way while the other half moved in the opposite direction. So when the western end deflected upward to a crest, the eastern end deflected downward to a trough. The middle of the bridge didn't appear to be moving, and the crests and troughs remained respectively one-fourth and three-fourths of the way across the span. At 11:00 a.m. on November 7, 1940, just four months after opening, the bridge collapsed.

The investigation that followed concluded that the design of the bridge allowed for aerodynamic forces that amplified the motion of the bridge until the suspension cables failed.

1. What kind of wave (transverse, longitudinal, or surface) was produced in the bridge as seen from the side?

2. What phenomenon of a wave is produced when the crests and the troughs don't seem to move horizontally?

3. Where on the bridge was constructive interference occurring?

17.10 DOPPLER EFFECT

Have you ever listened as an ambulance raced by you? The sound of the siren seemed to have a higher pitch as it approached but a lower tone as it moved away from you. You may have noticed the same thing at a car race or an airshow. Why does the tone—the frequency—seem to change as an ambulance goes by?

The frequency doesn't actually change, but we perceive it as changing. This apparent change in frequency of a wave due to the motion of the source or the receiver is called the **Doppler effect**. Austrian physicist Christian Doppler described this phenomenon in 1842. The relationship between wavelength and frequency causes this effect. We know that the speed of a wave depends only on the medium and that the wave speed formula shows an inverse relationship between frequency and wavelength. If we perceive the wavelength as decreased, then we perceive sound as a higher frequency, which we hear as a higher pitch. Conversely, if the wavelength increases, then the frequency decreases, and we perceive it as a lower tone.

In the image at right, an ambulance races past three observers. Each observer perceives a different pitch, or frequency, of the ambulance's siren *at the instant depicted in the image*. Observer A perceives the siren as a lower pitch than the real siren. The frequency is perceived as lower because as the ambulance moves away from him, each wave pulse begins from a position that is farther from him. Observer A observes these more widely spaced waves as having a longer wavelength and therefore a lower frequency. Observer B hears the actual frequency of the siren because there is no relative motion between him and the ambulance. Observer C observes a higher pitch than the actual siren. He hears this because as the ambulance approaches him, each successive wave starts from a closer position to him. Each of these waves is perceived as having a shorter wavelength and accordingly a higher frequency.

The Doppler effect occurs with all waves and has numerous applications. Astronomers use the Doppler effect to determine the rate at which objects in the universe are moving relative to Earth. In meteorology, radar technology has advanced tremendously by applying the Doppler effect. Weather radar can now determine how storms are moving relative to the radar site. This information provides warnings of possible tornadoes. Police use the Doppler effect to measure the speed of vehicles on the roadways to help keep people safe.

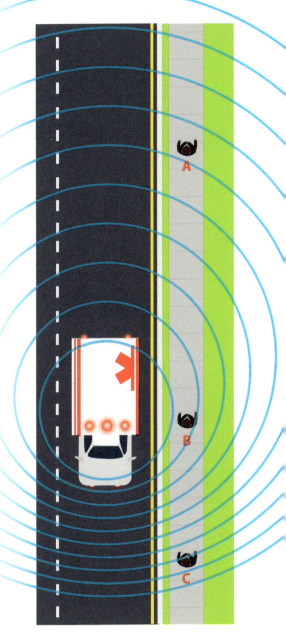

17C | REVIEW QUESTIONS

1. What happens to the energy in a wave when it reaches the surface of new material?
2. What is reflection?
3. Why do waves change directions when they enter a new material?
4. What is diffraction?
5. Compare constructive and destructive interference.
6. Define *Doppler effect*.

CHAPTER 17

17A PERIODIC MOTION

- Periodic motion occurs in cycles that repeat at regular time intervals called periods.
- The restoring force causes the oscillating motion of a spring system. The system transforms energy between kinetic and potential energies.
- The maximum displacement of an object undergoing periodic motion is called its amplitude.
- Simple harmonic motion is periodic motion in which the restoring force is proportional to the displacement.
- Periodic motion can be damped to decrease the amplitude or can be driven to increase it.
- Resonance occurs when the driving force causes an object to oscillate at rates of vibration specific to that object.

17A Terms

periodic motion	396
period	396
amplitude	399
simple harmonic motion	400
damping	400
resonance	401

17B WAVES

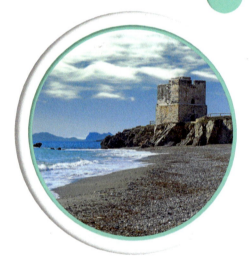

- A wave is a disruption that carries energy from one location to another.
- Mechanical waves require a medium through which to travel.
- Waves are described on the basis of their amplitude, wave height, wavelength, period, and frequency.
- We classify waves by the direction of the disruption compared with the direction of wave motion. Waves can be transverse, longitudinal, or surface.
- Frequency and period are reciprocals of each other.
- Wave speed is dependent on the medium, but it relates frequency and wavelength according to the formula $v = f\lambda$.

17B Terms

wave	403
mechanical wave	403
crest	404
trough	404
wavelength	404
transverse wave	406
longitudinal wave	406
compression	406
rarefaction	406
frequency	407

17C WAVE BEHAVIOR

- All waves can reflect (bounce off a surface), refract (bend due to a change in medium), or diffract (bend around an obstacle or through an opening).
- Waves may pass through each other unaffected. But they interfere with each other when parts of the waves overlap. The amplitudes of overlapping waves add together.
- At times when waves interfere, they can create a standing wave in which the crests and troughs don't appear to move. The length of the medium must be a multiple of ½λ for a standing wave to form.
- When a wave source or its detector moves, the detector perceives a change of frequency in a phenomenon called the Doppler effect.

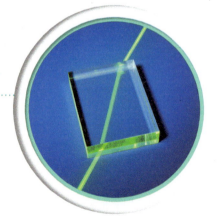

17C Terms

reflection	410
refraction	411
diffraction	412
interference	413
standing wave	414
Doppler effect	415

REVIEW

CHAPTER REVIEW QUESTIONS

Recalling Facts

1. Give three examples of periodic motion.
2. Define *period*.
3. As a spring system is oscillating, why doesn't the mass come to a stop when it reaches its rest position?
4. What do we call the force that moves an oscillating system back toward its equilibrium position?
5. What is a mechanical wave?
6. Classify seismic waves.
7. Describe a wave using the terms *crest, trough, amplitude, wave height, wavelength, period,* and *frequency*.
8. According to the law of reflection, if a wave approached a surface at 35°, at what angle would it reflect?
9. What is refraction?
10. How do waves interfere with each other?
11. What is a standing wave?

Understanding Concepts

12. Identify the period for each of the hands (second, minute, and hour) on a clock.
13. Using the spring system on pages 398–99 as a guide, sketch and explain the motion of a child on a swing. Include in your answer the following terms: *kinetic energy, potential energy, restoring force, velocity, amplitude, cycle,* and *period*.
14. Compare periodic and simple harmonic motion.
15. Your friend tells you that matter doesn't move as a wave passes. Is your friend correct? Explain.
16. Why can't a mechanical wave travel through space?
17. Compare wave height and amplitude.
18. Compare transverse and longitudinal waves.
19. What is the source for the wave formula $v = \frac{\lambda}{T}$?

Use the following information for Questions 20–21.

Radio waves travel at 3.00×10^8 m/s. Your favorite radio station has a frequency of 89.3 MHz.

20. How long are the radio waves from this station?
21. How much time passes between each wave arriving at your radio?
22. A classmate tells you that only mirrors reflect light. Is he correct? Explain.
23. Compare reflection, refraction, and diffraction.
24. Draw a standing wave with five nodes and four antinodes. Label the nodes and antinodes.
25. Explain how the Doppler effect occurs as a sound source approaches a receiver.

Critical Thinking

26. Using the illustration of a swing, explain the concepts of a damped oscillator and a driven oscillator.
27. You might have heard someone say that a statement *resonates* with them. How does that usage compare to the use of the word in physics?
28. Different types of earthquake waves travel through the earth at different speeds. Scientists use data from at least three seismic stations to determine the epicenter of an earthquake. In a hypothetical earthquake, P waves and S waves are the first waves to be detected. P waves arriving at one seismic station are 1191 m long and 4.16 waves arrive every second. S waves arriving at the same station are 715 m long and 4.16 waves arrive every second. If the station is 354 km from the epicenter, which waves arrive first and how much earlier?
29. From what you have learned in this chapter, why do you think it is a challenge to do spearfishing from a lakeshore?
30. Your classmate tells you that the ambulance sounds louder as it gets closer to you and quieter as it moves away because of the Doppler effect. Is he correct? Explain.

CHAPTER 18
Sound

BOOMING BUSINESS

You may have heard the window-rattling sonic boom of an aircraft breaking the sound barrier (left). But what exactly is this barrier? As the Second World War pushed aircraft performance limits ever higher, pilots noticed control problems as their planes approached the speed of sound, about 343 m/s. Some aeronautical engineers thought that these difficulties made supersonic flight impossible. But the engineers and aircraft designers kept at the problem. On October 14, 1947, test pilot Chuck Yeager (inset), at the controls of a Bell X-1, became the first man to fly faster than the speed of sound in level flight, breaking the sound barrier. The age of the high-performance jet had dawned. Amazingly, only six years passed between Yeager's historic flight and the first flight to reach Mach 2—twice the speed of sound. In this chapter, we'll examine the speed of sound and other sound characteristics.

18A Sound Waves	420
18B Hearing and Music	424
18C Using Sound Waves	430

18A Questions

- What is a sound wave?
- Does sound always travel at the same speed?
- How are the different characteristics of sound related?
- What makes one kind of sound distinct from others?

18A Terms

sound energy, acoustic spectrum, pitch, loudness, intensity, timbre

18A | SOUND WAVES

How fast is the speed of sound?

18.1 ACOUSTIC ENERGY

Structure of Sound Waves

The rumble of a passing jet is a form of energy—sound energy. **Sound energy** is a type of energy that can be transmitted as waves *through a medium*, either a gas, liquid, or solid. Some, but not all, sound energy can be detected by our ears. Sound waves are *longitudinal waves*. The atoms and molecules of the medium vibrate back and forth, parallel to the direction of the wave's motion. Sound waves are transmitted as an alternating series of compressions and rarefactions. Each compression is a region of high particle density. As a compression passes, the increased pressure forces them apart. The rarified region of a sound wave is its point of lowest particle density. As the rarefaction passes, the decreased pressure draws molecules in and evens out their distribution, and then the next compression pushes them together again. Alternating compressions and rarefactions distinguish the longitudinal waves of acoustic energy from other kinds of wave energy.

The Sound Medium

How is a sound wave created? Sound is a type of mechanical wave (see Chapter 17). To make a sound, something must vibrate in a physical medium, the substance through which the wave travels. Using a tuning fork is one of the simplest ways to make sound. When a tuning fork is tapped, the tines, or prongs, of the fork vibrate, making a sound with a single frequency or pitch. If you were to look closely, you might be able to see the tines vibrate. These vibrations push on the air, first in one direction, then in the other, sending out waves of compressions and rarefactions. Dipping the tines in water shows the periodic motion of the tines more clearly. The fork causes ripples in the water similar to those it causes in air.

We know that sound needs a physical medium to travel through because of an experiment on sound conducted by Robert Boyle in 1660. Boyle had suspected that air was necessary to transmit sound. To confirm his hypothesis, he suspended a watch with an alarm from a thread inside a sealed container. He then pumped the air out of the vessel to form a vacuum. He and his assistants waited for the sounding of the watch's alarm. Time passed, and they did not hear it. Then Boyle opened a valve and allowed air to reenter the chamber encasing the watch. As the air rushed in, the alarm grew louder and louder. Boyle had successfully demonstrated that sound needs a physical medium.

18.2 SPEED OF SOUND

Further experimentation showed that sound moves quite well through liquids and solids. In fact, those states of matter generally are better acoustic media than air. But how fast does sound travel in different media? From 1708 to 1709, William Derham first accurately measured the speed of sound through air. He provided assistants with synchronized watches and pistols and stationed them at various distances, some several kilometers away. When they fired their guns at specified times, Derham recorded the time the sound was heard. Knowing their distances and the length of time between the scheduled firing and the sound of its boom, Derham calculated the speed of sound. He averaged the results of several trials and came up with a value that was surprisingly accurate.

Scientists later tested Derham's value for the speed of sound under differing conditions. Their observations eventually indicated that the speed of sound is dependent on the density of air. But factors that don't affect the density of air, such as direction or time of day or year, had no impact. Today, the accepted speed of sound in dry air at 20 °C is 343 m/s. This value typically increases with temperature.

Table 1	
Speed of Sound	
Medium	Speed (m/s)
Oxygen (25 °C)	316
Air (0 °C)	331
Air (25 °C)	346
Air (35 °C)	352
Distilled water (25 °C)	1498
Salt water (25 °C)	1531
Glass (25 °C)	3980
Aluminum (25 °C)	6420

The speed at which sound travels through various media depends on the "stiffness" and temperature of the medium. In general, the stiffer the medium, the faster sound travels. Thus sound generally travels fastest in solids, slower in liquids, and slowest in most gases. But for gases, as the density of the gas increases, the speed of sound *decreases*. Thus, the speed of sound in air is higher on a hot summer's day than on a cold winter's day. The table at left shows the speed of sound in various media.

18.3 CHARACTERISTICS OF SOUND WAVES

Chapter 17 discussed some basic properties of all waves—wavelength, frequency, amplitude, and interference. God created us with sensory organs that can receive different types of wave-based information from our environment and transmit this information to our brains. For each type of wave, the basic properties listed above are interpreted differently. Our eyes sense electromagnetic waves, and our brains interpret the frequency, amplitude, and interference of these waves as color, brightness, and pattern. Our ears sense sound waves, and our brains interpret their properties as distinct sounds. Our ears can detect several measurable properties of sound waves, including frequency, loudness, intensity, and quality. All of these factors affect every sound we hear.

Pitch and Frequency

All of the possible sound waves form a continuous **acoustic spectrum** of energy. This spectrum consists of waves with differing frequencies, or vibrations per second. These vibrations correspond to the wavelength of the sound. The shorter the wavelength, the higher the sound's frequency. These frequencies include those that humans can hear, along with frequencies that are too low or too high to hear. **Pitch** is how high or low an audible tone sounds to the human ear. It describes how we perceive a sound's frequency. The higher a sound's frequency, the higher its pitch. Recall from Chapter 17 that frequency is measured in hertz (Hz). The bottom note on an eighty-eight-key piano has a frequency of 27.5 Hz, the highest note, about 4186 Hz. Young children can hear pitches ranging from about 20 Hz to 20 000 Hz. As a person grows older, he typically loses the ability to hear higher frequencies.

Do you think that different pitches travel at different speeds through a uniform medium? What would it sound like if the lower notes from the choir traveled faster than the higher notes? You would hear them at different times, and the choir wouldn't sound unified. Since this doesn't happen, we conclude that sounds of all pitches travel at the same speed. Experiments have confirmed this conclusion.

Loudness and Intensity

Loudness is our perception of the intensity of a sound wave. **Intensity** is a measure of the power contained in a sound wave and is indicated by the wave's amplitude. Intensity is the rate at which the sound wave transmits energy. The greater the amplitude of a sound wave, the more power it contains. Loudness involves interpretation by the human brain and is subjective. A sound of a certain intensity that seems painfully loud to one person may be just uncomfortably loud to another. Nevertheless, loudness and intensity are closely related since sounds of high intensity naturally tend to sound louder.

A sound's loudness is compared to the softest sound that an average young person can hear—the threshold of hearing. Loudness is measured in units called *decibels* (dB), which were named for Alexander Graham Bell. Like the seismic scale used for earthquakes, the increase in the loudness of a sound is not linear, but exponential. The threshold of hearing is assigned a value of 0 dB. Each 10 dB increase in loudness indicates a tenfold increase in our perception of the sound's intensity. A whisper is about 20 dB. Normal conversation is about 60 dB, or about 10,000 times the perceived intensity of a whisper. At 120 dB—the threshold of pain—sound becomes painfully loud and can permanently damage hearing. A 120 dB noise has about a trillion times as much power as the quietest sound a human can hear!

Timbre

Tuning forks produce sound with mainly one frequency. When a musical instrument plays the same note, the sound is different from that of the tuning fork because a musical instrument plays mixtures of acoustical tones. When a trumpeter, for example, plays a middle C, the instrument produces a mixture of frequencies that are multiples of middle C's frequency. This mixture gives the trumpet its unique sound. The particular sound of an instrument is its quality, or **timbre**—the particular combination of tones that makes any sound distinctive. We make judgments on whether a sound is pleasant on the basis of its quality.

	fireworks	140 dB
	jet engine	130 dB
threshold of pain	police siren	120 dB
extremely loud	trombone	110 dB
	helicopter	100 dB
	hairdryer	90 dB
very loud	truck	80 dB
loud	city traffic	70 dB
moderate to quiet	conversation	60 dB
	moderate rainfall	50 dB
faint	refrigerator	40 dB
		30 dB
		20 dB
		10 dB
		0 dB

18A | REVIEW QUESTIONS

1. Are sound waves transverse or longitudinal waves?
2. Which part of a sound wave has the lowest density of particles?
3. Why are sound waves classified as mechanical waves?
4. Rank gases, liquids, and solids from fastest to slowest by the rate at which sound travels in them.
5. Does sound travel through the atmosphere faster on a cold winter morning or a hot summer afternoon? Explain.
6. How long does it take for a soundwave to travel 5.0 km in salt water at 25 °C? Refer to Table 1 on the facing page.
7. How are pitch and frequency related?
8. What is the difference between loudness and intensity?
9. What gives a sound its particular timbre?

18B Questions

- How do humans produce sound?
- How do humans hear sound?
- How do musical instruments produce sound?

18B Terms

fundamental tone, overtone, harmonic

18B | HEARING AND MUSIC

Why does each kind of musical instrument produce a distinct sound?

18.4 THE HUMAN VOICE

Many creatures can make sounds through calls or voices, but God supplied only humans with unique physical features that permit us to do more than just make a few distinctive sounds. We can communicate complex thoughts using words, whether through speech or singing. Unlike the rest of God's creation, we are able to vocally praise Him.

No two people are exactly alike. Slight differences in the physical structures from one person to another make each person's voice sound slightly different. When singing, these differences allow singers to produce tones with different pitches, loudness, and timbre. The human voice also has amazing flexibility. The useful range of the human singing voice typically extends from just under 100 Hz (a low bass) to a little over 1000 Hz (a high soprano), or about four octaves. The exact range varies from person to person, depending on gender and the structure of the larynx. This variance is why choral music is subdivided into parts—soprano, alto, tenor, bass, and so on. How is the wondrous human voice produced? Let's find out!

PRODUCING SOUND

1 LARYNX

Vocal sounds originate in your *larynx*, a box-like structure located at the top of your *trachea*. It consists of nearly a dozen pieces of cartilage held together by ligaments to form a flexible container that supports your vocal cords.

2 VOCAL CORDS

Your *vocal cords* stretch across the upper portion of the larynx. Between them is an opening, the *glottis*. Muscles that surround the larynx control the tension in the vocal cords. During normal breathing, the vocal cords are relaxed and the glottis is open. During speaking or singing, the vocal cords tighten and partially close the glottis, allowing the vocal cords to vibrate with the passage of air. The tension in the cords controls the pitch of the sound that they produce.

3 THROAT AND SINUS PASSAGES

The back of your throat opens upward into the nasal and sinus passages. Sound produced by the vocal cords resonates within the *sinuses*. Together, the throat and sinuses form a resonating chamber for the voice.

4 TONGUE, TEETH, AND LIPS

The interior of your mouth forms another resonance chamber whose shape is completely flexible. Your tongue, teeth, and lips work together with facial muscles to form individual sounds.

5 DIAPHRAGM

The *diaphragm*, a large muscle that forms the floor of the chest cavity, controls the volume of air that passes over your vocal cords. This determines the power of your voice.

18.5 THE HUMAN EAR

Of course, being able to hear and appreciate speech or music is as important as producing it. God made us with an extraordinarily complex sensory organ that allows us to do so. Its incredible features give ample evidence of having been designed. The ear also doubles as a structure for providing us with a sense of balance. Let's examine how the ear receives and transforms sound energy.

THE EAR

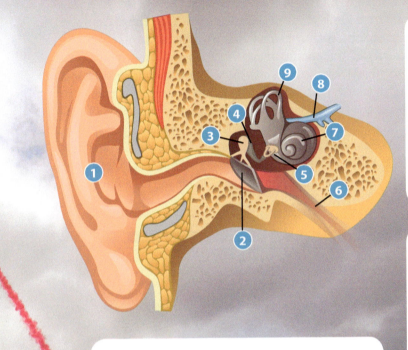

MIDDLE EAR

The middle ear consists of the eardrum and three bones called the *hammer* ③, the *anvil* ④, and the *stirrup* ⑤, each named for its shape. Together, these bones amplify (i.e., increase the force of) vibrations entering the inner ear.

The *eustachian tube* ⑥ connects the middle-ear cavity with the throat. It helps to equalize the air pressure on both sides of the eardrum.

INNER EAR

At the inner ear, sound vibrations are transferred to the liquid of the *cochlea* ⑦. The vibrations travel through the liquid to thousands of sensory hair cells and nerve endings. In the cochlea, the vibrations are converted into electrical impulses. Differences in structure allow only certain sensory cells to be moved by certain frequencies of vibration.

The sensory cells relay thousands of separate signals to the *auditory nerve* ⑧. These signals indicate the pitch, loudness, and quality of a sound. The auditory nerve conducts these signals to the brain, which interprets them as sound.

The pressure exerted by fluids upon sensory cells inside the *semicircular canals* ⑨ gives the brain information about the body's orientation. This information is critical for coordinating movement and maintaining balance.

OUTER EAR

Sound enters through the external ear. The outer ear helps us determine the direction of a sound's source by blocking sounds coming from behind.

The external *auditory canal* ① serves as a resonating chamber, amplifying frequencies from the middle range of the audible spectrum.

The *tympanic membrane* ②, or eardrum, separates the outer ear from the middle ear. This thin, flexible membrane converts acoustic energy to kinetic energy by vibrating in response to extremely tiny sound vibrations in the auditory canal.

18.6 MUSICAL INSTRUMENTS

Mankind's God-given drive to create music has led him to make thousands of different kinds of musical instruments. They range from well-known instruments, such as pianos, to the wildly exotic, like West African balafons (left). No matter their origin, all musical instruments are designed to vibrate, creating sound waves. The vibrations may be made by striking, plucking, bowing, or blowing into some part of the instrument. The unique timbre of each instrument is a combination of the tones produced by its design, the materials from which it is made, and the manner in which it is played. Today, instruments such as synthesizers and drum machines use electronics to create musical tones. Some of these tones sound very much like the tones made by more traditional instruments.

TONES

To show how an instrument can produce sound, let's examine what happens when a guitar string is plucked.

When a guitar string is plucked, it vibrates as one standing wave from end to end, producing a tone, called the **fundamental tone** (A), or simply, the fundamental. The string produces this same tone each time it is plucked. Players use their fingers to press down on the strings, changing the length of the vibrating portion. This changes the note produced by the fundamental.

If the string is tight enough, it also vibrates in other standing-wave patterns at the same time. This produces more than one crest along the string. These shorter, faster vibrations produce higher pitches called **overtones** (B and C).

The fundamental and its overtones that are whole-number multiples of the fundamental frequency are called **harmonics**. The fundamental is called the first harmonic. For example, if the fundamental frequency (first harmonic) is 440 Hz, then the second harmonic of that tone will be 880 Hz.

18B | REVIEW QUESTIONS

1. What is the normal range of frequencies for human voices?
2. Briefly describe how the human voice is produced.
3. Which structure in the ear converts sound energy to kinetic energy?
4. Which portion of the ear (outer, middle, or inner) plays a role in both hearing and balance?
5. How is the information in a sound wave transmitted to the brain?
6. What factors affect the timbre of a musical instrument?
7. (True or False) The fundamental tone of a plucked guitar string is produced by a combination of all the standing wave patterns in the plucked string.

When you pluck a guitar string, it typically produces several overtones at the same time. The fundamental tone is the loudest and determines the note that people hear. The quieter overtones contribute to the guitar's timbre.

The plucked string itself makes little noise. Instead, its vibrations are transferred to the guitar's wooden body, especially the top plate. The vibrating top plate can move the volume of air needed to make the guitar's sound audible.

The guitar's sound hole helps the top plate vibrate more freely. It also allows vibrations from the air inside the guitar to pass out of the instrument. All of these vibrating parts work together to produce the guitar's unique timbre.

18C Questions

- How can sound energy be used?
- Are there different kinds of sound energy?
- What is an ultrasound?

18C Terms

acoustic amplification, echolocation, sonar, infrasonic wave, ultrasonic wave, sonography

18C | USING SOUND WAVES

In what other ways do we use sound waves?

18.7 SOUND TECHNOLOGIES

Acoustic Amplification

Since sound is a form of energy, it makes sense that such energy can be put to good use. And indeed, not only people but animals too use sound energy for many purposes. In this section, we'll explore some of those uses.

Acoustic amplification is the process of making a sound louder. A megaphone is a simple kind of amplifier known as an *acoustic horn*. Normally when a person speaks, the sound waves of his voice spread out over a large area. A megaphone focuses that sound energy into one specific direction, making it sound louder. If you point the large end of a megaphone at the source of a sound and listen at the smaller end, the same thing happens. This is why people sometimes cup their ears to hear better. Amplification also happens in any enclosed space with hard surfaces, where sound reflection occurs easily.

○ HOW IT WORKS

Speakers

Stereo speakers are described with some strange terms. Woofers, midrange, tweeters—just what do these things mean? And how does a speaker work anyway? A speaker contains a cone (1) made of stiff material connected to a voice coil (2) made of wound copper wire. The voice coil is suspended inside a powerful circular magnet (3). An amplified electrical signal enters the voice coil, creating, in essence, an electromagnet (see Chapter 21). The electrical signal causes the voice coil's magnetic field to rapidly change polarity many times per second. As this rapidly changing field interacts with the field of the magnet, the voice coil vibrates back and forth. The vibrating speaker cone pushes against the air, creating sound waves.

The size of the cone relates to the frequency of the tones produced. Large cones are needed to produce deep, low-frequency sounds—the bass end of what humans can hear. These speakers are the "woofers" in a multi-cone speaker. "Tweeters" are the small cones that produce high-pitched, high-frequency sounds. Large speakers may have three or more cones, including mid-range cones. Such speakers are better able to re-create the variety of tones carried by an electrical signal. Single-cone speakers are limited in the range of frequencies that they can produce. They are often unable to produce some low- and high-end frequencies well.

Electronic amplification uses electronic technology to make sound louder. A sound receiver, such as a microphone, converts sounds into a relatively weak electronic signal. This signal is sent to an amplifier, which magnifies the electrical signal. The output signal is identical to the original, but its amplitude is much larger. The amplified signal is then sent to a speaker. There, the electronic signal is converted into the kinetic energy of the speaker's cone (see box on the facing page). In the performing arts, sound amplification has become essential as the size of performance halls has increased.

Sonar

Sound travels fast through most media, and its waves easily reflect off many materials. Because sound behaves this way, it can be used to measure the distance to an object. The time required for an outgoing pulse of sound to return as an echo can be used to determine the distance to the sound-reflecting surface. The direction from which the strongest echo returns gives the direction to the object. Using sound this way is called **echolocation**. Many animals, such as bats and dolphins, are able to use echolocation to move around and hunt when visibility is poor.

Bats emit a stream of sound in the form of high-pitched squeaks. The odd structures around the noses of some bats amplify the sound signal. Bat echolocation squeaks range from 20 kHz to 200 kHz, above the range of human hearing. Bat ears are designed to focus and amplify the returning signal.

The beam of sound reflects off objects, including prey.

The return echo gives the bat information about the size and type of prey along with its direction and distance from the bat.

Toothed whales such as dolphins focus their acoustic signals with a special dome-shaped structure in their heads called a *melon* (see below). Dolphins can even use powerful sound pulses to stun and disorient their prey before capture. The use of sound by these animals shows the marvels of God's design, equipping each animal with the tools it needs to succeed in its unique environment.

Scientists became interested in sound in water as early as 1822, when Jean-Daniel Colladon used a submerged bell to calculate the underwater speed of sound. This experiment spurred further interest in how sound travels in water. After the RMS *Titanic* struck an iceberg and sank in 1912, researchers considered ways to avoid future collisions with ice. They suggested using devices in the water to listen for the sounds that icebergs make. But it was the wartime need to detect submarines that spurred the development of sonar. The word **sonar** is an acronym for *so*und *na*vigation and *r*anging.

The earliest sonars could only receive, not transmit, underwater sounds. These *passive sonars* used underwater microphones called *hydrophones*. *Active sonars*, developed during and after World War I, produce sound as well as listen for it. Short pulses of sound—the "pings" often heard in submarine movies—echo back to the sending ship or submarine. These returning echoes are used to find the bearing and range to a target. Active sonar is a kind of echolocation.

Modern sonar also has nonmilitary uses, such as mapping the bottoms of oceans and lakes. The pulses of sound emitted by a ship downward into the water reflect if they hit a boundary between two very different media. Thus, a sandy ocean bottom will strongly reflect the sound upward toward the ship. A sonar screen displays an image of the surface that reflects the pulse. One form of sonar known as *side-scan sonar* is particularly good at creating images of submerged objects. Side-scan sonar has been an important tool for finding shipwrecks (left). Sonar is also used to keep floating deep-sea drilling rigs in position while drilling. By placing active sonar beacons on the ocean floor and hydrophones underneath the rig, an onboard computer controls thrusters that keep the massive vessel directly over the drill hole.

18.8 INFRASONIC AND ULTRASONIC WAVES

Infrasonic Waves

Infrasonic waves are sound waves of low to very low frequency. They are too low for humans to hear, but they can often be felt. For example, distant heavy machinery can produce infrasonic vibrations that make people feel nervous or restless. Recent studies reveal that infrasound is important to some forms of animal communication. Elephants, for instance, can reportedly communicate with each other up to 10–15 km away at frequencies of 12–35 Hz. Rather than using their ears, they seem to receive these waves through ground vibrations via their feet. Giraffes have been observed to vocalize at frequencies of 14–40 Hz. Scientists believe that they use these sounds to inform others of their distress or location. Infrasound can easily pass through jungle foliage and other obstructions. Cats use infrasonic sound not only to communicate but also to help other cats recover from injury. Research indicates that an increase of bone density, tissue repair, and pain relief occurs more rapidly in response to infrasonic purring near 20–50 Hz. So next time you're sick, don't shove your purring cat off the bed!

Ultrasonic Waves

Ultrasonic waves (also known as *ultrasound*) have frequencies above the range of human hearing—typically greater than 20 000 Hz. Ultrasonic waves are ideal for short-range sensors because they reflect well and do not strongly diffract around objects.

Ultrasound has many practical uses in a wide variety of fields. Perhaps the use of ultrasound with which you are most familiar is sonography. **Sonography** uses ultrasound to create images of objects found inside other objects. It can be used for doing short-range searches, such as in fish finders. It can "see" into buildings and other structures to search for weaknesses such as poor welds. But you probably associate it most often with medical imaging. Sonography allows medical personnel to examine the inside of a patient without doing surgery. A sonogram, the image produced on a monitor using sonography, can be used to track the progress of a baby in its mother's womb or to diagnose tumors, heart conditions, and other disease.

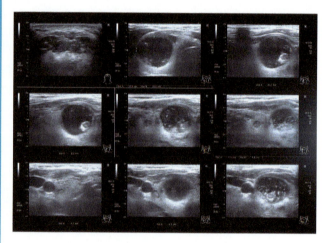

Ultrasound images of a thyroid

Ultrasound is not only useful for diagnosing disease—it can be used to treat it too. It is particularly useful because ultrasound treatment can be done without resorting to risky and costly surgeries. Ultrasound has been used to break up painful kidney stones. It is also commonly used during physical therapy. It's fascinating to see how many ways humans have found to use sound to help and serve others in need.

SERVING AS AN ACOUSTIC ENGINEER:
SOUNDS GREAT!

If you've ever enjoyed a good concert, then you likely owe thanks to Wallace Sabine. Before the twentieth century, whether a concert hall had good acoustics or not was largely a matter of chance. Some halls had great sound, while others were dreadful. Poorly designed concert halls produce a lot of reverberation—unwelcome lasting echoes that interfere with sound waves coming from a performance.

In 1895, Sabine was tasked with fixing the poor acoustics in a lecture hall at Harvard University's Fogg Art Museum. Over the next few years, Sabine and his assistants timed thousands of echoes of different frequencies. During these tests, he placed different kinds and amounts of sound-absorbing materials in different parts of the hall. All of this testing resulted in a mathematical model, the *Sabine formula*, for comparing the acoustics of any room to an ideal value. And so the field of acoustic engineering was born. Today, acoustic engineers merge advanced sound-absorbent materials with modern architecture. They help design concert halls and other venues that enhance, rather than ruin, the audience's listening experience.

PUTTING ULTRASOUND TO WORK

SONOGRAMS

Gone are the days of grainy, hard to interpret sonograms. Current ultrasound technology (below) can capture sharp, three-dimensional images of a developing baby.

NONDESTRUCTIVE TESTING

Back in the "old days," there was no way to peer into a welded joint to see whether the weld had been done properly. Today, engineers can use portable ultrasound technology (above) to "see" into a weld, ensuring that structures like pipelines are built safely.

VETERINARY MEDICINE

It isn't just humans who benefit from noninvasive diagnosis and treatment of disease. Veterinarians use ultrasound imaging for treating pets and livestock too.

MINI LAB

DEMONSTRATION: CATCH A WAVE

Essential Question:

How are sound waves produced by speech and singing different?

You can easily tell ordinary speech from singing when you hear it, but what makes one different from the other? During this demonstration, you'll "see" some waves produced by talking and singing and think about how they're different.

PROCEDURE

A Your teacher will use a tin can sound converter to convert sound waves into light waves. As you examine different kinds of sound-to-light patterns, answer the following questions on a separate sheet of paper.

1. Describe the pattern produced by ordinary speech.
2. Describe the pattern produced by singing a single note.

CONCLUSION

3. On the basis of what you have learned so far about music and how sound is produced, explain why the patterns produced by speech and music are different.

GOING FURTHER

4. What do you think the light pattern produced by a tuning fork would look like? State your answer in the form of a hypothesis.

18C | REVIEW QUESTIONS

1. When is sound amplification used?
2. Describe how a megaphone works.
3. Which part of a speaker actually produces sound waves?
4. Describe one similarity and one difference between echolocation in bats and dolphins.
5. What is the main difference between passive and active sonar?
6. Compare infrasonic and ultrasonic sound.
7. Describe two ways in which humans use ultrasound.

CHAPTER 18 REVIEW

18A SOUNDS WAVES

- Sound energy is a form of energy that is transferred as longitudinal waves through a physical medium.
- A sound wave consists of alternating regions of compressions and rarefactions.
- Sound travels at different speeds depending on its medium. Sound travels fastest in solids and slowest in gases.
- The speed of sound in dry air at 20 °C is 343 m/s.
- A sound's pitch is determined by its frequency. The higher the frequency, the higher the pitch.
- Loudness, measured in decibels, is a measure of how loud a sound seems to humans. Intensity is a measure of the power in a sound wave.
- The timbre of a sound is determined by its combination of the fundamental tone and overtones.

18A Terms

sound energy	420
acoustic spectrum	422
pitch	422
loudness	423
intensity	423
timbre	423

18B HEARING AND MUSIC

- The human voice is produced by vibrations of the vocal cords produced as air passes over them. Human sounds are modified and resonated by structures in the sinuses, throat, mouth, and face.
- Human hearing is the result of sound vibrations first being converted into kinetic energy by structures in the middle ear and then into nerve impulses in the inner ear.
- Musical instruments are designed to produce sound vibrations when struck, plucked, bowed, or blown. The distinctive sound of each instrument is caused by its combination of design, materials, and manner of playing.
- A note sounded by an instrument consists of a dominant fundamental tone coupled with one or more overtones.

18B Terms

fundamental tone	428
overtone	428
harmonic	428

CHAPTER 18

18C USING SOUND WAVES

- Amplification makes sound louder. It may be done mechanically, such as with a megaphone, or electronically as is done with an electronic speaker.
- Echolocation is the use of reflected sound for the purposes of navigating or hunting. Echolocation is common among bats and toothed whales.
- Sonar is the use of sound to detect objects underwater. It may be passive (listening only) or active (producing sound and listening for echo).
- Infrasonic sound has frequencies that are too low for humans to hear. Some animals, such as elephants and cats, can communicate with infrasonic sound.
- Ultrasonic sound has frequencies that are too high for humans to hear. Ultrasonic sound has many practical implications, such as sonography.
- Sonography uses ultrasonic sound to produce images of objects inside other objects. Sonography is commonly used to view developing babies prior to birth.

18C Terms

acoustic amplification	430
echolocation	431
sonar	432
infrasonic wave	433
ultrasonic wave	434
sonography	434

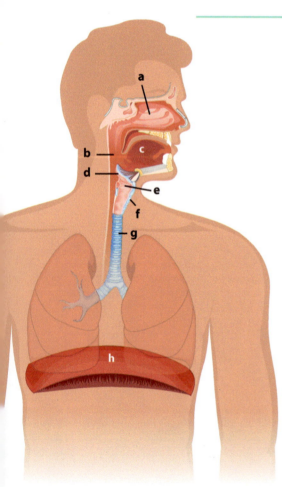

CHAPTER REVIEW QUESTIONS

Recalling Facts

1. Sketch a sound wave and label a compression and a rarefaction.
2. Which anatomical structures together act as a resonating chamber for the human voice?

For Questions 3–8, refer to the diagrams on this page to identify each indicated structure.

3. tympanic membrane
4. diaphragm
5. larynx
6. auditory canal
7. vocal cords
8. cochlea

9. Trace the path of a sound wave from entering the human ear to its conversion into a nerve impulse.
10. What determines whether a sound is considered infrasonic?
11. What advantages does infrasonic sound communication have for elephants?
12. Why is sonography a desirable means for diagnosing disease?
13. Is ultrasound used only for medical purposes? Explain.

Understanding Concepts

14. Can sound travel through deep space? Explain.
15. What will happen to the speed of a sound wave as it passes from air into water? Explain.

REVIEW

16. What will happen to the speed of sound as air temperatures drop from 30 °C to 20 °C? Explain.
17. You see a flash of lightning in the distance on a 25 °C day, and the sound of thunder reaches your ears 11 s later. How far away was the strike?
18. If the wavelength of a sound wave gets longer, what happens to the sound's pitch?
19. How will the characteristics of a sound change if the amplitude of its wave increases?
20. You and a friend hear a sonic boom, which your friend describes as an "intense" sound. What is wrong with your friend's description?
21. A person raises his voice from 60 dB to 80 dB. How much more sound wave power is he projecting?
22. A French horn and a trumpet can play notes of the same pitch and frequency but sound very different. Why is that so?
23. Which human muscle is particularly vital for producing sound? Explain.
24. In a standard guitar tuning, the lowest string, the E_2, is tuned to 82.41 Hz. What frequency is its third harmonic?
25. Small stereo systems with built-in speakers typically do not produce bass tones well. Why is this?
26. What characteristic of ultrasonic sound makes it particularly suited to locating underwater objects?

Critical Thinking

27. What will happen to the voice of a person with reduced lung capacity due to injury or disease? Explain.
28. What portion of a trumpet performs the same function as the body of a guitar? Explain.
29. Can a guitar be played without plucking its strings?
30. Why can active sonar be thought of as a type of echolocation?
31. Why would it not be a good idea for a submarine to use active sonar to find targets?
32. Describe a way that an acoustic engineer could serve others, specifically with regard to the church's mission of making disciples.

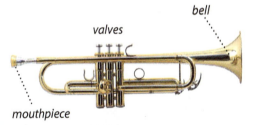

CHAPTER 19
Electricity

SIMPLY SHOCKING

The low-pitched hum intensifies, followed by the crackle of electricity. Suddenly, arcs of electricity leap to the man and the sparks seem to coincide with the rhythm of the music—because the sparks are creating the music! The man is standing on a device called a *Tesla coil*. Nikola Tesla, a Serbian-American engineer and inventor, developed the Tesla coil in an attempt to use static electricity. Tesla coils can generate electric sparks of more than a million volts!

Tesla worked tirelessly on inventions for wireless electric power transmission, wireless communication, and alternating current generation. Most of Tesla's inventions never became commercially successful, but they did lay the groundwork for research into the field of electricity. Today, Tesla coils are common in entertainment as part of light and music shows. Throughout this chapter we will learn about the static and current electricity that interested Tesla so much.

19A	Static Electricity	442
19B	Current Electricity	450
19C	Circuits	457

19A | STATIC ELECTRICITY

19A Questions

- What causes the electric force?
- What is an electric field?
- How can an electric force act at a distance?
- How can I put a charge on an object?

19A Terms

static electricity, electric force, electric field, grounding, law of conservation of charge

Why do I sometimes get shocked after walking across a carpet?

We are all familiar with static electricity, even if it is just from watching lightning storms, getting a shock from a doorknob on a winter day, or seeing socks cling to other clothing after coming out of the dryer. **Static electricity** is an accumulated electric charge on an object. How can static electricity result in both annoying static cling and potentially deadly lightning?

Static electricity for lightning, sparks, and static cling comes from a physical property of matter that God designed into the creation for our benefit. Recall from Chapter 3 that atoms consist of protons, neutrons, and electrons. Both protons and electrons are charged particles, with protons having a positive one charge and electrons having a negative one charge. Atoms are neutral because they have equal numbers of electrons and protons. But under some conditions, objects can gain or lose electrons, resulting in an object with a static electric charge. The coulomb (C) is the SI unit for electric charge.

19.1 ELECTRIC FORCE

You've heard the statement that opposites attract in personal relationships. But physicists refer to this idea when working with electricity. In Chapter 3 you learned about the law of electrostatic charges, which states that opposite electric charges attract each other, while like charges repel each other. This attraction or repulsion is a result of the **electric force**, which is the force between two charged objects. It is directly related to the charge on each object and inversely related to the square of the distance between them. The French physicist Charles-Augustin de Coulomb formalized the relationship in 1785. Coulomb's law is shown below.

$$F_e = k_c \frac{q_1 q_2}{r^2}$$

In this law, F_e is the electric force in newtons, k_c is Coulomb's constant, q_1 and q_2 are the charges in coulombs, and r is the distance in meters between the objects. The electric force is one of the forces that doesn't require the objects to be in contact. Recall that we refer to this as a field force or a force at a distance.

While Coulomb received credit for determining the formula for the electric force, many other scientists were working on determining the relationship between the variables. Early in the 1700s, scientists were convinced that there were parallels between the electric force and the gravitational force that you learned about in Chapter 12. They hypothesized that the electric force would be inversely related to the distance between the objects as was the gravitational force. Look at the two formulas side by side.

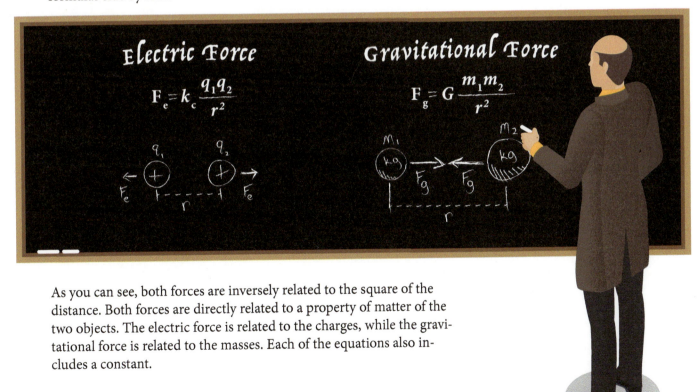

As you can see, both forces are inversely related to the square of the distance. Both forces are directly related to a property of matter of the two objects. The electric force is related to the charges, while the gravitational force is related to the masses. Each of the equations also includes a constant.

FACTORS THAT AFFECT THE ELECTRIC FORCE

Two objects with like charges each exert a repulsive electric force on the other.

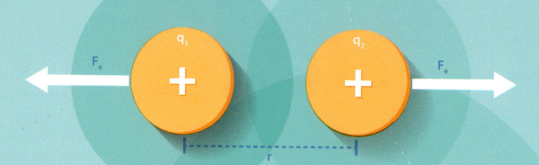

By doubling the charge on one of the objects, the force on each object is also doubled.

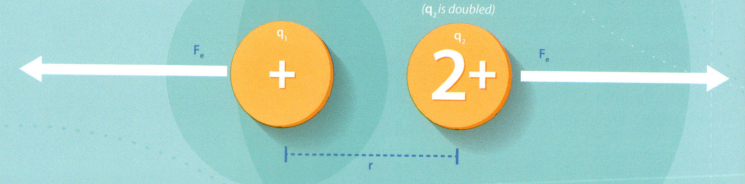

If the distance between the two original objects is doubled, the force on each object is reduced to one-fourth the original value.

Notice that the pairs of forces in each image are exactly equal but in opposite directions. The forces are examples of action-reaction pairs (see Chapter 12).

Force and charge are linearly related.

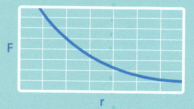

Force varies inversely with the square of the distance.

19.2 ELECTRIC FIELDS

So how does the electric force act at a distance? Forces at a distance, or field forces, were a challenge for scientists as they studied the electric force. British physicist Michael Faraday addressed this issue by hypothesizing that every charged object has an electric field surrounding it. The strength of the field indicates how much force the field would exert on a one-coulomb (1 C) charge placed in the field at that location. The strength of the field decreases the farther you get from the object. We depict field lines with arrows, similar to vector arrows. The direction of the field depends on the type of charge creating the field. Physicists agreed to use a small positive test charge for defining the direction of electric fields. Therefore, field arrows point away from positive objects and toward negative objects. So an **electric field** is a three-dimensional region around a charged object that will apply a force on other charged objects within that region. The interaction of the electric fields around two charged objects produces an electric force between those objects.

The similarities between gravity and electricity don't end with the mathematical models. Just as energy can be stored in a gravitational field as gravitational potential energy, so it is with electric fields. The *electric potential energy* of a charged particle depends on the charged particle's position in the electric field. We add gravitational potential energy to an object by working and moving it in a direction opposite the force. Electric potential energy works the same way. Any time we are working and moving the charge against the electric force, we are adding electric potential energy.

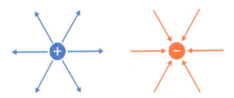

The image above shows the electric field lines pointing away from a positively charged object. The field lines point toward a negative object. Notice that the field lines are farther apart away from the object, indicating a decrease in the strength of the field.

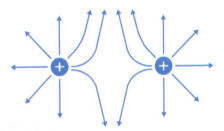

The interaction of two objects' field lines results in an electric force on each of the objects.

19.3 CHARGING OBJECTS WITH STATIC ELECTRICITY

By this point you understand that we can cause an object to be charged by somehow accumulating more protons or electrons than what typically occurs. To do this, we must be able to move positive or negative charges onto or off the object. We typically do this by moving electrons.

METHODS OF CHARGING

CHARGING BY FRICTION

Rubbing two objects together can transfer charge from one object to the other in a process called *charging by friction*. Walking across a carpeted floor often allows you to gain a negative electric charge as the friction between your feet and the carpet removes electrons from the carpet. This excess charge can leave your body as a spark between you and a doorknob or light switch. Electric force holds the paper cutouts on the balloon (left). Rubbing the balloon with a piece of wool or other material generated the static electricity that produced the electric force.

CHARGING BY CONDUCTION

If an isolated charged object comes in contact with a second uncharged isolated object, the excess charge will spread across the two objects in a process called *charging by conduction*. In the image below you can see an uncharged metal sphere and a metal sphere with excess electrons. If we bring the two spheres into contact with each other, the excess electrons, because they repel each other, spread out over the two objects. When we separate the two objects, they are each charged with the same type of charge as the originally charged object. The magnitude of charge on each object will depend on the size and shape of the objects.

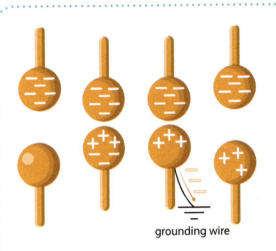
grounding wire

CHARGING BY INDUCTION

Charging by induction is a third way to charge an object. Again we have two isolated metal spheres (above), one with excess electrons. This time we bring the two spheres close but don't allow them to touch. The excess negative charge forces, or induces, the electrons in the other sphere to move to the opposite side of the sphere. If we then connect the uncharged sphere to the ground, we allow the negative charges to leave the sphere. **Grounding** is the act of providing a path for electric charge to move into the earth, or ground. When we disconnect the second sphere from the ground, we have a second charged sphere. Induced charges are opposite the original charge. The magnitude of the induced charge depends on the magnitude of the original charge and how close the two spheres become.

Whenever we charge an object, the electrons gained by one object are the same as the electrons lost by the other. This accounting of charge is required by the **law of conservation of charge**, which states that charge cannot be created nor destroyed but only transferred between objects.

ELECTRICITY

19.4 EVERYDAY STATIC ELECTRICITY

People often think of static electricity as being a hazard or at least a nuisance. And this is true when we think of the annoyance of static cling on our clothing or the potential hazard of a lightning strike. Static discharge can also be hazardous when dealing with flammable powders or gases. Airport ground personnel connect an aircraft and their trucks to the ground using a grounding wire during refueling operations. They do this because aircraft can build up a static charge when flying. The discharge of that excess charge could ignite the fuel vapors.

But static electricity can also be put to use to help people. Engineers have figured out how to use static electricity to do spray painting, laser jet printing, air filtration, and photocopying. So the next time you get frustrated by static cling, remember that static attraction also holds the words for your English paper on the page long enough for the laser printer to make them permanent!

Static cling forms the seal between plastic wrap and the container.

19A | REVIEW QUESTIONS

1. Define *static electricity*.
2. What does it mean that the electric force is a force at a distance?
3. What other force does the electric force act like?
4. What will happen to the electric force if we double both charges?
5. What is an electric field?
6. Draw the electric field lines around a positively charged sphere.
7. Explain how charging by conduction occurs.
8. (True or False) The induced charge on an object will always be the same type of charge as the object that induced the charge.

MINI LAB

OBSERVING ELECTROSTATIC CHARGE

The law of electrostatic charges states that like charges repel and opposite charges attract. One of the earliest instruments used to study electrostatics was the electroscope. The thin metal leaves of the electroscope shown below allow scientists to study how different charged objects interact. This lab uses clear plastic tape to act as the leaves of an electroscope. You will see the effects of charging by contact and by induction.

1. If two charged objects repel each other, what would you conclude about the objects?

Essential Question:

How can we observe accumulated charge on an object?

Equipment

clear plastic tape

balloon

PROCEDURE

A Fold approximately 1 cm of the end of the tape to form a handle. Now remove a piece of tape (approximately 8 cm) and tape it to your clean and dry desktop. Repeat with a second piece of tape, placing it directly on top of the first piece.

B Rapidly pull both pieces of tape off the desktop and then pull them apart. Holding one in each hand, move the pieces together with their nonadhesive sides facing each other. Record any observations of the tapes' behavior.

C Repeat Step **A**, but this time tape each piece to the desktop next to each other.

D Rapidly pull each piece of tape off the desktop. Holding one piece of tape in each hand, move the pieces together with their nonadhesive sides facing each other. Record any observations of the tapes' behavior.

ANALYSIS

2. Why did the tape behave as it did in each trial?

CONCLUSION

3. Can you determine the amount of charge on the objects with an electroscope? Explain.

GOING FURTHER

4. If you were to charge a balloon, how would you expect it to react if you placed it close to an electrostatically neutral object such as the wall?

ELECTRICITY 449

19B Questions

- What is current electricity?
- How is alternating current different from direct current?
- Does charge flow the same through all materials?
- What makes charge flow?

19B Terms

current electricity, electric current, electric circuit, direct current, alternating current, electrical conductor, electrical insulator, voltage, resistance, Ohm's law

19B | CURRENT ELECTRICITY

Why does the light turn on when I flip the switch?

19.5 ELECTRIC CURRENT

In the previous section, we studied static electricity—the accumulated charge on an object. This charge typically doesn't move and has limited usefulness. Most applications of electricity use **current electricity**, which is electricity involving moving electric charges. The movement of the electric charge through a complete loop is called an **electric current**, and the loop through which current electricity can flow is called an **electric circuit**. An *open circuit* is incomplete, preventing the movement of charge. A *closed circuit* is complete and allows current.

Direct Current

There are two types of current involved in current electricity: direct current and alternating current. Calculators, flashlights, and any other battery-powered systems all use direct current. The charges in **direct current** (DC) flow in one direction from the power source, through the components of the circuit, and then back to the power source. In most circuits, free electrons move through the circuit as the current. The electrons move from the negative end of the power source, through parts of the circuit, and end up at the positive end of the power source. Most physicists talk about *conventional current*, which is the movement that positive charges would have through the circuit. Therefore, the electrons moving in the circuit move in the direction opposite that of the conventional current.

UNDERSTANDING DIRECT CURRENT

It is easy to understand direct current if you compare it with water moving through a pump system.

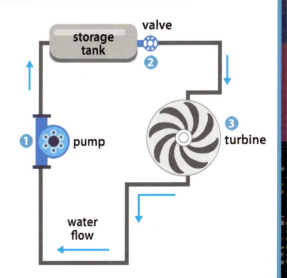

The pump ❶ does work as it lifts the water to the storage tank, storing gravitational potential energy.

The valve ❷ controls the flow of water.

The water does work through the turbine ❸ to convert the energy into other forms.

The water continues to flow until it returns to the pump ❶.

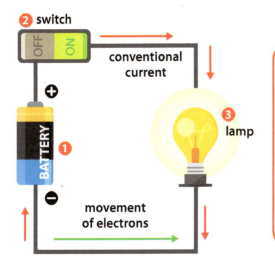

The battery ❶ stores energy. When needed, the battery transforms energy from chemical potential energy to electrical energy.

The switch ❷ controls the flow of electricity.

The charge carriers do work through the lamp ❸ as the electrical energy is converted to light and thermal energy.

The charge carriers continue moving through the circuit back to the battery ❶.

ELECTRICITY 451

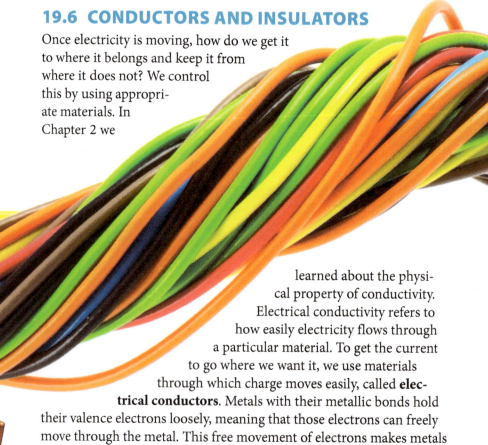

Alternating Current

While many small portable devices use direct current, much of the electrical equipment that we use in our homes operates by using alternating current. The electric charges in these circuits change direction repeatedly as the system operates. This changing, or alternating, of the direction of movement of the charge carriers is why we call this **alternating current** (AC). The alternating of the current happens in a regular pattern, making this a form of periodic motion. In the United States, most AC systems have 60 cycles of their forward and reverse current every second (60 Hz). The amazing part is that for the current to change direction, it must first stop, so the current stops 120 times every second, meaning that an electric device is unpowered 120 times every second.

There are many applications that need to change DC to AC or vice versa. Many items (e.g., cell phones, tablets, digital cameras) that operate using DC battery power are recharged using the AC in our homes. Their electrical cords usually have a rectifier that changes the alternating current into direct current. Gas-powered cars operate on DC power. A battery provides the DC electricity to start the car. Once the car is running, the alternator generates AC electricity that is converted to DC to operate the electrical systems and recharge the battery.

19.6 CONDUCTORS AND INSULATORS

Once electricity is moving, how do we get it to where it belongs and keep it from where it does not? We control this by using appropriate materials. In Chapter 2 we learned about the physical property of conductivity. Electrical conductivity refers to how easily electricity flows through a particular material. To get the current to go where we want it, we use materials through which charge moves easily, called **electrical conductors**. Metals with their metallic bonds hold their valence electrons loosely, meaning that those electrons can freely move through the metal. This free movement of electrons makes metals excellent electrical conductors.

If you look at the wires that run through your house, you will notice that plastic encases the metal wire. This design is used because charge moves poorly through plastic. A material through which charge flows poorly is an **electrical insulator**. We prevent current going to the wrong place by using insulators.

Some materials have conductivity values between those typical of conductors and insulators. These materials are called *semiconductors*; many of the metalloids are semiconductors. Unlike metals, the conductivity of semiconductors increases with increased temperature. Engineers can alter the properties of semiconductors by intentionally inserting other elements in the crystal lattice of a semiconductor. Many components in electronic equipment consist of semiconductors.

19.7 CURRENT, VOLTAGE, AND RESISTANCE

Current

The symbol for the flow of charge through a complete circuit—the electric current—is *I*, from the French for current intensity. The unit for current is the *ampere* (A). The naming of this unit was to honor the French physicist André-Marie Ampère, who some consider the Father of Electrodynamics. The current, or amperage, is the coulombs of charge that pass a point in a circuit each second, so 1 A = 1 C/s. We measure the amps in a circuit with an *ammeter*.

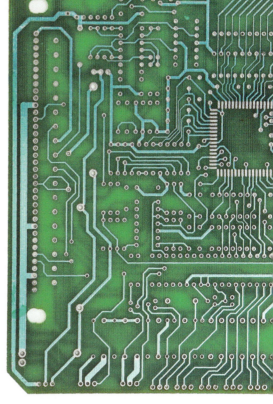

The green areas are the nonconducting portions of the circuit board. The light blue paths are the metallic circuits on the board.

Most people think that when an electrical switch is closed, charges start at the power source and then move through the circuit. But the charge carriers are always present throughout the conductors. These charge carriers are always in motion, but when the circuit is open, the movement is random with no net movement in any one direction. In a closed circuit, the net motion is in one direction, which we call current. But what makes the charges move through the circuit?

Voltage

The "force" that moves electric charge carriers through an electric circuit is the **voltage** (*V*). This is like water pressure in a water system. Voltage is the work done to move a coulomb of charge through a circuit. You can think of voltage as the energy converted from electric potential energy as each coulomb of charge moves through an electric field. Another term for voltage is the *electric potential difference*. The unit of voltage is the *volt* (V). Scientists named the volt after an early researcher into electric power, Italian physicist Alessandro Volta. We measure voltage in a circuit with a *voltmeter*. How do we get this "force," or voltage, into the circuit?

A power supply provides the voltage to a circuit. In DC circuits, the most common power source is a *battery*, a device that stores energy in the form of chemical potential energy. The chemical energy changes into electrical energy when we close the circuit. Most of us are familiar with AAA, AA, C, and D cell batteries. Each of these provides 1.5 volts of electric potential difference, though there is more energy stored in the larger cells. You may have also used a 9 V battery. Many applications include a number of these batteries used together. The common graphing calculator uses four AAA cells, providing 6 V (1.5 V × 4) of potential difference. Other sources of DC power are photovoltaic cells, which convert light into electrical energy, and hydrogen fuel cells, which convert the chemical energy released when hydrogen and oxygen react into electrical energy. Electrical generators are the typical source of AC power energy (see Section 20C).

Resistance

As charge carriers move through a circuit, they are constantly colliding with particles in the circuit. These collisions slow down, or resist, the movement of the charge carriers. Almost all materials have **resistance**—the property of matter indicating the degree to which the material slows down the flow of charge carriers. The symbol for resistance is the uppercase R, and its unit is the *ohm* (Ω). The ohm was selected as the unit of resistance to recognize the accomplishments of German physicist Georg Simon Ohm.

The amount of resistance that a particular wire has depends on the resistivity of its material, its length and diameter, and its temperature. Metals tend to have lower resistivity than other materials, which is why we use them as conductors. (See Table 1 for a list of the resistivity of common materials.) As the diameter of a wire increases, its resistance decreases. This relationship exists because there are more free electrons to carry the charge. As the length of the wire increases, the resistance increases because the charge carriers have to pass a greater number of atoms that they can bump into, slowing down the current. As the temperature of the wire increases, the resistance increases. This relationship is due to the increased kinetic

Table 1

The Resistivity of Common Materials

Material	Resistivity (at 20 °C) ($\Omega \cdot$m)
Silver	1.59×10^{-8}
Copper	1.68×10^{-8}
Gold	2.44×10^{-8}
Aluminum	2.65×10^{-8}
Silicon	6.40×10^{2}
Glass	$\approx 1 \times 10^{12}$
Quartz	7.5×10^{17}

energy of the charge carriers, resulting in more collisions. The greater number of collisions makes it harder for the charge carriers to move through the wire, resulting in an increased resistance.

In most applications, we want to reduce the resistance in a circuit because, like friction, it turns energy into less useful forms. To reduce the resistive losses in a circuit, engineers are working on materials that have zero resistance—*superconductors*. Currently, superconductors have zero resistance only at extremely cold temperatures. The goal is to make superconductors that can operate at higher temperatures.

19.8 OHM'S LAW

In 1827, Ohm published a paper that explained how current, voltage, and resistance relate to each other in conductors. Today we know this relationship as **Ohm's law**, which states that the current in a circuit is directly related to the voltage and inversely related to the resistance. The formula for Ohm's law is

$$I = \frac{V}{R},$$

where I is the current in amperes, V is the voltage in volts, and R is the resistance in ohms. Ohm's law allows scientists to analyze circuits to understand how various components will affect the conditions (current, voltage, and resistance) in the circuit. Let's consider a few examples of Ohm's law in action.

EXAMPLE 19-1: Determining Current

A simple DC circuit provides 6.0 V and has a resistance of 2.7 Ω. What is the current in the circuit?

Write what you know.

$V = 6.0 \text{ V}$

$R = 2.7 \text{ Ω}$

$I = ?$

Write the formula and solve for the unknown.

$$I = \frac{V}{R}$$

Plug in known values and evaluate.

$$I = \frac{6.0 \text{ V}}{2.7 \text{ Ω}} = 2.2 \text{ A}$$

EXAMPLE 19-2: Determining Voltage

An ammeter in a simple circuit indicates 3.5 A. The circuit has a resistance of 1.9 Ω. What is the voltage in the circuit?

Write what you know.

$I = 3.5$ A

$R = 1.9\ \Omega$

$V = ?$

Write the formula and solve for the unknown.

$$I = \frac{V}{R}$$

$$IR = \frac{V}{R}R$$

$$V = IR$$

Plug in known values and evaluate.

$$V = (3.5\text{ A})(1.9\ \Omega) = 6.7\text{ V}$$

EXAMPLE 19-3: Determining Resistance

A battery provides 12.0 V to a circuit and 0.891 A of current flow through the circuit. What is the total resistance in the circuit?

Write what you know.

$V = 12.0$ V

$I = 0.891$ A

$R = ?$

Write the formula and solve for the unknown.

$$I = \frac{V}{R}$$

$$IR = \frac{V}{R}R$$

$$\frac{\cancel{I}R}{\cancel{I}} = \frac{V}{I}$$

$$R = \frac{V}{I}$$

Plug in known values and evaluate.

$$R = \frac{12.0\text{ V}}{0.891\text{ A}} = 13.5\ \Omega$$

19B | REVIEW QUESTIONS

1. What is current electricity?
2. How is alternating current different from direct current?
3. Does charge flow the same through all materials?
4. What is a semiconductor?
5. Define *voltage*.
6. If you were to replace a copper wire with an identical wire made of one of the other materials found in Table 1 (p. 454), which material would make a wire with less resistance than the copper wire?
7. What is a superconductor?
8. A circuit has a total resistance of 35 Ω and has a 16 V power supply. What is the current through this circuit?

19C | CIRCUITS

How does current travel through a circuit?

19.9 BUILDING ELECTRIC CIRCUITS

To use current electricity, engineers build circuits to move an electric charge where they want it. The design of the circuit also gives the engineer the ability to control how much voltage and current are in each part of the circuit. Recall from Section 19B that current electricity is the flow of electric charge through an electric circuit. Energy is put into the circuit by the power supply, and the different components use that energy to do work. Any electrical device in a circuit that consumes electric energy is called an *electrical load*.

Since the goal of an electric circuit is to get the electric charge to move where it is wanted, we can also imagine that the engineer wants to avoid having the charge go where it isn't wanted. When the current takes an unintended path, called a **short circuit**, it is potentially hazardous. If the electricity moves directly to the ground, then the only problem is that the circuit fails to work properly. But many times the current in a short circuit moves through other materials or even people. Short circuits can cause equipment damage, fires, injuries, or even death. Therefore engineers have to be careful when designing and building electric circuits.

19C Questions

- What is the difference between a series circuit and a parallel circuit?
- What is a short circuit?
- Are all circuits set up the same?
- How can we safely use electricity?

19C Terms

short circuit, resistor, series circuit, parallel circuit, electric power, fuse, circuit breaker

COMMON COMPONENTS OF ELECTRIC CIRCUITS

Engineers use various components when building a circuit.
Let's look at a few of the most common ones along with their symbols.

POWER SUPPLY

battery   photovoltaic cell   AC generator

Common power sources are batteries, photovoltaic cells, and AC generators.

SWITCHES

open toggle switch   closed toggle switch   push switch

It is important to control whether there is current in the circuit. Closing a switch allows charge to move through the circuit. An open switch prevents current in the circuit.

RESISTORS

fixed   variable

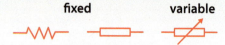

In some parts of a circuit we want to reduce the current, so we increase the resistance in that part of a circuit. The most common way to accomplish this is to install a **resistor**, an electrical device that converts electrical energy into other forms, such as thermal energy.

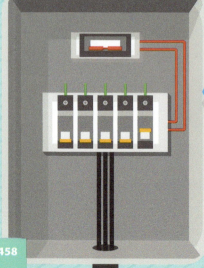

SAFETY DEVICES

ground   fuse   circuit breaker

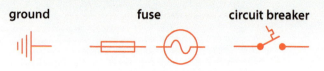

Electricity can be hazardous. Fuses, circuit breakers, and grounding provide needed safety for electric circuits. (See Subsection 19.12.)

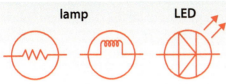

It is common to want a circuit to produce light. We can do this with a lamp or a light-emitting diode (LED). Notice the resistor in the first lamp symbol. Many lamps use resistance to convert electrical energy into thermal energy and light energy.

LIGHTS

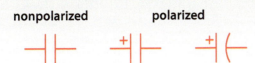

At times we want the circuit to store electrical energy, which we can accomplish with a *capacitor*.

CAPACITORS

While not a permanent part of most circuits, meters are often needed when building or analyzing circuits.

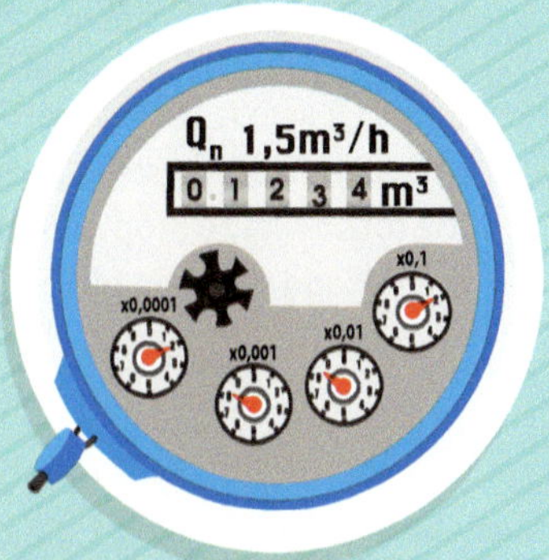

METERS

Let's draw a circuit using some of these components.

EXAMPLE 19-4: Drawing Electric circuits

Draw a circuit made up of a 15 V battery, an open toggle switch, two lamps, and a 5 Ω resistor, all in a single loop.

You can see that there is only one loop for the electricity to flow through. In this simple circuit, you can draw the components in any order, but it is easiest to just build the circuit in the order described. Make sure that you complete the circuit by returning to the battery.

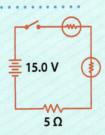

ELECTRICITY 459

19.10 SERIES AND PARALLEL CIRCUITS

As you can imagine, circuits can range from very simple to extremely complex. Circuits, like the one in Example 19-4, that provide only one path for the charge carriers to move through are called **series circuits**. A series circuit's total current must flow through each component of the circuit. The sum of the voltages lost in all the components must equal the total voltage supplied by the power source. Other circuits, called **parallel circuits**, have branches that provide multiple paths that current can take. In parallel circuits, the voltage drop across each branch equals the voltage supplied by the power supply, but the total current gets split among the multiple branches, and more of the current goes through the path with the least resistance.

COMPARING SERIES AND PARALLEL CIRCUITS

SERIES CIRCUIT

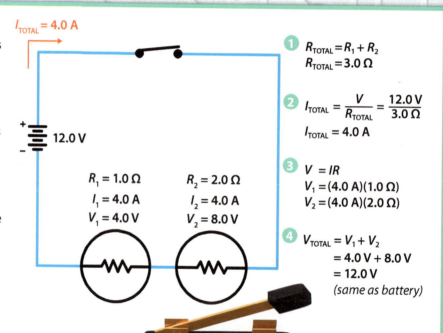

The series circuit on the right consists of a 12 V battery, a switch, and two lamps in series. Because the charge carriers have to travel through both lamps, we can find the total resistance ① (3.0 Ω). Ohm's law allows us to find the total current ② through the circuit. Since all of the current has to go through each lamp, we can use Ohm's law again to find the voltage drop ③ across each lamp. Finally, we check our work by seeing whether the total voltage drop ④ is equal to the voltage supplied by the battery.

① $R_{TOTAL} = R_1 + R_2$
$R_{TOTAL} = 3.0\ \Omega$

② $I_{TOTAL} = \dfrac{V}{R_{TOTAL}} = \dfrac{12.0\ V}{3.0\ \Omega}$
$I_{TOTAL} = 4.0\ A$

③ $V = IR$
$V_1 = (4.0\ A)(1.0\ \Omega)$
$V_2 = (4.0\ A)(2.0\ \Omega)$

④ $V_{TOTAL} = V_1 + V_2$
$= 4.0\ V + 8.0\ V$
$= 12.0\ V$
(same as battery)

Series circuits are often the simplest systems. Since the current has to flow through all the components, when one fails the whole system fails. The electric circuit within a flashlight is an example of a series circuit.

PARALLEL CIRCUIT

If we rearrange the circuit as in the image below, we can make a parallel circuit. We can treat the two paths as separate circuits. Since each path has only one lamp, we can find the current ❶ through each path of the circuit. Since the current in each path all returns to the battery, we can find the total current ❷ through the circuit. Notice that more current goes through the path with lower resistance. Electricity tends to follow the path of least resistance.

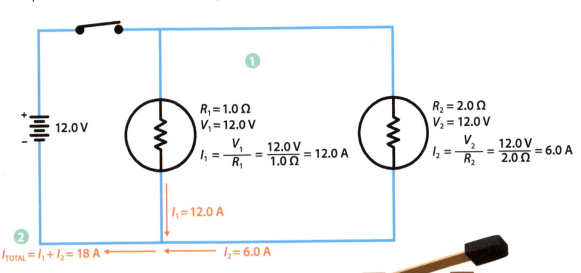

Parallel circuits have an advantage in that one branch of the circuit can operate without needing all the branches operating. When your house was built, the builder wired most of its circuits in parallel. If this were not true, you would have to turn on the toaster and the microwave to use the blender.

19.11 ELECTRIC POWER

Throughout the chapter, our focus has been on electrical energy and on the way that electrical systems and components can do work. You might remember learning about the relationship between work and power in Chapter 13. We defined power as the amount of work done per second. **Electric power** is the work done per second by an electrical system. The formula for power that we learned in Chapter 13—

$$P = \frac{W}{t}$$

—still applies, but it would be easier if we could use quantities that we measure in electric circuits, such as current, voltage, and resistance. Scientists have derived such a formula.

$$P = VI$$

In this formula, P is the power in watts, V is the voltage in volts, and I is the current in amps.

EXAMPLE 19-5: Determining Power

You and your friends meet weekly to watch one episode of your favorite show. If your plasma TV and streaming service draw 2.80 A and the electrical system is 120 V, how much electric power is needed for the TV and streaming device?

Write what you know.

$I = 2.80$ A

$V = 120$ V

$P = ?$

Write the formula and solve for the unknown.

$P = IV$

Plug in known values and evaluate.

$P = (120 \text{ V})(2.80 \text{ A}) = 336$ W

People often talk about the "power" company when referring to their utility service, but in reality, we should refer to it as an energy company. When your parents get the electric bill, the utility charges your household for the energy that your family used during the previous month. While scientists typically measure energy in joules, the energy companies usually charge us per kilowatt-hour (kW·h). How much total energy will you consume in watching your show?

EXAMPLE 19-6: Determining Energy Used

Each season of your favorite show has twenty-two episodes and each show runs for forty-eight minutes. How much energy—kilowatt-hours and joules—will you use to watch all five seasons?

Write what you know.

$P = 336$ W

$t = 5 \text{ seasons} \left(\frac{22 \text{ shows}}{\text{season}} \right) \left(\frac{48 \text{ min}}{\text{show}} \right) \left(\frac{1 \text{ h}}{60 \text{ min}} \right)$

$t = 88$ h

$E = ?$

Write the formula and solve for the unknown.

$P = \frac{E}{t}$

$Pt = \frac{E}{t} t$

$E = Pt$

Plug in known values and evaluate.

$E = (336 \text{ W})(88 \text{ h})$

$= 29\,568$ W·h

Now convert watt-hours to kilowatt-hours and joules:

$= 29\,568 \text{ W·h} \left(\frac{1 \text{ kW}}{1000 \text{ W}} \right) = 30$ kW·h

$29\,568 \frac{\text{J}}{\text{s}} \cdot \text{h} \left(\frac{3600 \text{ s}}{1 \text{ h}} \right) = 1.1 \times 10^8$ J

EXAMPLE 19-7: Determining Energy Cost

How much will it cost to watch all eighty-eight hours, assuming an average cost of $0.13/kW·h?

Write what you know.

$E = 29.568$ kW·h

$29.568 \text{ kW·h} \left(\frac{\$0.13}{\text{kW·h}} \right) = \$3.84$

19.12 ELECTRICAL SAFETY

As mentioned above, our houses are wired predominantly with parallel circuits. Parallel circuits do have one serious drawback. As you turn on each branch, you add more current to the circuit. It is quite possible that you will draw too much current, potentially causing your wiring to overheat. Most household circuits can carry either 15 A or 20 A. If you are using a curling iron, a hair dryer, and an electric space heater on the same circuit, you could easily exceed the current limit for that circuit.

But don't worry—the architects thought about this when designing the electrical systems for your house. They have included in each circuit one or more safety devices that protect the circuit from overheating due to excessive current. These devices include fuses, circuit breakers, ground wires, and ground fault interrupters. Let's see how these devices work.

CIRCUIT PROTECTION

A **fuse** is a safety device for electric circuits that opens the circuit by melting when an overheated condition occurs due to excessive current. It consists of a thin conductor that is in series with the rest of the circuit. If the current exceeds the limit for the circuit, the thin wire will melt, opening the circuit and stopping the current. The drawback in using fuses is that once they melt they need to be replaced before the circuit will work again.

A **circuit breaker** is a safety device consisting of an automatic switch that opens when there is too much current in the circuit. There are many different types of circuit breakers depending on how the switch is triggered. Most household circuit breaks operate by temperature or changes in a magnetic field. Circuit breakers are preferred over fuses because after a circuit breaker trips, or opens, it can be reset. Notice the circuit breakers in this household circuit breaker panel (left). Each circuit breaker indicates the current limit for that circuit.

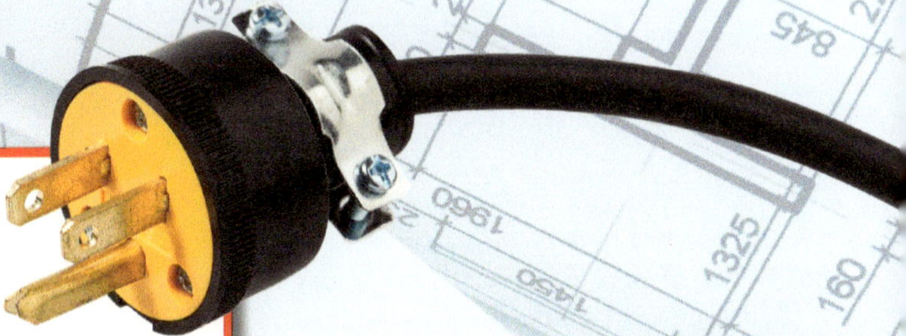

The third, u-shaped prong on many plugs today is the ground prong. It is connected to a wire that runs out of the house and into the earth. Most locations require electrical systems to be grounded. *Grounding* provides a safe path for the electricity in the event of a malfunction. It is also an important safety precaution when static discharge could be hazardous, like working with flammable materials or on sensitive electronic equipment.

A *ground fault interrupter* (GFI) is an additional safety device for electric circuits that have a high potential for electric shock. These are usually installed in kitchens, bathrooms, garages, and outdoors, where water and electricity are likely to come in contact with each other. And they don't mix! The GFI is a special outlet that has a circuit breaker built into it. You can recognize GFI outlets by the two push buttons. One allows you to test the GFI and the other resets the circuit breaker after it trips or after the system is tested.

God has given us the ability to develop electric circuits and systems. Every technology has its potential hazards. We continue to fulfill the Creation Mandate by developing safety systems that allow us to reap the benefits of current electricity while keeping people safe. Fuses, circuit breakers, and GFIs can help prevent mishaps.

HOW IT WORKS

Electric Cars

You may have noticed an increase in the number of electric cars on the road today. Many car companies are developing electric cars to meet consumer demand. Increases in gas prices, as well as concern for the environment, have driven the recent interest in electric cars. But most people don't realize that electric cars are not new. The first electric car was built in 1884, about twenty years before the first Model T. Most early cars were electric because they were simpler and more capable than gas-powered cars. Ironically, electric cars fell out of favor because of the invention of the electric starter for gasoline cars.

Chemical energy stored in large, high-voltage batteries provides the energy to power electric cars today. As chemical energy changes to electrical energy, the electrical energy powers one or more electric motors. Most electric cars can travel between 160 and 480 km on a charge. A car battery can be recharged by plugging the vehicle into a charging station. Many gas stations now have charging stations, and many cities have even installed charging stations in parking garages.

The greatest benefit of electric cars is that they are significantly more efficient than gasoline engines. Most gas-powered cars use less than 20% of the available chemical energy to move the vehicle. Electric cars, on the other hand, convert about 80% of the electrical energy into motion. Electric vehicles also emit no pollution when in operation. However, this advantage may just be shifting the pollution problem to the energy plant that produces the electricity.

Many nations are looking at electric cars as the solution to both the fossil fuel and pollution issues. They plan to prohibit the sale of gasoline-powered cars over the next three to four decades.

19C | REVIEW QUESTIONS

1. Draw the symbols for a fuse, a resistor, and a lamp.
2. Draw a series circuit consisting of a battery, a switch, two resistors, a lamp, and a capacitor.
3. A circuit consists of a 20 V battery with three resistors (10 Ω, 20 Ω, and 20 Ω) in series. If a total of 0.40 A of current leaves the battery, how much current flows through, and what is the voltage drop in the 10 Ω resistor?
4. What is the difference between a series circuit and a parallel circuit?
5. If your hair dryer consumes 1500 W of power in a 120 V system, what is the current through the hair dryer?
6. Compare fuses and circuit breakers.

CHAPTER 19 REVIEW

19A STATIC ELECTRICITY

- Static electricity is the accumulated charge on an object.
- An electric force is produced by two charged objects. The force is directly related to the charges on the objects and inversely related to the square of the distance between them.
- The electric force is a field force that is produced by the interaction of each object's electric field.
- Objects can be charged by friction, conduction (contact), or induction (by the electric force).

19A Terms

static electricity	442
electric force	443
electric field	445
grounding	447
law of conservation of charge	447

19B CURRENT ELECTRICITY

- Current electricity involves electric charge that is in motion. Typically electric current moves through closed loops called circuits.
- Electric current, the flow of electric charge, can move in one direction (direct current) or it can change direction (alternating current).
- Electric current moves easily through electrical conductors but poorly through electrical insulators.
- Electric current moves because of voltage, which can be thought of as the "force" on the electric charges.
- Electric current is slowed down by the resistance within a circuit.
- Ohm's law relates current, voltage, and resistance.

19B Terms

current electricity	450
electric current	450
electric circuit	450
direct current	451
alternating current	452
electrical conductor	452
electrical insulator	453
voltage	453
resistance	454
Ohm's law	455

19C CIRCUITS

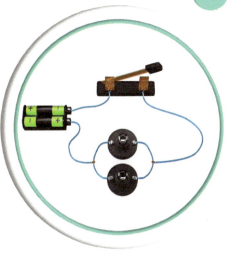

- We use current electricity by building electric circuits to move electricity from the power source to the components that will use the energy to do work.
- A short circuit is an unintended path that current moves through.
- Some circuits, called series circuits, have only one path for charge to flow through, while others have multiple paths, which we call parallel circuits.
- The power used in an electrical device depends on the current and voltage of that device.
- Because parallel circuits can increase the total current in a circuit, engineers have developed safety devices that will open a circuit if too much current is flowing. The most common of these devices are fuses and circuit breakers.

19C Terms

short circuit	457
resistor	458
series circuit	460
parallel circuit	460
electric power	462
fuse	464
circuit breaker	464

CHAPTER 19

CHAPTER REVIEW QUESTIONS

Recalling Facts

1. Describe the factors that affect the electric force.
2. What are the three methods for putting a static charge on an object?
3. Describe charging by friction.
4. What is an electric circuit?
5. What is electric current?
6. What is an open circuit?
7. What makes charge flow?
8. How is energy stored in a battery?
9. What characteristics of a wire determine the resistance in that wire?
10. What is the electrical load of a circuit?
11. What is a short circuit?
12. What type of meter measures current?
13. Define *series circuit*.

Understanding Concepts

14. Compare electric force and gravitational force.
15. What will happen to the electric force if we triple the distance between two objects?
16. What will happen to the electric force if we reduce the distance between the charges to one-fourth of the original value?
17. Use a graphic organizer to compare charging by conduction with charging by induction.
18. Compare static electricity and current electricity.
19. How are DC and AC similar to each other? How are they different?
20. Compare electrical conductors and electrical insulators.
21. What will happen to the resistance in a wire if we double its length? Explain.
22. Summarize Ohm's law.
23. A series circuit has a total resistance of 12 Ω and an ammeter in the circuit reads 2.7 A. What is the voltage of the circuit's battery?
24. Draw a series circuit consisting of a battery, a switch, and three lamps. Using a different color pencil, add a capacitor in parallel with the third lamp and a voltmeter in parallel with the second lamp.
25. Draw a circuit consisting of a battery, a switch, and two resistors that are in parallel with each other.
26. How does a parallel circuit differ from a series circuit?
27. In Resistor 1 of the circuit on the left, how much current flows through and what is the voltage drop?
28. On the basis of your answer to Question 27, how much current do you expect in Resistors 2 and 3?
29. Your friend claims that adding fuses and circuit breakers adds unnecessary cost to constructing a building. Do you agree or disagree? Explain.
30. What is the benefit of having ground fault interrupters in our houses?

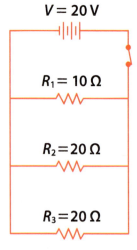

REVIEW

Critical Thinking

31. What will happen to the electric force between two objects if we triple one's charge, cut the other's charge in half, and cut the distance between the objects in half?

32. Draw the electric field lines around a system of charged objects consisting of a positively charged sphere and a negatively charged sphere.

33. Explain how the law of conservation of charge is an extension of the law of conservation of matter.

34. A circuit has a 12 V battery connected to a lamp and a resistor in series. If the lamp has a resistance of 2 Ω and can have no more than 1.1 A moving through it, what must the resistance of the resistor be?

35. If you leave a 75 W bulb on for a whole day (24 h), how much energy (in kW·h and J) will be used?

36. Your stereo system uses 11.8 A of electricity on a 120 V system. If electricity costs $0.13/kW·h, how much will it cost to run the stereo system continuously for a week?

37. The circuit breaker for the kitchen trips when your brother turns on the microwave. He asks, "Why do we have to have circuit breakers anyway?" How would you respond?

Use the Case Study below to answer Questions 38–40.

38. Do you think that the probes and wires in a Taser are made of conductors or insulators? Explain.

39. Tasers are high-voltage devices, but high current is dangerous to people. To provide high voltage but low current, should the resistance be high or low? Explain.

40. Justify the use of Tasers as part of fulfilling the Creation Mandate.

CASE STUDY: TASER®

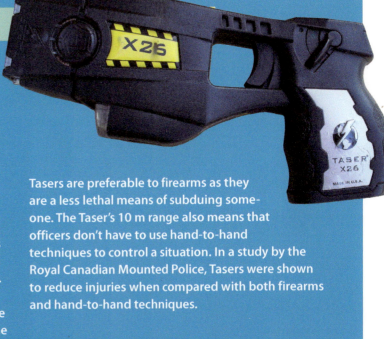

Police officers risk their lives every day as they work to protect others. These law enforcement officers often carry firearms to protect themselves and others. Police departments have turned to science and technology to find less lethal methods of neutralizing threats. Many officers now carry a Taser in addition to their firearm.

A Taser has two probes attached to the unit by conducting wires. When fired at a targeted person, the Taser launches the probes. At the end of the probes are barbs that hold the probes in the target's clothing. Once deployed, high-voltage, low-amp current flows through the wires and into the person. This current produces an involuntary contraction of the long muscles. In most cases, this will incapacitate the person while not harming him, keeping everyone safer than if the officer had needed to use his firearm.

Tasers are preferable to firearms as they are a less lethal means of subduing someone. The Taser's 10 m range also means that officers don't have to use hand-to-hand techniques to control a situation. In a study by the Royal Canadian Mounted Police, Tasers were shown to reduce injuries when compared with both firearms and hand-to-hand techniques.

CHAPTER 20
Magnetism

REAL MAGNETIC PERSONALITIES?

Homing pigeons are famous for their ability to find their way home after being released hundreds of kilometers away. They rely on a range of cues to do this. One such cue is an ability to detect the earth's magnetic field. Homing pigeons use this field to align themselves in the correct direction for their return flights. Amazingly, homing pigeons are not the only animals that can detect magnetic fields, a sensory ability called *magnetoreception*. Fruit flies, chickens, mole-rats, and many other animals can also detect magnetic fields. Although scientists have known about magnetoreception for decades, they have yet to learn how animals manage it. Like magnetoreception, magnetism itself seems like a very mysterious force. In this chapter we'll look at what magnetism is, how it works, and how human thriving depends on it every day.

20A	Magnets and Magnetism	472
20B	Electromagnetism	476
20C	Generating and Using Electricity	480

20A Questions

- What is a magnet?
- What are the properties of magnets?
- What are magnetic fields?

20A Terms

magnet, magnetism, magnetic pole, magnetic field, magnetic domain, ferromagnetism

20A | MAGNETS AND MAGNETISM

Why do magnets stick to some materials and not to others?

20.1 MAGNETS

We've all seen a magnet. You probably have a few on your fridge at home. But just what is a magnet, and what makes it stick to some metal objects? A **magnet** is any material or object that can produce a *magnetic field*. The properties of magnets and magnetic fields and the phenomena that they produce are together known as **magnetism**. The existence of magnetism has been known for thousands of years, though what causes it was a mystery. The ancient Greeks noted that certain stones called *lodestones* could attract metal objects. Later, the Chinese carved lodestones into spoon-like shapes whose "handles" pointed south. They used these as crude compasses for navigation. Around AD 1100, they developed compasses that used magnetized needles. In 1600, an English physician, William Gilbert, published *De Magnete*. Based on the results of many experiments and observations, Gilbert's book was the first comprehensive treatment of the properties of magnets.

20.2 MAGNETIC PROPERTIES

Magnetic Poles

In 1269, French scholar Petrus Peregrinus de Maricourt observed that a spherical lodestone had two separate spots where bits of iron collected. From this observation, de Maricourt reasoned that the magnetic properties of the lodestone were concentrated at these two locations. Today these two areas of concentrated magnetic force are called **magnetic poles**. Magnets have two poles, a north pole and a south pole. You probably already know that the opposite poles on two magnets, a north on one and a south on the other, will attract each other, while like poles will repel each other. Because this tug or push between two magnets can occur at a distance, magnetism is considered a field force (see Chapter 12).

Magnetic Fields

A magnet's **magnetic field** is the region surrounding the magnet where it can exert a magnetic force. Magnetic fields are invisible, of course, but there is a way to observe them (see right). Iron filings sprinkled onto a sheet of paper laid atop a magnet will align themselves in a distinct pattern. The filings reveal the shape and strength of the field. The well-defined, closely spaced lines nearest the magnet indicate where its greatest magnetic effects are felt. The two poles are also clearly seen.

Field Direction

It's not apparent from looking at the iron filings, but magnetic fields also have a direction. A compass needle placed within a magnetic field will always orient itself so that the north-pointing tip will point *away from* the north pole of the magnet and *toward* the south pole. This north-to-south direction of the field explains why two south poles will push each other away (right), while the opposite poles of two magnets will attract each other.

20.3 EARTH'S MAGNETIC FIELD

Compass needles, of course, always align themselves in a north-south direction. That's because Earth itself has a magnetic field. The earth's magnetic field is thought to originate deep inside the earth, probably due to the flow of liquid iron in the outer core. It is thought that the movement of the charged particles of the fluid creates the magnetic field. This magnetic field extends far out into space. Near Earth, the field deflects or traps the high-energy charged particles of the solar wind. Without a magnetic field, these particles would bombard Earth's atmosphere and surface, damaging or destroying life. Creation scientists point to Earth's magnetic field as just one of its many characteristics that are favorable for life. This is strong evidence in favor of a Creator.

You might be surprised to learn that Earth's magnetic North Pole is not truly a *north* magnetic pole. In reality, Earth's magnetic "North" Pole is a *south* magnetic pole. This may require some mental adjustment to understand, but it has little impact on the use of magnetic compasses.

Earth also has *two* North Poles (and two South Poles too)! The *geographic* North Pole is at the "top" of the earth at the axis of Earth's spin. This is sometimes referred to as "true north." All lines of longitude intersect at the geographic North and South Poles. The *magnetic* North Pole is where the earth's concentrated magnetic field lines are vertical at the earth's surface. The position of the magnetic North Pole is not fixed. Its position is constantly changing. As of 2018, the magnetic North Pole was located to the northwest of the Canadian Arctic and drifting westward toward Siberia. Since compasses are attracted to the magnetic North Pole, not the geographic North Pole, corrections must be made when using a compass to navigate. The difference in direction between the magnetic North Pole and geographic North Pole at a given location is called *magnetic declination*.

20.4 FACTORS AFFECTING MAGNETISM

The Domain Model of Magnetism

Each atom within a magnetic material creates a tiny magnetic field of its own. This magnetic force can cause groups of neighboring atoms to align their fields in the same direction. Such a group of atoms whose magnetic fields are aligned is called a **magnetic domain**. Within a magnetic object, domains may also be aligned in the same direction, or they may be arranged so that their fields are oriented in different directions. The degree of alignment affects how strongly magnetic a material is.

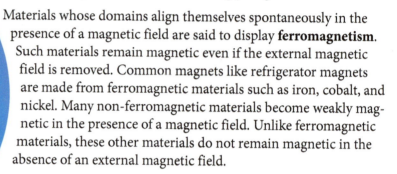

Materials whose domains align themselves spontaneously in the presence of a magnetic field are said to display **ferromagnetism**. Such materials remain magnetic even if the external magnetic field is removed. Common magnets like refrigerator magnets are made from ferromagnetic materials such as iron, cobalt, and nickel. Many non-ferromagnetic materials become weakly magnetic in the presence of a magnetic field. Unlike ferromagnetic materials, these other materials do not remain magnetic in the absence of an external magnetic field.

Temporary Magnets

Some materials can be temporarily magnetized in the presence of a strong magnetic field. For example, a common nail or sewing needle normally has unaligned domains. Repeatedly running a strong magnet along the length of a nail or needle aligns its domains, creating a magnetized nail or needle. But if the strong magnetic field is removed, the temporary magnet's domains will soon return to their random orientations. Even some fluids, called ferrofluids, can be temporarily magnetized, often resulting in bizarre shapes (see above left).

Magnets and Temperature

Magnetism begins at the particle level of matter, and temperature affects how particles move. It stands to reason then that temperature has an effect on magnetism. The higher the temperature of a material, the more random the motion of its particles becomes. In a magnetic material, this means that particles become increasingly less likely to keep their magnetic fields aligned. Iron, for example, abruptly loses its magnetic properties when heated to 770 °C.

20A | REVIEW QUESTIONS

1. What is a magnet?
2. Sketch Earth and its magnetic field lines. Indicate its *true* North and South Poles on the diagram and show the direction of the field.
3. The direction of a magnetic field is always from the _____ pole to the _____ pole.
4. What is magnetic declination?
5. What is a magnetic domain?
6. How are the domains in strongly magnetic and weakly magnetic materials different?
7. Why does iron lose its magnetic properties at temperatures above 770 °C?

MINI LAB

MAGNETIC FIELDS

You learned in the previous section that iron filings can show the shape of a magnetic field around a magnet. Magnets come in different shapes, including bars, horseshoes, and flat discs. Do all of these shapes produce similar magnetic fields? Let's find out!

Essential Question:

Does the shape of a magnet affect the shape of its magnetic field?

PROCEDURE

A Choose a magnet and place it flat on a desktop or table. Lay a sheet of paper over the magnet. If the edges of the paper droop too much, use something to prop them up.

B Sprinkle a small amount of iron filings onto the paper above the magnet. Use enough filings to clearly reveal the shape and extent of the magnet's magnetic field.

C On a separate sheet of paper, identify the shape of the magnet you used and sketch its magnetic field.

D Repeat Steps **A** – **C** for two additional but differently shaped magnets.

Equipment

magnets (3), (three different shapes)

sheet of paper

iron filings

ANALYSIS AND CONCLUSIONS

1. Did the magnetic fields for your magnets all have the same shape? If not, why do you think this was so?

2. Were you able to identify both poles on each magnet? If not, why do you think you could not?

GOING FURTHER

3. How do you think the field lines appear when looking straight down at one pole?

4. From time to time, the sun experiences an event called a *coronal mass ejection*. During these events, larger than normal amounts of energetic particles head toward Earth. At such times, astronauts aboard the International Space Station must take shelter in a heavily shielded compartment within the ISS. Why do you think this is so?

MAGNETISM 475

20B Questions

- Is there a link between magnetism and electricity?
- How is electromagnetism used?

20B Terms

right-hand rule, solenoid, electromagnet, electromagnetic induction

20B | ELECTROMAGNETISM

How do electric and magnetic fields interact?

20.5 ELECTRIC CURRENT AND MAGNETISM

Magnetism from Electricity

By the early 1800s, scientists believed that electricity and magnetism were linked, but they hadn't yet shown how. In 1820, while getting ready for a classroom demonstration, Danish physicist Hans Christian Oersted noticed that a nearby compass needle moved when he sent an electric current through a wire. He concluded that the flowing current had produced a magnetic field. Further experiments with the wire and compass showed that a current moving through a wire creates a circular magnetic field around the wire.

The direction of the magnetic field depends on which direction the electric current flows through the wire. The direction of the field can be found by using the **right-hand rule**. If you grasp a wire so that your right thumb points in the direction of the current flow, the wire's magnetic field lines will point in the same direction as your fingers wrapped around the wire. Changing the direction of the current changes the direction of the field.

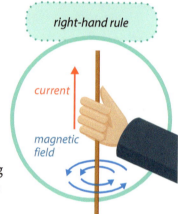

right-hand rule

Hans Christian Oersted's current experiment

wire with current

battery

moving compass needle

Shortly after Oersted's discovery, the French physicist André-Marie Ampère announced one of his own. He found that two parallel wires carrying current in the same direction were magnetically attracted (see below). After replacing one of the wires with a magnet, the remaining wire would move either away from or toward the magnet. The direction of movement hinged on the direction of current flow in the wire and the orientation of the magnet. Ampère furthermore learned that an increase in the electrical current also increased the strength of the magnetic field around the wire.

André-Marie Ampère's current experiment

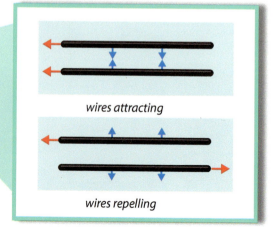

wires attracting

wires repelling

Later, scientists found that the magnetic field caused by a current moving through a wire is concentrated inside a loop of the wire. This arrangement produces distinct magnetic north and south poles, similar to those of a bar magnet. If the wire is further formed into a coil, the magnetic field strength is multiplied by the number of loops. Such a coil of wire that acts as a magnet is called a **solenoid** (below left). In 1823, Englishman William Sturgeon observed that placing a ferromagnetic bar inside a solenoid (below right) greatly increased the force of its magnetic field. Today, this coupling of a strong magnet inside an electrified wire coil is called an **electromagnet**. Sturgeon's solenoid with a pure iron core produced enough magnetic force to lift a piece of iron twenty times as heavy as the iron core itself! These days, electromagnets are critical parts of many devices. We'll look at some of them at the end of this section.

solenoid

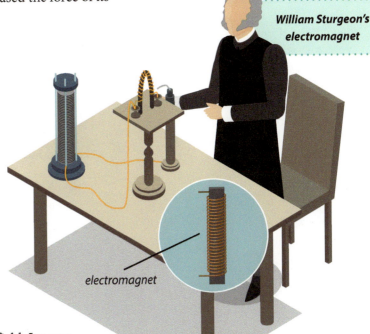

William Sturgeon's electromagnet

electromagnet

Electricity from Magnetism

Electrical current in a wire creates a magnetic field. It turns out that the reverse is true as well—a magnetic field can induce an electrical current in a wire. If a bar magnet is moved through a coil of wire, a current is produced in the wire, but only while the magnet is moving. When the magnet stops moving, the current stops too. Sliding the wire coil along the magnet has the same effect as moving the magnet. Increasing either the number of coils or the speed of motion increases the current produced. How does a magnetic field produce an electrical current? A magnetic field affects the motion of electrical charge carriers in a conductor. When the concentration of magnetic field lines near a conductor increases or decreases, a force is exerted on the electrons in the conductor. The highly mobile electrons move through the conductor, creating a current if the conductor is part of an electrical circuit. The faster the magnetic field strength changes, the more the electrons move. This effect is called **electromagnetic induction**. But even when a magnetic field surrounds a conductor, if the field does not change or the conductor does not move through the field, no current is produced.

20.6 USING ELECTROMAGNETISM

Since Sturgeon's discovery, electromagnets have been put to many uses where an easily controlled but variable force is needed. Let's look at a few of these.

ELECTROMAGNETISM AT WORK

Heavy Metal. If you visit a junkyard, you might see a crane using a large flat disk to lift pieces of cars and trucks. This disk is part of an electromagnet. When direct current flows through the disk, a strong magnetic force is produced. Pieces of metal can be released from the magnet by turning the current off.

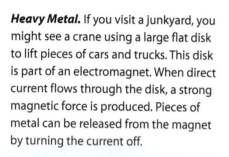

People Movers. To solve big-city commuter problems, a few countries have experimented with high-speed, magnetically levitated (maglev) trains. These trains have no engine and hover 1–10 cm over the guide rails using electromagnets. Since the train is not in direct contact with the rails, friction is greatly reduced, allowing the train to approach speeds of 500 kph!

Staying Focused. Particle accelerators—large machines used in particle physics research (see pp. 48–49)—use electromagnets to increase the speed of charged particles, to focus beams of charged particles, and to bend their paths, keeping them from colliding with the walls of the accelerator.

Taking a Peek Inside. MRI machines contain special electromagnets called *superconducting magnets*. These contain coiled wires made of niobium and either titanium or tin. Once cooled by liquid helium to temperatures below 20 K (−253 °C), resistance in the coiled wires is eliminated. They can then conduct large amounts of electric current and produce powerful magnetic fields. The strong magnetic forces cause hydrogen atoms in our bodies to generate radio waves, which the computer system turns into an image.

Music Makers. Electromagnets are used in pickups for electric musical instruments. The vibrating strings of a guitar, for example, create disturbances in the magnetic fields of the pickup's magnets. The disturbed fields induce an electric current that is then amplified and converted to sound energy by a speaker.

guitar pickup

Sorting It Out. Electromagnets are also useful for sorting metals. If a mixture of scrap iron and aluminum is carried on a conveyor through a magnetic field, the iron pieces strongly deflect to one side, while the aluminum passes on unaffected. Smelters use such sorters to keep iron contaminants out of recycled aluminum.

20B | REVIEW QUESTIONS

1. What is the relationship between magnetism and electricity?
2. An electric current in a wire flows from left to right. Use the right-hand rule to determine whether the field lines at the top of the wire point *toward* or *away from* you.
3. Describe a solenoid and explain why its shape is significant.
4. Compare solenoids and electromagnets.
5. Give a brief description of three uses for electromagnets.

20C Questions

- How is electricity generated?
- How is electrical energy transmitted?

20C Terms

generator, transformer

20C | GENERATING AND USING ELECTRICITY

Where does the electricity in my house come from?

20.7 GENERATING ELECTRICITY

Generators

Our modern world runs on electricity. Industrial machines, computers, home appliances—the list of things that run on electricity is a long one! You probably don't even think twice about flipping a switch to light up a dark room or pushing a button to heat a snack in the microwave. But such conveniences were unimaginable just a few generations ago. Though scientists had known about electricity for centuries, for much of that time it was mostly a novelty. It took the tireless efforts of some rather brilliant engineers and inventors to devise the means of producing cheap, reliable electric energy—and new ways to put that energy to work.

The electricity that comes from an outlet in your house is in the form of alternating current (AC). It is produced by electrical **generators** (left) that convert mechanical energy into electrical energy. A generator can produce AC in one of two ways: by rotating wire coils within a stationary magnetic field or by rotating a magnetic field within stationary wire coils. The rotating portion of the generator is called the *rotor*, while the stationary portion is called the *stator*. Mechanical energy to turn the rotor is provided by a *prime mover*, such as a steam turbine (see page 335), falling water, or an engine. As the rotor turns within the stator, the orientation of the magnetic field relative to the wires constantly changes. This causes the magnitude and direction of the current within the wire to constantly change, producing alternating current.

Direct current (DC) generators are much like AC generators. To produce current that flows in only one direction, DC generators include a structure called a *commutator*. The commutator functions much like a gate that allows current to flow from the generator in only one direction.

Electric Motors

We use the electricity in our homes in various ways, perhaps to produce heat in a curling iron or light from a lamp. Many times we need the electricity to power a motor (left), such as you would find inside an electric fan, drill, mixer, or vacuum cleaner. Electric motors are machines that convert electrical energy into rotary mechanical motion. How a motor works differs depending on the kind of current supply. Both AC and DC motors are constructed much like their corresponding generators. The operation of a motor is also much like that of a generator, but in reverse order. Instead of using magnetic fields or wire coils to turn a rotor inside a stator, producing electricity, motors use electricity to cause the interaction between the magnetic fields of the magnet and the wire coil's current to turn a rotor. The rotor is usually connected to some sort of tool to do work, such as a drill bit, drone propeller, or conveyor belt.

20.8 TRANSMITTING ELECTRICITY

Transformers

The first widely used generators produced direct current. First built in the 1870s, these generators were used to power the first electric street lights. The efficient transmission of electricity through power lines requires high voltages. But the voltages produced by early DC generators could not be easily changed. The solution proved to be AC current and the use of transformers. A **transformer** is a device that is used to increase or decrease the voltage of an AC current.

As you read earlier, a steady direct current flowing through a wire produces a magnetic field. This field, though, doesn't change, which means that it can't produce any electricity in nearby wires by electromagnetic induction. An alternating current, on the other hand, is constantly changing magnitude and direction. It produces a magnetic field whose magnitude and direction constantly change as well. The changing magnetic field of an AC conductor can generate current in other nearby conductors through induction.

Suppose two coils of insulated wire are placed next to each other. If alternating current is supplied to one coil, its alternating magnetic field induces an alternating current in the other coil. This process isn't very efficient. But if the insulated wires are wrapped around opposite sides of a rectangular iron frame (above), the strong ferromagnetic effect of the iron core greatly increases the efficiency of the induction between the coils. This is the basic design of a transformer.

If the coils described above each have the same number of wraps, the voltage in each coil is the same. But if they have different numbers of wraps, then their voltages differ as well. The difference in voltage is proportional to the difference in the number of wraps between the input and output coils. In the figure at left, the left-hand, or input, coil of the transformer has more wraps than the right-hand, or output, coil. The changing magnetic current in the input coil produces a magnetic field whose strength is related to the number of wraps of wire—the more wraps, the greater the field strength. The changing magnetic field in the iron core induces a current in the output coil. The voltage in the output coil, however, is lower because it has fewer wraps than the input coil.

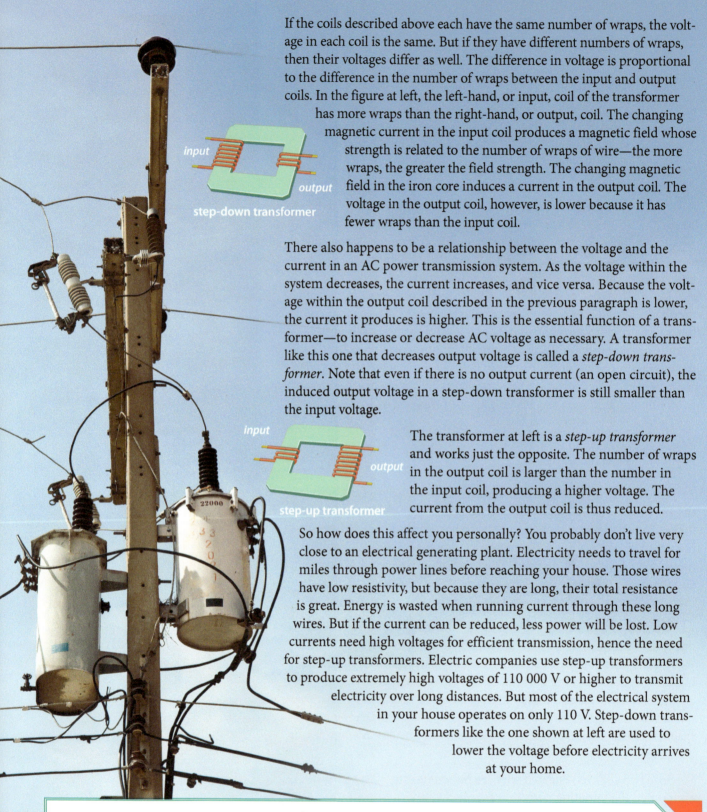

There also happens to be a relationship between the voltage and the current in an AC power transmission system. As the voltage within the system decreases, the current increases, and vice versa. Because the voltage within the output coil described in the previous paragraph is lower, the current it produces is higher. This is the essential function of a transformer—to increase or decrease AC voltage as necessary. A transformer like this one that decreases output voltage is called a *step-down transformer*. Note that even if there is no output current (an open circuit), the induced output voltage in a step-down transformer is still smaller than the input voltage.

The transformer at left is a *step-up transformer* and works just the opposite. The number of wraps in the output coil is larger than the number in the input coil, producing a higher voltage. The current from the output coil is thus reduced.

So how does this affect you personally? You probably don't live very close to an electrical generating plant. Electricity needs to travel for miles through power lines before reaching your house. Those wires have low resistivity, but because they are long, their total resistance is great. Energy is wasted when running current through these long wires. But if the current can be reduced, less power will be lost. Low currents need high voltages for efficient transmission, hence the need for step-up transformers. Electric companies use step-up transformers to produce extremely high voltages of 110 000 V or higher to transmit electricity over long distances. But most of the electrical system in your house operates on only 110 V. Step-down transformers like the one shown at left are used to lower the voltage before electricity arrives at your home.

20C | REVIEW QUESTIONS

1. How can a generator produce AC current?
2. Why is it necessary for rotation to occur within a generator?
3. How can a generator that otherwise produces AC current be modified to produce DC current?
4. Describe the relative number of input and output coils in a step-up transformer.

CHAPTER 20 REVIEW

20A MAGNETS AND MAGNETISM

- A magnet is any material or object that can produce a magnetic field.
- An object's magnetic field is the region over which it can exert a magnetic force. The field connects two areas of concentration, called poles.
- Magnetic fields have direction, pointing away from the magnet's magnetic north pole and toward its magnetic south pole.
- The earth also has a magnetic field that helps to protect it from the solar wind. Earth's magnetic poles drift over time.
- A magnetic domain consists of a group of atoms whose magnetic fields are aligned. The greater the alignment among an object's domains, the stronger its magnetic field.
- Ferromagnetic materials have domains that remain aligned even in the absence of an external magnetic field.

20A Terms

magnet	472
magnetism	472
magnetic pole	472
magnetic field	473
magnetic domain	474
ferromagnetism	474

20B ELECTROMAGNETISM

- An electric current flowing through a wire produces a magnetic field. The direction of the field can be determined by using the right-hand rule.
- The loops of wire in a solenoid increase the strength of the current's induced magnetic field.
- Placing a ferromagnetic bar within a solenoid increases the strength of the solenoid's magnetic field, creating an electromagnet.
- A moving magnetic field produces an electric current in a wire via electromagnetic induction.

20B Terms

right-hand rule	476
solenoid	477
electromagnet	477
electromagnetic induction	477

20C GENERATING AND USING ELECTRICITY

- Electrical generators produce electric current by rotating either wire coils inside a magnetic field or a magnetic field within wire coils.
- A prime mover provides the mechanical energy needed to turn the rotor of an electrical generator.
- The long-distance transmission of AC current requires very high voltages.
- The voltage of an AC current is increased by a step-up transformer prior to transmission and decreased by a step-down transformer prior to residential use.

20C Terms

generator	480
transformer	481

MAGNETISM 483

CHAPTER 20

CHAPTER REVIEW QUESTIONS

Recalling Facts

1. What do scientists currently believe causes Earth's magnetic field?
2. Permanent magnets are made of what kind of material?
3. What type of material is added to a solenoid to create an electromagnet?
4. What is electromagnetic induction?
5. How is electromagnetism used in a particle accelerator?
6. Why does part of a generator have to rotate?
7. What function does a commutator serve?

Understanding Concepts

8. What is the distinction between a magnet and a magnetic field?
9. Is Earth's magnetic North Pole an actual magnetic north pole? Explain.
10. How does Earth's magnetic field make life on the planet possible?
11. Topographic maps usually include an indication of the angular difference, called the magnetic declination, between true north and magnetic north on the map. Why is the declination shown on a forty-year-old map likely to be incorrect?
12. Describe how a temporary magnet can be made.
13. Sketch a wire and place an arrow next to it to indicate the direction of an electric current flowing through the wire. Add magnetic field lines to your sketch to show the magnetic field around the wire.
14. Apart from converting it into an electromagnet, how can the strength of a solenoid's magnetic field be increased?
15. How do magnets cause electromagnetic induction?
16. Briefly describe three uses for electromagnets.
17. Examine the images at left. If you wanted to increase the voltage in the wire on the right side, which transformer would you use and why?
18. Why is the voltage stepped up prior to sending electricity through high-voltage power lines?

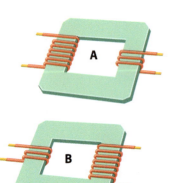

Critical Thinking

19. On the basis of current scientific theory, would you consider the earth a permanent magnet or an electromagnet? Explain.
20. What do you suppose happens to the poles and magnetic fields of a magnet if you break it in half?
21. In what sense is a permanent magnet not really permanent?
22. Why would an electric motor not make a good prime mover for an electrical generator?
23. Explain how wire coils, or windings, might be used for both the rotor and stator of an electric motor.

REVIEW

WORLDVIEW SLEUTHING: THE WAR OF THE CURRENTS

As you've already seen, there are two types of electrical current that can be generated and used, alternating current (AC) and direct current (DC). Your toaster and waffle iron run on AC. Your battery-powered flashlight and smartphone operate on DC. What you might *not* be familiar with is why AC is used in your home instead of DC.

In the late 1800s, two brilliant inventors waged a battle of sorts over which type of current would provide electrical energy to the nation. Thomas Edison pushed for the adoption of DC, while Nikola Tesla argued for AC. This struggle came to be known as "The War of the Currents."

TASK

It's 1885, and the city council of Tenebraria is debating whether to use AC or DC to power its new street lights. You will assume the role of either Edison (DC) or Tesla (AC) and create a presentation for the council that argues for your particular system.

PROCEDURE

1. Start your research on the issue by doing a keyword search for "war of the currents." Identify factors in the debate such as the costs, benefits, hazards, and required infrastructure for using each type of current.

2. Plan your presentation and collect any required materials. Be sure to cite your sources. Remember that for this particular task, you will be attempting to persuade your audience to choose your system—even if your research leads you to regard the other system as the better choice!

3. Show your presentation to a classmate or friend for feedback.

4. Present your findings to your classmates.

CONCLUSION

Deciding whether to adopt DC or AC for widespread power generation and distribution wasn't a simple task. Each system has its benefits and drawbacks. And as in other instances of deciding how to derive the most benefit from scientific discoveries, God's command to love and serve others can be obscured by a person's desire to further his own self-interest.

Edison   *Tesla*

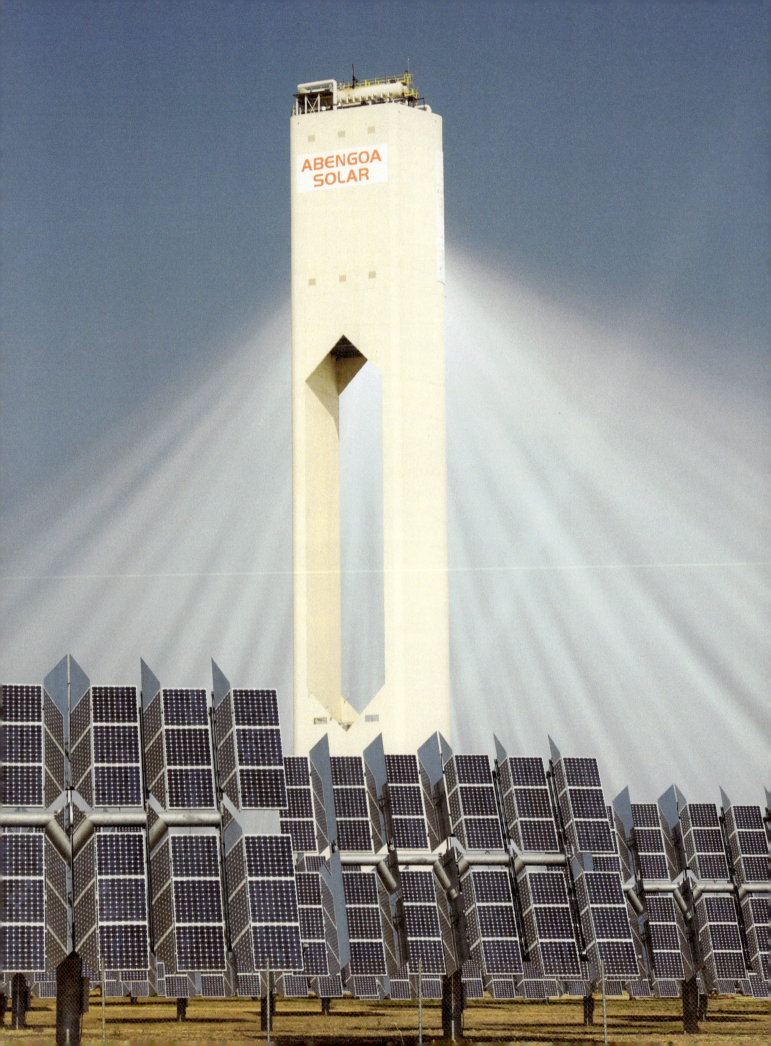

CHAPTER 21
Electromagnetic Energy

SOLAR POWER TOWER

Some people support solar power research because of their concern for the environment. Others like the prospect of using an available renewable energy resource. There may even be some people that like the PS10 Solar Power Tower because of its sheer beauty.

Towering 115 m over the plains of Seville in Spain, the PS10 solar power plant became the first commercial concentrating power tower in 2004. The tower collects the solar energy reflected from 624 mirrors that surround it. This reflected energy superheats water to 275 °C at a pressure of 50 PSI. The superheated water is then used to generate electrical power. The plant produces a total of 23.5 GWh of electrical energy each year.

The energy that powers the PS10 solar power plant travels through the vacuum of space in the form of electromagnetic waves. Throughout this chapter you will learn about electromagnetic waves and how we use them.

21A Electromagnetic Waves   488

21B The Electromagnetic Spectrum   496

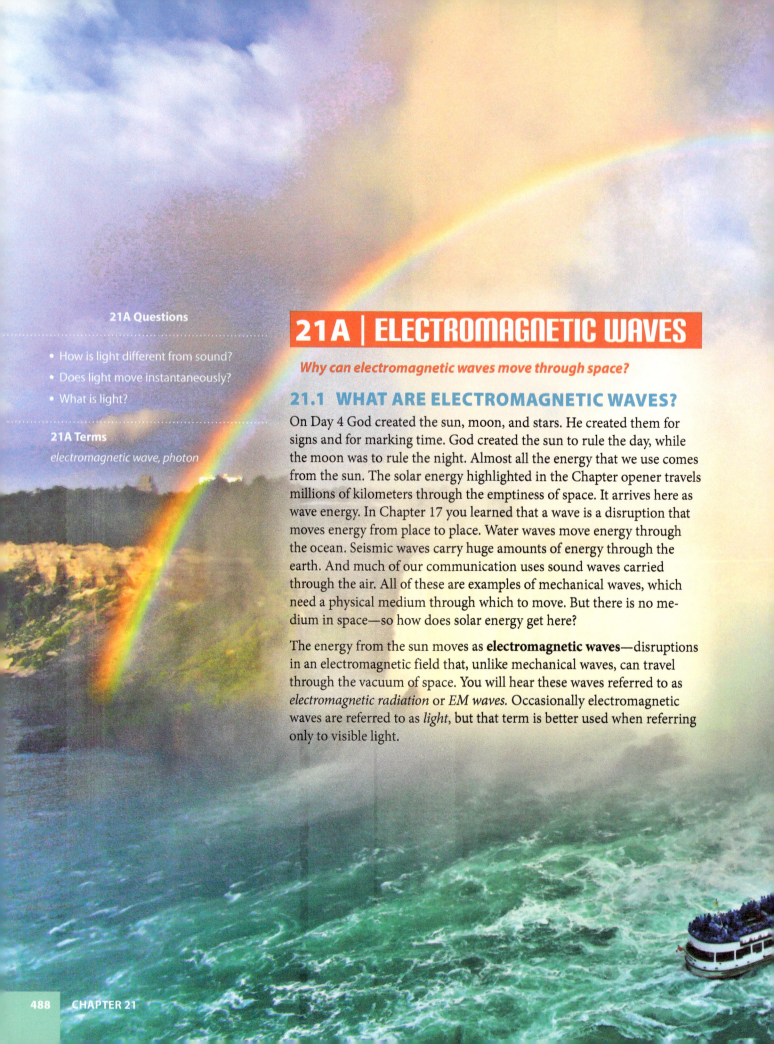

21A Questions

- How is light different from sound?
- Does light move instantaneously?
- What is light?

21A Terms

electromagnetic wave, photon

21A | ELECTROMAGNETIC WAVES

Why can electromagnetic waves move through space?

21.1 WHAT ARE ELECTROMAGNETIC WAVES?

On Day 4 God created the sun, moon, and stars. He created them for signs and for marking time. God created the sun to rule the day, while the moon was to rule the night. Almost all the energy that we use comes from the sun. The solar energy highlighted in the Chapter opener travels millions of kilometers through the emptiness of space. It arrives here as wave energy. In Chapter 17 you learned that a wave is a disruption that moves energy from place to place. Water waves move energy through the ocean. Seismic waves carry huge amounts of energy through the earth. And much of our communication uses sound waves carried through the air. All of these are examples of mechanical waves, which need a physical medium through which to move. But there is no medium in space—so how does solar energy get here?

The energy from the sun moves as **electromagnetic waves**—disruptions in an electromagnetic field that, unlike mechanical waves, can travel through the vacuum of space. You will hear these waves referred to as *electromagnetic radiation* or *EM waves*. Occasionally electromagnetic waves are referred to as *light*, but that term is better used when referring only to visible light.

In the 1600s, Sir Isaac Newton discovered that he could form a rainbow by using a prism to disperse sunlight. Newton didn't know that light was related to electrostatic or magnetic forces. It was André-Marie Ampère who in 1820 suggested a theory that combined these forces as electromagnetism. It wasn't until 1864 that the Scottish physicist James Clerk Maxwell published a set of comparatively simple mathematical equations that describe the relationship between electricity and magnetism. He also showed that both electric and magnetic fields interact in wavelike fashion and suggested that light actually moves as a wave disturbance in electric and magnetic fields.

21.2 MAKING AND MOVING ELECTROMAGNETIC WAVES

In Chapter 17, you learned that a vibrating, or oscillating, object produces mechanical waves. Electromagnetic waves are no different, but in this case the waves are produced by vibrating charged particles, typically electrons. To understand how electromagnetic waves are produced and how they move through space, we have to connect a few things that you learned in previous chapters. Recall from Chapter 19 that electric fields are the regions around charged objects that can exert an electric force on other charged objects. And in Chapter 20 you learned that moving charges and changing electric fields both produce magnetic fields. You also learned that a changing magnetic field produces electric fields. Maxwell showed that the interaction between electric and magnetic fields produces electromagnetic waves. As an electron vibrates, it produces a changing electric field and a changing magnetic field. The interaction of these fields causes the movement of waves that carry energy away from the source. In the image on the right, notice that the waves in both the electric and magnetic fields are transverse waves and that they are also perpendicular to each other. The electric field (blue) is oscillating vertically while the magnetic field (red) moves horizontally (the wave is moving from left to right).

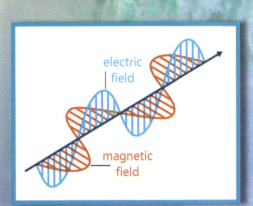

ELECTROMAGNETIC ENERGY 489

21.3 PROPERTIES OF ELECTROMAGNETIC WAVES

Speed of Electromagnetic Waves

All the properties of waves that we discussed in Chapter 17 apply to electromagnetic waves. One property that intrigued scientists was the speed at which these waves moved. Initially, some scientists thought that light moved instantaneously. But in 1676 Danish astronomer Ole Rømer proved that light didn't move instantaneously. While studying the moons of Jupiter, Rømer noticed that the time for light to travel to Jupiter and back to Earth varied depending on the position of the two planets in their orbits. This indicated that light moved at a finite speed. Due to the high rate at which light moves, experiments for measuring the speed of light were initially limited to the field of astronomy.

American pysicist Albert Michelson was one of the scientists fascinated by the question of how fast light moved. He spent fifty years working on land-based experiments to refine measurements of the speed of light. The image below shows the setup for Michelson's most successful experiment. In the early 1930s, Michelson's team set up a 1.6 km long vacuum tube with a light source, mirrors (including an eight-sided rotating mirror), and sensors. This setup allowed for a beam of light to move the length of the tube ten times. Michelson's team measured the speed of light to within 0.004% of the currently accepted value.

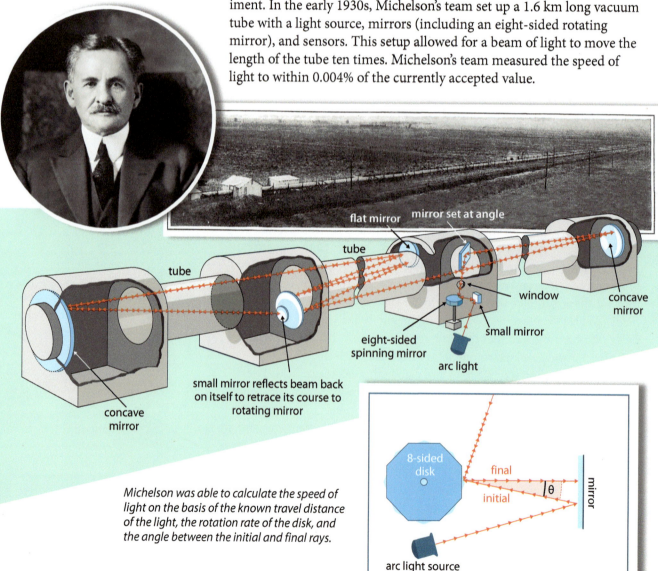

Michelson was able to calculate the speed of light on the basis of the known travel distance of the light, the rotation rate of the disk, and the angle between the initial and final rays.

There are many types of electromagnetic waves besides visible light, but they all move at the speed of light, a constant indicated by the letter c (as seen in Einstein's famous equation $E = mc^2$). Today we know that the speed of light in a vacuum is 299 792 458 m/s. We know it exactly because scientists redefined the meter on the basis of the speed of light. For our purposes, we will use the rounded value of 3.00×10^8 m/s to make calculations easier. Light travels slower in other materials.

Wavelength and Frequency

Just like with mechanical waves, we can measure the wavelength and frequency of electromagnetic waves. In Chapter 17 you learned the relationship between wave speed, wavelength, and frequency. This relationship is true for electromagnetic waves as well.

Since wave speed equals the product of frequency and wavelength, we can write the formula for the speed of electromagnetic waves as

$$c = f\lambda,$$

where c is the speed of the electromagnetic wave in a vacuum in meters per second, f is the frequency of the wave in hertz, and λ is the wavelength in meters.

Since c is constant in a vacuum, frequency and wavelength are inversely proportional to each other, meaning that higher-frequency waves have shorter wavelengths and vice versa.

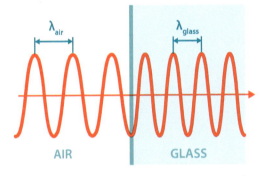

The vibrations of the source determine the frequency of the wave and the frequency doesn't change as light enters a new medium. So when an EM wave enters a denser material, its speed slows, causing the wavelength to shorten.

EXAMPLE 21-1: Determining Wavelength of Electromagnetic Waves

What is the wavelength of a 105.1 MHz radio wave? of a 3.00×10^{18} Hz gamma ray?

Write what you know.

$c = 3.00 \times 10^8$ m/s

$f_{radio} = 105.1 \; \text{MHz} \left(\dfrac{10^6 \text{ Hz}}{1 \text{ MHz}} \right) = 1.051 \times 10^8$ Hz

$f_{gamma\,ray} = 3.00 \times 10^{18}$ Hz

$\lambda_{radio} = ?$

$\lambda_{gamma\,ray} = ?$

Write the formula and solve for the unknown.

$c = f\lambda$

$\dfrac{c}{f} = \dfrac{\cancel{f}\lambda}{\cancel{f}}$

$\lambda = \dfrac{c}{f}$

Plug in known values and evaluate.

$\lambda_{radio} = \dfrac{3.00 \times 10^8 \; \frac{m}{s}}{1.051 \times 10^8 \; \frac{1}{s}} = 2.85$ m

$\lambda_{gamma\,ray} = \dfrac{3.00 \times 10^8 \; \frac{m}{s}}{3.00 \times 10^{18} \; \frac{1}{s}} = 1.00 \times 10^{-10}$ m

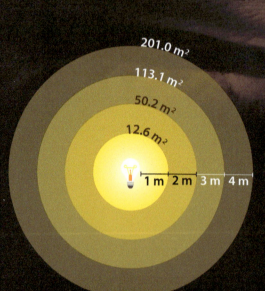

As light radiates from its source, the energy is spread over an increasing area. The energy per square meter is inversely proportional to the distance from the source.

Energy

Recall that the energy of a wave is related to its amplitude, but since electromagnetic waves do not displace matter, we can't measure the amplitude of the waves directly. But German physicist Max Planck determined that the energy of a particle of light—a *photon*—is directly related to the frequency of the wave. Planck's equation is written as

$$E = hf,$$

where E is the energy of the wave, h is Planck's constant, and f is the frequency of the wave. Since we know that frequency and wavelength are inversely related, we know that energy and wavelength must also be inversely related. Therefore, high-energy waves have high frequencies and short wavelengths, while waves with long wavelengths have low frequencies and relatively low energies.

Intensity

In Chapter 18 you learned about the intensity of sound waves. Electromagnetic waves also vary in their intensity. The intensity of electromagnetic waves coming from a spherical source, like a star or a light bulb, is related to the energy in the wave as well as to how far the observer is from the source. The intensity from this source is an inverse square law just like the law of universal gravitation and Coulomb's law. So when an observer moves farther away from a source of electromagnetic energy, the intensity of the wave decreases. If the distance is doubled, the intensity decreases to one-fourth of its original value. If we triple the distance, the intensity is decreased to one-ninth of its original value.

21.4 NATURE OF ELECTROMAGNETIC WAVES

We've been learning about electromagnetic waves, and we have also referred to the term *photons* as particles of light. So does electromagnetic energy move as a wave or a particle? Let's look at the two major models of electromagnetic energy.

MODELS OF ELECTROMAGNETIC WAVES

WAVE MODEL

In 1801, English scientist Thomas Young proved that light behaved as a wave. He set up a light source shining on a screen with two thin slits. Young hypothesized that if the light acted as particles, he would see two bright bands on a second screen, one for each slit. If the light acted as waves, each slit would diffract the light, and the two light sources (slits) would produce light waves that would interfere with each other, creating repeated bright and dark bands. Young's results showed the wave nature of light. Electromagnetic waves exhibit the same behaviors as mechanical waves.

PARTICLE MODEL

Isaac Newton was a firm believer that light consisted of particles. Evidence for light's particle nature came in 1887 as German scientist Heinrich Hertz noticed that certain metals would emit electrons when electromagnetic energy struck them. Further investigation showed that particular metals would emit electrons only if the light were of a high enough frequency. Once that frequency was reached, higher intensities of light would produce more electrons.

The first image at left shows light with a frequency that is too low to produce electrons, regardless of its intensity. The second image shows a higher frequency of light of the same intensity producing many electrons. In the third image, the frequency is still high enough to cause the emission of electrons, but fewer are emitted due to the lower intensity of the light.

If light were waves, we would expect sufficiently intense light of any frequency to emit electrons. But if light were particles, then only particles with sufficient energy (frequency) would transfer enough kinetic energy to the electrons to eject them. This behavior is known as the *photoelectric effect*.

ELECTROMAGNETIC ENERGY

Wave-Particle Duality

So is light a wave or a particle? Albert Einstein solved the problem in his 1905 Noble Prize-winning paper on the photoelectric effect. He explained that all electromagnetic energy travels as massless particles called **photons**. A photon is a bundle of electromagnetic wave energy. So in reality, electromagnetic energy travels as both particles and waves—scientists call this behavior the *wave-particle duality*.

But you may be wondering how light can be both a wave and a particle. Remember that the goal of science is to develop models to explain or describe the world around them. Good models are workable—they explain observations and make accurate predictions. Our current model of light is that light exhibits both wave and particle characteristics. Is light actually both a wave and a particle? Maybe, but we at least know that our current model is the most workable one developed. As we continue to refine our model of light, we learn more about God's created world and can use our knowledge to help others.

21A | REVIEW QUESTIONS

1. Define *electromagnetic wave*.
2. How is light different from sound?
3. Who is credited with the mathematical formulation of electromagnetism?
4. What is the wavelength of a microwave with a frequency of 6.72×10^9 Hz?
5. What is the frequency of a radio wave with a wavelength of 2.92 m?
6. Ultraviolet waves have higher frequencies than infrared waves. Which ones have more energy? Explain.
7. If I move a lamp farther from my desk, why does the light get dimmer?
8. What evidence is there for both the wave nature and particle nature of light?

MINI LAB

TESTING SUNSCREEN

Too much of a good thing! This statement can be said about many things and is especially true of exposure to ultraviolet waves. We need short-term exposure to produce vitamin D, but too much can result in a painful sunburn. In this lab activity, you will investigate the effectiveness of various sunscreens.

Essential Question:

How well do sunscreens block UV waves?

Equipment

sunscreen

petri dishes (3)

ultraviolet detection beads (18)

1. Where, within the range of electromagnetic energies, do we find ultraviolet waves?

PROCEDURE

A Place six UV detection beads in each of the three petri dishes and cover.

2. What do you think the ultraviolet detection beads are modeling?

B Using two sunscreens with different SPF ratings, prepare two of the petri dishes by applying a thin layer of sunscreen to the exterior sides and the top of the lid. Wash any sunscreen off your hands.

C Place all three petri dishes in direct sunlight.

3. Predict how much color change will be seen for the beads in each of the three petri dishes. Write your prediction as a hypothesis.

ANALYSIS

D After fifteen minutes, remove petri dishes from the sunlight and open all three dishes. Record your observations.

4. How did your results compare with your hypothesis?

5. What was the purpose of the untreated petri dish?

CONCLUSION

6. Was there a significant difference in color change on the basis of the SPF rating of the sunscreen applied?

GOING FURTHER

7. Would you expect UV detection beads to change color on a cloudy day? Explain.

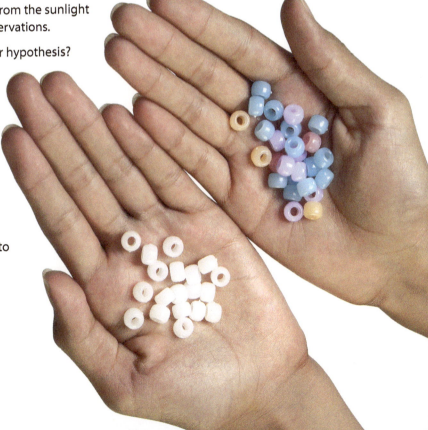

ELECTROMAGNETIC ENERGY 495

21B Questions

- What is a spectrum?
- What are the different types of electromagnetic waves?
- What distinguishes one electromagnetic wave from another?
- How do we use the various types of electromagnetic waves?

21B Terms

electromagnetic spectrum, radio wave, microwave, infrared wave, visible light, ultraviolet wave, x-ray, gamma ray

21B | THE ELECTROMAGNETIC SPECTRUM

How can we use electromagnetic energy?

21.5 TYPES OF ELECTROMAGNETIC WAVES

Electromagnetic waves come in many forms. We classify these waves on the basis of their frequency, wavelength, and energy. The collection of all the electromagnetic waves is called the **electromagnetic spectrum**.

ELECTROMAGNETIC SPECTRUM

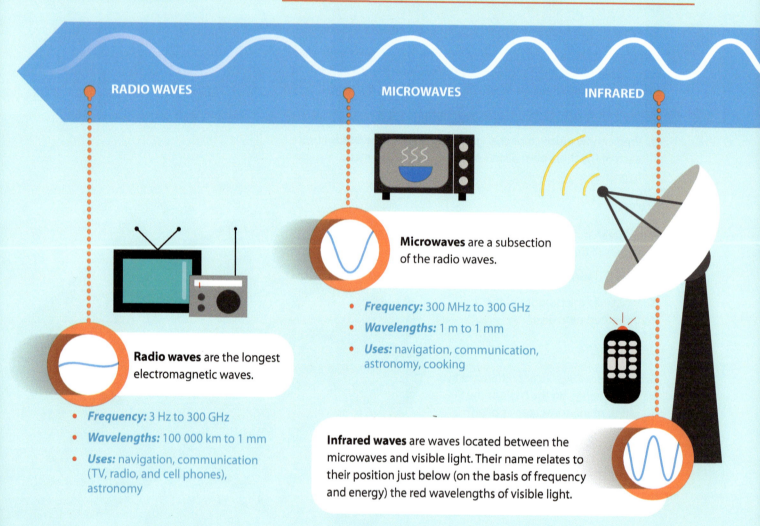

RADIO WAVES

Radio waves are the longest electromagnetic waves.

- *Frequency:* 3 Hz to 300 GHz
- *Wavelengths:* 100 000 km to 1 mm
- *Uses:* navigation, communication (TV, radio, and cell phones), astronomy

MICROWAVES

Microwaves are a subsection of the radio waves.

- *Frequency:* 300 MHz to 300 GHz
- *Wavelengths:* 1 m to 1 mm
- *Uses:* navigation, communication, astronomy, cooking

INFRARED

Infrared waves are waves located between the microwaves and visible light. Their name relates to their position just below (on the basis of frequency and energy) the red wavelengths of visible light.

- *Frequency:* 300 GHz to 430 THz (terahertz, 10^{12})
- *Wavelengths:* 1 mm to 700 nm
- *Uses:* astronomy, medical imaging, and wireless devices (Some animals such as snakes, bats, and insects can detect waves in the infrared spectrum.)

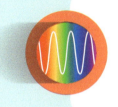

Visible light is the collection of electromagnetic waves that we can **see**. We think of these waves as special, but in reality it is the special design of our eyes that makes it possible for us to see these waves.

- *Frequency:* 430 THz to 790 THz
- *Wavelengths:* 700 nm to 380 nm

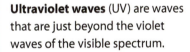

VISIBLE SPECTRUM | ULTRAVIOLET | X-RAYS | GAMMA RAYS

X-rays are high energy waves that were discovered accidentally by Wilhelm Röntgen (see Chapter 8).

- *Frequency:* 30 PHz to 30 EHz (exahertz, 10^{18})
- *Wavelengths:* 10 nm to 10 pm (picometers 10^{-12})
- *Uses:* medical imagery, transportation security, nondestructive inspection

Ultraviolet waves (UV) are waves that are just beyond the violet waves of the visible spectrum.

- *Frequency:* 790 THz to 30 PHz (petahertz, 10^{15})
- *Wavelengths:* 380 nm to 10 nm
- *Uses:* medical treatment, dentistry, killing bacteria

Gamma rays are the highest-energy electromagnetic waves that are emitted during radioactive decay (Ch. 8) and from sources in space.

- *Frequency:* above 30 EHz
- *Wavelengths:* 10 pm down to 1 fm (femtometers, 10^{-15})
- *Uses:* medical treatment, astronomy

ELECTROMAGNETIC ENERGY

21.6 USING ELECTROMAGNETIC WAVES

As you have seen, there are many different types of electromagnetic waves. You can imagine that there are many ways that we use them in our daily lives.

Communication

Radio waves are used for transmitting radio signals. We refer to a particular radio station by either its call letters or frequency. You may like listening to music on your favorite radio station 93.5. Or you may listen to a ball game on 1030. These numbers refer to the frequency of the radio waves carrying the signal. The number 93.5 refers to 93.5 MHz on the FM band, which ranges from 88.0 to 108.0 MHz. The number 1030 also refers to a frequency, 1030 kHz on the AM band (530 to 1600 kHz). AM and FM refer to the methods by which the radio signal is produced.

A radio system needs a signal producer and an antenna, which changes the electrical signal into a radio wave. A radio receiver also needs an antenna, in this case to change the radio wave back into an electrical signal. The speaker in the receiver then changes the electrical signal into a sound wave that we can hear. In both AM and FM systems, the information is encoded in the radio wave by modulating, or changing, some characteristic of the wave. In an AM system, the information is encoded in the wave by changing the amplitude of the wave, a process known as *amplitude modulation* (AM). By changing the frequency of the wave—*frequency modulation* (FM)—information is encoded in an FM radio wave.

Radio waves are also used to transmit television signals and navigational information.

unmodulated wave

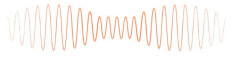
amplitude modulation (AM)

frequency modulation (FM)

Telephone

It used to be that people had telephones only in their homes and offices. If you needed to make a call while away from those locations, you had to look for a phone booth and then hope that you had enough change to make your call. Today almost everyone has a phone at hand all the time. The cell phone you carry around uses a variety of electromagnetic waves to accomplish the tasks you ask of it. It can take photographs in the visual spectrum. It uses Bluetooth® (high-frequency radio waves) to communicate wirelessly. It transmits information, using radio waves, for using the internet, sending text messages, and, of course, making phone calls.

Cooking

The microwave oven has provided us a tremendously convenient method of cooking and reheating food. The oven produces microwaves that cause polar molecules (of water in particular) to rotate as they try to align with the varying fields. These molecules collide with other molecules, increasing the kinetic energy (thus, temperature) of food. Microwave ovens cook food much quicker than conventional methods do. We also use infrared waves for heating food. Many restaurants and large kitchens have heat lamps that produce infrared waves.

Medicine

Many electromagnetic waves are used in the field of medicine. The first that probably comes to mind is the x-ray, used to observe within the body. Gamma rays are used also, especially in the treatment of cancer. Ultraviolet waves can be used to treat some skin conditions, while infrared waves are used to reduce some side effects of chemotherapy.

WORLDVIEW SLEUTHING: AUTONOMOUS VEHICLE SENSORS

In the United States, almost six million car accidents happen every year, costing over $200 billion. Distracted, impaired, or fatigued drivers cause many of these accidents. As of 2018, forty-six automotive and technology companies were working on autonomous vehicles, which drive without human input. The greatest challenge to an autonomous vehicle is providing it enough information to drive safely. Autonomous vehicles (see image opposite page) use a variety of sensors that collect information from different parts of the electromagnetic spectrum.

TASK

You write for a technology-based website and are tasked with writing an article about the variety of sensors that autonomous vehicles use to drive safely without human input.

PROCEDURE

1. Do the necessary research to develop your article. Do an internet search using the keywords "autonomous vehicles" and "sensors of self-driving cars."
2. Plan your article and collect any required materials. Remember to cite your sources.
3. Show your article to a classmate or friend for feedback.
4. Submit your article.

CONCLUSION

Autonomous vehicles can't get distracted, be impaired, or fall asleep and so could prevent the majority of car accidents. God gives us the technological ability to solve problems to protect human life for His glory.

21B | REVIEW QUESTIONS

1. What is the electromagnetic spectrum?
2. List the seven major bands in the electromagnetic spectrum in order from lowest to highest frequency.
3. What distinguishes one type of light from another?
4. Which types of electromagnetic waves have wavelengths between visible light and gamma rays?
5. What types of electromagnetic waves can we use for cooking?
6. How do microwaves heat food?
7. Pick two types of electromagnetic energy and explain their use in medicine.

CHAPTER 21

21A ELECTROMAGNETIC WAVES

- Electromagnetic waves are disruptions in the electromagnetic field that carry energy, even through the vacuum of space.
- Electromagnetic waves are produced by vibrating charged particles that create transverse waves in electric and magnetic fields.
- All electromagnetic waves move at the speed of light. The speed of light in a vacuum is 299 792 458 m/s.
- The energy of a photon is directly proportional to the frequency of the wave.
- Electromagnetic energy exists as wave bundles, called photons, and exhibits both wave and particle behaviors, reflecting the wave-particle duality of our models.

21A Terms

electromagnetic wave	488
photon	494

21B THE ELECTROMAGNETIC SPECTRUM

- Electromagnetic waves come in a variety of forms and are classified on the basis of their wavelength, frequency, and energy.
- The collection of all electromagnetic waves is called the electromagnetic spectrum.
- Electromagnetic waves have been used in a variety of ways—communication, navigation, medicine, cooking, astronomy, and transportation.

21B Terms

electromagnetic spectrum	496
radio wave	496
microwave	496
infrared wave	496
visible light	497
ultraviolet wave	497
x-ray	497
gamma ray	497

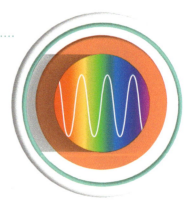

CHAPTER REVIEW QUESTIONS

Recalling Facts

1. Who first introduced the concept of electromagnetism?
2. Describe the waves in both the electric and magnetic fields as an electromagnetic wave moves.
3. Do electromagnetic waves move instantaneously? Explain.
4. Radio waves have longer wavelengths than x-rays. Which ones have more energy? Explain.
5. The observation that specific frequencies of light have enough energy to eject electrons from metals supports the particle model of light. What do scientists call this behavior?
6. Define *photon*.
7. Name the band of electromagnetic waves that has the highest energy and the one with the lowest energy.
8. List the seven major bands in the electromagnetic spectrum in order from shortest to longest wavelength.
9. What types of electromagnetic waves do cell phones use?

Understanding Concepts

10. Why can electromagnetic waves move through space?
11. Compare mechanical and electromagnetic waves.
12. Explain how an electromagnetic wave is produced.
13. How many waves pass by per second if waves of blue light are 480 nm long?

REVIEW

14. How much distance exists between crests of an x-ray if 4.88×10^{18} waves pass you every second?

15. Thomas Young's experiment supported which theory of light? Explain.

16. Explain what is meant by *wave-particle duality*.

17. What does *infrared* mean?

18. What makes visible light special?

19. What type of light is considered to be just *beyond* the visible light portion of the electromagnetic spectrum?

20. Pick two bands of electromagnetic waves from the spectrum and compare them.

21. Compare AM and FM waves.

22. What does a radio antenna do?

Critical Thinking

23. Why do you think that Michelson needed the light to travel ten times through the vacuum tube when measuring the speed of light?

24. Your classmate tells you that Planck's equation ($E = hf$) implies that all the photons of a particular frequency of light have the same energy. How could a source of that light emit more energy?

25. If you move from 10 m from a light source to 2.5 m (one-fourth of the original distance), how much more intense will the light be? Explain.

26. Why do you think that physicists accept the wave-particle duality as their current model of light?

27. From a biblical perspective, explain why a Christian can accept the uncertainty about the nature of light (i.e., whether it is a wave, a particle, or both).

28. Respond to the following statement: Due to their harmful effect on living organisms, we should avoid all ultraviolet waves.

Use the information in the Case Study below to answer Questions 29–31.

29. Where are infrared and ultraviolet light on the electromagnetic spectrum?

30. Many people consider the visible spectrum to be specific wavelengths and frequencies of light. How could we change the definition on the basis of how birds, snakes, and other animals see?

31. How do you think a naturalistic scientist and a creationist would explain the ability of different organisms to sense different frequencies?

CASE STUDY: SEEING IS BELIEVING

By this point, you know that visible light has wavelengths that range from 380 nm to 700 nm. However, some animals would disagree with that. Some animals can sense infrared waves, and others can actually see ultraviolet waves.

Pit vipers, pythons, and some insects can detect infrared waves. The pit viper has a pit just behind each nostril that senses thermal energy associated with infrared waves. The combination of its eyes, sensing in the visible range, and its infrared sensors make vipers tremendous hunters.

Insects, birds, and even some reptiles and mammals can see light in the ultraviolet range. The structure of a bird's eye includes four color sensing cones, including one that senses UV light. Recently scientists have conducted studies to see whether UV vision varies in different species of birds. They found that the smaller the bird's eye, the better its UV vision is.

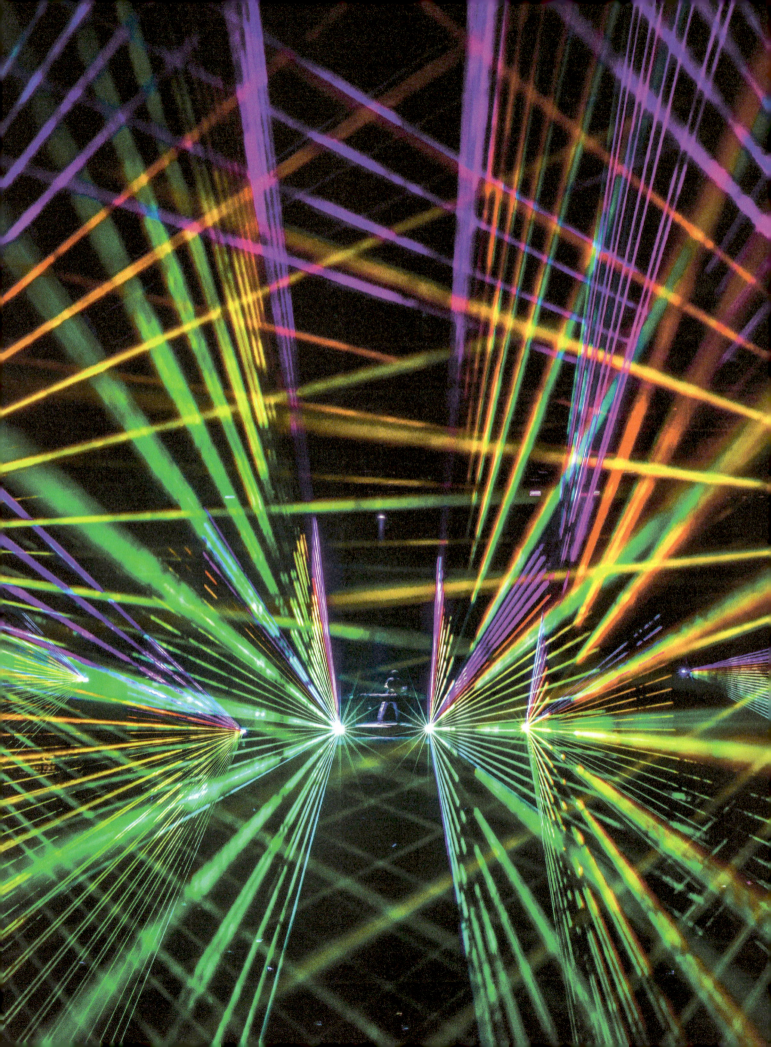

CHAPTER 22
Light and Optics

LIGHT SHOW

"Let there be light." With those words, recorded for us in Genesis 1:3, God created electromagnetic energy. One portion of that spectrum of energy is especially important to humans—visible light. Seeing light and color allows us to appreciate beauty within both the natural world and in various art forms, such as painting, photography, the performing arts, or laser light shows (facing page). In this chapter, we'll look at what visible light is, how it behaves, and how it can be used to benefit people.

22A	Light Behavior	504
22B	Color	507
22C	Reflection and Mirrors	510
22D	Refraction and Lenses	514

22A | LIGHT BEHAVIOR

What does light do?

22A Questions

- What is visible light?
- What is a luminous object?
- How is light modeled?
- Do all materials transmit light equally well?

22A Terms

luminous object, illuminated object, ray, transparent material, translucent material, opaque material

22.1 VISIBLE LIGHT

As you learned in Chapter 21, visible light is just a very small portion of the electromagnetic spectrum. Still, it's the only portion that humans can see, though some animals can see electromagnetic energy with longer or shorter wavelengths. In fact, when we say "visible light," we really mean light that is visible to *humans*. That's why it's worth looking at visible light in more detail.

VISIBLE LIGHT

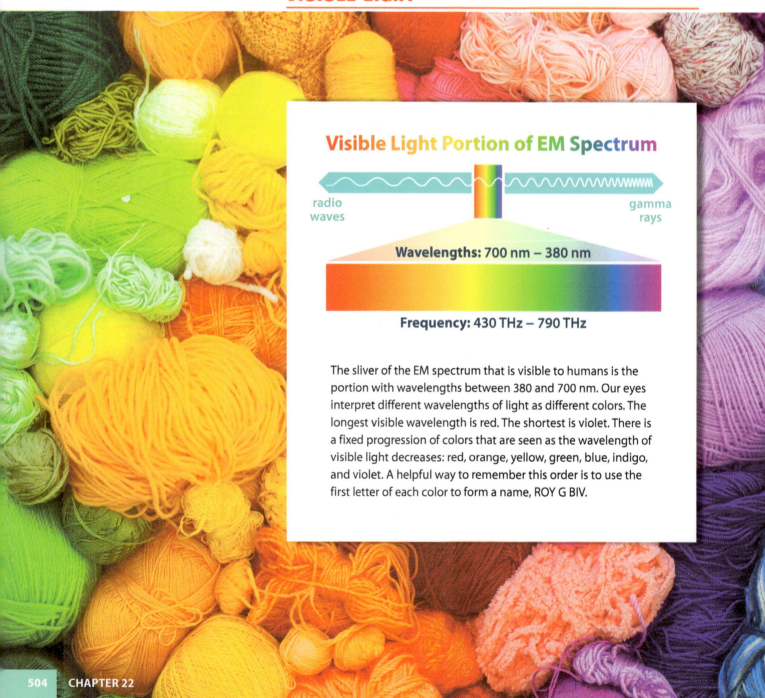

Visible Light Portion of EM Spectrum

radio waves — gamma rays

Wavelengths: 700 nm – 380 nm

Frequency: 430 THz – 790 THz

The sliver of the EM spectrum that is visible to humans is the portion with wavelengths between 380 and 700 nm. Our eyes interpret different wavelengths of light as different colors. The longest visible wavelength is red. The shortest is violet. There is a fixed progression of colors that are seen as the wavelength of visible light decreases: red, orange, yellow, green, blue, indigo, and violet. A helpful way to remember this order is to use the first letter of each color to form a name, ROY G BIV.

22.2 LUMINOUS OR ILLUMINATED?

You've probably had the unpleasant experience of stumbling about in a darkened room. There are limited options for fixing the problem. You could flip a switch to turn on a lamp, or you might resort to using a flashlight or even a lit candle. Lamps, flashlights, and candles are examples of **luminous objects**—those that can produce visible light. Most objects, of course, do not produce visible light. We can still see them, though, *if* they are **illuminated objects**, meaning that they are reflecting some light. We stumble in a darkened room because with no light to reflect, objects remain invisible to us.

Visible light can be produced in different ways. Let's take a look at some of these.

PRODUCING LIGHT

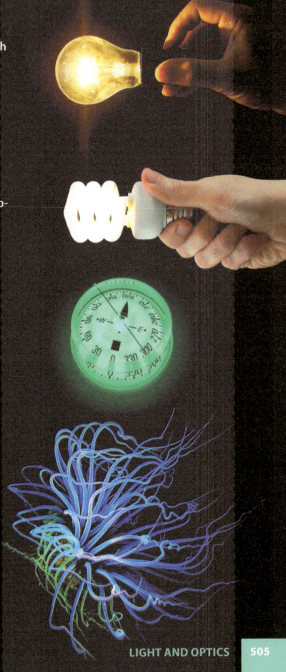

Incandescence is light produced by objects that glow as a result of high temperature. Old-fashioned incandescent light bulbs use electrical resistance to heat a thin filament inside a glass bulb. The hot filament converts some energy into visible light. Stars are also incandescent, though their energy is a product of nuclear fusion. The colors of stars are related to their temperatures. Blue stars are the hottest, red the coolest.

Fluorescence occurs when a material gives off visible light after absorbing electromagnetic energy. Fluorescent light bulbs contain mercury vapor. An electric current causes the mercury gas to emit ultraviolet (UV) light, which humans can't see. The UV light strikes a powdered coating inside the bulb, causing the coating to glow.

Phosphorescence is very much like fluorescence, but in this case the process is slower. As a result, a phosphorescent material glows for a longer period of time after absorbing energy.

Some chemical reactions give off energy in the form of light. This is called *chemiluminescence*. It's what happens when you snap a glow stick. Some living things can produce light in this manner, in which case it is called *bioluminescence*.

ray

22.3 MODELING LIGHT WITH RAYS

When a light source emits light, the waves radiate outward from the source in many directions, like ripples from a pebble tossed into a lake. Normally when we consider how light interacts with other things, we're interested only in a portion of the light energy being emitted, not the entire wavefront. So to model how light behaves, scientists think in terms of light *rays*. Recall from mathematics that a ray is a half-line that has one endpoint and continues infinitely in one direction. A **ray** of light has its endpoint at the light's source, and its direction is perpendicular to the light's wavefront. This means that the ray points in the direction of the wave's travel. We'll use the ray model of light throughout the rest of this chapter while looking at light behavior.

22.4 TRANSPARENCY

When a ray of light strikes a material, several things can happen. It may be reflected by the material, or its energy may be absorbed. The material may allow the ray to pass through mostly unhindered, or it may bend the ray's path of travel as it passes. We'll look at reflection and absorption in a bit, but for now let's consider how materials can be classified on the basis of how well they transmit light.

Materials through which light can easily pass are said to be **transparent materials**. Transparent materials include, among other things, glass, cellophane wrap, and air. We should be especially thankful for that last one, otherwise we wouldn't be able to see anything! Transparent materials do not transmit all wavelengths of light equally well. Color filters make use of this fact. They allow only certain wavelengths of light to pass while blocking others.

Translucent materials do not allow light to pass easily. Some of the light is scattered, either at the surface or within the material itself. As a result, you may be able to see a dim shape through a translucent material, but you won't see a detailed image.

Opaque materials do not allow visible light to pass. We cannot see through an opaque material because it is neither transparent nor translucent. But materials that are opaque to visible light may still be transparent to other forms of electromagnetic energy. X-rays, for example, can pass through many materials that are opaque to visible light. This makes x-ray scanning possible.

transparent (air)

translucent (kite)

opaque (girl)

22A | REVIEW QUESTIONS

1. What is visible light?
2. Which color of visible light has the longest wavelength? Which has the shortest?
3. What is the difference between a luminous object and an illuminated object?
4. What is bioluminescence?
5. Why are rays used to model light?
6. The word *translucent* comes from the Latin meaning "shine through." Why would it be incorrect today to describe a clear glass window as translucent?
7. Does a material described as opaque necessarily block all forms of electromagnetic energy? Explain.

22B | COLOR

How many colors are there?

22.5 SHADES OF COLOR

In 1972, a color television set cost at least three times as much as a black-and-white set. Yet that same year was the first in which more color TV sets were sold in the United States than black-and-white sets. Why were people willing to spend so much more money on a color TV? Simple—the human eye is designed to see color and to see it well. Thus, a color viewing experience is far more enjoyable than one consisting solely of shades of gray.

Just how many shades of color are there? Remember ROY G BIV? Those colors, of course, represent only the general *progression* of colors in the visible spectrum. Within and between each color band of the spectrum are thousands of different shades of color, each one representing a minuscule change in wavelength. Humans can distinguish several million different colors. That's why two students can spend hours debating the exact color of someone's shirt! In theory, though, an *infinite* number of colors exist, since the wavelengths of the electromagnetic spectrum are not a set of distinct values but rather a continuous range of values.

22B Questions

- How sensitive to color are human eyes?
- What is the difference between primary and secondary colors?
- What are additive and subtractive colors?

22B Terms

primary color, secondary color, additive color, subtractive color

22.6 ADDITIVE COLORS

Take another look at the visible spectrum on page 504. Try to find brown on the spectrum. Do you give up? Brown doesn't exist on the visible spectrum because it is not caused by a single wavelength of light. Brown, along with its many shades, is an example of a color produced by a mixture of different visible light wavelengths. The most familiar example of this is white, which is a mixture of all the visible wavelengths of light.

Scientists have learned that the human eye and brain can distinguish three colors of light called primary colors (see top left). **Primary colors** are those that can be mixed to produce most other colors. These primary colors of light are red, blue, and green. The colors that they produce when mixed are called **secondary colors**. Yellow (red + green), magenta (red + blue), and cyan (blue + green) are the secondary colors produced by combining two of the primary colors at equal intensities. All these colors are collectively referred to as **additive colors** because their wavelengths blend together to produce various hues.

When we say "mixed," it's important to remember that perceiving mixed light is a two-part process. The combining of different wavelengths of light is the first part. The second part is how we see the mixed wavelengths. Our eyes contain millions of light-sensitive cells. Those called *cones* are responsible for seeing color and exist in three forms. Each form is sensitive to a different portion of the visible spectrum. When stimulated by light, cones send signals to the brain via the optic nerve. A particular mix of number and type of signals is interpreted as a shade of color.

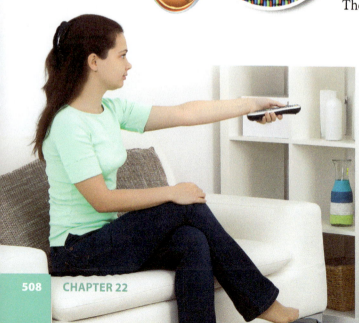

TVs, computer monitors, and cell phone screens take advantage of the eye and brain's ability to mix colors. A color screen consists of a repeating pattern of very fine red, green, and blue dots (left). To display a color other than these three, mixtures of different intensities of these dots are used. A close look at a cyan object on a television screen with a magnifying glass will reveal only green and blue dots. There's really no cyan there at all!

22.7 SUBTRACTIVE COLORS

Illuminated objects don't produce light. They can only reflect it. The color that we perceive of an illuminated object or material depends on what wavelengths of light it reflects. When a substance reflects all wavelengths of visible light, it appears white. When it absorbs all wavelengths, it appears black. In between are all the other colors that we can see, each based on which wavelengths are absorbed and which are reflected.

Pigments are substances that absorb some wavelengths of light and reflect others. They are used to create paints, inks, and dyes. The colors that pigments produce are known as **subtractive colors** because the pigments are absorbing and thus removing some wavelengths of light from a mixture rather than adding them. Pigments give color to objects and materials, such as glazes used to color ceramics, or melanin, the pigment that colors human skin. Different pigments can be mixed, like what we see in the finely powdered solids found in paints. The more pigments that are mixed, the greater the range of wavelengths that the mixture absorbs. You may have seen this when using paints. If you mix red, green, and blue paints together, you'll likely end up with something close to dark brown or gray. Ideally you would have black, but no primary pigment absorbs a given color completely. The image above shows the primary subtractive colors (cyan, magenta, and yellow) and the secondary colors (red, blue, and green) obtained when they are mixed in equal proportion.

If you study the subtractive primary colors closely, you'll see a relationship to the additive primary colors (see top of facing page). Secondary subtractive colors are the same as additive primary colors. But how can this be? Let's answer that question by examining why we see blue when the pigments cyan and magenta are mixed. Recall from Subsection 22.6 that magenta-colored *light* can be produced by mixing the primary additive colors of red and blue light. Magenta *pigment* on the other hand reflects those two colors but absorbs green light. Similarly, cyan pigment reflects green and blue light but absorbs red light. Thus when we mix the subtractive colors magenta and cyan, the only primary additive color of light that is reflected by both pigments is blue. The other primary additive colors are absorbed, so blue is what we see.

22B | REVIEW QUESTIONS

1. How many colors are there?
2. What are the three primary additive colors of light?
3. What cells in the human eye are responsible for detecting color?
4. What is a pigment?
5. Compare additive and subtractive colors.
6. The color green is seen when looking at white light through overlapping cyan and yellow filters. Is green a primary or secondary color in this example? Explain.

22C Questions

- What is reflection?
- How do mirrors produce images?
- What are mirrors used for?

22C Terms

diffuse reflection, specular reflection, incident ray, reflected ray, law of reflection, plane mirror, virtual image, real image, concave mirror, focal point, focal length, convex mirror

22C | REFLECTION AND MIRRORS

How do mirrors produce images?

22.8 REFLECTION

We see objects when they either emit their own light or reflect light from another source. Most things we see are not luminous—we see them by reflected light. There are two kinds of reflection, and each depends on the kind of reflective surface present. The most common type of reflection, called **diffuse reflection**, happens when rough or uneven surfaces, especially at the microscopic level, reflect light rays in all directions (white arrows below). Fabric, brick, concrete, paper, and most natural materials reflect light diffusely.

Smooth, highly polished surfaces, especially metal and glass, reflect light in a different manner. **Specular reflection** (blue arrows below) occurs when a smooth surface reflects light rays in mostly the same direction. This type of reflection is also called *regular reflection*. A perfectly specular reflection is mirror-like, reflecting all rays and wavelengths uniformly.

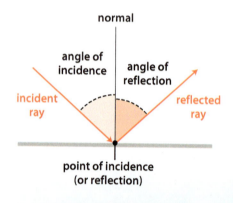

A light ray approaching a reflective surface is called an **incident ray**. The outgoing ray is the **reflected ray**. When a light ray reflects off a surface, the angle of the incident ray, called the *angle of incidence*, equals the angle of the reflected ray, called the *angle of reflection*. This principle is called the **law of reflection** (see left). The law of reflection holds true for light rays in both diffuse and specular reflection. Physicists measure the angles of incidence and reflection from an imaginary line that is perpendicular to the surface at the point of incidence. This line is called the *normal*. These angles are measured from the normal because it can always be defined, even on a curved or uneven surface. It would be nearly impossible to measure these angles from the surface itself.

510

22.9 MIRRORS AND IMAGES

Plane Mirrors and Virtual Images

You've probably fussed over yourself in front of a mirror. When we hear the word *mirror*, we usually think of flat mirrors known as **plane mirrors**. In a plane mirror (right), you see what appears to be an upright image of yourself an equal distance away on the opposite side of the mirror. But that image exists only as a construction made by your eyes and brain from reflected light rays. Physicists call this a **virtual image** (see below right) because it is seen in a position where no light rays actually originate—behind the plane of the mirror.

Every point on a luminous or illuminated object emits or reflects light rays in all directions. Light rays that fall on a plane mirror are reflected according to the law of reflection. But only the reflected rays from a mirror enter our eyes, not the incident rays. Since our visual sense assumes that light rays travel in straight lines from their source, we interpret reflected rays as coming from behind the mirror in straight lines through the mirror to our eyes. As our eyes gather all the light rays reflected from the mirror, our minds construct an image just as if we were looking at the object itself. There is one catch—although the image appears to be reversed, it isn't. It is a "mirror image" of the real object. A mirror image (right) shows the right side of an object where we would expect to see the left side if we were looking directly at it, and vice versa. When you raise your right hand, your image raises its "left" hand (if it were a real person). But it is the hand on the right side of the *image* that is in fact raised. The illusion is created by subconscious expectations when looking at the human figure. In all other ways, though, plane mirror images preserve the sizes, shapes, and distances of reflected objects.

Real Images and Concave Mirrors

Real Images

Specular reflections from flat mirrors can produce only virtual images. This is because light rays diverge, or spread apart, from their sources and continue to diverge after they are reflected. The opposite of a virtual image is a real image. A **real image** is formed when the rays of light from a source are made to converge or come together by a focusing optical device, such as a special mirror or lens. Where the rays intersect, they form an image, which is a representation of a point on the source's surface. This image exists even if there is no observer to see it, or instrument, like a camera, to record it. If the optical device arranges all the image points from the source in their correct relationship, an image of the entire source can be reconstructed. When the image points fall on a flat surface, such as a movie screen, the image can be viewed. Thus, real images can be projected; virtual images cannot.

Concave Mirrors

When it comes to mirrors, real images can be produced only by a special curved mirror called a **concave mirror**. A concave mirror is slightly dished in on the reflective side. The deepest point of the dished surface is the center of the mirror. An imaginary line normal to the mirror's surface at this point is called the *principal optical axis*. The optical axis is helpful for understanding how concave mirrors form images. Let's examine what happens when reflected light rays from an object fall on the mirror.

CONCAVE MIRRORS

Light rays reflecting from an object ① diverge slightly from every point on the object. Multiple rays from each point strike the mirror at different locations.

The curved surface of the mirror reflects the incident rays according to the law of reflection ②. The image is produced at the point where all the reflected rays intersect ③.

An important point related to concave mirrors is the **focal point** ④, the point in front of the mirror where the reflected rays from all incident rays that are parallel to the optical axis will intersect. Depending on the mirror's shape and the distance of the object from the mirror, the reflected rays may intersect at a single point. If the rays do intersect, a real image of the original object is produced.

The focal point lies on the optical axis at a distance from the mirror's center called the **focal length** of the mirror. Because the image in this example can be projected and exists apart from our minds, it is a real image. But the image is reversed—it appears upside down and reversed left to right, unlike a virtual "mirror image," which is upright with right and left sides reflected as right and left.

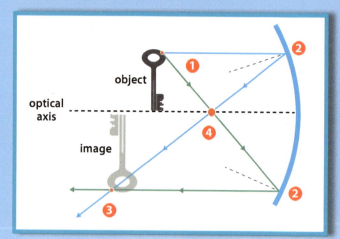

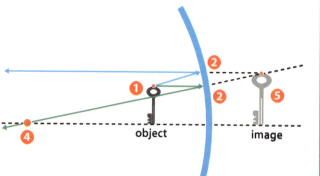

A concave mirror can produce an upright, enlarged, virtual image ⑤ when the mirror reflects light from an object positioned closer to the mirror than the mirror's focal point. Magnifying makeup mirrors use this principle.

512 CHAPTER 22

Using Concave Mirrors

Concave mirrors are used in reflecting telescopes (facing page at bottom) to collect parallel rays of light from a star and direct them to the focal point. Near this point there is usually a smaller plane mirror that reflects the rays to the side or rear of the telescope. There the image may be photographed or viewed by an observer. Concave mirrors may also reflect light in the opposite direction. If a light bulb is placed at the focal point of a concave mirror, the incident rays of light hit the mirror and reflect outward in a tight beam of light. This arrangement is used to produce concentrated beams of light in flashlights, searchlights, and automobile headlights.

Convex Mirrors

The surface of a mirror may also bulge outward. This kind of mirror is a **convex mirror**. It cannot produce a real image because all light rays reflecting off a convex surface diverge even more than from a plane mirror. Every image seen in a convex mirror is an upright, virtual image. Because of the exaggerated divergence of the reflected light rays, convex mirror images appear smaller than the objects that they reflect, and the reflected image includes more background area. For this reason, convex mirrors are used to provide wide-angle views. They are useful for security monitoring, in the passenger-side side view mirrors of cars, and at blind corners of hallways and street intersections.

HOW IT WORKS

Lasers

You've probably seen a laser pointer in action or maybe have seen a laser light show. But how does a laser work, and how is laser light different from other light forms? Ordinary light sources produce light waves of many different wavelengths that radiate out in all directions. Laser light, on the other hand, consists of a single wavelength of light in a tightly concentrated beam. Scientists call this type of light *coherent light*. Coherent light can remain focused over extremely long distances, even from Earth to objects in orbit and back. Laser light can be produced by adding energy, such as electrical energy, to the atoms of certain crystals or gases. Electrons in the energized atoms temporarily move to higher energy levels. When they return to their original energy levels, they emit the extra energy in the form of photons. Since the amount of energy gained and emitted is the same each time, the wavelength of light produced is constant. High-energy lasers are used in many industrial processes, such as laser cutting and engraving (below).

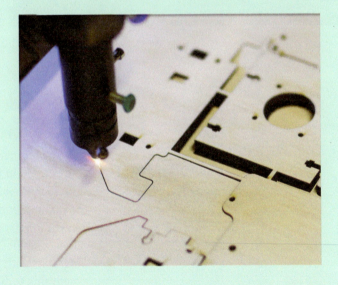

22C | REVIEW QUESTIONS

1. Compare specular and diffuse reflection.
2. State the law of reflection.
3. From what reference point are angles of incidence and reflection measured?
4. Why is the image produced by a plane mirror described as a virtual image?
5. How is a real image different from a virtual image?
6. Which kind of mirror can produce both real and virtual images?
7. What is the focal length of a mirror?
8. State one use for a concave mirror and one for a convex mirror.
9. Which atomic model explains why a laser produces a single wavelength of light?

22D Questions

- Is the speed of light always the same?
- Can light rays be bent?
- How does fiber optic cable work?
- What do different kinds of lenses do?
- How are lenses used to correct poor vision?

22D Terms

index of refraction, total internal reflection, lens, converging lens, diverging lens

Refractive Indexes

Material	n
Air	1.000 277
Water	1.330
Ethanol	1.361
Glass	1.458
Diamond	2.417

22D | REFRACTION AND LENSES

How do glasses help people who can't see well?

22.10 LIGHT REFRACTION

As you learned in Chapter 21, the value that we normally think of as the speed of light is valid only when light travels in a vacuum, such as space. But like other kinds of waves, when light waves pass from one medium into another, their speed changes. This change in speed can cause the waves to refract, or bend. The more a light wave speeds up or slows down as it passes from one medium into another, the greater the bending of its path will be. One measure of how much bending will occur is a medium's **index of refraction**, which is the ratio of the speed of light in a vacuum to the speed of light in the medium. The formula for finding a medium's index of refraction is

$$n = \frac{c}{v},$$

where n is the index of refraction, c is the speed of light in a vacuum, and v is the speed of light in the material. Since the two speeds are measured in the same units, the units cancel, leaving n as a unitless value. Because the speed of light in a vacuum will always be the higher of the two compared speeds, the value of n is always greater than 1 for all forms of matter. Note also that a higher value for n indicates a slower speed of light in a particular medium. The table on the left shows the index of refraction for various media.

MINI LAB

BENDING LIGHT

Rainbows are caused when white light from the sun is refracted and reflected by tiny water droplets in the atmosphere. The different colors within white light don't bend at the same rate, so they separate as they bend. You've probably seen a prism used to produce the same kind of separation. But once the colors are separated, is it possible to recombine them into white light again? Your teacher will use a light ray box and several kinds of lenses to help you think about this question.

Essential Question:

Can a rainbow be undone?

1. Do the colors of light separated by the prism follow the progression described in Subsection 22.1?
2. On the basis of what you observe, which color of light is bent the most and which the least? How can you tell?
3. Do you think it will be possible to recombine the separated colors back into white light using a lens? If so, which lens do you think will work?
4. Describe how each kind of lens affects the separated colors.

CONCLUSION

Different kinds of lenses differ in how they bend light rays. As you read further, you'll learn about these different lenses and see how they can be used to benefit people.

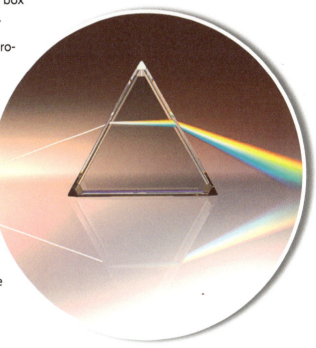

In the example on the right, you can see how light bends as it passes through different media. The ray first bends as it passes from air into water because water's index of refraction is greater than air's. The ray is bent again as it passes into glass because glass has a greater index of refraction than water. Because the boundaries in this example are parallel, the ray returns to its original direction of travel as it passes out of the glass and back into air.

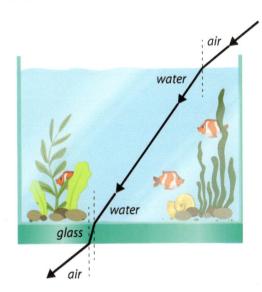

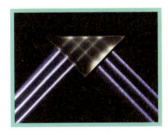

Total Internal Reflection

In some situations, a ray of light *can't* pass from one medium to another. Instead, the ray reflects off the boundary between the two media and remains within the first medium. This phenomenon is called **total internal reflection** (left). It happens when a light ray's angle of incidence exceeds a certain critical value when going into a medium with a lower index of refraction.

Humans have put total internal reflection to very good use. The phenomenon is what allows electronic high-speed data to be converted to pulses of light that can be transmitted via fiber optic cable. Such cables are constructed in a manner that "traps" a light beam as it travels the length of the transparent fiber. At the receiving end, the pulses are converted back into an electronic signal. Compared with sending signals through copper wires, fiber optic cable is faster, safer, and does a better job of maintaining signal strength.

22.11 LENSES

The most common and useful application of optical refraction is a lens. A **lens** is a disk of transparent material that refracts light to produce a real or virtual image. An object may be magnified so that it appears larger than it actually is, or it may be reduced, depending on the purpose of the lens. Lenses can be classified as converging or diverging. A **converging lens**, also called a *convex lens*, collects incoming rays of light and focuses them at a point, just like a concave mirror does. A **diverging lens**, or *concave lens*, spreads rays of light apart similar to what a convex mirror does. Lenses are used in many kinds of optical instruments such as telescopes, microscopes, binoculars, rifle scopes, and cameras.

LENSES

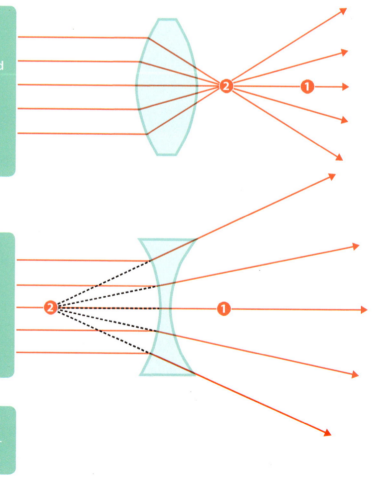

CONVERGING LENS

A converging lens is thicker in the center and thinner at the edges. Many of the terms that you learned for mirrors apply to lenses. The imaginary line through the center of a lens is the optical axis ①.

Parallel light rays from an object pass through the lens and converge at the focal point ②, located on the optical axis on the opposite side of the lens.

DIVERGING LENS

A diverging lens is thinner at the center and thicker at the edges. Parallel rays entering a diverging lens are bent *away* from the lens's optical axis ①. This means that a diverging lens's focal point ② is actually on the same side of the lens as the incoming rays.

The distance to a focal point from the center of either lens is the focal length. The focal length shortens as the curvature of the lens increases.

22.12 CORRECTING VISION

In our fallen world, many people suffer from poor vision. Happily, lenses allow for some vision problems to be corrected. The human eye itself contains a convex lens. The lens is designed to focus light rays onto the eye's retina. But in some people, a slightly misshapen eye causes the retina to lie slightly in front of or behind the lens's focal point. If the focal point of the lens is in front of the retina, the person experiences nearsightedness. If it is behind the retina, the person suffers from farsightedness. Lenses can correct both conditions.

Correcting Nearsightedness

Nearsightedness is corrected with diverging lenses. A diverging lens is used to spread out rays of incoming light before they enter the eye. As those rays then pass through the eye's lens, they are focused farther back than they normally would be. The two lenses together, acting as a lens group, have a longer focal length than the eye's lens alone. The new focal length results in a focused image falling on the retina, and the affected person can then see clearly.

Correcting Farsightedness

Farsightedness is corrected similarly, but with converging lenses. The rays that enter a farsighted person's eyes do not converge enough. The result is that the focal length of the person's lens is too long. Light entering the person's eye strikes the retina before the rays can intersect and produce a sharp image. The eye's focal length needs to be shortened. A converging lens combined with the eye's lens creates a lens group with a shorter focal length than either lens alone.

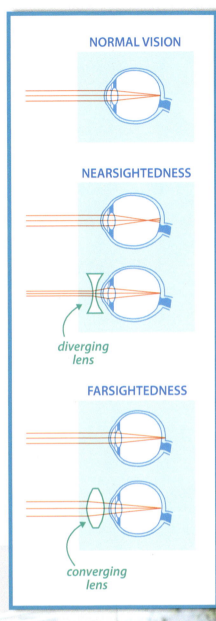

22D | REVIEW QUESTIONS

1. What is the index of refraction?
2. Light travels faster in medium *A* than in medium *B*. Which medium has a higher index of refraction?
3. What is total internal reflection?
4. Compare converging and diverging lenses.
5. Which type of lens is used to correct nearsightedness?
6. Explain how a biblical worldview justifies using knowledge of lenses to correct poor vision.

CHAPTER 22

22A LIGHT BEHAVIOR

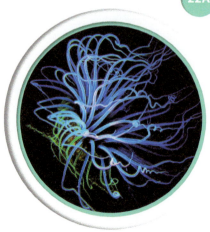

- Visible light is the portion of the electromagnetic spectrum that humans can see.
- Luminous objects produce light; illuminated objects can only reflect light.
- A light ray models a portion of a light source's energy beginning at the source and traveling in a direction perpendicular to the light's wavefront.
- Transparent materials transmit light easily. Translucent materials do not transmit light easily, scattering some of it from or within the material. Opaque materials do not transmit light.

22A Terms
luminous object	505
illuminated object	505
ray	506
transparent material	506
translucent material	506
opaque material	506

22B COLOR

- An infinite number of colors is possible, but the general progression of colors from longest wavelength to shortest is red, orange, yellow, green, blue, indigo, violet.
- Primary colors can be mixed to produce most other colors.
- Additive colors are produced by blending different wavelengths of light together. Subtractive colors are produced when pigments absorb certain wavelengths of light and reflect others.

22B Terms
primary color	508
secondary color	508
additive color	508
subtractive color	509

22C REFLECTION AND MIRRORS

- Diffuse reflection occurs when an object reflects light rays in many different directions. Specular reflection happens when an object reflects light rays in mostly the same direction.
- The law of reflection states that the angle of a reflected light ray is equal to its angle of incidence.
- A virtual image is one that appears to be in a location where none actually exists. A real image occurs where reflected light rays intersect.
- Plane mirrors produce upright, virtual images.
- A concave mirror can produce an inverted, real image or an upright, enlarged, virtual image depending on where the reflected object is positioned relative to the mirror's focal point.
- Convex mirrors produce only upright, reduced, virtual images.

22C Terms
diffuse reflection	510
specular reflection	510
incident ray	510
reflected ray	510
law of reflection	510
plane mirror	511
virtual image	511
real image	511
concave mirror	512
focal point	512
focal length	512
convex mirror	513

REVIEW

22D **REFRACTION AND LENSES**

- Like other kinds of waves, light waves can be refracted.
- A material's index of refraction is a ratio of the speed of light in a vacuum to the speed of light in that material. An index of refraction is a unitless number.
- Total internal reflection occurs at a boundary between two media (the second medium has a lower index of refraction) when the angle of an incident ray exceeds a critical value, resulting in the ray being reflected into the first medium.
- Lenses are disks made of transparent materials that are used to refract light.
- Converging lenses focus light rays onto a single point called a focal point. Diverging lenses spread light.
- The distance along the lens's optical axis from the center of the lens to its focal point is the lens's focal length.

22D Terms

index of refraction	514
total internal reflection	515
lens	516
converging lens	516
diverging lens	516

CHAPTER REVIEW QUESTIONS

Recalling Facts

1. Why is visible light usually discussed in the context of *human* vision?
2. In order of decreasing frequency, name the colors of the visible spectrum.
3. What is the range of wavelengths within the visible light spectrum?
4. (True or False) The longest wavelength of visible light is violet.
5. Which type of light source relies on a heated filament?
6. The endpoint of a light ray is at the light's _____ , and its direction is _____ to the light's wavefront.
7. State the three secondary additive colors and the combination of primary colors that produces each.
8. State the secondary subtractive color produced by each of the following combinations.
   a. cyan and yellow
   b. magenta and cyan
   c. yellow and magenta
9. What is the reference line used for measuring angles of incidence and reflection?
10. Which side of a concave mirror is reflective?
11. Describe the image created by convex mirrors.
12. Why does light refract?
13. You are looking at a lens with two convex surfaces. Is this a converging or diverging lens?
14. What causes nearsightedness?

LIGHT AND OPTICS

CHAPTER 22

Understanding Concepts

15. Give two examples of luminous objects and two of illuminated objects.
16. Compare incandescent and fluorescent lightbulbs.
17. Relate chemiluminescence and bioluminescence.
18. Identify each of the following examples as either transparent (TP), translucent (TL), or opaque (O).
    a. sunglasses
    b. fog
    c. wax paper
    d. wood paneling
    e. air
19. Why is it incorrect to say that the visible light spectrum consists of the colors red, orange, yellow, green, blue, indigo, and violet?
20. How does a color television screen produce colors other than red, blue, and green if those are the only colors emitted by the screen's pixels?
21. Why does a mix of different colors of paint tend to produce a brown or gray color?
22. Light reflecting from you strikes a nearby rock. Why don't you see your image reflected on the rock?
23. An incident ray strikes a convex mirror at an angle of incidence of 53°. What is the angle of reflection?
24. Under what conditions does a concave mirror produce a real image? a virtual image?
25. Why would a convex mirror not be useful for creating a searchlight?
26. Why can a convex mirror never produce a real image?
27. Why is a material's index of refraction always greater than 1?
28. The index of refraction for material A is 1.005, and that of material B is 2.150. In which material does light travel faster? Explain.
29. Explain how the reflected image in the photo at left is possible.
30. Why do we use a diverging lens to correct nearsightedness?

Critical Thinking

31. How would color vision be affected if humans lacked the cones that are sensitive to green light?
32. During a ceramics project, you run out of red glaze. A friend tells you that you can make red by adding together blue and green glazes. Is your friend giving you sound advice? Explain.
33. A website claims that a plane mirror creates a virtual image because the right side of a person appears as the left side of the person reflected in the mirror. Is this correct? Explain.
34. Chad focuses a pair of binoculars until he sees a sharp image of a mountain peak in the distance. He then passes the binoculars to Ellen, but to her the image appears blurry. How is this possible?
35. Is visible light special? Explain your position from a biblical standpoint.

REVIEW

Use the Ethics box below to answer Question 36.

36. Write a three-paragraph response on the ethics of sharing photos on social media. In the first paragraph, discuss principles from Scripture that may guide Christians in this area. In the second paragraph, discuss the possible outcomes of heeding or ignoring the guidelines of Scripture. In the third paragraph, state the biblical motivation involved in posting images online and give your opinion on whether additional safeguards need to be put in place to protect the intellectual property rights of social media users.

ETHiCS — WHO OWNS YOUR PHOTOS?

Lenses make digital photography possible. Gone are the days when amateur photographers had to send their rolls of film to a developer and wait for days to see their pictures. Now we can digitally store thousands of images on our cell phones and have instant access to them. We also live in an age where hundreds of millions of people regularly share snapshots of their daily lives on social media.

Have you ever thought about what happens to your photos after you upload them to a social media site? Once you share them, are you still in control over what happens to them and who sees them? Should you be?

To prepare for answering Question 36, start by looking up the *terms of service* for several social media platforms. Look for the part that discusses what the service can do with your content once you have posted it. Terms of service vary from company to company and from country to country.

APPENDIX A

UNDERSTANDING SCIENTIFIC TERMS

You may find science a little intimidating because of all the long, unfamiliar words that scientists use. However, you can often unravel the meaning of these words by breaking them down into simple parts that *do* have meaning to you. When you see a difficult scientific word, look at the entries in this appendix to help you understand that term. These word parts may come at the beginning, at the end, or in the middle of the term, depending on their meaning.

For example, if you ran into the word *magnetohydrodynamics*, you could separate it into three parts—*magneto*, *hydro*, and *dynamics*. *Magneto* means "magnetic," *hydro* means "water" or "fluid," and *dynamics* means "study of power." So magnetohydrodyamics has to do with the study of the magnetic properties of moving fluids.

Changes in the sun's surface are due to the movement of the charged particles that make up the sun.

a, **an** (Gk.)—not, without
ab (L.)—away from
ac, **ad**, **ag** (L.)—to, toward
acou (Gk.)—hearing
aer, **aero** (Gk.)—air
alter (L.)—change
amal (Gk.)—soft
amphi, **ampho** (Gk.)—on both sides
ant, **anti** (Gk.)—opposite, against
ante (L.)—before
aqua (L.)—water
audio (L.)—hear
aut, **auto** (Gk.)—self
bar, **baro** (Gk.)—weight, pressure
bi (L.)—two, twice, double
bio, **bios**, **biot** (Gk.)—life
calc, **calci** (L.)—calcium
calor (L.)—heat
centi (L.)—a hundred
centr, **centri**, **centro** (Gk.)—center
chem, **chemi**, **chemo** (Gk.)—chemistry
chrom, **chromo** (Gk.)—color
chron, **chrono** (Gk.)—time

cline (Gk.)—sloping
co, **com**, **con** (L.)—with, together
cupr (Gk.)—copper
cycl, **cyclo** (Gk.)—circle, wheel
de (L.)—loss, removal
deci (L.)—tenth
di (Gk.)—two
div (Gk.)—apart
duce, **duct** (L.)—to lead
dyna (Gk.)—power
eco (Gk.)—house
electro (L.)—electricity
en, **end**, **endo** (Gk.)—within, inner
equ, **equa**, **equi** (L.)—equal
ex, **exo** (Gk.)—out, outside, without
extra (L.)—outside, more, beyond, besides
fissi (L.)—split, divide
flam (L.)—fire
fund (L.)—basis
fusi (L.)—to join together
grad (L.)—step, walk, slope
graph, **grapho**, **graphy** (Gk.)—to write

grav (L.)—heavy
gyro (Gk.)—spinning
halo (Gk.)—salt
hemi (Gk.)—half
hetero (Gk.)—other, different
homeo, homio, homo (Gk.)— like, same, resembling
hydr, hydra, hydro (Gk.)—water, fluid
hyper (Gk.)—over, beyond
hypo (Gk.)—under, beneath
ic (Gk.)—of, relating to
inter (L.)—between
is, iso (Gk.)—equal
ism (Gk.)—belief, process of
kine, kinema, kinemato, kines, kinesi, kinet, kineto (Gk.)— move, moving, movement
log, logo, logus, logy (Gk.)—word, study of
macr, macro (Gk.)—large
magneto (Gk.)—magnetic
medi, media, medio (L.)—middle
met, meta (Gk.)—between, with, after, change
meter, metry (Gk.)—measure
micro (Gk.)—small
mill, mille, milli, millo (L.)— one thousand
mit (L.)—to send
mono (Gk.)—one
morph, morpha, morpho (Gk.)—form, shape
multi (L.)—many
nan, nani, nano, nanus (Gk.)— dwarf, one billionth
nomy (Gk.)—the science of
nuc, nucle, nucleo (L.)—central part
ocul, oculi, oculo, oculus (L.)—eye
opt, opti, opto (Gk.)—eye, vision

organ (Gk.)—living
orth, ortho (Gk.)—upright, perpendicular
ox, oxy (Gk.)—oxygen
par, para (Gk.)—beside
pause (Gk.)—to stop
pend (L.)—hanging
peri (Gk.)—around, near
phon, phono (Gk.)—sound
phos, phot, photo (Gk.)—light
phyt, phyto, phytum (Gk.)—plant
poly (Gk.)—many
post (L.)—after
pre (L.), **pro** (Gk.)—before, in front of
prot, prote, proto (Gk.)—first, original
pyro (Gk.)—fire
radi, radia, radio (L.)—spoke, ray
retro (L.)—backward
sal (L.)—salt
scient (L.)—knowledge
scope, scopy (Gk.)—to see, watch
sect (L.)—to cut
seism (Gk.)—earthquake
semi (L.)—half
sol (L.)—sun
son (L.)—sound
spec (L.)—to see, look at
sphere (Gk.)—ball, globe
stasis (Gk.)—to stand still
sub (L.)—below, under
super (L.)—above, over
syn (Gk.)—together
tele (Gk.)—distant
terra (L.)—earth
tetr, tetra (Gk.)—four
therm, thermo (Gk.)—heat
top, topo (Gk.)—place

tran, trans (L.)—across, through
trop, tropae, trope, tropo (Gk.)—turn, change
uni (L.)—single, one
vacu (L.)—empty
vari, vario (L.)—difference
vect (L.)—to carry
volu (L.)—bulk, amount

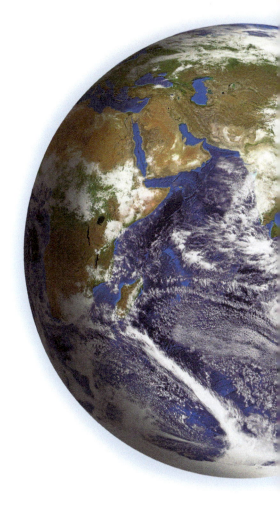

UNDERSTANDING SCIENTIFIC TERMS

APPENDIX B

MATH HELPS

This appendix provides you the necessary helps for solving the many kinds of physical science problems found in this textbook. It's not intended to be all-inclusive. We assume that you understand or are learning the methods for solving single-variable algebraic equations, including using the order of operations and solving equations for an unknown quantity.

Rounding Rules

We round numbers all the time. Sometimes it's convenient to round, for example, the number of residents in your town to the nearest 100 people or the distance around the earth to the nearest 1000 kilometers. Rounding is useful for several reasons. It helps us grasp large numbers when the specific number is not needed but the approximate size is. Rounding also allows us to report measured scientific data correctly.

For the purposes of measuring scientific data and performing calculations, scientists have agreed on certain rounding rules. These ensure that calculation solutions are properly rounded to show the needed precision of numerical quantities. We suggest that you use the following rules for problems in this textbook to closely imitate those used by scientists.

1. Identify the place value that you are going to round to. This is the rounded place value.

2. If the digit to the right of the rounded place value is 0–4, the digit in the rounded place value remains unchanged ("rounded down"). If the digit to the right is 5–9, the digit in the rounded place value is increased by 1 ("rounded up").

    a. If the rounded place value is to the *right* of the decimal point, all digits to the *right* of the rounded place value will be dropped after rounding.

       Examples: 105.6̲39 g → 105.6 g

       82.4̲7 mm → 82.5 mm

       32.9̲5 cm → 33.0 cm

    b. If the rounded place value is to the *left* of the decimal point (or assumed decimal point), all digits between the rounded place value and the decimal point become zeros and the decimal point (if one is shown) is omitted.

       Examples: 1 6̲94 m → 1700 m

       19̲.6 °C → 20 °C

3. Notice in two of the examples above that rounding to a particular place value may affect other digits to the left of the rounding place value. When a 9 is rounded up, the 9 becomes 0 and 1 is added to the digit to the left.

Mathematics and Measurement

When measured quantities appear in equations as terms or factors, follow the same arithmetic principles that you have learned for solving any equation. But acting as a scientist, you have to think about some extra things before arriving at a solution to a problem.

Measured quantities include *units*. Units, such as meters (m), seconds (s), and meters per second (m/s), are treated as parts of a measurement. So any arithmetic operation applies to both the number and the unit factors in a measurement.

Measured quantities also have *precision* (see Subsection 1.9). Precision is the exactness of a measurement. Scientists exert a lot of effort to ensure that their data properly represents the exactness that their instruments can read. Scientists alert other scientists to the precision of their measurements by the number of *significant figures* (SF) they include in their reported data.

On the following pages you will find some rules for doing basic arithmetic operations using measurements.

Determining Significant Figures

When we are reading measured data collected by someone else, we have to be able to identify the significant figures. (Significant figures do not apply to counts or definitions—only to measured data.) Scientists follow an established set of rules to determine which digits in a measurement are significant.

Rule 1: All nonzero digits are significant.

Rule 2: All zeros between significant figures are significant.

Rule 3: All ending zeros to the right of the decimal point are significant.

Examples:

a. 1.694 mL → 4 SF (all four digits are nonzero—Rule 1)

b. 51.021 m → 5 SF (four nonzero digits and a zero between SF—Rule 2)

c. 22.7500 g → 6 SF (four nonzero digits and two ending zeros to the right of the decimal point—Rule 3)

d. 0.097 N → 2 SF (the zeros are not ending zeros—Rule 3)

The fourth example can be confusing, but it is much easier to understand if you change the number to scientific notation.

$$0.097 \text{ N} = 9.7 \times 10^{-2} \Rightarrow \text{only two SF}$$

One other case in which determining SF can be difficult is when ending zeros occur to the left of the decimal point.

Example:

How many SF does the measurement 7600 m have? You know that two of the digits must be significant, but you as the reader cannot determine which, if either, of the zeros are significant. Does this number have two, three, or four SF? Again, the person reporting the value can use scientific notation to avoid confusion.

$7600 \rightarrow 7.6 \times 10^3$ (two SF)

$7600 \rightarrow 7.60 \times 10^3$ (three SF)

$7600 \rightarrow 7.600 \times 10^3$ (four SF)

Adding and Subtracting Physical Science Data

Math Rule 1: Added or subtracted data must have the same units. Even if the kinds of data are the same, if their units are not the same, then you are adding apples and oranges. For example, you can't add the lengths 3.1 m and 45 cm together without first converting one of the measurements to the other's units.

Math Rule 2: The result of adding or subtracting data can't be more precise than the least precisely measured data used in the calculation.
After adding or subtracting, round the result to the decimal place of the estimated digit in the least precise measurement.

EXAMPLE 1: Adding Data: Length

Add 3.1 m and 45 cm.

Unit Conversion:

$$45 \text{ cm} \left(\frac{1 \text{ m}}{100 \text{ cm}}\right) = 0.45 \text{ m}$$

Add data in column:

$$\begin{array}{r} 3.\underline{1} \text{ m} \\ +0.4\underline{5} \text{ m} \\ \hline 3.\underline{55} \text{ m} = 3.6 \text{ m} \end{array}$$

The estimated digits in each piece of data and the sum are underlined. The sum contains two underlined digits that result from the summing operation. These must be rounded to the place having the least precise estimated value, in this case the 0.1 m place.

Multiplying and Dividing Physical Science Data

Math Rule 3: The result of multiplying or dividing data can't have more SF than the data with the fewest SF used in the calculation.
After multiplying or dividing, round the results to the same number of significant figures as the data with the fewest SF.

EXAMPLE 2: Dividing Data: Density

Calculate the density of a sample of quartz with a mass of 27.55 g and a volume of 10.4 cm³.

Given data: $m = 27.55$ g (four SF), $V = 10.4$ cm³ (three SF)

Calculation:

$$d = \frac{m}{V}$$

$$= \frac{27.55 \text{ g}}{10.4 \text{ cm}^3} \left(\frac{\text{four SF}}{\text{three SF}}\right)$$

$$= 2.6\underline{4}9038462 \frac{\text{g}}{\text{cm}^3}$$

$$= 2.65 \frac{\text{g}}{\text{cm}^3} \text{ (three SF allowed)}$$

Since 10.4 cm³ has only three SF, only three SF are allowed in the quotient.

Math Rule 4: The result of multiplying or dividing a measured quantity and a pure number has the same number of decimal places, or the same precision, as the measured quantity used in the calculation.

Examples:

a. (7)(2.35 cm) = 16.45 cm (not 16.5 cm)

b. 2.63 cm ÷ 5 = 0.53 cm (not 0.526 cm)

Notice that the number of SF in the measured data do not determine the number of SF in the result when multiplying or dividing by pure numbers. Math Rule 4 ensures that you preserve the precision of the original measurement, even if you change the number by multiplying or dividing by a pure number.

Proportionalities in Physical Science

As you study physical science, you will discover that many measurable quantities change or vary in a reliable, predictable way with other quantities. For example, the gravitational potential energy of a ball increases with its height above the ground. If its height is doubled, then its potential energy is doubled as well. Similarly, if you double your running speed around a cross-country track, you halve the time it takes to run the course. These kinds of relationships are called *proportionalities*.

Proportionality

A minute has 60 seconds, right? How many seconds are in 2 minutes? Correct, 120 seconds. The number of seconds is *directly proportional* to the number of minutes. The word *proportional* means "having a constant ratio." In the case of seconds and minutes, the ratio is always

$$\frac{60 \text{ s}}{1 \text{ min}} \text{ or } \frac{1 \text{ min}}{60 \text{ s}}$$

—they are directly proportional.

If two numbers, A and B, are directly proportional, then they can be set equal to each other by including a proportionality constant factor—k. The equation is

$$A = kB,$$

which means that

$$\frac{A}{B} = k,$$

and k is a constant for all ratios of these two quantities. For our example, suppose A is time in minutes and B is time in seconds. The proportionality constant k is equal to 1 min/60 s, as shown below.

$$A = \left(\frac{1 \text{ min}}{60 \text{ s}}\right) B$$

You'll work with many quantities that are directly proportional. You will study direct proportions that involve momentum, force, acceleration, mechanical advantage, and other concepts.

You'll also find that other pairs of quantities are *inversely proportional*. If the value of one increases, the other decreases in proportion.

The equations for these proportions set up as shown below.

$$A = \frac{k}{B}$$

If you multiply both sides of the equation by B, you end up with

$$AB = k.$$

So if k is a constant, as A increases, B must decrease in the same proportion to maintain the equality.

Examples of inverse proportions include the pressure and volume of a confined gas (Boyle's law), electric current and resistance in Ohm's law, frequency and period, and wavelength and frequency of a wave.

Lastly, there are other kinds of proportionalities that are neither entirely direct nor inverse relationships. For example, the kinetic energy of an object increases much faster than expected when its speed increases. If a baseball's speed is doubled, it will have four times as much kinetic energy than it did at the slower speed. Similarly, the force of gravity and the electric force decrease much faster with distance than would be the case with a simple inverse proportion. Double the distance between two massive objects and the gravitation force is only one-fourth its value at the closer distance. These kinds of proportionalities include an extra math operation in the direct or inverse relationship. Scientists give the latter case a special name—the *inverse square law*.

GRAPHING

Scientists often like to compare two or more groups of numbers visually to see whether they are related in some way. When scientists plot data to compare different quantities, they make a graph. The simplest graphs compare two changing quantities, called *variables*. Usually, one of the variables is determined by the scientist or else changes in a regular way (e.g., time). This is called an *independent variable*. Its value doesn't depend on anything in the data. The other variable is expected to change in some way related to the independent variable. Its value depends on the first variable, so it is called the *dependent variable*. That makes sense! The values of the independent and dependent variables are called the *coordinates of the data*. To plot the data, an ordered pair of coordinates is used. You have graphed ordered pairs in the form (x, y) in a math class.

Scientists usually plot the independent variable on the horizontal axis (x-axis), with increasing values to the right. The dependent variable is typically plotted on the vertical axis (y-axis), increasing upward. These are not hard and fast rules, and many graphs are arranged differently to more clearly see the relationships in the plotted data. Useful graphs include a title describing the graph's purpose and labels identifying the quantities and units used on each axis. Graphs are often depicted on a scaled grid with the x- and y-axis scales shown so that estimates of the values of the plotted variables can be made.

Scatterplots

Graphs come in different forms. A simple plot of points on a graph is called a *scatterplot*. This is the starting point for many graphs. Scientists like to detect trends in the data and then create a mathematical equation (model) that describes the trend. They draw a *best-fit curve* through the pattern of dots. The kind of curve drawn depends on how the data changes. We still call it a best-fit curve even if it doesn't actually curve. Curves that are straight lines are called *linear graphs*. These are fairly rare in nature. Most trends in nature range from slightly curved to really wavy! These graphs, logically enough, are called *nonlinear graphs*.

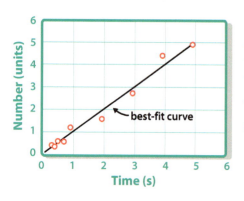

A scatterplot with a direct proportion

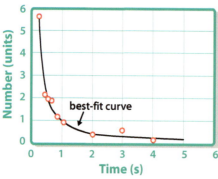

A scatterplot with an inverse proportion

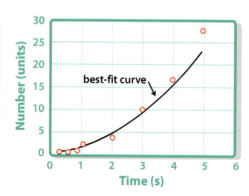

A scatterplot with a square proportion

Slope

One of the most important things to be learned from a graph is the rate at which the dependent variable changes in comparison to the independent variable. This rate is indicated by the *slope* of the graph. A graph with a steep slope shows that things are changing quickly. One with a horizontal slope shows that the dependent variable is not changing at all. A graph that rises from left to right has a positive slope; a dropping curve has a negative slope.

To calculate the slope (m) of a curve, use the slope formula,

$$m = \frac{\Delta y}{\Delta x} = \frac{y_2 - y_1}{x_2 - x_1},$$

where m is the slope, Δy is the change in the y variable, and Δx is the change in the x variable.

EXAMPLE 3: Calculate the Slope Between Two Points

What is the slope of a line between (–4, 8) and (5, 2)?

Given data: $x_1 = -4, y_1 = 8, x_2 = 5, y_2 = 2; m = ?$

Calculation:

$$m = \frac{\Delta y}{\Delta x} = \frac{y_2 - y_1}{x_2 - x_1}$$

$$m = \frac{2 - 8}{5 - (-4)} = \frac{-6}{9} = -\frac{2}{3}$$

The graph slopes downward, losing two units vertically for every three units moved horizontally.

Estimating Data

You can sometimes use scatterplots and trend lines to obtain information that is not in the measured data set. If you follow the trend line between two data points, these values are estimated—not measured. Obtaining unmeasured data this way is called *interpolation*. Scientists also often try to predict the values of the dependent variable beyond the range of the measured independent variable data points. This method of analysis is called *extrapolation*. Extrapolating data assumes that the trend will continue in the same fashion as observed within the measured data.

Other Types of Graphs

In a bar graph, dependent data is plotted as vertical bars at each independent data value. Area graphs fill in the area of a graph. When several dependent variables are plotted on a single graph, the areas can be compared to show their relationships visually. Pie charts are another type of area graph. They are especially useful for showing percentages of a whole.

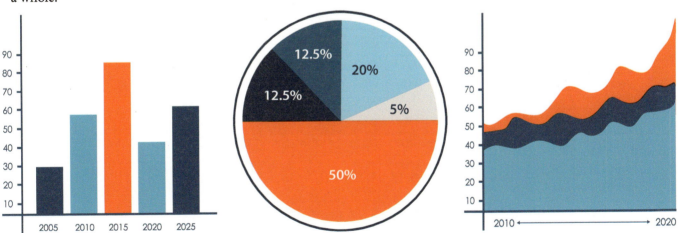

A bar graph (left), pie chart (center), and an area graph (right)

MATH HELPS

APPENDIX C

FUNDAMENTAL AND DERIVED UNITS OF THE SI

Fundamental Units

Dimension	Name	Symbol	Definition
length	meter	m	The meter is the length of the path traveled by light in a vacuum during a time interval of 1/299 792 458 of a second.
mass	kilogram	kg	The kilogram is the unit of mass. It is defined on the basis of the Planck constant (6.626 070 15 × 10^{-34} J·s), the meter, and the second.
time	second	s	The second is the duration of 9 192 631 770 cycles of the radiation associated with a specific transition of the cesium-133 atom.
electrical current	ampere	A	The ampere is the current required for one coulomb of charge to pass through a conductor every second. The coulomb is defined on the basis of the elementary charge, the electric charge carried by a single proton.
temperature	kelvin	K	The kelvin, the unit of thermodynamic temperature, is defined on the basis of the Boltzmann constant (1.380649 × 10^{-23} J/K), the kilogram, the meter, and the second.
amount of substance	mole	mol	A mole is the amount of material that contains 6.022 140 76 × 10^{23} entities. This value is the Avogadro constant.
luminous intensity	candela	cd	The candela is the luminous intensity, in a given direction, of a source that emits monochromatic radiation of frequency 540 × 10^{12} hertz and that has a radiant intensity in that direction of 1/683 watt per steradian.

Note: The unit definitions are a contribution of the National Institute of Standards and Technology and come directly from the NIST website.

Commonly Used Derived Units

Dimension	Name	Symbol	Definition
force	newton	N	$\frac{kg \cdot m}{s^2}$
energy, work, heat	joule	J	$N \cdot m = \frac{kg \cdot m^2}{s^2}$
power	watt	W	$\frac{J}{s} = \frac{kg \cdot m^2}{s^3}$
pressure	pascal	Pa	$\frac{N}{m^2} = \frac{kg}{m \cdot s^2}$
electrical charge	coulomb	C	$A \cdot s$
voltage	volt	V	$\frac{J}{C} = \frac{kg \cdot m^2}{A \cdot s^3}$
electrical resistance	ohm	Ω	$\frac{V}{A} = \frac{kg \cdot m^2}{A^2 \cdot s^3}$
frequency	hertz	Hz	$\frac{cycles}{s} = s^{-1}$
particle mass	atomic mass unit	u	1/12 of a carbon-12 nuclide

APPENDIX D

METRIC PREFIXES

Prefix	Meaning (Origin)	Symbol	Factor	Example	Application
exa-	six (Gk.)	E	10^{18}	exabyte (EB)	amount of data created every second
peta-	five (Gk.)	P	10^{15}	petahertz (PHz)	frequency of ultraviolet radiation
tera-	monster (Gk.)	T	10^{12}	terawatt (TW)	average annual power generation in the United States
giga-	giant (Gk.)	G	10^{9}	gigabyte (GB)	hard-drive storage capacity
mega-	great (Gk.)	M	10^{6}	megajoule (MJ)	energy to heat 10 L of water from 0 °C to 100 °C
kilo-	thousand (Gk.)	k	10^{3}	kilometer (km)	distances on Earth
hecto-	hundred (Gk.)	h	10^{2}	hectometer (hm)	wavelength of radio waves
deca-	ten (Gk.)	da	10^{1}	decapascal (daPa)	pressure of sound
(base)			10^{0}		
deci-	tenth (L.)	d	10^{-1}	decibel (dB)	sound loudness
centi-	hundredth (L.)	c	10^{-2}	centimeter (cm)	distances in a laboratory
milli-	thousandth (L.)	m	10^{-3}	millivolt (mV)	EKG signal from heart at skin
micro-	small (Gk.)	µ	10^{-6}	microsecond (µs)	time for light to travel 1 km
nano-	dwarf (Gk.)	n	10^{-9}	nanometer (nm)	size of atoms
pico-	small (Sp.)	p	10^{-12}	picometer (pm)	long gamma-ray wavelength
femto-	fifteen (Dan.)	f	10^{-15}	femtogram (fg)	mass of a virus
atto-	eighteen (Dan.)	a	10^{-18}	attosecond (as)	period required to image an orbiting electron

APPENDIX E

COMMON ABBREVIATIONS AND SYMBOLS

Abbreviations

Unit	Abbreviation	Dimension
ampere	A	current
atmosphere	atm	pressure
atomic mass unit	u, Da	mass
bar	bar	pressure
candela	cd	light intensity
coulomb	C	electrical charge
day	d	time
decibel	dB	relative power
degree	°	temperature or angle
degree (Celsius)	°C	temperature
degree (Fahrenheit)	°F	temperature
gram	g	mass
hertz	Hz, s^{-1}	frequency
hour	h	time
joule	J	energy or work
kelvin	K	temperature
kilogram	kg	mass
kilowatt-hour	kW·h	energy
liter	L, ℓ	volume
meter	m	length/distance
millibar	mbar	pressure
minute	min	time
molarity	M	solution concentration
mole	mol	quantity (particles)
newton	N	force
ohm	Ω	electrical resistance
pascal	Pa	pressure
second	s	time
volt	V	electrical potential difference
watt	W	power
year	y	time
–	×	magnification power
–	α, $^4_2$He	alpha particle

Unit	Abbreviation	Dimension
–	AC	alternating current
–	AM	amplitude modulation
–	*AMA*	actual mechanical advantage
–	(aq)	chemical state: aqueous
–	β, $^{0}_{-1}$e	beta particle
–	CAT/CT	computed axial tomography
–	DC	direct current
–	e^-	electron
–	EM	electromagnetic
–	*EPE*	elastic potential energy
–	FM	frequency modulation
–	γ	gamma ray
–	(g)	chemical state: gas
–	GFI	ground-fault interrupter
–	*GPE*	gravitational potential energy
–	*IMA*	ideal mechanical advantage
–	IR	infrared
–	IUPAC	International Union of Pure and Applied Chemistry
–	*KE*	kinetic energy
–	(l)	chemical state: liquid
–	LED	light-emitting diode
–	*MA*	mechanical advantage
–	n	neutron
–	NIST	National Institute of Standards and Technology
–	p^+	proton
–	*PE*	potential energy
–	pH	acidity/alkalinity
–	(s)	chemical state: solid
–	SF	significant figure(s)
–	sp gr	specific gravity
–	SI	Système International d'Unités
–	STP	standard temperature and pressure
–	UV	ultraviolet

Formula Symbols and Quantities

Symbol	Quantity
A	area; mass number
a	acceleration
c	speed of light
c_{sp}	specific heat
d	density, distance
$\Delta \mathbf{x}$	displacement
F	force
$\mathbf{F}_g$	gravitational force, weight
$\mathbf{F}_{in}$	input force
$\mathbf{F}_{out}$	output force
f	frequency
G	gravitational constant
g	gravitational acceleration
h	height
h	Planck's constant
h_I	image height
h_o	object height
I	electrical current
k	constant (unspecified)
λ	wavelength
L	length
ℓ	lever arm; liter

Symbol	Quantity
ℓ_{in}	input lever arm
ℓ_{out}	output lever arm
L_f	heat of fusion
L_v	heat of vaporization
m	mass
N_{in}	number of turns, input
N_{out}	number of turns, output
P	power; pressure
p	linear momentum
Q	charge; heat
R	electrical resistance
r	radius
s	speed
T	period
T_C	Celsius temperature
T_F	Fahrenheit temperature
T_K	Kelvin temperature
Δt	time interval
V	volume; voltage, electrical potential difference
v	velocity
W	work
Z	atomic number

APPENDIX F

CREATING GRAPHIC ORGANIZERS

Several times in this textbook you are asked to demonstrate that you understand a concept or a group of concepts by creating a *graphic organizer*—a visual way to represent data. Use this appendix as a guide to help you create different kinds of graphic organizers to show how concepts in physical science compare, how they are structured, or how they move through a process.

COMPARING CLASSES OF LEVERS

	First-Class Lever	Second-Class Lever	Third-Class Lever
Order of parts	F_{out} – fulcrum – F_{in}	fulcrum – F_{out} – F_{in}	fulcrum – F_{in} – F_{out}
Comparing lengths of input arm (ℓ_{in}) and output arm (ℓ_{in})	any	$\ell_{in} > \ell_{out}$	$\ell_{in} < \ell_{out}$
Mechanical advantage	any	> 1	< 1

Table/T-Chart

A table is often used in science to organize data, but it can also be used to organize descriptions and other information. A table can be a helpful way to compare two or more concepts in physical science. For example, the three classes of levers are compared in the table at left. If you are comparing only two concepts, this kind of table with only two columns can be called a *T-chart* because of its shape.

When you create a table, place the concepts that you are comparing in the top row. The characteristics under consideration should go in the leftmost column. Fill in the cells, considering how each characteristic or category differs for the concepts that you are comparing.

Hierarchy Chart

A *hierarchy chart* is a diagram that shows the relationships between several concepts. For example, a company has an organizational chart that does the same thing. Hierarchy charts are especially useful in physical science when studying classification.

When you create a hierarchy chart, think of the different categories that a larger category includes. The example below shows how all the objects in the solar system can be classified as planets, small solar system bodies, dwarf planets, the sun, or moons.

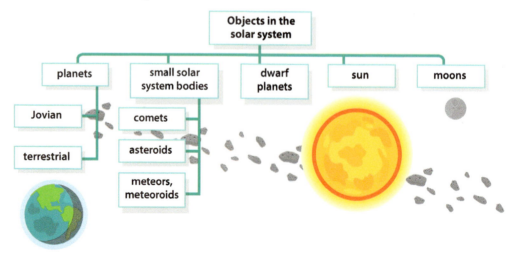

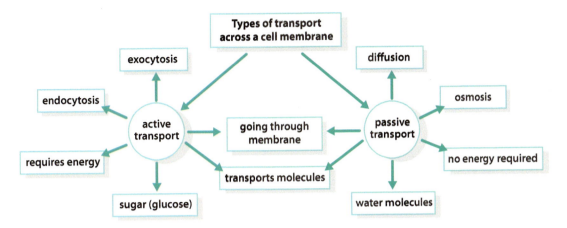

Concept Map

A *concept map* (above) is a freeform graphic organizer that relates concepts, sub-concepts, and their related characteristics. It's similar to an outline in visual form. Because concept maps are so flexible, they can be more difficult to create.

To make a concept map, start with the biggest concept and put it in a circle. Use lines or arrows to connect this main concept with smaller concepts. You can use linking words to help a viewer see the connection between two concepts and read the concept map like a sentence. The further you get from the main idea, the more specific the information should become.

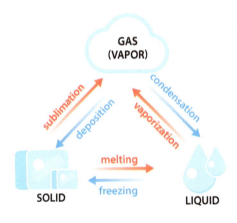

Process Map/Flowchart

A *process map* (left) illustrates a specific chain of events from the first in a physical science process. Because of its purpose, it almost always includes arrows to indicate the direction of the process.

When you create a process map, think of the process that you are illustrating and the different steps that are part of this process. Make sure that you put them in order! You can show a process in a line of events such as a timeline, or, if the process repeats itself, you can show it in a circle as a cyclical process that doesn't have a beginning or an end.

Concept Definition Map

The last type of graphic organizer is a *concept definition map*. The graphic organizer on the right takes a concept and considers four questions: What is it? What is it not? What are its properties? What are some examples?

Think through these four questions for a concept when you create a concept definition map. Then put the concept in the middle of the map and organize these four questions in boxes labeled with these questions around the concept.

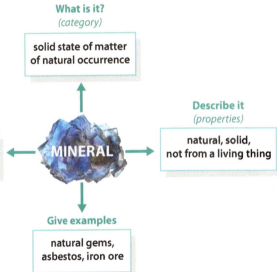

CREATING GRAPHIC ORGANIZERS   537

GLOSSARY

A

absolute zero The coldest temperature possible.

acceleration (a) The rate of change in velocity.

accuracy The comparison of a measurement to an accepted or expected value.

acid A substance that produces hydrogen ions in an aqueous solution.

acoustic amplification The process of making a sound louder.

acoustic spectrum The continuum of all possible sound waves.

activation energy The minimum energy needed for a chemical reaction to occur.

active sonar An underwater device that produces short pulses of sound that echo back to the sending object; used to find the bearing and range of submerged objects.

actual mechanical advantage (*AMA*) The mechanical advantage that accounts for all mechanical losses sustained through using a machine.

additive color A color that is produced by combining the wavelengths of different colors.

alchemists Scientists who were interested in turning low-value materials such as lead into high-value substances like gold.

alcohol A substituted hydrocarbon in which a hydroxyl group (OH) replaces a hydrogen atom.

aldehyde A substituted hydrocarbon in which the replacement by an oxygen atom of a pair of hydrogen atoms at the end of a hydrocarbon chain forms a carbonyl group (C=O).

alkali metal An element in Group 1 of the periodic table, having one valence electron that it can easily lose to form a 1+ cation, making it extremely reactive; the most reactive of all the metals.

alkaline earth metal An element in Group 2 of the periodic table, having two valence electrons that it tends to lose easily to become a 2+ cation, making it very reactive.

alkane A saturated hydrocarbon that has only single bonds between its carbon atoms.

alkene An unsaturated hydrocarbon with at least one double bond between carbon atoms.

alkyne An unsaturated hydrocarbon with at least one triple bond between carbon atoms.

allotrope One of multiple different forms of the same element in the same state.

alloy A solid solution made of two or more elements, including at least one metal.

alpha decay A nuclear decay that results in the emission of an alpha particle.

alpha particle A helium nucleus (α or ^4_2He) that is emitted from a nucleus when a radioactive isotope experiences alpha decay.

alternating current (AC) Electric current in which the charge carriers change direction periodically.

amino acid A class of organic compounds that serve as the building blocks of proteins.

ammeter A meter used to measure electric current through a circuit.

amorphous solid A solid that consists of a mass of particles with no discernible pattern.

ampere (A) The fundamental SI unit of electric current.

amplitude The maximum displacement from the equilibrium position during periodic motion.

amplitude modulation (AM) The process of putting information into a radio wave by changing, or modulating, the amplitude of the wave.

anion A negatively charged ion.

antenna A device that converts electrical energy to radio waves and vice versa.

antinode The position on a standing wave that has maximum displacement.

anvil One of the three bones, along with the hammer and stirrup, of the middle ear that transmit energy from the outer ear to the inner ear.

aqueous solution A water-based solution.

Archimedes's principle The principle that states that an immersed object displaces an amount of fluid equal to its volume and that the weight of the displaced fluid is equal to the buoyant force acting on the object.

aromatic hydrocarbon An unsaturated hydrocarbon that contains at least one benzene structure.

artificial transmutation A manmade nuclear change.

atom The building block of all matter; consists of protons, electrons, and (usually) neutrons.

atomic mass The mass of an atom expressed in atomic mass units; the weighted average of the masses of all the naturally occurring isotopes of an element.

atomic number The unique number of protons found in every atom of a particular element.

atomic radius The distance from the center of an atom's nucleus to its outermost energy level.

B

balanced forces Simultaneous forces whose pushes and pulls cancel each other out.

base A substance that produces hydroxide ions in an aqueous solution.

battery A power source for DC electrical systems consisting of two or more electrochemical cells.

benzene A six-carbon unsaturated hydrocarbon ring with the electrons from the C–C bonds equally distributed among the carbon atoms. It is the key feature in aromatic hydrocarbons.

Bernoulli's principle The principle that states that the fluid pressure of a flowing fluid decreases as its speed increases.

beta decay A nuclear decay that results in the emission of a beta particle.

beta particle An electron (β or $_{-1}^{0}e$) emitted from a nucleus when a radioactive isotope experiences beta decay.

binary compound A compound made from only two elements.

biochemistry The chemistry of living organisms and their processes.

biomass energy Chemical potential energy obtained from renewable organic materials.

block and tackle A system of fixed and movable pulleys connected by ropes.

Bohr model The atomic model developed by Niels Bohr in which electrons travel in distinct spherical regions called *energy levels* at fixed distances from the nucleus.

boiling The relatively fast form of vaporization in which the energy within a liquid creates higher pressure within the liquid than the air pressure outside the liquid.

boiling point The temperature at which a liquid starts to boil.

boiling point elevation The increase in the boiling point of a solvent due to the presence of dissolved solute.

bonding pair Two shared electrons in a covalent bond.

Boyle's law A gas law stating that the pressure and volume of a sample of gas at a constant temperature are inversely proportional.

branched chain An organic molecule with carbon atoms connected to each other in such a way as to create more than one chain of carbons.

Brownian motion The random movements of microscopic particles due to collisions.

buffer A weak acid or base in solution that resists changes in pH when a moderate amount of either an acid or base is added.

buoyancy The tendency of an object to float when immersed in a fluid.

buoyant force The upward force caused by the displacement of a fluid.

C

caloric theory The now-obsolete theory that stated that heat was an invisible self-repelling fluid.

calorimeter A device that enables scientists to measure the thermal energy transferred in reactions and between systems.

capacitor A device that stores electric charge.

carbohydrate An organic compound comprised of carbon, hydrogen, and oxygen atoms that is the basic energy source for living organisms; includes sugars and starches.

carbonyl group A functional group comprised of an oxygen atom double bonded to a carbon atom (C=O); a substituent in an aldehyde or ketone.

carboxide One of a group of compounds made of carbon and oxygen.

catalyst A substance that helps a reaction happen faster but is not used up in the reaction.

cation A positively charged ion.

centripetal acceleration Acceleration that causes an object to move along a circular path.

centripetal force A force that accelerates an object toward the center of a circular path.

chain reaction A self-sustaining nuclear fission process in which neutrons produced in one fission reaction trigger more fission events.

charging by conduction The process by which one charged object can produce a second charged object by the two objects being placed in contact with each other and the excess charge being shared.

charging by friction The process by which an object can gain excess charge while being rubbed by another object.

charging by induction The process by which one charged object can produce a second charged object by allowing the electric force to move excess charge onto the second object while the two objects are not in contact with each other.

Charles's law A gas law stating that the volume and absolute (Kelvin) temperature of a sample of gas at constant pressure are directly proportional.

chemical bond An electrostatic attraction that forms between atoms when they share or transfer valence electrons.

chemical change A change that alters the chemical composition of a substance.

chemical equation A combination of chemical formulas and symbols that models a chemical reaction.

chemical equilibrium The state that occurs when forward and reverse reactions each happen at the same rate.

chemical formula A shorthand way of identifying a chemical compound consisting of the element symbols and subscripts that represent the number of atoms of each element.

chemical property A property of a substance that describes how its chemical identity changes in the presence of another substance or under certain conditions.

chemical reaction A process that rearranges the atoms in one or more substances into one or more new substances.

chemiluminescence The emission of light energy from chemical reactions.

chemistry The study of the composition, structure, and properties of matter, and the changes that take place in matter.

circuit breaker An electrical safety device consisting of an automatic switch that opens when there is too much current in a circuit.

circular motion Movement along a circular path.

classical elements Earth, air, water, fire, and aether; considered by the Greeks to be the basic elements of which all materials were formed.

classical mechanics The study of how and why large objects move; also called *Newtonian mechanics*.

closed circuit A complete electric circuit that allows charge to flow.

cochlea The inner-ear organ that converts kinetic energy to electrical impulses.

coefficient A number placed in front of a chemical formula within a chemical equation that shows how many units of a reactant or product are needed to balance the chemical equation.

cohesion The attraction of like particles to each other.

colligative property A physical property of solutions that depends only on the concentration of dissolved particles and not on the identity or properties of the solute.

collision model The model that states that for a reaction to occur, the reactant particles must collide with each other, with the proper alignment, and with enough energy.

colloid A heterogeneous mixture in which the dispersed particles are between 1 nm and 1 μm in diameter and will not settle out.

combined gas law A gas law that states that the ratio of the product of a gas sample's pressure and volume to its absolute (Kelvin) temperature is a constant.

combustion An exothermic chemical reaction in which a fuel reacts with oxygen.

commutator A structure in DC motors and generators that converts current between DC and AC.

compound A pure substance consisting of atoms of two or more elements that are chemically combined in a fixed ratio.

compound machine A machine that combines two or more simple machines.

compressibility The property of a substance indicating how easily its particles can be pushed closer together.

compression A region of high density and pressure in a longitudinal (compression) wave.

compression force Any force that pushes two objects together.

compression wave See *longitudinal wave*.

computed tomography Medical imaging technology that uses x-rays to take millions of images that are combined into cross-sectional images of the body.

concave lens See *diverging lens*.

concave mirror A curved mirror with the reflective side on the inside of the curve.

concentration The actual amount of solute dissolved in a given amount of solution.

condensation The change of state from a gas (vapor) to a liquid.

conduction Movement of electric charge or thermal energy through an object or from object to object through direct contact.

conductivity The physical property indicating how easily a material transfers thermal energy or electric charge.

constructive interference An adding of overlapping waves that creates a wave with a larger amplitude.

contact force A force that acts only when one object touches another.

convection Movement of thermal energy as fluids move.

convection current The cycle of warmer fluid rising while cooler fluid sinks.

conventional current The direction in a DC circuit that positive charges would flow; decided by agreement as the standard current direction.

converging lens A lens that causes light rays to come together (converge); also known as a *convex lens*.

converging rays Rays of light that come together after being reflected or refracted.

conversion factor Two quantities with different units that are equivalent to each other and that are written as a fraction.

convex lens See *converging lens*.

convex mirror   A curved mirror with the reflective side on the outside of the curve.

coulomb (C)   The derived SI unit for electric charge; 1 C = 1 A·s

covalent bond   A chemical bond formed as a result of two atoms sharing valence electrons.

Creation Mandate   God's command that directs us to exercise wise and good dominion over His creation to the glory of God and for the benefit of fellow humans.

crest   The highest point of a wave.

critical mass   The smallest mass of fissionable material that can sustain a chain reaction.

crystal lattice   An extensive three-dimensional structure of atoms or ions built up by repeating subunits, such as ions in ionic compounds, or individual atoms or molecules, as in metals or covalent solids.

crystalline solid   A solid with particles arranged in regular, repeating patterns.

current electricity   Electricity involving moving electric charges.

D

damped oscillator   A system that is designed to oscillate but includes a force to reduce the amplitude.

damping   Reducing the amplitude of periodic motion by applying a force that works against the motion.

data   The collection of observations made by scientists during investigations.

decibel (dB)   The unit for measuring relative sound intensity.

decomposition reaction   A chemical reaction in which a single reactant breaks down into two or more products.

density   The mass per unit volume of an object.

deoxyribonucleic acid   See *DNA*.

deposition   The change in state directly from a gas (vapor) to a solid without condensing first.

derived unit   A unit that is a mathematical combination of fundamental SI units.

destructive interference   An adding of overlapping waves that creates a wave with a smaller amplitude.

diaphragm   A large muscle that forms the floor of the chest cavity and controls the volume of air that passes over the vocal cords.

diatomic element   An element found in nature as a molecule consisting of two atoms of that element.

diatomic molecule   A molecule made of two atoms.

diffraction   The bending of waves around an obstacle or through an opening.

diffuse reflection   The reflection off a rough or uneven surface that reflects light rays in all directions.

diffusion   The process of spreading out and mixing due to random particle motion.

direct current (DC)   Electric current in which electric charges move in only one direction.

displacement (Δx)   A vector quantity that describes a change in position.

dissociation   The physical separation of the ions in a solid ionic compound by the action of a solvent.

distance (*d*)   How far an object moves during a time interval.

distillation   A process for separating mixtures that uses the different boiling points of the materials.

diverging lens   A lens that causes light rays to separate (diverge); also known as a *concave lens*.

diverging rays   Rays of light that move apart after being reflected or refracted.

DNA   The nucleic acid responsible for most cellular reproduction, growth, and development; abbreviation for *deoxyribonucleic acid*.

Doppler effect   The perceived change in the frequency of a wave due to the motion of the wave's source or of the receiver.

dosimeter   A device used to measure exposure to radiation.

double (covalent) bond   A covalent bond formed by two pairs of shared valence electrons.

double-replacement reaction   A chemical reaction in which two compounds swap cations or anions with each other; also known as a *double-displacement reaction*.

driven oscillator   A system that is designed to oscillate at a constant amplitude by including a force to overcome frictional losses.

ductility   The physical property indicating how easily a material can be pulled into a wire.

dynamics   The branch of physics that studies forces and how they can change an object's motion.

E

echolocation   The process that uses the time interval and direction of an echo to determine the position of an object.

efficiency   A comparison of the amount of usable energy remaining after a process with the original amount of energy that went into the process.

electric circuit   The loop through which current electricity can flow.

electric current (*I*)   The movement of electric charge through a complete loop.

electric field A three-dimensional region around a charged object that will apply a force on other charged objects within that region.

electric force The field force between two charged objects.

electric potential difference See *voltage*.

electric potential energy Energy that is stored by a charged object in an electric field. The quantity depends on the magnitude of the charges and the object's position in the field.

electrical conductor A material through which electric charge moves easily.

electrical insulator A material through which electric charge does not move easily.

electrical load An electrical device in a circuit that consumes electrical energy.

electrical power The work per second done or produced by an electrical system.

electrical resistance (R) An object's opposition to the movement of charge carriers.

electromagnet The combination of a strong magnet and an electrified wire coil.

electromagnetic force The force produced by static and moving charges.

electromagnetic induction The principle that voltage is produced in a conductor whenever the conductor and a magnetic field are changed relative to each other.

electromagnetic radiation See *electromagnetic wave*.

electromagnetic spectrum The entire range of all electromagnetic waves.

electromagnetic wave A disruption in an electromagnetic field that carries energy, even through the vacuum of space.

electromagnetism The theory that electricity and magnetism are parts of the same fundamental force, called the *electromagnetic force*.

electron The smallest of the main subatomic particles, located outside the nucleus, having a negative charge and a mass that is 1/1836th that of a proton.

electron dot notation A representation of an atom consisting of its chemical symbol with surrounding dots representing its valence electrons.

electronegativity A measure of an element's ability to attract and hold electrons when bonded to other atoms.

electroscope An instrument that detects the presence and relative magnitude of static charge.

element A pure substance that consists of atoms with the same atomic number.

EM wave See *electromagnetic wave*.

endothermic reaction A chemical reaction that absorbs more thermal energy than it releases.

energy The ability to do work.

energy level In the Bohr model of the atom, the regions located at fixed distances from the nucleus of an atom in which electrons are found.

ethics A system of moral values; a theory of proper conduct.

eustachian tube The canal that connects the middle-ear cavity with the throat to allow the equalization of pressure on both sides of the eardrum.

evaporation The relatively slow form of vaporization in which liquid particles obtain sufficient energy to change to the gaseous state through the random collisions of particles.

exothermic reaction A chemical reaction that releases more thermal energy than it absorbs.

F

family A column of elements in the periodic table having similar valence electron arrangement, resulting in similar chemical properties; also known as a *group*.

fat A lipid comprised of carbon, hydrogen, and oxygen atoms; serves as long-term energy storage for plants, animals, and humans.

ferromagnetism The physical property in which a material's magnetic domains spontaneously align to an external magnetic field.

field force A force that acts between objects that are not touching; also called *force at a distance*.

filtration A process for separating solids from fluids by using a barrier that allows only the fluid to pass through.

first-class lever A lever in which the fulcrum is between the input and output forces.

first law of thermodynamics The law that states that energy cannot be created or destroyed but only transferred between objects or transformed; also known as the *law of conservation of energy*.

fission A nuclear reaction in which a large nucleus splits into smaller nuclei.

flammability The chemical property indicating how well a material burns in the presence of oxygen.

fluid Any substance that can flow.

fluorescence The emission of light after an object absorbs high-energy electromagnetic waves.

focal length The distance from the center of a lens or mirror to its focal point.

focal point The point on an optical axis at which all reflected or refracted light rays from incident rays that are parallel to the optical axis converge.

force (F) A push or pull on an object.

force diagram See *free-body diagram*.

forensic science A kind of scientific investigation that attempts to reconstruct the scenario of a crime.

formula unit The smallest whole-number ratio of ions within an ionic compound.

fossil fuel A fuel that has formed from the remains of plants and animals that lived in the past, including coal, petroleum (oil), and natural gas.

frame of reference A coordinate system used to describe the motion of an object.

free-body diagram A sketch that shows an object and the forces acting on it; also called a *force diagram*.

free fall The motion of an object that falls due to gravity alone, with no other forces acting on it.

freezing The change of state from a liquid to a solid.

freezing point depression The lowering of the freezing point of a solvent due to the presence of a dissolved solute.

frequency The number of waves or cycles that occurs per second.

frequency modulation (FM) The process of putting information into a radio wave by changing, or modulating, the frequency of the wave.

friction A contact force that works against the motion of objects trying to move past each other.

fulcrum The point about which a lever pivots or rotates.

functional group An atom or group of atoms that replaces a hydrogen atom to form a substituted hydrocarbon; also called a *substituent*.

fundamental force Any one of the four forces that appear to underlie all the other known forces: gravity, strong force, weak force, and electromagnetic force.

fundamental tone The longest (lowest frequency) standing wave produced by a vibration of a structure.

fundamental units The seven units (meter, kilogram, second, ampere, Kelvin, mole, and candela) in the SI from which all other units are derived.

fuse An electrical safety device that opens the circuit by melting when an overheated condition occurs due to excessive current.

fusion A nuclear reaction in which small nuclei combine to form a more massive nucleus.

G

gamma decay A nuclear decay that results in the emission of gamma rays.

gamma ray A high-energy photon (γ) that is emitted from a nucleus when a radioactive isotope experiences gamma decay.

gas The state of matter in which particles are far apart, move rapidly, and have little interaction with each other.

Gay-Lussac's law A gas law stating that the pressure and absolute (Kelvin) temperature of a sample of gas at a constant volume are directly proportional.

gear A simple machine that consists of a wheel with teeth on its perimeter that mesh with similar teeth of other gears to do work.

Geiger counter A device designed to measure ionizing radiation.

generator A device that converts mechanical energy into electrical energy by electromagnetic induction.

genetic damage Any damage done to DNA in cells that can affect growth and reproduction of the cells; can be passed to offspring if it occurs in reproductive cells.

geocentric model An obsolete model of the universe with an unmoving Earth at its center and with all planets, the sun, and other celestial objects moving about it.

geographic North Pole The "top" of the earth at the axis of Earth's spin, where the lines of longitude intersect.

geographic South Pole The "bottom" of the earth at the axis of Earth's spin, where the lines of longitude intersect.

geothermal energy Thermal energy that originates deep within the earth's interior.

glycogen A polymer made of glucose monomers that animals and people use for short-term energy storage.

gravitational potential energy (GPE) Potential energy due to an object's position within a gravitational field.

gravity A field force that acts between the masses of any two objects.

ground-fault interrupter A safety device consisting of an outlet with a built-in circuit breaker.

grounding The act of providing a path for electrical charge to move into the earth.

group See *family*.

H

half-life The time in which half the atoms of a radioactive sample will probably decay.

halogen An element in Group 17 of the periodic table having seven valence electrons. It easily gains an electron, forming a 1− anion, which causes it to be highly reactive.

hammer One of the three bones, along with the anvil and stirrup, of the middle ear that transmit energy from the outer ear to the inner ear.

harmonics The fundamental tone and its overtones.

heat Movement of thermal energy from an area of higher temperature to one of lower temperature.

heat of fusion The heat required to melt a quantity of a solid.

heat of vaporization The heat required to vaporize a quantity of liquid.

heliocentric model The modern model of the solar system with the sun as the gravitational center.

hertz (Hz) The derived SI unit of frequency;

$$1 \text{ Hz} = 1 \frac{\text{cycle}}{\text{s}} = \frac{1}{\text{s}} = \text{s}^{-1}$$

heterogeneous mixture A mixture that does not have a uniform appearance since the combined substances are unevenly distributed.

homogeneous mixture A mixture that has a uniform appearance throughout; also known as a *solution*, especially of a solid dissolved in a liquid.

hydraulics The branch of physics concerned with the forces within and work done by liquids.

hydrocarbon An organic compound consisting of only hydrogen and carbon atoms.

hydroelectric energy Electrical energy generated by the movement of water.

hydronium ion An ion (H_3O^+) formed by the combination of a water molecule and a hydrogen ion.

hydroxide ion An ion (OH^-) formed by oxygen and hydrogen.

hydroxyl group A functional group made of an oxygen atom covalently bonded to a hydrogen atom (OH).

hypothesis An initial, testable explanation of a phenomenon that stimulates and guides scientific investigation.

I

ideal mechanical advantage (IMA) The mechanical advantage for an ideal machine (100% efficient) with no friction.

illuminated object An object that is visible because it reflects light from an external source.

incandescence The emission of light due to an object's high temperature.

incident ray A light ray approaching a surface; an incoming ray.

inclined plane A simple machine that consists of a plane whose opposite ends are at different heights.

indicator A substance that changes color in the presence of an acid or base.

inertia The tendency of matter to resist changes in its motion.

inert substance A substance that has almost no chemical reactivity except under extreme conditions.

infrared wave (IR) An electromagnetic wave that is just below (less energetic than) red visible light; used in astronomy, medical imaging, and wireless devices.

infrasonic sound Sound having frequencies below the range of human hearing.

inhibitor A substance that slows the rate of a reaction by reducing the effectiveness of catalysts.

inner transition metal An element from either of two rows usually placed below the periodic table; a member of either the lanthanide or actinide series. It typically has two valence electrons.

instantaneous speed Speed at a particular point in time.

intensity A measure of the power contained in a wave; often refers to sound or electromagnetic waves.

interference The combining of waves where they overlap.

International System of Units See *SI*.

International Union of Pure and Applied Chemistry (IUPAC) The international organization responsible for standardization in chemistry.

ion A charged atom or group of atoms that has gained or lost electrons, producing an unequal number of protons and electrons.

ionic bond A chemical bond formed when atoms transfer valence electrons.

ionization energy The minimum amount of energy required to remove the first electron from the outermost shell of a neutral atom in its gaseous state.

ionizing radiation Radiation that is energetic enough to knock electrons out of atoms or molecules. This radiation is the most damaging to living organisms.

isomer Any of a group of compounds that have the same molecular formula but different structures.

isotope An atom of an element that has a different number of neutrons compared with other atoms of that element, resulting in a different mass number.

isotope notation A symbol that distinguishes between different isotopes. It is written in the form $^A_Z X$ where A is the mass number, Z is the atomic number, and X is the element symbol.

IUPAC See *International Union of Pure and Applied Chemistry*.

J

joule (J) The derived SI unit of work or energy;

$$1 \text{ J} = 1 \text{ N·m} = 1 \frac{\text{kg·m}^2}{\text{s}^2}$$

K

ketone A substituted hydrocarbon in which a pair of hydrogen atoms bonded to a carbon atom other than those at the end of the carbon chain are replaced by an oxygen atom, forming a carbonyl group (C=O).

kinematics The study of how things move.

kinetic energy (KE) Energy that an object possesses due to its motion.

kinetic-molecular model The model that says that all particles are in constant motion and that thermal energy is the sum of the kinetic energies of these particles.

L

larynx The box-like structure located at the top of the trachea that supports the vocal cords.

law A model, often expressed as a mathematical equation, that describes phenomena under certain conditions.

law of acceleration The law that states that the acceleration of an object is directly proportional to the net force acting on the object and is inversely proportional to its mass; also called *Newton's second law of motion*.

law of action-reaction The law that states that for every action force, there is an equal and opposite reaction force; also called *Newton's third law of motion*.

law of cause and effect The basic law that states that every effect has a specific, identifiable cause, and for every cause there is a definite and predictable effect.

law of conservation of charge The principle that states that charge cannot be created or destroyed but only transferred between objects.

law of conservation of energy The law that states that energy cannot be created or destroyed but only transferred between objects or transformed; also known as the *first law of thermodynamics*.

law of conservation of matter The law that states that matter can neither be created nor destroyed, but can only change form.

law of conservation of momentum The law that states that within a closed system, the total momentum remains constant.

law of definite proportions The law that states that the masses of chemical substances combine in definite, characteristic integer ratios when forming compounds.

law of electrostatic charges The law that states that opposite electrical charges attract each other, while like charges repel each other.

law of inertia The law that states that objects at rest remain at rest and objects in motion continue in a straight line at a constant velocity unless acted on by a net external force; also called *Newton's first law of motion*.

law of octaves The principle published by chemist John Newlands that stated that the properties of the forty-nine then-known elements repeated every eighth element, as in a musical octave.

law of reflection The law that states that when a light ray reflects off a surface, the angle of the incident ray equals the angle of the reflected ray.

law of torques The law that states that in a lever system in rotational equilibrium, the sum of the torques must be zero.

law of universal gravitation The law that states that the strength of gravity varies in direct proportion to the masses of the objects involved and inversely to the square of the distance between their centers of mass.

Le Châtelier's principle The principle that states that a chemical system in equilibrium will adjust its equilibrium position in such a way as to reduce the effect of any changes made to the system.

lever A simple machine that consists of a rigid bar that turns about a pivot point (the fulcrum).

Lewis structure A system for modeling the covalent bonds between atoms in a molecule and any un-bonded electrons in the molecule.

light Electromagnetic energy in the visible light range.

lipid Organic compounds that provide long-term energy storage in living organisms; includes fats, oils, waxes, and steroids.

liquid The state of matter in which particles are close together but able to move around.

longitudinal wave A wave in which the disruptions are parallel to the direction of wave travel.

loudness Human perception of the intensity of a sound wave.

luminous object An object that is producing visible light.

luster The physical property indicating how well a substance reflects light.

M

macromolecule A very large molecule.

magnet Any material or object that can produce a magnetic field.

magnetic declination The angular difference between the directions to the geographic and magnetic North Poles.

magnetic domain A group of atoms whose individual magnetic fields are aligned.

magnetic field The region surrounding a magnet or current-containing wire that can apply a magnetic force on magnetic materials.

magnetic North Pole The magnetic pole in the Northern Hemisphere. It is currently located beneath the Arctic Ocean and is moving toward Siberia. This pole is actually the south pole of Earth's magnetic field.

magnetic pole One of the two regions of concentrated magnetic fields in a magnet; traditionally called the *north* and *south poles*.

magnetic South Pole The magnetic pole in the Southern Hemisphere. It is currently located beneath the Antarctic Ocean south of Australia. This pole is actually the north pole of Earth's magnetic field.

magnetism The properties of magnets and magnetic fields and the phenomena that they produce.

malleability The physical property indicating how easily a material can be hammered or pressed into sheets.

mass The measure of the amount of matter in an object.

mass number (*A*) The total number of particles found in the nucleus of a particular isotope of an element.

materials science An interdisciplinary science that studies the properties of materials to discover new materials, new uses for existing materials, and changes to existing materials to meet the needs of people.

matter Anything that has mass and occupies space.

measurement Data that is based on numbers or quantities; includes a number and a unit; also known as *quantitative data*.

mechanical advantage (*MA*) The amount by which a simple machine multiplies an input force to produce an output force.

mechanical energy The sum of the kinetic energy and all the forms of potential energy in a system.

mechanical wave A wave that carries energy through a physical medium.

mechanics The study of motion.

melting The change of state from a solid to a liquid.

melting point The temperature at which a solid turns to a liquid.

metal An element that is typically dense, solid, ductile, malleable, highly conductive, and chemically reactive, especially in the presence of nonmetal elements. Metals are located on the left end of the periodic table.

metallic bond The attraction between metal atoms and their collectively shared valence electrons.

metalloid An element with characteristics between those of metals and nonmetals; also called *semiconductors*; located between metals and nonmetals on the periodic table.

metric prefixes A set of prefixes representing powers of ten that are attached to metric units to create smaller or larger units.

microwave A wave from the upper end of the radio wave band; used for navigation, communication, and cooking.

mixed group Any of Groups 13–16 in the periodic table; so named because they contain metals, nonmetals, and metalloids. These groups are often named for the first element in the family.

mixture A physical combination of two or more substances in a changeable ratio.

model A workable explanation or description of a phenomenon.

molarity (*M*) A measure of solution concentration; the number of moles of solute per liter of solution.

molar mass The mass of one mole of a substance.

mole (mol) SI fundamental unit for the quantity of matter in a substance; a count equal to 6.022×10^{23}.

molecular formula A formula that shows the exact composition of a molecule. It is comprised of element symbols along with subscripts for the number of atoms of each element.

molecule A particle consisting of two or more atoms covalently bonded together.

momentum (*p*) A property of a moving system that is equal to its velocity times its mass.

monomer A simple molecule that can link with other monomers to form large molecules called *polymers*. For example, glucose monomers combine to form the polymer starch.

N

natural science The study of objects and phenomena of the physical world.

net force The vector sum of all the forces acting on an object.

neutral buoyancy The condition when an object submerged in a fluid neither ascends nor descends because its density is equal to that of the fluid.

neutralization A double-replacement reaction between an acid and a base that produces a salt and water.

neutron A subatomic particle found in the nucleus of most atoms that has no electrical charge and a mass slightly greater than that of a proton.

newton (N) The derived SI unit of force; $1\text{ N} = 1\ \dfrac{\text{kg}\cdot\text{m}}{\text{s}^2}$

Newton's first law of motion See *law of inertia*.

Newton's second law of motion See *law of acceleration*.

Newton's third law of motion See *law of action-reaction*.

noble gas An element in Group 18 on the periodic table having eight valence electrons that fill the outer energy level. (Helium is an exception with only two.) With a full outer energy level, it is inert (i.e., nonreactive).

node The position on a standing wave that has no displacement.

nonmetal An element that typically has four or more valence electrons and that does not exhibit the general properties of metals; located on the right side of the periodic table.

nonpolar bond A covalent bond in which electrons are shared equally.

nonrenewable energy resource An energy resource that is not replaced naturally.

normal line A line that is perpendicular to a surface (e.g., the direction of the normal force or the line from which a light ray's angle of incidence is measured).

normal force The force that acts in a direction that is perpendicular to the surface where two objects make contact.

north magnetic pole The end of a magnet from which magnetic field lines leave.

nuclear change Any change that alters the composition of the nucleus in an atom.

nuclear decay See *radioactive decay*.

nuclear energy Energy produced by nuclear reactions.

nuclear model The atomic model developed by Ernest Rutherford in which an atom is made up of a tiny, dense, positively charged central nucleus surrounded by negatively charged electrons.

nuclear stability The likelihood of an isotope to not undergo radioactive decay.

nuclear waste Any waste that is radioactive.

nucleic acid A biochemical polymer that encodes, stores, and provides instructions for cellular processes.

nucleotide Any of a group of biochemical molecules that act as the monomers to make nucleic acids. Each consists of a sugar, a nitrogen base, and a phosphate group.

O

octet rule The principle that states that atoms generally are most stable when they have eight electrons in their valence energy level.

ohm (Ω) The derived SI unit for electrical resistance; $1\ \Omega = 1\ \frac{V}{A} = 1\ \frac{kg \cdot m^2}{s^3 A^2}$

Ohm's law The law that states that the current in a circuit is directly related to the voltage and inversely related to the resistance.

opaque material A material through which visible light cannot pass.

open circuit An incomplete electric circuit that prevents the movement of charge.

orbit The circular path of an electron around the nucleus in the Bohr model.

orbital The region within an atom where an electron will most probably be located; part of the quantum-mechanical model of the atom.

organic chemistry The study of the composition, structure, and properties of carbon-containing compounds.

organic compound A covalently bonded compound containing carbon.

overtone A shorter, faster vibration (higher pitch) in addition to the fundamental tone produced by a vibrating structure.

oxidation A loss of electrons in a chemical reaction.

oxidation-reduction reaction See *redox reaction*.

oxidation state A positive or negative number showing the electric charge on an element when it forms a compound.

oxide A compound that contains oxygen, often formed in an oxidation reaction.

P

parallel circuit A circuit with multiple paths that electric current can take.

particle model of matter A model that states that all physical matter exists in the form of particles.

pascal (Pa) The derived SI unit of pressure; $1\ Pa = 1\ \frac{N}{m^2} = 1\ \frac{kg}{m \cdot s^2}$

Pascal's principle The principle that states that pressure applied to a fluid is transmitted throughout the fluid.

passive sonar A system of underwater microphones that can only receive, not produce, underwater sounds in order to detect a submerged object.

peer review The assessment process for scientific research in which other scientists review and respond to research.

period (1) A row in the periodic table of the elements; also called a *series*. (2) The time interval (T) for one complete cycle of periodic motion.

periodicity The idea that properties of elements repeat in regular patterns in relation to some basic characteristic such as atomic mass or atomic number.

periodic law The law that states that the properties of the elements vary with their atomic numbers in a regular, repeated pattern.

periodic motion Motion that repeats in equal time intervals.

periodic table of the elements A table of the chemical elements arranged to display their periodic properties in relation to their atomic numbers.

periodic trend The change of a particular property along rows or columns of the periodic table.

pH A unitless number between 0 and 14 that tells how acidic or basic a substance is.

phase diagram A diagram summarizing the temperature and pressure conditions under which a pure substance exists as a solid, liquid, or gas.

phenomenon An observable or measureable event, object, process, or property.

phosphorescence A slow emission of absorbed electromagnetic energy, similar to fluorescence.

photoelectric effect The dislodging of electrons from metals caused by light of sufficiently high frequency, supporting the particle model of light.

photon A wave bundle, or particle, of electromagnetic energy.

physical change Any change in matter that does not alter its chemical or nuclear composition.

physical property Anything about a substance that can be observed or measured without altering its chemical composition.

physical science The study of nonliving matter and energy.

physics The study of matter and energy and the interactions between them.

pigment A substance that absorbs some wavelengths of light and reflects others; used to create paints, inks, and dyes.

pitch How high or low an audible tone sounds to the human ear; related to the concept of wave frequency.

plane mirror A flat mirror.

plasma A gas-like state of matter, formed at very high temperatures, that consists of high-energy ions and free electrons.

plum pudding model The atomic model developed by J. J. Thomson that views atoms as spheres of positively charged material with embedded electrons.

polar covalent bond A covalent bond in which electrons are shared unequally.

polarity The unequal distribution of electric charge in a covalent bond.

polar molecule A molecule with negatively charged and positively charged regions as a result of polar covalent bonds and molecular geometry.

polyatomic ion A group of covalently bonded atoms that together have gained or lost electrons and act as a single ionized particle.

polyatomic molecule A molecule formed by two or more atoms, including diatomic molecules.

polymer A large molecule formed by linking smaller molecules, called *monomers*. For example, nucleic acid, a polymer, is formed by linking nucleotide monomers.

potential energy (PE) Stored energy that can be used later.

power The rate of doing work.

precipitate A solid that forms when two solutions react.

precision The degree of exactness of a measurement; can indicate the closeness or repeatability of measurements.

pressure (P) A measurement of the amount of force acting upon a unit area.

primary color One of the three colors (red, blue, and green) of visible light that the human eye can sense; mixed to produce other colors.

prime mover The agent that provides the mechanical energy to turn the rotor of a generator.

principle of uniformity The basic law that says that nature acts the same today as it did yesterday and that we can fully expect it to act the same way tomorrow.

product A substance that is formed during a chemical reaction.

projectile motion The two-dimensional motion of any flying object whose path is determined by the influence of an external force only, such as gravity.

protein A biochemical polymer made of amino acids. Proteins are the building blocks for muscle, blood, skin, and hair in humans and animals.

proton A subatomic particle found in the nucleus of an atom and having a positive charge and a mass slightly less than that of a neutron.

pulley A simple machine that consists of a wheel and axle system with a groove around the perimeter of the wheel in which a rope, cable, or belt moves with the wheel as it rotates.

pure substance A material made of only one kind of element or compound.

Q

qualitative data Data that is based on qualities of an object or phenomenon.

quantitative data Data that is based on measureable dimensions of an object or phenomenon; also known as *measurement*.

quantum-mechanical model The currently accepted atomic model in which electrons are found in orbitals that are positioned around a nucleus that contains protons and (usually) neutrons.

quantum mechanics The branch of physics that explores the behavior of matter and energy at the atomic and subatomic level.

R

radiation Movement of energy in the form of electromagnetic waves.

radioactive decay The naturally occurring, spontaneous change of an unstable isotope to a more stable one by emitting particles or energy or both; also known as *nuclear decay* or *radioactivity*.

radioactivity See *radioactive decay*.

radiometric dating A method for mathematically estimating the age of an object by measuring the present amount of a radioactive isotope and comparing that with the starting amount of that isotope in the object.

radiotracer A radioactive isotope used in nuclear medicine to study how an isotope moves through or collects in a certain organ or system.

radio wave The longest and lowest energy type of electromagnetic wave; used for navigation and communication.

rarefaction A region of lower density and pressure in a longitudinal (compression) wave.

ray (1) A half-line. (2) A model of light in which the endpoint represents the light's source and the light travels in one direction.

reactant A substance that enters into a chemical reaction.

reaction rate The speed of a reaction.

reactivity The chemical property indicating to what degree a material reacts with another substance.

real image An image that forms at the point where converging rays of light intersect.

redox reaction A chemical reaction in which electrons are transferred; oxidation and reduction occurs.

reduction A gain of electrons in a chemical reaction.

reflected ray The ray that bounces off a surface.

reflection The bouncing of waves off a surface.

refraction A change in wave direction due to a change in a wave's speed as it enters a new medium.

regular reflection Reflection that occurs when a smooth surface reflects light rays in mostly the same direction; also known as a *specular reflection*. A perfectly regular reflection is mirrorlike, reflecting all rays and wavelengths uniformly.

relativistic mechanics The study of the motion of objects whose speeds are near the speed of light.

renewable energy resource An energy resource that is easily replaced by natural methods.

resistor An electrical device that slows down electric current by converting electrical energy into other forms, such as thermal energy.

resonance An increase in the amplitude of a vibration due to additional wave input.

retrograde motion The apparent motion of some planets in which they appear to slow down, stop, reverse their direction, and then resume their normal motion.

reversible reaction A reaction in which the products can react together to re-form the original reactants.

ribonucleic acid See *RNA*.

right-hand rule A mnemonic device used to remember the direction of the magnetic field induced by an electric current moving through a wire.

ring An organic molecule made by connecting the two ends of a carbon chain.

RNA A nucleic acid that plays a critical role in protein synthesis; abbreviation for *ribonucleic acid*.

rolling friction The frictional force between two objects when one is rolling relative to the other.

rotor The rotating portion of a generator.

S

salt An ionic compound formed by a combination of cations and anions.

saturated hydrocarbon A hydrocarbon that has only single bonds between its carbon atoms.

saturated solution A solution that contains the maximum amount of solute that it can hold at a given temperature.

scalar A measurable quantity that consists of magnitude (size) only.

science The systematic study of the universe that produces observations, inferences, and models, including the products that it creates through this systematic study.

scientific inquiry An ongoing, orderly, cyclical approach used to investigate the world.

screw A simple machine that consists of an inclined plane wrapped around a cone or cylinder in a spiral pattern.

secondary color A color of light that is produced when primary colors are mixed.

second-class lever A lever in which the output force is between the input force and the fulcrum.

second law of thermodynamics The law that states that energy can flow from a colder object to a warmer object *only* if something does work.

semiconductor A material with conductivity between those of conductors and insulators.

series circuit A circuit with only one path that electric current can take.

short circuit An unintended path for an electric current.

SI A standardized system of measurement units used for science. SI stands for *International System of Units* (from the French *Système International d'Unités*).

simple harmonic motion Periodic motion that is caused by a restoring force that is proportional to the displacement of the system.

simple machine A basic mechanical device that changes the magnitude, direction, or distance traveled of the force used when doing work.

single (covalent) bond A covalent bond formed by one pair of shared valence electrons.

single-replacement reaction A chemical reaction in which one element in a compound is replaced by another element; also known as a *single-displacement reaction*.

sliding friction The frictional force between two objects when one is sliding past the other; also called *kinetic friction*.

soda lake A lake with a high pH (basic) due to the presence of carbonates.

solar energy Energy produced by the sun.

solenoid A coil of current-carrying wire used as a magnet.

solid The state of matter in which particles vibrate in fixed positions, giving a substance a fixed shape and volume.

solubility The maximum amount of solute that can dissolve in a given amount of solvent at a given temperature.

solute In a solution, the dissolved material or the material in lesser abundance.

solution A mixture with a uniform appearance throughout; also called a *homogeneous mixture*.

solvation The process of dissolving a solute into a liquid solvent.

solvent In a solution, the dissolving material or the material in greater abundance.

somatic damage Any damage to cells that are not involved in reproduction, thus harming the organism but not any future offspring.

sonar A device for detecting submerged objects by using sound; an acronym for <u>so</u>und <u>na</u>vigation and <u>r</u>anging.

sonography Technology that uses ultrasound to create images of objects found inside other objects.

sound energy A type of mechanical wave energy that can be detected by the human ear.

south magnetic pole The end of a magnet into which magnetic field lines return.

specific heat (c_{sp}) The energy required to raise the temperature of 1 g of a substance 1 °C.

specular reflection See *regular reflection*.

speed A scalar quantity indicating the rate at which an object moves.

speed of light The speed at which all electromagnetic waves travel in a vacuum (299 792 458 m/s).

standing wave A wave that is moving even though the locations of the crests and troughs appear to be stationary.

state of matter A physical form of matter determined by the forces between and energy of its particles. The three most common states of matter on Earth are solid, liquid, and gas.

static electricity The accumulated electric charge on an object.

static friction The frictional force between two objects that are touching but not moving relative to each other.

stator The stationary portion of a generator.

step-down transformer A transformer that is designed to decrease an output voltage.

step-up transformer A transformer that is designed to increase an output voltage.

stirrup One of the three bones, along with the anvil and hammer, of the middle ear that transmit energy from the outer ear to the inner ear.

straight chain An organic molecule consisting of a single continuous series of any number of carbon atoms bonded to each other.

strong acid An acid in which all or most of the molecules produce hydrogen ions in an aqueous solution.

strong base A base that readily produces large numbers of hydroxide ions in an aqueous solution.

strong force An attractive force that holds protons and neutrons together in a nucleus.

structural formula A drawing depicting the composition and arrangement of atoms in a molecule.

sublimation The change in state directly from a solid to a gas (vapor) without melting first.

substituent See *functional group*.

substituted hydrocarbon A hydrocarbon in which at least one of the hydrogen atoms has been replaced with another atom or group of atoms. The replacement atom or group of atoms is called a *functional group*.

subtractive color A color that forms because of pigments absorbing some wavelengths of light that strike an object.

superconductor A material with zero resistance.

supersaturated solution A solution that contains more than the maximum amount of solute that it can normally hold at a given temperature.

surface wave A wave that occurs along the interface between two media.

suspension A heterogeneous mixture of a fluid and large particles (< 1 μm) that will settle out over time.

synthesis reaction A chemical reaction in which two or more reactants combine into a single, more complex product.

system A portion of a larger motion that we are interested in studying.

T

temperature (T) The measure of the hotness or coldness of a substance; proportional to the average kinetic energy of the particles within the substance.

tension A pulling force that is transmitted through a rope, chain, or similar object.

theory A model that explains a related set of phenomena; can be used to predict unobserved aspects of the phenomena.

thermal conductor A material through which thermal energy moves easily.

thermal energy The sum of the kinetic energies of all the particles within an object.

thermal expansion The property of many materials to increase in volume when heated and contract when cooled.

thermal insulator A material through which thermal energy does not easily move.

thermodynamics The study of thermal energy and heat and how they relate to work and other forms of energy.

thermometric property Any property that changes predictably with changes in temperature.

third-class lever A lever in which the input force is between the output force and the fulcrum.

third law of thermodynamics The law that states that entropy would be at its minimum value at absolute zero. Therefore, absolute zero can never be achieved.

tidal energy Mechanical energy in rising and falling tides.

timbre The distinctive sound of an instrument; also called *quality*.

torque A force that tends to cause a rotation about a pivot point.

toxicology The scientific study of poisons and their negative effects on organisms.

traction The frictional force between a vehicle's tires and the road; responsible for accelerating the vehicle.

trajectory The curved path of a projectile.

transformer A device that uses electromagnetic induction to increase or decrease the voltage of an AC current.

transition metal Any elements in Groups 3–12 of the periodic table, typically having one or two valence electrons, which it easily loses, resulting in cations with charges of 1+ or 2+.

translucent material A material through which light passes but the light is scattered and does not transmit a clear image.

transmutation Any process that converts one isotope into another by changing the number of protons or neutrons.

transparent material A material through which light passes without scattering, transmitting a clear image.

transverse wave A wave in which the disruptions move perpendicular to the direction of wave travel.

triads A model of periodicity developed by Johann Döbereiner that is based on groups of three elements with similar properties.

triple (covalent) bond A covalent bond formed by three pairs of shared valence electrons.

triple point The temperature and pressure at which the solid, liquid, and vapor states of a substance exist in equilibrium.

trough The lowest point of a wave.

tympanic membrane The thin, flexible membrane that separates the outer ear from the middle ear and converts acoustic energy to kinetic energy.

Tyndall effect The scattering of light by the particles in a colloid.

U

ultrasonic sound Sound having frequencies above the range of human hearing.

ultraviolet wave (UV) An electromagnetic wave that is just beyond (more energetic than) violet visible light; used for medical treatment, dentistry, and killing bacteria.

unbalanced forces A collection of forces on an object that don't cancel out and thus cause an acceleration.

unsaturated hydrocarbon A hydrocarbon that has at least one double or triple bond between its carbon atoms.

unsaturated solution A solution that contains less than the maximum amount of solute that it could hold at a given temperature.

urban heat island Term used to describe the difference in temperature of an urban area compared with that of the surrounding region.

V

valence electron Any electron in the outermost energy level of a neutral atom. Unpaired valence electrons are usually involved in chemical bonding.

vaporization The change of state from a liquid to a gas (vapor).

vector A measurable quantity with both magnitude and direction.

vector addition The mathematical process of combining vectors that takes into account both the magnitude and direction of the vectors.

velocity (v) A vector quantity indicating the rate at which an object's position changes.

virtual image   An image produced by diverging rays. The image is formed at the point from which the diverging rays would have originated. Because the rays don't actually intersect, virtual images cannot be projected onto a screen.

viscosity   A measure of a fluid's resistance to flowing.

visible light   Electromagnetic waves that humans can see.

vocal cords   Folds of tissue in the throat that when vibrated produce the sound waves that humans use to communicate.

volt (V)   The derived SI unit for electric potential difference; $1\ V = 1\ A \cdot \Omega = 1\ \dfrac{kg \cdot m^2}{s^3 A}$

voltage   The "force" that moves electric charge carriers through an electrical circuit; also called *electric potential difference*.

voltmeter   A meter used to measure voltage.

volume (V)   The space enclosed or occupied by an object.

W

water displacement   A method for determining the volumes of irregularly shaped objects on the basis of Archimedes's principle.

watt (W)   The derived SI unit of power; $1\ W = 1\ \dfrac{J}{s} = 1\ \dfrac{kg \cdot m^2}{s^3}$

wave   A disruption that carries energy from one location to another.

wave height   The vertical distance between the trough and crest of a wave.

wavelength (λ)   The distance between two identical points on successive waves.

wave-particle duality   The currently accepted model of light indicating that light has both wave and particle characteristics.

wave pulse   A single wave disruption.

wave train   A series of wave pulses.

weak acid   An acid that produces few hydrogen ions in an aqueous solution.

weak base   A base that produces few hydroxide ions in an aqueous solution.

weak force   The extreme short-distance force that holds the subatomic particles together inside protons and neutrons.

wedge   A simple machine consisting of two inclined planes attached at an acute angle and used to spread a material apart as it is forced into the material.

weight   The force of gravity acting on the matter in an object.

wheel and axle   A simple machine consisting of a wheel with a rod running through its axis that acts as the pivot point.

wind energy   Energy from the wind; can be used to turn turbine blades that are attached to generators.

wind shear   A significant change in wind speed, direction, or both in a small geographic area; a significant hazard for aircraft.

work   The energy transferred to a system by an external force when it acts on the system to move it.

workability   The basis upon which a model is assessed, taking into account how well it explains or describes a set of observations and how well the model makes predictions.

worldview   How a person views the world; a set of basic beliefs, assumptions, and values that arises from an overarching narrative about the world and that produces individual and group action.

X

x-ray   (1) A form of electromagnetic energy with wavelengths between 10 pm (picometers, 10^{-12}) and 10 nm that is used for medical imagery, transportation security, and nondestructive inspection. (2) A photograph taken with x-rays.

Y

Young double-slit experiment   An early experiment that demonstrated that electromagnetic waves interfere with each other, supporting the wave model of light.

INDEX

Boldface page numbers denote the location of the definition of key terms.

A

absolute zero, 38, **350**–51, 367
acceleration, **261**–63
  calculation of, 262, 279
acceleration, law of, **278**–79
accuracy, **17**
acid, **224**–25, 227
  strong, **228**
  sulfuric, 227
  weak, **228**, 231
acid concentration, 228, 229
acid indigestion, 237–38
acid strength, 228
acidity, 228–31
acoustic amplification, **430**–31
acoustic energy, **420**–23, 430
acoustic engineering, 434
acoustic spectrum, **422**
action-reaction, law of, **280**–81
activation energy, **160**, 163
actual mechanical advantage (AMA), 299, **301**
  calculation of, 300–301
air convection, 357
air pollution, 221
alchemist, 70
alcohol, 133, 134, **135**
  isopropyl, 133
aldehyde, 134–**35**
algae, blue-green, 223
alkali metal, 76, 77, 80, **82**
alkaline earth metal, **82**
alkalinity, 226, 228–31
alkane, 128
alkene, 129
alkyl halide, 134
alkyne, 130
allotrope, 93
alloy, 82–83, **205**
alpha decay, 177, 179, **180**–81
alpha particle, 53, 177, 179, 180–81, 186, 194–95
alpha radiation survey meter, 195
alternating current (AC), 441, 451, **452**, 454, 458
amalgam, 205

amine group, 139
amino acid, **139**
ammeter, 453, 459
amorphous solid, 37
ampere, 15, 453, 455
Ampère, André-Marie, 453, 476, 489
amplitude, **399**, 401, 404–5, 407, 413
amplitude modulation (AM), 498
analemma, 397
angle of incidence, 510, 515
angle of reflection, 510
anion, **60**, 84, 100
antacid, 237–38
antinode, 414
anvil (ear), 427
aqueous solution, **224**
Archimedes, 376
Archimedes's principle, **376**
Aristotle, 27, 50, 246
aromatic hydrocarbon, **131**
artificial leaf, 322–23
artificial transmutation, 185–86
atom, 28, **29**, 50–62
  ionization of, 60
  properties of, 57–58, 61–62
  structure of, 56–62
atomic mass, **61**–62
atomic model, 50–55
atomic number, 33, **57**
  and atomic mass, 61
  and mass number, 58
  in isotope notation, 59
atomic radius, **89**–90
atomic structure, 86–88
auditory canal, 427
auditory nerve, 427
autonomous vehicle, 499

B

backscatter x-ray, 193
balance, 18
ballistic material, 44
base, **226**–27
  strong, **229**
  weak, **229**, 231
base concentration, 228

base strength, 228
battery, 451–52, 454, 458, 460–61, 466
Becquerel, Henri, 177
belt of stability, 178
benzene ring, **131**
Bernoulli, Daniel, 389
Bernoulli's principle, **388**–**89**
Berzelius, Jacob, 71, 125
beta decay, 177, **180**
beta particle, 177, 180
binary compound, **109**
binary covalent compound, 115–16
binary ionic compound, 109
binary salt, 232
biochemistry, 136–140
bioluminescence, 505
Black, Joseph, 364
Blackett, Patrick, 186
block and tackle, 311–12
  calculation of mechanical advantage of, 312
blood sugar, 138
Bohr model, **54**, 86–88
Bohr, Niels, 54
boiling, **42**
boiling point, **42**, 211
boiling point elevation, 217, 218
  and maple syrup, 221
bond
  chemical, **96**–105
  covalent, **99**, 101–4
  double, 102, 126, 129–31, 135, 139
  ionic, **100**, 104–5
  metallic, **100**
  nonpolar, 103
  polar covalent, 103
  single, 102, 126–29, 131, 139
  triple, 102, 129–30
  types, 99–105
Bose-Einstein condensate (BEC), 38, 367
Boyle, Robert, 71, 381, 421
Boyle's law, 380–82 (**381**)
  calculation with, 382
Brahe, Tycho, 246, 247
branched chain, 126–27
Brownian motion, 28

Brown, Robert, 28
buckyball, 93
buffer, **231**
buffered system, 231
bulletproof technology, 44
buoyancy, 375, 377–78
    neutral, 378
buoyant force, **375**–76

C

calicheamicin, 131
caloric, 364–65
caloric theory, 364–65
calorie, 365
Calorie (kilocalorie), 365
calorimeter, 360
calorimetry, 360–61
candela, 15
capacitor, 459
carbohydrate, **138**
carbon, 126–27
carbon-14, 183, 193
carbonyl group, 134–35
carboxide, 115
carboxyl group, 139
Castle Bravo nuclear weapon test, 189
catalyst, **163**
cathode ray, 52, 176
cation, **60**, 82–83, 100
cause and effect, law of, 6–7
Celsius, Anders, 350
Celsius scale, 348, 350–53
centripetal acceleration, **263**
centripetal force, **288**
Chadwick, James, 57
chain reaction, **187**, 190
change
    chemical, **41**
    physical, **41**
charge carrier, 451–55, 460
Charles, Jacques, 382
Charles's law, 382–**83**
    calculation with, 383
chemical bond, **96**–105
chemical change, **41**
chemical equation, **150**–54, 155
    balancing of, 150–52, 155
    parts of, 150
chemical equilibrium, 165–68

chemical formula, **108**–17
chemical property, **41**
chemical reaction, 41, **148**–68
    and activation energy, 160
    and conservation of energy, 161
    and equilibrium, 164–68
    and reaction rate, 162–63
    classification of, 156–58
    energy in, 159–61
    evidence of, 148–49
    reversible, 164
chemiluminescence, 505
chemistry, **5**
    organic, 124–40
Chernobyl nuclear power plant, 195
chloroethane, 134
cholesterol, 139
circuit breaker, 458, **464**–65
circular motion, **263**–64
classical mechanics, 247
closed circuit, 450, 453
coal, 337, 338
cochlea, 427
coefficient, **150**
coherent light, 513
Colladon, Jean-Daniel, 432
colligative property, **217**–18
collision model, **162**–63
colloid, **203**
color, 503–4, 507–9, 515
    additive, **508**–9
    primary, **508**–9
    secondary, **508**–9
    subtractive, **509**
color-glass condensate (CGC), 38
combined gas law, **385**–86
    calculation with, 386
combustion, 146, **157**, 160, 164
commutator, 481
compass, 472–73, 476
compound, **33**, 96–98, 108
    binary, **109**
    binary covalent, 115–16
    binary ionic, 109
    ionic, 110–12
    organic, **125**. *See also* organic chemistry
compressibility, 37
compression, **406**, 420–21
compression force, 40
compression wave, 406

computed tomography (CT), 192
concave lens, 516–17
concave mirror, **512**–13, 516
    uses of, 513
concentrated solution, 212
concentration
    acid, 228, 229
    base, 228, 229
    of a solution, **212**–15, 217–18, 223
condensation, **43**
conduction, **356**
conductivity, electrical, 40, 225
conductor, electrical, **452**–55, 464
cone (eye), 508
conservation of energy, law of, **161**, 325
conservation of matter, law of, 42–**43**
conservation of momentum, law of, 259
constructive interference, 413–14
contact force, **271**
convection, 356–**57**
convection current, 357
conventional current, 451
converging lens **516**–17
conversion factor, 18–20
convex lens, 516–17
convex mirror, **513**
cooling curve, 363
Copernicus, Nicolas, 12
corpuscle, 52
coulomb, 442–43, 445, 453
Coulomb, Charles-Augustin de, 443
Coulomb's constant, 443
Coulomb's law, 443
Count Rumford, 365
covalent bond, **99**, 101–4
crash test, 268–69
Creation Mandate, **8**, 9–10, 66, 140, 196, 343, 465, 469
crest, **404**–6, 414
critical mass, **187**
Crookes tube, 52
crystal, 100
crystal lattice, 33
crystalline compound, 33
crystalline solid, 33, 37
current electricity, 441, **450**–62, 464–65
    calculation of, 453–56
cycle, 399, 404, 407, 408

D

Dalton, John, 51
damping, **400**–401
data, 14
de Laval nozzle, 389
decay chain, 180
decibel, 423
decomposition reaction, **156**
definite proportions, law of, **28**, 51
degree, 348, 350, 351–53
Democritus, 27, 50
density, **29**, 30, 31, 37, 377–79
    and buoyancy, 377–78
density stack, 379
deoxyribonucleic acid, **140**, 194
deposition, **43**
Derham, William, 421
derived units, 15
destructive interference, 413
diabetes, 123, 140
diamond, 33, 69, 93
    anvil cell, 68, 69
diaphragm, 425
diatomic molecule, **101**–2
diffraction, **412**
diffraction grating, 412
diffuse reflection, 410, **510**
diffusion, 28
dilute solution, 212
direct current (DC), **451**–52, 454
displacement (position), **249**–52, 254, 257, 260, 263, 264, 398–400, 405, 408, 413
    calculation of, 252
displacement (volume), 375–76
dissociation, **207**, 229
dissolving process, 206–7
dissolving rate, 209
distance, 248, **249**–54
distillation, in separating mixtures, 211
diverging lens, **516**–17
DNA, **140**, 194
Döbereiner, Johann, 73
Doppler effect, **415**
dosimeter, 195
double covalent bond, 102
double-replacement reaction, **157**
Dubrow, Geoff, 140
ductility, 40
dynamics, **270**–90

E

ear, 426–27
    inner, 427
    middle, 427
    outer, 427
Earth's magnetic field, 473
echolocation, **431**–32
Edison, Thomas, 485
efficiency, 300–302 (**301**), 311, 312
    calculation of, 301–2
Einstein, Albert, 189, 247, 324, 491, 494
electric car, 466
electric charging, 446–47
    by conduction, 447
    by friction, 446
    by induction, 447
electric circuit, **450**–65
electric components, 451, 453, 455, 457, 458–59, 460, 462
electric current, 441, **450**–62, 464–65
    calculation of, 453–56
electric field, **445**, 453, 489
electric force, **443**–46
electric motor, 481
electric potential energy, 445, 453
electric power, **462**–63
electrical conductor, **452**–55, 464
electrical generation, 480–81
electrical insulator, 452–**53**
electrical load, 457
electrical potential difference. *See* voltage
electrical resistance, 354
electrical safety, 458, 464–65
electrical transmission, 481–82
electricity, 441–48, 450–66
    generation of, 335–36
electromagnet, 430, **477**–79
electromagnetic energy, 357, 487–94, 496–99, 501
electromagnetic force, 290
electromagnetic induction, **477**, 481
electromagnetic radiation. *See* electromagnetic wave
electromagnetic spectrum, **496**–99, 504
electromagnetic wave, 403, 408, 487–94 (**488**), 496–99
    and communication, 498–99
    energy of, 492
    frequency of, 491–93, 496–98
    intensity of, 492
    models of, 493–94
    speed of, 408, 490–91
    uses of, 498–99
    wavelength of, 491–92, 496–97
electromagnetism, 476–79
    uses of, 478–79
electron, 52, **56**
    and atomic number, 57
    in atomic models, 52–55
    in ions, 60
electron dot notation, **88**
electronegativity, 89–**90**
    and bonding, 99, 104
    and polarity, 103
electroscope, 449
electrostatic charges, law of, **52**, 449
element, **33**, 70–90, 93
    classification, 80–85
    in Greek thought, 70
    organization, 70–79
EM radiation. *See* electromagnetic wave
EM spectrum, 504
emission spectrum, 54
emulsifier, 377
endothermic reaction, **160**–61, 163, 167
energy, 296, **324**–43
    and work, 324–25, 326, 327, 328, 329, 333
    biomass, **340**
    calculating, 359
    changes in, 332–36
    chemical potential, 328, 335–36, 340
    conservation of, 325
    conserving of, 343
    elastic potential, 328, 333–34
    geothermal, **340**
    gravitational potential, 329–31, 332–34, 340
    gravitational potential, calculation of, 329–30
    gravitational potential, factors affecting, 329
    hydroelectric, **340**, 342
    interconvertability with matter, 324–25
    kinetic, 36, **326**–28, 330, 332–34
    kinetic, calculation of, 327–28
    kinetic, factors affecting, 326–27
    mechanical, **332**–34, 364, 365

energy (continued)
    nuclear, 188–89, **339**
    potential, **328**–30, 331, 332–34
    solar, **341**
    sound, **420**–23, 430tidal, **341**
    transfer of, 328, 335, 341
    wind, **341**
energy level, **54**, 76, 78, 84, 86–89
energy resource, 337–43
    nonrenewable, **337**–39
    renewable, **337**, 340–42
    renewable, advantages and disadvantages of, 342
    renewable, types of, 340–41
energy transformation, 335–36
    efficiency of, 335–36
equilibrium, 164–68
    chemical, **165**–68
    factors affecting, 166–67
    of a system, 304
    physical, 165
equilibrium position, 398–400, 404
etching, 224
ethane, 128
ethene, 130
ethics, **9**–10
    and antacids, 238
    and fast food, 143
    and helmet laws, 293
    and nuclear power generation, 199
    and photos on social media, 521
    and pollution, 21
    and pseudoephedrine, 118
    and radar detectors, 267
    and radiation, 66
    and scientific data, 23
ethyne, 130
eustachian tube, 427
evaporation, **42**
    in separating mixtures, 210
exothermic reaction, **160**–61, 163, 166, 167, 168
eye (human)
    and light, 504, 507–8, 511, 517

F

Fahrenheit, Daniel Gabriel, 350
Fahrenheit scale, 348, 350–53
Fall, the, 8, 299
family, **76**–77, 82–84, 86–88
Faraday, Michael, 131, 445
farsightedness, 517
fast food, ethics of, 143
fat, 139
ferrofluid, 474
ferromagnetism, **474**, 477, 481
fiduciary point, 350
field force, **271**, 277, 282, 443, 445, 472
filtering, in separating mixtures, 211
first law of thermodynamics, **366**
fission, 185, **186**–87, 188, 189
flammability, 41
Flood, the
    and fossil fuels, 338
fluid, 37, 371–78 (**372**), 380–89
    properties of, 372–78
fluid pressure, 372–74
fluorescence, 505
focal length, **512**, 516–17
focal point, **512**–13, 516–17
food chemist, 140
force, **270**–72, 296–97, 299–302, 304, 306–15, 317
    balanced, 272–74, 277, 280
    centripetal, **288**
    classification of, 270–72
    compression, 40
    contact, **271**
    electromagnetic, 290
    field, **271**, 277, 282
    friction, 286–87, 399, 400, 401
    fundamental, 289–90
    gravitational, 443
    input, 299–301, 306–07, 310–12
    normal, **281**
    output, 300–01, 306–07, 309–10
    strong nuclear, 290
    types of, 282–90
    unbalanced, 272–73, 277, 280
    weak nuclear, 290
force diagram, 273, 281
forensic anthropology, 8
forensic science, 3, 8
formula
    molecular, 128, 132
    structural, 128
formula unit, **100**
fossil fuel, **337**–338
    and the Flood, 338
    formation of, 338

frame of reference, **248**–49
free fall, **262**–63, 280, 283, 284
free-body diagram, **273**–74
freezing, **42**
freezing point, 42
freezing point depression, 218
frequency, **407**–9, 415, 421, 422–23, 427–28, 430, 433, 434
frequency modulation (FM), 498
friction, **286**–87, 399, 400, 401
    and weight, 286
    rolling, 287
    sliding, 287
    static, 287
    types of, 287
Frisch, Otto, 186
Fukushima accident, 339
fulcrum, **303**, 306–07
functional group, **134**–35
fundamental force, 289–90
fundamental tone, **428**–29
fundamental units, 15
fuse, 458, **464**–65
fusion, 185, **188**–89

G

Gabeira, Maya, 394–95, 407
Galileo, 247, 275–76
gamma decay, 177, **181**
gamma ray, 177, 181, 188, **497**, 499
gas (vapor), 27, **35**, 37
gas laws, 380–86
Gay-Lussac, Joseph, 384
Gay-Lussac's law, **384**–85
    calculation with, 385
gear, **310**, 316, 317, 318
    calculation of mechanical advantage of, 310
gecko, 24–25
Geiger counter, 195
generator, 454, 458, **480**–81
genetic damage, **194**
genocide, 8
geocentric model, 12
geographic pole, 473
Gilbert, William, 472
glottis, 425
glucose, 33, 96, 138
glue, 121
glutamine, 136

glycogen, 138
gravitational acceleration, 282–85
gravitational constant, 282
gravitational field, 282–83
gravitational force, 443
gravitational potential energy, 329
    calculation of, 329
gravity, **282**–85, 288
    acceleration due to, 282–85
    and weight, 284–85, 288–89
Greek prefixes, and binary covalent compounds, 116
greenhouse gas, 106
ground fault interrupter (GFI), 465
grounding, **447**, 448, 458, 465
group, **76**–77, 82–84, 86–88

H

Hahn, Otto, 186
half-life, **182**–84, 193
    calculations with, 184
Halifax explosion, 146, 147
halogen, 81, **84**, 134
hammer (ear), 427
harmonic, **428**
heat, 347, **355**–67
heat engine, 364
heat of fusion, 362
heat of vaporization, 363
heating curve, 362
heliocentric model, 12
Hertz, 407, 408
Hertz, Heinrich, 493
heterogeneous mixture, **34**, 202–3
hexane, 133
homogeneous mixture, **34**, 202, 204–5
horsepower, 298
hydraulics, **387**–88
hydrocarbon, **128**–132, 134–35, 137
    aromatic, **131**
    classification, 128–31
    saturated, **128**
    substituted, **134**–35, 137
    unsaturated, **129**–31
Hydrogen Seven, 101
hydronium ion, **224**
hydrophone, 432
hydroxide ion, **226**
hydroxyl group, 133, 134, 135
hypothesis, **13**

I

ideal mechanical advantage (IMA), 299–**300**, 310–11
    calculation of, 300–301
illuminated object, **505**, 509, 511
imagineer, 264
incandescence, 505
incident angle, 410
incident ray, **510**–13
inclined plane, **313**–15
index of refraction, **514**–15
indicator, **225**, 234–35
inert, 84
inertia, **275**–77, 278
    law of, **276**–77
infrared wave, **496**, 499, 501
infrasound, 433
inhibitor, **163**
inner transition metal, 80–81, 82–**83**
insulator, electrical, **452**–53
intensity, 422, **423**
interference, **413**–14
International System of Units, **14**–16
    prefixes, 16
International Union of Pure and Applied Chemistry, 75
invisibility cloak, 35
ion, **60**, 97, 100, 105
    hydronium, **224**
    hydroxide, **226**
    polyatomic, **112**
ionic bond, **100**, 104–5
ionic compound, 110–12
ionization, 228, 229
ionizing radiation, 175, 194, 195
irradiated food, 174–75
isomer, **132**–33
isooctane, 96
isopentane, 132
isoprene, 96
isopropyl alcohol, 133
isotope, **58**
    and differences in physical properties, 59
    and mass number, 58
    naming of, 58
    stability, 178, 180–82, 185, 188
isotope name, 58
isotope notation, 59
isotopes, and radiation, 178–84, 188, 192–93
IUPAC, 75

J

joule, 296
Joule, James Prescott, 296, 365

K

Kekulé, Friedrich, 131
kelvin, 15, 351, 353
Kelvin scale, 348, 350–51, 353
Kepler, Johannes, 246, 247, 275
ketone, 134–**35**
kilogram, 15
kilowatt-hour, 463
kinematics, **246**–64
kinetic energy, 36, **326**–28, 330, 332–34, 348–49, 356, 362–63, 366
    calculation of, 327–28
    factors affecting, 326–27
    in a spring system, 398–99
kinetic-molecular model, 28, 365–66
Koff, Clea, 8

L

Lake Natron, 222–23
Large Hadron Collider, 38, 48–49, 168
larynx, 424, 425
laser, 513
latex, 95
Lavoisier, Antoine, 71, 364
law, **11**
    of acceleration, **278**–79
    of action-reaction, **280**–81
    of cause and effect, 6
    of conservation of charge, **447**
    of conservation of energy, 161, 325
    of conservation of matter, 42–**43**, 180, 181, 189
    of conservation of momentum, 259
    of definite proportions, **28**, 51
    of electrostatic charges, **52**, 449
    of inertia, **276**–77
    of octaves, 74
    of reflection, 410, **510**–12

law (continued)
    of torques, 304–5
    of universal gravitation, 282
Le Châtelier's principle, **168**
LED, 459
lens, **516**–17
    and vision correction, 517
    converging, **516**–17
    diverging, **516**–17
Leucippus, 27
lever, **303**–4, 306–8, 317
    classes of, 306–7
    first-class, 306, 308
    second-class, 307
    third-class, 307
Lewis structure, **101**, 106
LHC. See Large Hadron Collider
light, 488, 489, 490–94, 497
    visible, **497**
light behavior, 504–6
light-emitting diode (LED), 459
Linnaeus, Carl, 350
lipid, **139**
liquid, 27, **35**, 37
litmus, 225
litmus paper, 225, 227
lodestone, 472
longitudinal wave, **406**, 420
Lord Kelvin, 350
loudness, 422, **423**, 424, 427
luminous object, **505**, 510, 511
luster, 40

M

machine, 299–304, 306–17
    compound, **316**–18
    simple, **299**–300, 303–4, 306–15
macronutrient, 136, 138
maglev train, 478
magnet, **472**–74, 476–79, 481
    superconducting, 479
    temporary, 474
magnetic declination, 473
magnetic domain, **474**
magnetic field, 471–74 (**473**), 476–77, 479, 480–82, 489
    direction, 473
    Earth's, 473
magnetic pole, **472**–73, 477
magnetic properties, 472–73, 474

magnetically levitated train (maglev), 478
magnetism, 471–74 (**472**), 476–79
    and electric current, 476–82
    and temperature, 474
    domain model, 474
    separating mixtures, 210
magnetoreception, 471
malleability, 40
Mars Climate Orbiter, 19
mass, 18, 26, **29**, 30
    of a solution, 216
mass number, **58**
    and isotope notation, 59
materials scientist, 24, 25, 26, 35
matter, **26**–43
    change in state of, 42–43, 361–63
    changes in, 39–43
    classifying, 32–34
    conservation of, 42–43
    states of, 35–38, 42–43
Maxwell, James Clerk, 489
measurement, **14**–18
    limits of, 17–18
mechanical advantage, **299**–302, 303, 306, 307, 309–17
    actual, 299, **301**
    and compound machines, 316–17
    calculation of, 300–302
    ideal, 299–**300**, 310–11
    of gear, 310
    of inclined plane, 313–14
    of lever, 306–07
    of pulley system, 310–12
    of wheel and axle, 309
mechanical energy, **332**–34, 364, 365
mechanical wave, **403**, 421
mechanics, **246**–64, 270–90
    classical, 247
    Newtonian, 247
    quantum, 247
    relativistic, 247
medium, 420–22
Meitner, Lise, 186
melting, **42**
melting point, **42**
Mendeleev, Dmitri, 74–75
metal, **80**–84, 88
metallic bond, **100**
metalloid, **81**, 84, 93
meter, 15

methanal, 134, 135
methane, 106, 128
methyl cyanoacrylate (CA), 121
metric system, **14**–16
    prefixes, 16
Michelson, Albert, 490
microburst, 371
micronutrient, 143
microwave, **496**, 499
mirror image, 511–13
mixed group, **84**
mixture, **34**, 202–11
    heterogeneous, **34**, 202–3
    homogeneous, **34**, 202, 204–5
    separation of, 210–11
model, **11**–13, 27–29, 36, 39, 55, 65, 106–7, 150, 247, 364–66, 493–94
    atomic, 50–55
    collision, **162**–63
    workability, 12, 55, 247, 494
modeling, 11–13
molar mass, **153**
molarity, **215**
mole, 15, **152**–53
molecular formula, 128, 132
molecule, 28, **29**, 99
    polar, 103
mole-mass conversion, 153–54
moment arm, 304, 306
momentum, **258**–59
    in a spring system, 398–99
monomer, **136**, 139, 140
Moseley, Henry, 75
motion, 261–64
    circular, **263**–64
    graph of, 247–58, 262–64
    in two dimensions, 251, 256
    periodic, **396**–401, 404, 407
    projectile, **264**
    retrograde, 246
    rotational, 304, 310, 317
    simple harmonic, **400**
MRI, 479
multiple bond, 102

N

Nagaoka, Hantaro, 53
natural gas, 337, 338

natural rubber, 96
natural science, 4–5
nearsightedness, 517
neopentane, 132
net force, 272–74, 277–79
    calculation of, 274, 288–89
neutral atom, 57
neutralization, **236**–38
neutralization reaction, 236
neutrino, 188
neutron, **57**
    and mass number, 58
Newlands, John, 74
newton, 270
Newton, Isaac, 247, 276, 278, 282, 489, 493
Newton's law of universal gravitation, 247
Newton's laws of motion, 247, 275–81
    first law, 276–77
    second law, 278–79
    third law, 280–81
Newtonian mechanics, 247
noble gas, 81, **84**
node, 414
Nomex, 137
non-matter, 26
nonmetal, **81**, 84, 87
nonpolar bond, 103
normal (line), 410, 411
normal force, **281**
nuclear change, 175–89, 189
    effects of, 194
    uses of, 191–94
nuclear decay, 176–84, (**178**)
nuclear energy, 188–89, **339**
nuclear medicine, 192
nuclear model, **53**–54
nuclear power plant, 186–87
    and clean energy, 199
    Chernobyl, 195
nuclear radiation, 175
nuclear stability, 178
nuclear waste, 196
nuclear weapons tests
    Castle Bravo, 189
    Trinity, 187
    Tsar Bomba, 187
nucleic acid, **140**
nucleotide, **140**

nucleus, 38, 86–90
    and atomic number, 57
    and atomic structure, 56–58
    and mass number, 58
    in atomic models, 53–54
numerical prefixes, 128

O

octane, 129
octet rule, **97**, 101–2, 104, 110
Oersted, Hans Christian, 476
ohm, 454–55
Ohm, Georg Simon, 454
Ohm's law, **455**–56, 460
    calculations with, 455–56
oil, 337, 338
oil (fat), 139
opaque material, **506**
open circuit, 450
orbit, of electrons, 53–55
orbital, 55
order in our world, 4–8
organic chemistry, 124–40
organic compound, 124–40 (**125**)
oscillation, 400–401, 414
oscillator
    damped, 400
    driven, 401
overtone, **428**–29
oxidation, **158**
oxidation state, **113**–14

P

parabolic curve, 264
parallel circuit, **460**–61, 464
particle
    alpha, 53, 177, 179, 180–81, 186, 194–95
    beta, 177, 180
    subatomic, 50, 56–57
particle accelerator, 478
particle model (of EM waves), 493
particle model (of matter), **28**, 36–37
pascal, **373**–74
Pascal, Blaise, 373, 374
Pascal's principle, **374**, 388
peer review, 13
pendulum, 397, 400

pentane, 132
percent by mass, 213
percent by volume, 214
Peregrinus, Petrus, 472
period (in motion), **396**, 399, 404, 407–9
period (in the periodic table), **78**–79, 86–90
periodic law, 74–76 (**75**)
periodic motion, **396**–401, 404, 407
periodic table, 6, 70, 72–79 (**75**), 86–90
periodic trend, 86–90
periodicity, 73–74
PET scan, 193
pH, **230**–31
pH scale, 230–31
phase, 35
phase diagram, 43
phenomenon, 4
phosphorescence, 505
photoelectric effect, 493
photon, 54, 492, 493–**94**
photosynthesis, 223
    artificial, 323
physical change, **41**
physical equilibrium, 165
physical property, **39**–41
physical science, 4–**5**
physics, **5**
pigeon, homing, 471
pigment, 509
piping engineer, 389
pitch, 421, **422**, 424–25, 427, 428, 430, 431
Planck constant, 492
Planck, Max, 492
Planck's equation, 492
plane mirror, **511**, 513
plasma, **35**, 37
Plato, 364
plum pudding model, **52**–53
point of incidence, 510
polar covalent bond, 103
polar molecule, 103
polarity, **103**–4
pollution, air, 221
polyatomic ion, **112**
polyatomic molecule, 102
polyethylene, 130
polymer, **136**–40
polymer, synthetic, 137

position versus time graph, 257–58, 262–63
positron, 188
post-transition metal, 81, 84
potential energy, **328**–30, 331, 332–34
    in a spring system, 398–99
power, **298**
    calculation of, 298, 462
power supply, 454, 457, 458, 460
precipitate, 148, 157
precision, **17**
pressure, 37, **372**–75, 380–86, 388–89
    atmospheric, 372, 373–74, 382, 386
    calculation of, 373
prime mover, 480
principal optical axis, 512, 516
principle of uniformity, 6–7
product, **150**
projectile motion, **264**
propan-2-ol, 134
propane, 128, 129
propanol, 133
property
    chemical, **41**
    physical, **39**–41
protein, **139**, 140
proton, **56**
    and atomic mass, 61
    and atomic number, 57
    and mass number, 58
    in ions, 60
proton pump inhibitor, 238
PS10 Solar Power Tower, 487
pseudoephedrine, 118
p-t graph, 257–58, 262–63
pulley, **310**–12
pure substance, **33**
PVA glue, 121

Q

qualitative data, 14
quantitative data, 14
quantity, dimensionless, 300
quantum, 54
quantum mechanical model, **55**
quantum mechanics, **54**, 247
quark-gluon plasma (QGP), 38

R

radar detector, 267
radiation, 66–67, 356–**57**
    and worldview, 196
    detection of, 194–95
    ionizing, 175, 194, 195
    nuclear, 175
radio wave, **496**, 498–99
radioactive decay, 176–84, (**178**)
radioactive isotope, 84. *See also* isotope, stability
radioactivity, 176–78, 196
radiocarbon dating, 59
radiometric dating, 183
radiotracer, **193**
radon, 66–67
rainbow, 515
rarefaction, **406**, 420–21
rate of dissolving, 209
ray (light), **506**, 510–13, 515–17
reactant, **150**
reaction
    chemical, 41
    combustion. *See* combustion
    decomposition, **156**
    double replacement, **157**
    endothermic, **160**–61, 163, 167
    exothermic, **160**–61, 163, 166, 167, 168
    neutralization, 236
    redox, 158
    reversible, **164**
    single-replacement, **157**
    synthesis, **156**, 158, 167, 168
reaction rate, 162–**63**
    factors affecting, 162–63
reactivity, 41, 224
real image, **511**–13
redox reaction, 158
reduction, **158**
reflected angle, 410
reflected ray, **510**–12
reflection, **410**, 510–13, 515–16
    diffuse, 410, **510**
    specular, 410, **510**–11
refraction, **411**–12
    of light, 514–17
regular reflection, 510–11
relativistic mechanics, 247

relativity, general theory of, 283
resistance, **454**–56, 458, 459, 460–62
    calculation of, 455–56
resistor, **458**–59
resonance, **401**
resonance chamber, 425
rest position, 398–400
restoring force, 398–400, 406
retrograde motion, 12, 246
reversible reaction, **164**
ribonucleic acid, 140
right-hand rule, **476**
ring, 127–28, 131
    benzene, **131**
RNA, 140
Rochester cloak, 35
rod (eye), 508
Rømer, Ole, 490
Röntgen, Wilhelm, 176
rotor, 480–81
rubber, 95
Rutherford, Ernest, 53, 54, 177

S

Sabine, Wallace, 434
safety, electrical, 458, 464-95
salt, **232**–33, 236–37
saturated fat, 139
saturated hydrocarbon, **128**
saturated solution, **212**
saturation, of a solution, 212–13
Saturnian model, 53
scalar, **252**
scale, 18
science, **4**
    forensic, 3, 8
    natural, 4–5
    physical, 4–**5**
scientific data, reporting, 23
scientific inquiry, **12**–13
scientific notation, 19
screw, **315**
second, 15
second law of thermodynamics, **366**
semicircular canal, 427
semiconductor, 453
series circuit, **460**, 464
setpoint, 354
short circuit, **457**

SI, **14**–16
    prefixes, 16
sifting, in separating mixtures, 211
significant figures, 18, 19
simple harmonic motion, **400**
single covalent bond, 101
single-replacement reaction, **157**
sinus, 425
smoke detector, 179
Snell's law, 411
Socrates, 364
soda lake, 223
solar energy, 487, 488
solar wind, 473
solenoid, **477**
solid, 27, **35**, 37
    amorphous, 37
    crystalline, 37
solubility, **208**–9, 212
    and pressure, 209
    and temperature, 208–9
solubility curve, 208, 209
solute, **204**, 206–7
solution, 34, 202–18 (**204**)
    aqueous, **224**
    dilute, 212
    in equilibrium, 212
    saturated, **212**
    saturation, 212–13
    supersaturated, 208, **212**
    unsaturated, **212**
solution concentration, **212**–15, 217–18
    using molarity, 215
    using percent by mass, 213
    using percent by volume, 214
solvation, 206–7
    endothermic process, 207, 208
    exothermic process, 207, 208
    state of being hydrated, 207
    state of being solvated, 207
solvent, **204**, 206–7
somatic damage, **194**
sonar, 431–**32**
    active, 432
    passive, 432
    side-scan, 432
sonogram, 434–35
sonography, **434**–35

sound, 419–35
    and musical instruments, 423, 428–29
    speakers, 430–31
    speed of, 419, 421–22, 432
sound barrier, 419
sound energy, **420**–23, 430
sound wave, 420–23, 428, 430–34
specific heat, **359**–62
specular reflection, 410, **510**–11
speed, **253**–54, 261–63
    calculation of, 254
spring, 397, 398–99, 401
spring system, 398–400, 405
standing wave, **414**
starch, 138
state of matter, 35–38, 42–43
    change in, 42–43, 361–63
static electricity, **442**–48
stator, 480–81
stirrup (ear), 427
straight chain, 126–27
Strassmann, Fritz, 186
strength
    acid, 228
    base, 228, 229
strong acid, **228**
strong base, **229**
strong force, **178**
strong nuclear force, 290
structural formula, 128
Sturgeon, William, 477
subatomic particle, 50, 56–57
subcritical mass, 187
sublimation, **43**
substituent, 134
substituted hydrocarbon, **134**–35, 137
sugar, 138, 140
sulfuric acid, 227
superconductor, 455
supercritical mass, 187
superfluid, 38
supersaturated solution, 208, **212**
surface wave, 406
suspension, **202**
suspension (car), 401
swamp gas, 106
switch, electrical, 451, 453, 458, 460, 464

synthesis reaction, **156**, 158, 167, 168
synthetic polymer, 137
system, **248**, 258, 259

T

Tacoma Narrows Bridge, 414
Taser, 469
temperature, **348**–56, 359–60, 362–63, 367
    conversion of, 352–53
temperature change, calculating, 360
tension, **280**, 288
Tesla coil, 441
Tesla, Nikola, 441, 485
theory, **11**
thermal conductor, **357**
thermal energy, 348–**49**, 355–57, 359, 361–66
    calculations with, 359–61
thermal expansion, 349, 354
thermal insulator, **357**
thermistor, 354
thermodynamics, **364**–67
    first law, **366**
    second law, **366**
    third law, **367**
thermometer, 348–50, 354
thermometric property, **349**
    viscosity, 378
thermoscope, 350
thermostat, 349, 354
third law of thermodynamics, **367**
Thompson, Benjamin, 365
Thomson, Joseph John, 52–53
Thomson, William, 350
throat, 425, 427
timbre, **423**, 424, 428–29
time interval, 249, 254, 257, 261
torque, **304**, 305
    calculation of, 304
torques, law of, 304–5
total internal reflection, **515**–16
toxicology, 153
traction, 277, 287
trajectory, **264**
transformer, **481**–82
    step-down, 482
    step-up, 482

transition metal, 80–81, 82–**83**
translucent material, **506**
transparency, 506
transparent material, **506**, 516
transverse wave, **406**
triad, 73
Trinity nuclear weapon test, 187
triple covalent bond, 102
triple point, 43, 350–51
trough, **404**–6, 414
Tsar Bomba nuclear weapon test, 189
Twaron, 137
2-butene, 129
tympanic membrane, 427
Tyndall effect, 203

U

ultrasound, 434-35
ultraviolet wave, **497**, 499
unified atomic mass unit, 61
uniformity, principle of, 6–7
unit conversions, 18–20
universal gravitation, law of, 282
unsaturated fat, 139
unsaturated hydrocarbon, **129**–31
unsaturated solution, **212**
urban heat island, 347, 365
UV wave, **497**, 499

V

valence electron, **76**, 78, 80–84, 87–90, 97
 and bonding, 99–105
valence energy level, 97
vapor pressure, 217
vaporization, **42**
vector, 251, **252**, 256, 261
vector addition, 256
velocity, **254**–59, 261–64
 calculating, 255–56, 330
velocity versus time graph, 257, 262–63
Villard, Paul, 177
vinyl acetate (VA), 121
virtual image, **511**–13, 516
viscosity, 37, **378**, 387, 389
visible light, **497**, 503–6, 508, 509
 progression of colors, 504
 wavelengths and frequencies of, 504
vision correction, 517

vital force, 125
vocal cord, 425
voice, human, 424–25
volt, 453
Volta, Alessandro, 453–54
voltage, **453**–56, 457, 460, 462, 466
 calculation of, 455–56
voltmeter, 454, 459
volume, **29**, 30, 31, 37
 of a solution, 216
v-t graph, 257, 262–63

W

water displacement, 31
watt, 298
Watt, James, 298
wave, 395, 400, **403**–15
 compression, 406
 electromagnetic, 403, 408
 infrasonic, **433**
 longitudinal, **406**, 420
 mechanical, **403**, 421
 sound, 420–23, 428, 430–34
 standing, **414**
 surface, 406
 transverse, **406**
 ultrasonic, **434**-35
wave height, 404–5
wave model (of EM waves), 493
wave pulse, 403, 413, 415
wave speed, 408–9, 415
 calculations with, 408–9
wave train, 403
wavelength, **404**, 407–9, 413, 414–15
 calculations of, 409, 491
wave-particle duality, 494
weak acid, **228**, 231
weak base, **229**, 231
weak nuclear force, 290
wedge, **315**
weight, 18, **30**, 270, 272, 281, 284–85, 286, 288–89
 calculation of, 285
wheel and axle, **309**–12, 316, 317
wind shear, 371
Wöhler, Friedrich, 125
work, **296**–317
 calculation of, 297, 308
workability, **12**, 55, 247, 494
worldview, 7–8
 and radiation, 196

X

Xpogo stick, 332–34
x-ray, 176–77, 192–94, 196, **497**, 499

Y

Yeager, Chuck, 419
Young, Thomas, 493

PHOTO CREDITS

Key: (t) top; (c) center; (b) bottom; (l) left; (r) right; (i) inset, (bg) background

COVER

Picsfive/Shutterstock.com

FRONT MATTER

i Picsfive/Shutterstock.com; **ii–iii** alexeys/iStock/Getty Images Plus/Getty Images; **iv** Pattadis Walarput/iStock zz/Getty Images Plus/Getty Images; **v** Mr. Klein/Shutterstock.com; **vi** DRogatnev/Shutterstock.com; **vii** Pavel L Photo and Video/Shutterstock.com; **vii**bg PhotoSky/Shutterstock.com; **viii** Mix3r/Shutterstock.com; **ix**tr Aflo Co. Ltd./Alamy Stock Photo; **ix**tri © iStock.com/PragasitLalao; **ix**tl Soloviova Liudmyla/Shutterstock.com; **ix**c Sergio33/Shutterstock.com; **ix**bl Cultura Images/Eugenio Marongiu/Media Bakery; **x**t Serhii Hrebeniuk/Shutterstock.com; **x**b Elenathewise/iStock/Getty Images Plus/Getty Images; **x**bi Mikhail Japaridze/TASS/Getty Images; **xi**ct Ian Cuming/Ikon Images/Getty Images Plus/Getty Images; **xi**l GIPhotoStock/Science Source; **xi**c NotionPic/Shutterstock.com; **xi**cb Jose Luis Pelaez Inc/DigitalVision/Getty Images; **xi**r "Elektrodynamisk-högtalare" by Svjo/Wikimedia Commons/CC BY-SA 3.0

CHAPTER 1

2 Matthew Horwood/Alamy Stock Photo; **4** Adobe Stock/Patrizio Martorana; **5**bg Alexander Limbach/Shutterstock.com; **5** (earth, leaf) shaimaadesigns/Shutterstock.com; **5** (light bulb, rocket, atom model) derter/Shutterstock.com; **5** (beaker) DRogatnev/Shutterstock.com; **6, 7**bg New Line/Shutterstock.com; **6** (broccoli) Kert/Shutterstock.com; **6, 7** (sun, moon) 21kompot/Shutterstock.com; **6** (snowflake) Kichigin/Shutterstock.com; **6** (microscope) DRogatnev/Shutterstock.com; **6** (dominos) Africa Studio/Shutterstock.com; **6** (nautilus shell) aaltair/Shutterstock.com; **7** (sunflower) Ian 2010/Shutterstock.com; **7** (scientist) Andrew Brookes/Cultura Exclusive/Getty Images; **7** (earth) LuckyVector/Shutterstock.com; **7** (robot) MONOPOLY919/Shutterstock.com; **8**l Bryan Chan/Los Angeles Times/Getty Images; **8**r Richard T. Nowitz/Science Source; **9, 22** © iStock.com/shapecharge; **10**t manusapon kasosod/Shutterstock.com; **10**b © iStock.com/stocknroll; **11** Paper Boat Creative/DigitalVision/Getty Images; **11**ti asharkyu/Shutterstock.com; **11**ci Marina Sun/Shutterstock.com; **11**bi Gorodenkoff/Shutterstock.com; **12** both, **21**c ValentinaKru/Shutterstock.com; **13** Matej Kastelic/Shutterstock.com; **14, 21**b Barrett & MacKay/All Canada Photos/Getty Images; **15** (ruler) Neveshkin Nikolay/Shutterstock.com; **15** (kilogram) donatas1205/Shutterstock.com; **15** (stopwatch) Stepan Bormotov/Shutterstock.com; **15** (ampere meter) Stefan Rotter/Shutterstock.com; **15** (thermometer) Tomas Ragina/Shutterstock.com; **15** (coal) Trum Ronnarong/Shutterstock.com; **15** (candle) Aksenova Natalya/Shutterstock.com; **16** (dam) Yao yilong/Imaginechina/AP Images; **16** (radio) Yuliyan Velchev/Shutterstock.com; **16** (jet) Digital Storm/Shutterstock.com; **16** (candy bar) burnel1/Shutterstock.com; **16** (cup) Kimberly Hall/Shutterstock.com; **16** (hair) Steve Gschmeissner/Science Photo Library/Getty Images; **16** (sign) photocritical/Shutterstock.com; **17**t hwanchul/Shutterstock.com; **17**i Andrey_Kuzmin/Shutterstock.com; **17**bl © iStock.com/MicroStockHub; **17**br iofoto/Shutterstock.com; **18** sciencephotos/Alamy Stock Photo; **20** © iStock.com/StefaNikolic; **21**t GoncharukMaks/Shutterstock.com; **23** Syda Productions/Shutterstock.com

CHAPTER 2

24 Anant Kasetsinsombut/Shutterstock.com; **24**i Robert Clark/National Geographic/Getty Images; **26, 45**t Soloviova Liudmyla/Shutterstock.com; **27** MatiasDelCarmine/Shutterstock.com; **28**t Trevor Clifford Photography/Science Source; **28**cl Christopher Meade/Shutterstock.com; **28**cr Matias DelCarmine/Shutterstock.com; **28**b Dmitry Lobanov/Shutterstock.com; **29** yitewang/Shutterstock.com; **30** Red monkey/Shutterstock.com; **31** prill/123RF; **32** Prasit Rodphan/Shutterstock.com; **33** (chlorine gas) Ian Miles-Flashpoint Pictures/Alamy Stock Photo; **33** (liquid mercury) Alexey V Smirnov/Shutterstock.com; **33** (diamond) SPbPhoto/Shutterstock.com; **33** (sugar cubes) Rob Stark/Shutterstock.com; **33** (table salt) etorres/Shutterstock.com; **34**t Agustin Vai/Shutterstock.com; **34**bl Turtle Rock Scientific/Science Source; **34**bc, **45**c Sergio33/Shutterstock.com; **34**br Africa Studio/Shutterstock.com; **35** University of Rochester; **36**t Sergiy Kuzmin/Shutterstock.com; **36**b, **45**b Cultura Images/Eugenio Marongiu/Media Bakery; **37**lt Sebastian Janicki/Shutterstock.com; **37**rt Cromagnon/Shutterstock.com; **37**rc DONOT6_STUDIO/Shutterstock.com; **37**lb Turtle Rock Scientific/Science Source; **37**rb Zimiri/Shutterstock.com; **38** Peter Macdiarmid/Getty Images News/Getty Images; **38**ti CERN; **38**bi SPL/Science Source; **39** Stefano Montesi/Corbis News/Getty Images; **40** Flegere/Shutterstock.com; **40**ri Phil Degginger/Alamy Stock Photo; **40**li manfredxy/Shutterstock.com; **41** © iStock.com/YongEe; **41**i Turtle Rock Scientific/Science Source; **42–43**bg HAKKI ARSLAN/Shutterstock.com; **44** Aflo Co. Ltd./Alamy Stock Photo; **44**i © iStock.com/PragasitLalao; **46** MSSA/Shutterstock.com

CHAPTER 3

48 CERN; **50**t ORNL/Science Source; **50**b NLM/Science Source; **51** Pictorial Press Ltd/Alamy Stock Photo; **51**bg Public Domain; **52**t World History Archive/Alamy Stock Photo; **52**b © iStock.com/abzee; **53**t Public Domain; **53**b RGB Ventures/SuperStock/Alamy Stock Photo; **54** Sueddeutsche Zeitung Photo/Alamy Stock Photo; **54**bg Marina Sun/Shutterstock.com; **55**l magnetix/Shutterstock.com; **55**r EugenP/Shutterstock.com; **56**l, **57**t, **64** Leigh Prather/Shutterstock.com; **56**r tele52/Shutterstock.com; **57**b martan/Shutterstock.com; **58**t Carolyn Franks/Shutterstock.com; **58**bl hans.slegers/Shutterstock.com; **58**br VVZann/iStock/Thinkstock; **59** Vadiar/Shutterstock.com; **60–61**bg Darwin Brandis/Shutterstock.com; **63** TaraPatta/Shutterstock.com;

65bg brkart/Shutterstock.com; 67t Andrey_Kuzmin/Shutterstock.com; 67br Kamenetskiy Konstantin/Shutterstock.com; 67bl Grigorita Ko/Shutterstock.com

CHAPTER 4

68 Jeffrey Hamilton/DigitalVision/Getty Images; 70bg Sylverarts Vectors/Shutterstock.com; 71t Science History Images/Alamy Stock Photo; 71cr Quagga Media/Alamy Stock Photo; 71cl "Lavoisier decomposition air"/Wikimedia Commons/Public Domain; 71b © iStock.com/gameover2012; 73 INTERFOTO/Alamy Stock Photo; 74t The History Collection/Alamy Stock Photo; 74ct "Newlands periodiska system 1866"/Wikimedia Commons/Public Domain; 74cb Heritage Image Partnership Ltd/Alamy Stock Photo; 74b RIA Novosti/Science Source; 75, 91 Historic Images/Alamy Stock Photo; 80, 81l Charles D. Winters/Science Source; 81r Turtle Rock Scientific/Science Source; 82 Alexandre Dotta/Science Source; 83t Bjoern Wylezich/Shutterstock.com; 83b SPL/Science Source; 84 Bjoern Wylezich/Shutterstock.com; 85 totojang1977/Shutterstock.com; 93tl Anatoly Maslennikov/Shutterstock.com; 93bl, r magnetix/Shutterstock.com

CHAPTER 5

94 Pulsar Imagens/Alamy Stock Photo; 96tl molekuul_be/Shutterstock.com; 96tc chromatos/Shutterstock.com; 96tr Adobe Stock/raimund14; 96b Tono Balaguer/Shutterstock.com; 97t SPL/Science Source; 97b Sulfur-sample by Ben Mills/Wikimedia Commons/Public Domain; 99t 32 pixels/Shutterstock.com; 99b, 100, 119 Mushakesa/Shutterstock.com; 102 Asier Romero/Shutterstock.com; 103 Mark Taylor/Minden Pictures; 104 Tom Kolossa/EyeEm; 105 uchar/E+/Getty Images; 106–7 © iStock.com/Instants; 108 Damsea/Shutterstock.com; 109 © iStock.com/LightFieldStudios; 114 dashingstock/Shutterstock.com; 115bg daizuoxin/Shutterstock.com; 116 shironosov/Shutterstock.com; 117 H. Mark Weidman Photography/Alamy Stock Photo; 118 © iStock.com/gradyreese; 121 © iStock.com/cyano66

CHAPTER 6

122 Stoked/Comstock/Media Bakery; 124–25 Monkey Business Images/Shutterstock.com; 126–27bg Djem/Shutterstock.com; 128–31bg Phutcharapan Mdr/Shutterstock.com; 129l Balaiban Mihai/Shutterstock.com; 129r tab62/Shutterstock.com; 130t Adobe Stock/Samuel B.; 130c Adobe stock/nationkp; 130b donikz/Shutterstock.com; 131tl "Calicheamicin gamma 1 3D spacefill"/Jynto/Wikimedia Commons/Public Domain; 131tr Simon Jarratt/Corbis/VCG/Getty Images; 131b Oleksandr Lysenko/Shutterstock.com; 133 metwo/Shutterstock.com; 135t yurakrasil/Shutterstock.com; 135c Gregory Davies/Science Source; 135b, 141t popcorner/Shutterstock.com; 136t Edilus/Shutterstock.com; 136b, 141b Antonina Vlasova/Shutterstock.com; 137i Turtle Rock Scientific/Science Source; 137 Nadezda Murmakova/Shutterstock.com; 138–39 ifong/Shutterstock.com; 140r Stockforlife/ducu59us/Shutterstock.com; 140lbg Meranda19/Shutterstock.com; 143 Markus Mainka/Shutterstock.com

CHAPTER 7

146, 169 TOMAS HULIK/AFP/Getty Images; 146i Bettmann/Getty Images; 148–49bg Mix3r/Shutterstock.com; 149 DenisMArt/Shutterstock.com; 150–51bg ananaline/Shutterstock.com; 152 Oleksiy Mark/Shutterstock.com; 152–53bg Djem/Shutterstock.com; 153 Robert F. Sisson/National Geographic/Getty Images; 155 jeehyun/Shutterstock.com; 156–57 (people) Sentavio/Shutterstock.com; 158 Charles D. Winters/Science Source; 158–59bg Valdecasas/Shutterstock.com; 160l vectortatu/Shutterstock.com; 160r Brumarina/Shutterstock.com; 161t, 170t Volodymyr Goncharuk/Shutterstock.com; 161b amenic181/Shutterstock.com; 162–63, 170b Belozersky/Shutterstock.com; 172 ArturHenryk/Shutterstock.com; 173t Turtle Rock Scientific/Science Source; 173b age fotostock/Alamy Stock Photo

CHAPTER 8

174 DOE/Science Source; 174i Tata Donets/Shutterstock.com; 176t Science & Society Picture Library/SSPL/Getty Images; 176b, 197t Fototeca Storica Nazionale/Hulton Archive/Getty Images; 177t Library of Congress/Science Faction/Getty Images; 177c "Becquerel plate" by Henri Becquerel/Wikimedia Commons/Public Domain; 177bl RGB Ventures/SuperStock/Alamy Stock Photo; 177br Archives de l'Academie des Sciences/EMILIO SEGRE VISUAL ARCHIVES/AMERICAN INSTITUTE OF PHYSICS/Science Source; 178 "Isotopes and half-life" by BenRG/Wikimedia Commons/Public Domain; 179t Mile Atanasov/Shutterstock.com; 179b Andrey_Popov/Shutterstock.com; 183 Juniors Bildarchiv GmbH/Alamy Stock Photo; 185 FABRICE COFFRINI/AFP/Getty Images; 185i, 197c THOMAS MCCAULEY, LUCAS TAYLOR/CERN/Science Photo Library/Getty Images; 186–87bg IndustryAndTravel/Shutterstock.com; 187t "Trinity Detonation T&B"/United States Department of Energy/Wikimedia Commons/Public Domain; 187b "Trinitite-detail6" by Shaddack/Wikimedia Commons/CC 3.0; 188 Diana Hlevnjak/Shutterstock.com; 189 Album/Alamy Stock Photo; 191, 197b Eva-Katalin/E+/Getty Images; 192ti cyano66/iStock/Getty Images Plus/Getty Images; 192t Sentavio/Shutterstock.com; 192ci Alfred Pasieka/Science Photo Library/Getty Images; 192c Golden Sikorka/Shutterstock.com; 192b, 193c Vectorpocket/Shutterstock.com; 193t Wellcome Dept. of Cognitive Neurology/Science Photo Library/Getty Images; 193br Golden Sikorka/Shutterstock.com; 193bri David Kasza/Shutterstock.com; 193bl Science History Images/Alamy Stock Photo; 194i lvcandy/DigitalVision Vectors/Getty Images; 194 jhorrocks/iStock/Getty Images Plus/Getty Images; 195bg Yann Arthus-Bertrand/Getty Images; 195t P_Wei/iStock/Getty Images Plus/Getty Images; 195b MediaProduction/iStock/Getty Images Plus/Getty Images; 196t Satakorn/Shutterstock.com; 196b andresr/E+/Getty Images; 198, 199 MediaProduction/E+/Getty Images

CHAPTER 9

200 YOSHIKAZU TSUNO/Staff/Getty Images; 202tl Denis Semenchenko/Shutterstock.com; 202tc © iStock.com/doug4537; 202tr LittleMiss/Shutterstock.com; 202bl Vandathai/Shutterstock.com; 202br StockFood Ltd./Alamy Stock Photo; 203tl Charles D. Winters/Science Source; 203bl,

219t fotog/Getty Images; **203**bg d3sign/Getty Images; **203**r (whipped cream) Nattika/Shutterstock.com; **203**r (mayonnaise) stockcreations/Shutterstock.com; **203**r (paint) jannoon028/Shutterstock.com; **203**r (marshmallow) © iStock.com/subjug; **203**r (opal) Alexander Hoffmann/Shutterstock.com; **203**r (vase) Simon Curtis/Alamy Stock Photo; **204**t Andrey_Kuzmin/Shutterstock.com; **204**b © iStock.com/chictype; **205**tl Aleksandra Gigowska/Shutterstock.com; **205**tc SPL/Science Source; **205**tr F. JIMENEZ MECA/Shutterstock.com; **205**bl (coke) DenisMArt/Shutterstock.com; **205**bl (alcohol) © iStock.com/jfmdesign; **205**bl (gatorade) © iStock.com/EasyBuy4u; **205**br Olena Yakobchuk/Shutterstock.com; **206** Foodcollection RF/Getty Images; **208**t Martyn F. Chillmaid/Science Source; **208**b David GABIS/Alamy Stock Photo; **209**t Mariyana M/Shutterstock.com; **209**b Garsya/Shutterstock.com; **210–11** T.W. van Urk/Shutterstock.com; **210**i Turtle Rock Scientific/Science Source; **211**ti gritsalak karalak/Shutterstock.com; **211**bi udaix/Shutterstock.com; **212**, **219**b Turtle Rock Scientific/Science Source; **213** design56/Shutterstock.com; **214** waltereicsy/Shutterstock.com; **215** Richard J Green/Science Source/Getty Images; **216** Adie Bush/Cultura RF/Getty Images; **217**t annick vanderschelden photography/Moment/Getty Images; **217**b (fudge) chris kolaczan/Shutterstock.com; **217**b (caramel) MaraZe/Shutterstock.com; **217**b (nougat) vandycan/Shutterstock.com; **217**b (toffee) © iStock.com/cislander; **218**t MaraZe/Shutterstock.com; **218**b imageBROKER/Alamy Stock Photo; **220**l Turtle Rock Scientific/Science Source; **220**cl Venturelli Luca/Shutterstock.com; **220**cr Turtle Rock Scientific/Science Source; **220**r Joe Belanger/Shutterstock.com; **221** Serhii Hrebeniuk/Shutterstock.com

CHAPTER 10

222 Gerry Ellis/Minden Pictures/Getty Images; **224–25**bg kubais/Shutterstock.com; **224** Dario Lo Presti/Shutterstock.com; **225**t, **226**b grmarc/Shutterstock.com; **225**c PHOTO RESEARCHERS, INC./Science Source/Getty Images; **225**b, **239**t traveliving/Shutterstock.com; **226–27**bg Fotaro1965/Shutterstock.com; **226**t Unuchko Veronika/Shutterstock.com; **227**t Turtle Rock Scientific/Science Source; **227**b Anucha Tiemsom/Shutterstock.com; **230–31**, **239**c elenabsl/Shutterstock.com; **232** Olya Detry/Shutterstock.com; **233**t Turtle Rock Scientific/Science Source; **233**ti NMeM/Royal Photographic Society/SSPL/Getty Images; **233**cl GIPhoto Stock/Science Source; **233**cr Africa Studio/Shutterstock.com; **233**bl Long Bao/Shutterstock.com; **233**br Haluk Köhserli/iStock/Getty Images Plus/Getty Images; **234** MaraZe/Shutterstock.com; **235**l Lijuan Guo/Shutterstock.com; **235**r Sakarin Sawasdinaka/Shutterstock.com; **237** szefei/Shutterstock.com; **238** OBprod/Shutterstock.com; **239**b StefaniaArca/Shutterstock.com; **240**l TREVOR CLIFFORD PHOTOGRAPHY/Science Source; **240**r MARTYN F. CHILLMAID/Science Source

CHAPTER 11

244 kuri2000/iStock Getty Images Plus/Getty Images; **246–47**bg GoodStudio/Shutterstock.com; **246** Science Source/Getty Images; **247**t "Justus Sustermans - Portrait of Galileo Galilei, 1636"/Wikimedia Commons/Public Domain; **247**c Hulton Archive/Getty Images; **247**bl Ian Dagnall/Alamy Stock Photo; **247**br CERN/Science Source; **248–49**, **265**t Nadya_Art/Shutterstock.com; **250–51** Alex Oakenman/Shutterstock.com; **252** DRogatnev/Shutterstock.com; **253**, **265**c coldsnowstorm/E+/Getty Images; **254–55** Xinhua/Alamy Stock Photo; **256** alazur/Shutterstock.com; **258** Hulton Deutsch/Corbis Historical/Getty Images; **259** all Macrovector/Shutterstock.com; **260** © iStock.com/mihtiander; **260**ti dnd_project/Shutterstock.com; **260**bi rzstudio/Shutterstock.com; **261** Joe Raedle/Getty Images; **263**, **265**b Rick Neves/Shutterstock.com; **266** Eric Gevaert/Shutterstock.com

CHAPTER 12

268 Benoist/Shutterstock.com; **270–71** 7Horses/Shutterstock.com; **271**i BlueRingMedia/Shutterstock.com; **272**i GingerArt/Shutterstock.com; **272–73**, **291**t Squared pixels/E+/Getty Images; **276**, **277** Wayne0216/Shutterstock.com; **277**li studioloco/Shutterstock.com; **277**ri wavebreakmedia/Shutterstock.com; **277**c, ri ESB Professional/Shutterstock.com; **278–79**bg okimo/Shutterstock.com; **278** miniwide/Shutterstock.com; **280**t, **281**t, **291**c Alena Che/Shutterstock.com; **280–81**bg Evgenia L/Shutterstock.com; **280**b, **282**i, **283**i NotionPic/Shutterstock.com; **281**b Good_Stock/Shutterstock.com; **282**, **283** D1min/Shutterstock.com; **282–83**bg PremiumArt/Shutterstock.com; **284** Graiki/Moment/Getty Images; **286**, **291**b Kekyalyaynen/Shutterstock.com; **287** IR Stone/Shutterstock.com; **289** gualtiero boffi/Shutterstock.com; **290** Darryl Leniuk/The Image Bank/Getty Images; **293** Buena Vista Images/The Image Bank/Getty Images

CHAPTER 13

294 NurPhoto/Getty Images; **296** AFP/Getty Images; **297**t, bl, **319**t Mascha Tace/Shutterstock.com; **297**br Good_Stock/Shutterstock.com; **298** Avalon_Studio/iStock/Getty Images Plus/Getty Images; **299** Jaren Kane/500px/Getty Images; **300–301** ProStockStudio/Shutterstock.com; **302**tl Stoked | Thinkstock/Media Bakery; **302**bl Konrad Wothe/Minden Pictures/Getty Images; **302**r ESB Basic/Shutterstock.com; **303** kali9/E+/Getty Images; **305** Kartinkin77/Shutterstock.com; **306** Monty Rakusen/Cultura/Getty Images; **306**i Serg64/Shutterstock.com; **307**t, **319**b Fabrice LEROUGE/ONOKY/Getty Images; **307**ti Pirotehnik/iStock/Getty Images Plus/Getty Images; **307**c loco75/Stock/Getty Images Plus/Getty Images; **307**b Image Source/Getty Images; **308**t, **308**bl GIPhotoStock/Science Source; **308**bcl Sam72/Shutterstock.com; **308**bcr dei-sin/iStock/Getty Images Plus/Getty Images; **308**br T.Dallas/Shutterstock.com; **309** iMoved Studio/Shutterstock.com; **313**, **320** kali9/E+/Getty Images; **315**i Ilya Bolotov/Shutterstock.com; **314–15** Patryce Bak/The Image Bank/Getty Images; **316–17** AlexSava/iStock/Getty Images Plus/Getty Images; **318**l Ian Hooton/Science Source; **318**r Ted Kinsman/Science Source

CHAPTER 14

322 Danita Delimont/Gallo Images/Getty Images; **322**i REUTERS/Jim Drury; **324–25** Gabe Rogel/Aurora Creative/Getty Images; **326**, **345** Mascha Tace/Shutterstock.com; **327**l OZaiachin/Shutterstock.com; **327**r Vadym Zaitsev/Shutterstock.com; **328**, **344**t YellowPaul/Shutterstock.com;

329 Mykhailo Bokovan/Shutterstock.com; 331 Westend61/Getty Images; 332–34, 344c Barcroft/Barcroft Media/Getty Images; 335, 336t CW craftsman/Shutterstock.com; 336b Jiw Ingka/Shutterstock.com; 337 nattapon supanawan/Shutterstock.com; 338–39 Daniel Schoenen/Getty Images; 340–41bg city hunter/Shutterstock.com; 340t, 344b Jacob Maentz/Corbis Documentary/Getty Images; 340c "Adam Beck Complex"/Ontario Power Generation/Wikimedia Commons/CC BY 2.0; 340b Puripat Lertpunyaroj/Shutterstock.com; 341t RoschetzkyIstockPhoto/iStock/Getty Images; 341b Doug McLean/Shutterstock.com; 342–43bg Gary Saxe/Shutterstock.com; 342t "Elwha Dam"/DancingBear/Wikimedia Commons/Public Domain; 342c "Elwha Dam with hole"/NPS/Wikimedia Commons/Public Domain; 342b "Elwha Dam Finished May2013"/Zandcee/Wikimedia Commons/CC BY-SA 3.0; 343l Anton Prado PHOTO/Shutterstock.com; 343r © iStock.com/pepifoto

CHAPTER 15

346 TheJim999/Shutterstock.com; 348–49 Andriy Blokhin/Shutterstock.com; 349t, 368t GIPhotoStock/Science Source; 349bi David R. Frazier Photolibrary, Inc./Alamy Stock Photo; 350 Scala/Art Resource, NY; 351 Fouad A. Saad/Shutterstock.com; 352 Ekaterina Pokrovsky/Shutterstock.com; 354t Heymo/Shutterstock.com; 354bl bbeltman/E+/Getty Images; 354br Charles D. Winters/Science Source; 355 bogdan kosanovic/E+/Getty Images; 356–57, 368c Pavel L Photo and video/Shutterstock.com; 357ti Martyn F. Chillmaid/Science Source; 357bi Elmar Krenkel/imageBROKE/Getty Images; 358 Creative-Family/iStock/Getty Images Plus/Getty Images; 360 Andrei Nekrassov/Shutterstock.com; 360–61bg a_Taiga/iStock/Getty Images Plus/Getty Images; 362, 363t Turtle Rock Scientific/Science Source; 363bl GIPhotoStock/Science Source; 363bc BlueRingMedia/Shutterstock.com; 363br no_limit_pictures/E+/Getty Images; 364, 368b Gary Saxe/Shutterstock.com; 366–67bg gerenme/E+/Getty Images; 366 Kateryna Kon/Shutterstock.com; 367 NASA/JPL-Caltech

CHAPTER 16

370 JC Patricio/Moment/Getty Images; 372 Lissandra Melo/Shutterstock.com; 374t, 390t Valentyn Hontovyy/Shutterstock.com; 374bl Berke/Shutterstock.com; 374br Chutima Chaochaiya/Shutterstock.com; 375 Cavan Images/Alamy Stock Photo; 376 sub job/Shutterstock.com; 377 Image Source/Getty Images; 378 Jonathan Bird/Photolibrary/Getty Images; 379 Martyn F. Chillmaid/Science Source; 380, 381 LineTale/Shutterstock.com; 382 Doug Allan/Science Source; 383, 390c Take Photo/Shutterstock.com; 384 age fotostock/Alamy Stock Photo; 387 Yevhenii Dorofieiev/Unitone Vector/iStock/Getty Images Plus/Getty Images; 388–89, 390b Elenathewise/iStock/Getty Images Plus/Getty Images; 389i Mikhail Japaridze/TASS/Getty Images; 391 "Aircraft venturi 1" by YSSYguy/Wikimedia Commons/Public Domain

CHAPTER 17

394 Ron Dahlquist/Perspectives/Getty Images Plus/Getty Images; 394i David Livingston/Getty Images Entertainment/Getty Images; 396–97 mikroman6/Moment/Getty Images; 397i, 416t Frank Zullo/Science Source; 400–401 NH/Shutterstock.com; 400i Sunwand24/Shutterstock.com; 401i kvsan/Shutterstock.com; 402 Andy Roberts/Caiaimage/Getty Images; 403, 416c Rod Jones/Alamy Stock Photo; 404–5 Kite_rin/Shutterstock.com; 405i Image by Chris Winsor/Moment/Getty Images; 410–11bg TierneyMJ/Shutterstock.com; 410t Andrew Mayovskyy/Shutterstock.com; 410b Beth Van Trees/Shutterstock.com; 411tl Olga Popova/Shutterstock.com; 411tr, 412cl, 416b GIPhoto Stock/Science Source; 411bl Frans Lanting/MINT Images/Science Source; 411br Tefi/Shutterstock.com; 412–13bg TierneyMJ/Shutterstock.com; 412t Esa Hiltula/Alamy Stock Photo; 412cr tamara_kulikova/iStock/Getty Images Plus/Getty Images; 412b dropStock/Shutterstock.com; 413t Ted Kinsman/Science Source; 413b Nicholas Toh/Shutterstock.com; 414 AP Photo; 415 VectorShow/Shutterstock.com

CHAPTER 18

418, 437t SVSimagery/Shutterstock.com; 418i "Chuck Yeager X-1 (color)" by Jack Ridley/Wikimedia Commons/Public Domain; 420 Syda Productions/Shutterstock.com; 421t ilbusca/DigitalVision Vectors/Getty Images; 421b Classic Image/Alamy Stock Photo; 422–23b granata68/Shutterstock.com; 423t trgrowth/Shutterstock.com; 424, 437b Speed Kingz/Shutterstock.com; 425l, 438bl Vecton/Shutterstock.com; 425r stockshoppe/Shutterstock.com; 426–27 Avpics/Alamy Stock Photo; 427i, 438br Sedova Elena/Shutterstock.com; 428t commerceandculturestock/Moment/Getty Images; 428–29b Roman Babakin/Shutterstock.com; 429ti, bi Brian A Jackson/Shutterstock.com; 430t Jose Luis Pelaez Inc/DigitalVision/Getty Images; 430b "Elektrodynamisk-högtalare" by Svjo/Wikimedia Commons/CC BY-SA 3.0; 431 Paul Colley/iStock/Getty Images Plus/Getty Images; 432–33, 438t bearacreative/Shutterstock.com; 432li Golden Sikorka/Shutterstock.com; 432ri U.S. Navy; 434tl Christian Beirle González/Moment Select/Getty Images Plus/Getty Images; 434bl Public Domain; 434r Richman Photo/Shutterstock.com; 435t Sidekick/E+/Getty Images; 435li BorupFoto/iStock/Getty Images Plus/Getty Images; 435ri "Ultrasonic pipeline test"/Davidmack/Wikimedia Commons/Public Domain; 435b Monty Rakusen/Cultura/Getty Images; 439t Horatiu Bota/Shutterstock.com; 439b BLOOMimage/Getty Images

CHAPTER 19

440 Barcroft Media/Getty Images; 442, 467t Ted Kinsman/Science Source; 445 Kan Taengnuanjan/EyeEm/Getty Images; 446–47bg Comaniciu Dan/Shutterstock.com; 446b Dave King/Dorling Kindersley/Science Source; 447b Sonsedska Yuliia/Shutterstock.com; 448t vanbeets/iStock/Getty Images Plus/Getty Images; 448b Dave King/Dorling Kindersley/Getty Images Plus/Getty Images; 449 SCIENCE PHOTO LIBRARY/Science Source; 450–51 Sean Pavone/Shutterstock.com; 451i, 454t VectorShow/Shutterstock.com; 452t, 453b Vasilius/Shutterstock.com; 452b, 467c Science Source/Science Source; 453t Laborant/Shutterstock.com; 454–55b 9dream studio/Shutterstock.com; 457 kadm/iStock/Getty Images Plus/Getty Images; 458t In-Finity/Shutterstock.com; 458ct sumkinn/Shutterstock.com; 458cb Scrap4vec/Shutterstock.com; 458b fullvector/Shutterstock.com; 459t RedlineVector/Shutterstock.com; 459c Alexandr III

/Shutterstock.com; **459**b Webspark/Shutterstock.com; **460–61**bg stockchairatgfx/Shutterstock.com; **460, 461, 467**b haryigit/Shutterstock.com; **462–63** Tim Allen/iStock/Getty Images Plus/Getty Images; **464–65**bg Mr Twister/Shutterstock.com; **464**t NOPPHARAT STUDIO 969/Shutterstock.com; **464**b Yentafern/Shutterstock.com; **465**t PitukTV/Shutterstock.com; **465**c ony Freeman/Science Source; **465**b chonticha stocker/Shutterstock.com; **466**t Bettmann/Getty Images; **466**b Nerthuz/Shutterstock.com; **469** "Police issue X26 TASER-white"/Junglecat/Wikimedia Commons /CC BY-SA 3.0

CHAPTER 20

470 OLI SCARFF/AFP/Getty Images; **472** New Africa/Shutterstock.com; **472**i (magnet) New Africa/Shutterstock.com; **472**i (William Gilbert) "William Gilbert 45626i"/Wikimedia Commons/Public Domain; **472**i (Petrus Peregrinus) INTER FOTO/Alamy Stock Photo; **472**i (globe) ixpert/Shutterstock.com; **472**i, **483**t (compass) Alex Staroseltsev/Shutterstock.com; **473**t CORDELIA MOLLOY/Science Photo Library/Getty Images; **473**c pippeeContributor/Shutterstock.com; **473**b VectorMine/Shutterstock.com; **474** hoch2wo/Alamy Stock Photo; **475** Peter Sobolev/Shutterstock.com; **478–79**, **483**c Bernd Mellmann/Alamy Stock Photo; **478**i BanksPhotos /iStock/Getty Images Plus/Getty Images; **479**ti stockphoto-graf/Shutterstock.com; **479**bi Studio GL/Shutterstock.com; **480** leezsnow/E+/Getty Images; **481**l, **483**b Sugrit Jiranarak /Shutterstock.com; **481**r Wpadington/Shutterstock.com; **482** suradech sribuanoy/Shutterstock.com; **485**l Chronicle/Alamy Stock Photo; **485**r "Tesla3" by Napoleon Sarony/Wikimedia Commons/Public Domain; **485**i (toaster) Makc/Shutterstock.com; **485**i (flashlight) paradesign/Shutterstock.com

CHAPTER 21

486 Kevin Foy/Alamy Stock Photo; **488–89** JS`s favorite things/Moment Open/Getty Images; **489**i, **500**b Vector Mine/Shutterstock.com; **490**t Bettmann/Getty Images; **490**b "Michelson speed of light measurement 1930" by H. H. Dunn/Wikimedia Commons/Public Domain; **492, 500**t Carlos Fernandez/Moment/Getty Images; **493**l INTERFOTO /Alamy Stock Photo; **493**r GL Archive/Alamy Stock Photo; **494** S.Borisov/Shutterstock.com; **495** Richard Hutchings /Science Source; **496–97** VectorMine/Shutterstock.com; **498**t Designua/Shutterstock.com; **498**b Zapp2Photo /Shutterstock.com; **501** Ikhsan Yohanda/iStock/Getty Images Plus/Getty Images; **501**i Ted Kinsman/Science Source

CHAPTER 22

502 WIN-Initiative/Getty Images Plus/Getty Images; **504**i VectorMine/Shutterstock.com; **504–5** lazyllama/Shutterstock.com; **505**tr, **505**ctr H.waritha/Shutterstock.com; **505**cbr fishmonger/Shutterstock.com; **505**br, **518**t Katia Vittoria Di Gioia/EyeEm/Getty Images; **506** Lane Oatey/Getty Images; **507** Nella/Shutterstock.com; **508**tl Sakurra/Shutterstock.com; **508**tr Alexander Tsiaras/Science Source; **508**bl Andrey_Popov/Shutterstock.com; **508**br David Tadevosian/Shutterstock.com; **509** Bence Katona-Kovacs/Shutterstock.com; **510** chirajuti/Shutterstock.com; **511, 518**b New Africa /Shutterstock.com; **512** tompics/iStock/Getty Images Plus /Getty Images; **513**l Josfor/Getty Images/iStock/Getty Images Plus/Getty Images; **513**r Pressmaster/Shutterstock.com; **514** waldru/Shutterstock.com; **515**t Ian Cuming/Ikon Images/Getty Images Plus/Getty Images; **515**bl GIPhoto Stock/Science Source; **515**br NotionPic/Shutterstock.com; **516** Sergey Merkulov/Shutterstock.com; **517, 519** andresr/E+/Getty Images; **520** Damsea/Shutterstock.com; **521** pixelfit/E+/Getty Images

APPENDIXES

522 "The sun is an MHD system that is not well understood-2013-04-9 14-29"/NASA/SDO/Wikimedia Commons /Public Domain; **523** Becart/iStock/Getty Images Plus /Getty Images; **524** Peter Dazeley/The Image Bank/Getty Images Plus/Getty Images; **531** all Kinjalben Apoorva Patel /DigitalVision Vectors/Getty Images; **536**l, r Nadya_Art /Shutterstock.com; **537** Albert Russ/Shutterstock.com

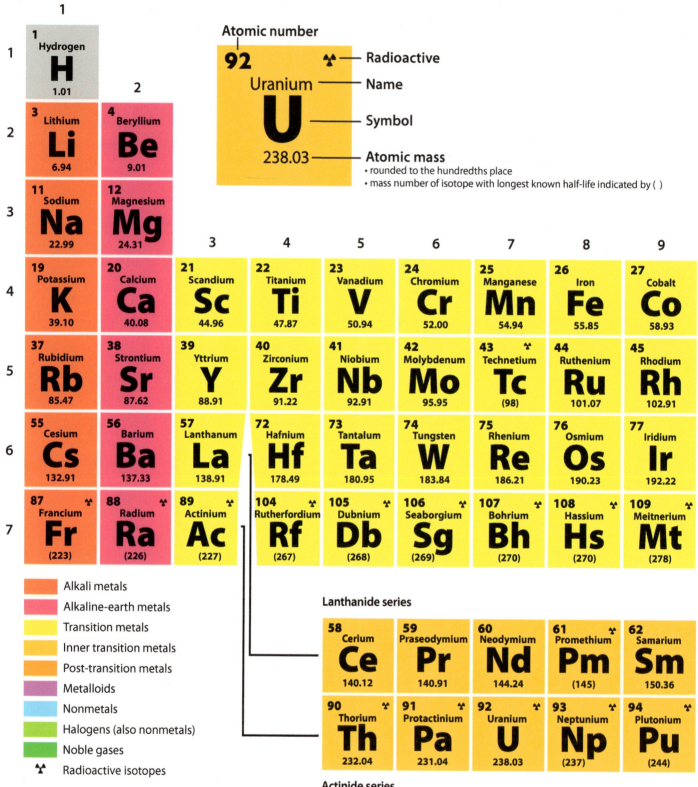

PERIODIC TABLE OF THE ELEMENTS

						18
						2 Helium **He** 4.00

13	14	15	16	17	
5 Boron **B** 10.81	6 Carbon **C** 12.01	7 Nitrogen **N** 14.01	8 Oxygen **O** 16.00	9 Fluorine **F** 19.00	10 Neon **Ne** 20.18
13 Aluminum **Al** 26.98	14 Silicon **Si** 28.09	15 Phosphorus **P** 30.97	16 Sulfur **S** 32.06	17 Chlorine **Cl** 35.45	18 Argon **Ar** 39.95

10	11	12						
28 Nickel **Ni** 58.69	29 Copper **Cu** 63.55	30 Zinc **Zn** 65.38	31 Gallium **Ga** 69.72	32 Germanium **Ge** 72.63	33 Arsenic **As** 74.92	34 Selenium **Se** 78.97	35 Bromine **Br** 79.90	36 Krypton **Kr** 83.80
46 Palladium **Pd** 106.42	47 Silver **Ag** 107.87	48 Cadmium **Cd** 112.41	49 Indium **In** 114.82	50 Tin **Sn** 118.71	51 Antimony **Sb** 121.76	52 Tellurium **Te** 127.60	53 Iodine **I** 126.90	54 Xenon **Xe** 131.29
78 Platinum **Pt** 195.08	79 Gold **Au** 196.97	80 Mercury **Hg** 200.59	81 Thallium **Tl** 204.38	82 Lead **Pb** 207.24	83 Bismuth **Bi** 208.98	84 Polonium **Po** (209)	85 Astatine **At** (210)	86 Radon **Rn** (222)
110 Darmstadtium **Ds** (281)	111 Roentgenium **Rg** (282)	112 Copernicium **Cn** (285)	113 Nihonium **Nh** (286)	114 Flerovium **Fl** (289)	115 Moscovium **Mc** (290)	116 Livermorium **Lv** (293)	117 Tennessine **Ts** (294)	118 Oganesson **Og** (294)

63 Europium **Eu** 151.96	64 Gadolinium **Gd** 157.25	65 Terbium **Tb** 158.93	66 Dysprosium **Dy** 162.50	67 Holmium **Ho** 164.93	68 Erbium **Er** 167.26	69 Thulium **Tm** 168.93	70 Ytterbium **Yb** 173.05	71 Lutetium **Lu** 174.97
95 Americium **Am** (243)	96 Curium **Cm** (247)	97 Berkelium **Bk** (247)	98 Californium **Cf** (251)	99 Einsteinium **Es** (252)	100 Fermium **Fm** (257)	101 Mendelevium **Md** (258)	102 Nobelium **No** (259)	103 Lawrencium **Lr** (266)

PERIODIC TABLE OF THE ELEMENTS